MICROPLASTICS IN ANIMAL AND HUMAN PHYSIOLOGY

MICROPLASTICS IN ANIMAL AND HUMAN PHYSIOLOGY

TOXICITY, MECHANISMS, AND MITIGATION

Edited by

RAJESH KUMAR
Department of Biosciences, Himachal Pradesh University, Shimla, India

ARUP GIRI
Department of Zoology, Baba Mastnath University, Asthal Bohar, Rohtak, Haryana, India

RANBIR CHANDER SOBTI
Department of Biotechnology, Punjab University, Chandigarh, India

ELSEVIER

ACADEMIC PRESS
An imprint of Elsevier

Academic Press is an imprint of Elsevier
125 London Wall, London EC2Y 5AS, United Kingdom
50 Hampshire Street, 5th Floor, Cambridge, MA 02139, United States

ISBN: 978-0-443-45635-0

For information on all Academic Press publications visit our website at https://www.elsevier.com/books-and-journals

Publisher: Andre Wolff
Acquisitions Editor: Jessica Mack
Editorial Project Manager: Sweta Bharti
Production Project Manager: Aishwarya R
Cover Designer: Bakyalakshmi S

Typeset by Integra Software Services

Contents

Contributors

Nouara Abdelli Department of Biological Sciences, College of Science, King Faisal University Al-Ahsa, Saudi Arabia

Mohammad Ayaz Ahmad Department of Mathematics, Physics & Statistics Faculty of Natural Sciences, Turkeyen Campus University of Guyana, Greater Georgetown, Guyana

Gangadhar Andaluri Department of Civil and Environmental Engineering, Temple University, Philadelphia, PA, United States

Kim Andrews Department of Mathematics & Natural Sciences, Prince Mohammad Bin Fahd University, AL Khobar, Saudi Arabia

Nguyen Tuan Anh Department of Environment and Natural Resources, Nong Lam University of Ho Chi Minh City – Gialai Campus, Pleiku City, Vietnam

Federica Arrigo Department of Veterinary Science, University of Messina, Messina, Italy

Elisabetta Benedetti Department of Life, Health and Environmental Sciences, University of L'Aquila, L'Aquila, Italy

Ajmal Rashid Bhat Department of Chemistry, Rashtrasant Tukadoji Maharaj Nagpur University, Nagpur, India

Neelam Bhatti Department of Life Sciences, Sant Baba Bhag Singh University, Jalandhar, India

Shuvra Debnath Faculty of Engineering and Information Technology, University of Technology Sydney, Ultimo, NSW, Australia

Dimple Dimple Department of Zoology, Baba Mastnath University, Asthal Bohar, Rohtak, Haryana, India

Karim Ebrahimpour Department of Environmental Health Engineering, University of Medical Sciences, Isfahan, Iran

Caterina Faggio Department of Chemical, Biological, Pharmaceutical and Environmental Sciences, University of Messina Viale Ferdinando Stagno d'Alcontres, Messina; Department of Eco-sustainable Marine Biotechnology, Stazione Zoologica Anton Dohrn, Naples, Italy

Roberta Maria Ferri Department of Life, Health and Environmental Sciences, University of L'Aquila, L'Aquila, Italy

Mohammad Mehdi Fowzi Department of Civil Engineering, University of Isfahan, Isfahan, Iran

Arup Giri Department of Zoology, Baba Mastnath University, Asthal Bohar, Rohtak, Haryana, India

Aysel Caglan Gunal Department of Biology Education, Faculty of Gazi Education, Gazi University, Ankara, Turkey

Younis Ahmad Hajam Department of Life Sciences, Sant Baba Bhag Singh University, Jalandhar, India

Nesrein M. Hashem Department of Animal and Fish Production, Alexandria University, Alexandria, Egypt

Xinlei Huang Faculty of Engineering and Information Technology, University of Technology Sydney, Ultimo, NSW, Australia

Valentina Iezzi Department of Clinical Medicine, Public Health, Life and Environmental Sciences, University of L'Aquila, L'Aquila, Italy

Khudrotul Nisa Indriyasari Mathematics and Natural Sciences Doctoral Program, Universitas Airlangga, Surabaya, Indonesia

Md Saiful Islam Department of Civil and Environmental Engineering, Temple University, Philadelphia, PA, United States

Rabiul Islam School of Pharmacy, BRAC University, Dhaka, Bangladesh

Ehsan Jafarpisheh Department of Civil Engineering, University of Isfahan, Isfahan, Iran

Divya Jyoti Lovely Professional University, School of Bioengineering and Biosciences, Phagwara, India

Nahid Khoshnamvand Department of Environmental Health Engineering, Lorestan University of Medical Sciences, Khorramabad, Iran

Rajesh Kumar Department of Biosciences, Himachal Pradesh University, Shimla, India

Eko Prasetyo Kuncoro Department of Biology, Universitas Airlangga, Surabaya, Indonesia

Nguyen Minh Ky Faculty of Environment and Natural Resources, Nong Lam University, Ho Chi Minh City, Vietnam

M. Amin Mir Department of Mechanical Engineering, Prince Mohammad Bin Fahd University, Dammam, Saudi Arabia

Adel Mokammel Ardabil University of Medical Sciences, Social Determinants of Health Research Center, Ardabil, Iran

Cristiana Roberta Multisanti Department of Veterinary Sciences, University of Messina, Messina, Italy

Poonam Mundlia Department of Zoology, Baba Mastnath University, Asthal Bohar, Rohtak, Haryana, India

Salvina Chowdhury Neha School of Pharmacy, BRAC University, Dhaka, Bangladesh

Nguyen Thi Nhan Master Program in Water Technology, Reuse, and Management, Vietnamese German University, Ben Cat, Binh Duong Province, Vietnam

Huynh Do Hieu Nhu Master Program in Water Technology, Reuse, and Management, Vietnamese German University, Ben Cat, Binh Duong Province, Vietnam

Oyegunle Emmanuel Oke Department of Animal Physiology, Federal University of Agriculture, Abeokuta, Nigeria

Giuseppe Piccione Department of Veterinary Science, University of Messina, Messina, Italy

Ruby Ruby Department of Zoology, Baba Mastnath University, Asthal Bohar, Rohtak, Haryana, India

Auroshree Sadhu Ecotoxicology Research Laboratory, Department of Zoology, The University of Burdwan, Burdwan, West Bengal, India

Shubhajit Saha Ecotoxicology Research Laboratory, Department of Zoology, The University of Burdwan, Burdwan, West Bengal, India

Suvash C. Saha School of Mechanical and Mechatronic Engineering, University of Technology Sydney, Sydney, NSW, Australia

Nisha Sharma Department of Physical Science, Sant Baba Bhag Singh University, Jalandhar, India

Ranbir Chander Sobti Department of Biotechnology, Punjab University, Chandigarh, India

Agoes Soegianto Department of Biology, Universitas Airlangga, Surabaya, Indonesia

Oksana Stoliar Ternopil Volodymyr Hnatiuk National Pedagogical University, Ternopil, Ukraine

Toha Tasnem School of Pharmacy, BRAC University, Dhaka, Bangladesh

Victoria Anthony Uyanga Department of Animal Sciences, Purdue University, West Lafayette, IN; Department of Agriculture and Environmental Sciences, Lincoln University of Missouri, Jefferson City, MO, United States

Pinar Arslan Yüce Department of Biology, Faculty of Science, Çankırı Karatekin University, Çankırı, Turkey

Noushi Zaidi Department of Chemistry, College of Science, King Faisal University, Al-Ahsa, Saudi Arabia

CHAPTER

1

Microplastics: A threat to the animal world

Mohammad Mehdi Fowzi[a], *Md Saiful Islam*[b], *and Gangadhar Andaluri*[b]

[a]**Department of Civil Engineering, University of Isfahan, Isfahan, Iran**
[b]**Department of Civil and Environmental Engineering, Temple University, Philadelphia, PA, United States**

1.1 Introduction

Once considered an emerging organic pollutant, microplastics (MPs) are recognized as a silent but serious threat to wildlife across multiple ecosystems (Khanjani and Mohammadi, 2025; Nguyen et al., 2023). When MPs are <5 mm in diameter, they originate from diverse sources. They can adsorb a range of environmental contaminants, including heavy metals, polycyclic aromatic hydrocarbons (PAHs), polychlorinated biphenyls (PCBs), per- and polyfluoroalkyl substances (PFAS), bisphenols, phthalates, tire wear particles (6-PPD-q), pharmaceuticals and personal care products, 1,4-dioxane, and antibiotics, thereby increasing their environmental toxicity and potential for accumulation within biological systems (Gheisari et al., 2025; Islam et al., 2025; Massahi et al., 2024). These tiny particles are generally classified into two main types: primary MPs, which are intentionally manufactured into small sizes, such as toothpaste, facial scrubs, and soaps (Singh and Mishra, 2023; Chengappa et al., 2023), and directly enter the environment, and secondary MPs, which are formed through the breakdown of larger plastic materials over time. Owing to their small size and resistance to degradation, MP particles can persist in the environment for long periods and continuously enter various ecosystems. Key sources of MPs in environmental matrices include plastic waste, agricultural mulching films, wastewater treatment plants, atmospheric fallout, compost, and leachate (Bank, 2022; Fowzi et al., 2025; Gheisari et al., 2025). Fig. 1.1 compares the characteristics and origins of primary and secondary MPs. Primary MPs are intentionally manufactured and commonly found in personal care products such as toothpaste and cosmetics (Zorin et al., 2025). In contrast, secondary MPs arise from the breakdown of

Microplastics in Animal and Human Physiology
https://doi.org/10.1016/B978-0-443-45635-0.00007-0

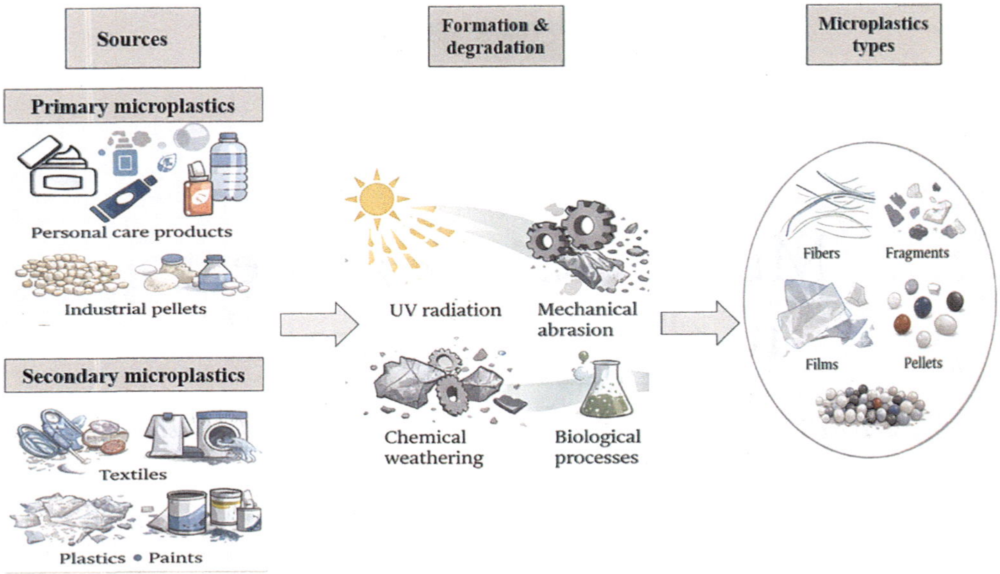

FIG. 1.1 Schematic classification of microplastics. Classification based on their origin, shape, sources, and formation processes, illustrating the distinction between primary and secondary microplastics and their transformation pathways in the environment.

larger plastic items, such as bottles and packaging, through physical, chemical, and biological degradation. The figure emphasizes their differences in origin, shape, and size (Gunawardhana et al., 2024).

Once released into the environment, MPs can interact with a wide range of animal species across different habitats. Research has documented their presence in numerous terrestrial and aquatic organisms, including fish, birds, mammals, farmed animals, and other forms of wildlife (Lackner and Branka, 2024; Sarkar et al., 2023). There is evidence that >200 marine animal species have encountered MPs at various life cycle stages (Gall and Thompson, 2015). Studies have revealed MPs in the gastrointestinal tracts (GITs), livers, and muscles of marine organisms (Haave et al., 2021). Exposure to these particles by terrestrial and aquatic organisms has been associated with behavioral alterations, impaired feeding activity, metabolic disturbances, and adverse effects on reproductive and neurological health, ultimately facilitating their entry into the food web and posing broader ecological and health risks (Hamidian and Feizi, 2025). However, the full extent of their impact on wildlife remains limited (Porcino et al., 2023).

Therefore the main objective of this study was to explore the environmental pathways of MPs, their interactions with animal species in aquatic and terrestrial systems, and the broader ecological consequences of their accumulation. Recent case studies and proposed mitigation strategies are also reviewed to provide a comprehensive understanding of this growing environmental challenge.

1.2 Sources, pathways, and fate of MPs in animal habitats

MPs originate from diverse sources, including the fragmentation of larger plastic debris, synthetic fibers released during textile washing, microbeads from personal care and cleaning products, and polymer-based carriers in pharmaceutical formulations. These particles can enter the environment through wastewater, atmospheric fallout, and mismanaged plastic waste, ultimately accumulating in soil, water, and biota (Ashrafy et al., 2023; Browne, 2015). These tiny particles can be introduced into a range of habitats through several routes, such as water bodies, stormwater runoff, atmospheric transport, and movement by living organisms. After entering the environment, they may accumulate in sediments, be resuspended by water currents or anthropogenic disturbances, and be taken up by animals, facilitating their further dispersal across ecosystems (Ashrafy et al., 2023).

One key factor determining the fate of MPs in aquatic environments is their density (Duis and Coors, 2016). Some MPs, such as expanded polystyrene (PS) (0.01–0.04 g/cm^3), low-density polyethylene (PE) (0.89–0.93 g/cm^3), and high-density PE (0.94–0.98 g/cm^3), have densities lower than those of seawater and can float on the surface. Marine animals living at or near the surface, including fish, may mistake these particles for food and ingest them. The color of MPs also plays a critical role, as brightly colored or food-like particles may be mistaken for prey, increasing the likelihood of ingestion by aquatic organisms (Islam et al., 2023; Kabir et al., 2021). Conversely, MPs such as polyvinyl chloride (PVC) (1.16–1.58 g/cm^3), polyurethane (1.2 g/m^3), polyester (1.24–2.3 g/m^3), and polytetrafluoroethylene (2.1–2.3 g/m^3) have densities greater than those of seawater and tend to sink, making them more likely to be consumed by organisms living in deeper parts of the ocean (Andrady, 2011). Moreover, biofilm formation on MP surfaces can reduce their density, enabling these particles to rise back to the water surface. Similarly, physical wear and degradation processes may alter the density of plastic debris. Moreover, turbulence driven by wind mixing in upper water layers can cause MPs to float to the surface and sink beneath it.

1.3 MP impacts on animals and ecosystems

MPs have a growing ecological impact on both individual species and broader ecosystems. The consumption of MPs by a wide range of organisms, from microscopic plankton to land-dwelling animals, can lead to both physical damage and chemical exposure. These effects include blockages in the digestive system, reduced energy availability, and the release of harmful substances from plastic materials (Sharda et al., 2024). Research has demonstrated that MPs can negatively influence the growth, developmental processes, behavioral patterns, and survival rates of aquatic organisms (Chae and An, 2017). Although MPs are seldom directly lethal, prolonged exposure may harm individual organisms and disrupt essential ecological functions, such as bioturbation and the transfer of carbon to deeper ocean layers (Galloway et al., 2017).

1.3.1 Aquatic organisms

The occurrence of MPs in aquatic environments is increasingly concerning because of their toxic effects. The small size of MPs, especially primary MPs, allows them to disperse extensively across diverse aquatic habitats (Ashrafy et al., 2023; Avio et al., 2017; Eriksen et al., 2013). Each year, approximately one to two million tons of plastic waste make their way into the ocean, contributing significantly to marine pollution. Ingestion of plastic debris has been reported in >1280 marine species, including over 400 fish species (Santos et al., 2021; Savoca et al., 2021). MPs can remain in the GIT of aquatic organisms for several days or even weeks before being excreted. This prolonged retention may allow their transport to new locations or facilitate their transfer through the food chain (Ghosh, 2025). Among these, fish are among the most affected groups, as they are more likely to ingest MPs because of their feeding habits and their widespread distribution across polluted aquatic environments (Merrill et al., 2023; Savoca et al., 2021). As shown in Table 1.1, various studies have reported the presence of different types of MPs in aquatic organisms. The table summarizes the polymer types, detected locations within the organisms, and corresponding references, highlighting the widespread bioaccumulation of MPs across different species and tissues.

1.3.1.1 *Effects of MPs on fish*

MPs pose significant threats to fish in aquatic ecosystems. Fish may ingest MPs either through direct consumption or indirectly via contaminated prey, leading to the accumulation of harmful chemical additives, such as phthalates and flame retardants, as well as environmental pollutants, such as heavy metals and PCBs, within their tissues (Arienzo et al., 2021; Avio et al., 2015; Ibrahim et al., 2025). Studies have demonstrated a range of adverse effects in fish, including decreased food intake, changes in feeding behavior, weakened immune

TABLE 1.1 Summary of aquatic organisms and microplastic types in organs and their accumulation in the liver, gills, gut, intestine, blood, gonads, and gastrointestinal tract.

Aquatic organism species	Microplastic types	Detected location and effect in organisms	References
Danio rerio	Polystyrene	Tissue accumulation (liver, gills, and gut)	(Lu et al., 2016)
Portunus trituberculatus	Polystyrene	Intestine	(Chen et al., 2024)
Pomatoschistus microps	Polyethylene	–	(Fonte et al., 2016)
Zebrafish (*Danio rerio*), protozoan	Polyethylene	Intestinal tissues	(Zhang et al., 2024)
Clarias gariepinus	Polyvinyl chloride	Gills, gastrointestinal tract	(Jabeen et al., 2018)
Gadus morhua	Polystyrene	Blood samples, gonadal tissue, and liver tissue	(Fernández-Míguez et al., 2023)
Zooplankton	Polyethylene, polystyrene, polyamide	–	(Lusher et al., 2015)

responses, and signs of cytotoxicity, neurotoxicity, liver stress, blood clotting, altered molecular transport, and altered fat metabolism (Ghosh, 2025). Physiologically, exposure to MPs has been linked to reduced feeding activity, gill dysfunction, and immune system suppression. Moreover, neurobehavioral disturbances, such as impaired social interactions and cognitive dysfunction, along with reproductive impairments linked to disruptions in the hypothalamic–pituitary–gonadal axis, have been reported in fish and are often associated with altered neurotransmitter activity and endocrine regulation (Hasan et al., 2024). However, it should be noted that the risk from contaminated MPs is often lower than that from consuming natural prey, which generally contains higher pollutant levels (Ibrahim et al., 2025). Studies suggest that MP ingestion does not significantly increase fish exposure to hydrophobic organic chemicals because of limited transfer rates and low bioaccumulation potential. Nonetheless, other pollutants, such as metals and certain persistent chemicals attached to MPs, might further increase overall risk (Ibrahim et al., 2025; Koelmans et al., 2016).

1.3.1.2 *Effects of MPs on marine mammals*

Owing to increased anthropogenic activities, such as fishing and shipping, the amount of plastic entering the ocean has significantly increased. Marine mammals, which play a vital role in shaping the structure and function of marine ecosystems and act as sentinels of ocean health, are now facing increasing conservation concerns, with many species at risk (Zantis et al., 2021). Ingestion of plastic debris has been documented in 123 species of aquatic mammals, including approximately 36% of seal species, 59% of whale species, and approximately 63% of all known cetaceans (Nabi et al., 2022).

As with fish, MPs can bioaccumulate in marine mammals through direct ingestion or via the food web (Nabi et al., 2022). In the GIT of these animals, gut fluids may increase the release of pollutants (PAHs, PCBs, PFAS, and heavy metals), thereby increasing their uptake into vital organs (Lee et al., 2019). Current studies have mostly reported MPs in the GIT and feces. There is concern that nanoplastics and MPs can penetrate cellular membranes, enter the bloodstream, cross the blood–brain barrier, and accumulate in organs such as the brain, liver, kidneys, placenta, and reproductive organs (Prietl et al., 2014). This raises the possibility of fetal exposure via transplacental transfer. Toxic accumulation of MPs and associated chemicals may lead to various health issues in marine mammals, including reproductive disorders and endocrine disruption, immune suppression, liver and kidney dysfunction, neurological damage, congenital disabilities, and cancer (Nabi et al., 2022). Moreover, synergistic interactions among multiple pollutants carried by MPs can amplify these effects, leading to severe outcomes, including neurotoxicity, cytotoxicity, immunotoxicity, and infertility. Alarmingly, some marine mammals lack key detoxifying enzymes (e.g., paraoxonase 1), increasing their vulnerability to organophosphate toxicity (Meyer et al., 2018).

Case studies highlight the measurable impacts of MPs on various aquatic animal species. In bivalves such as *Atactodea striata*, exposure to 10–1000 items/L of 63–250 μm PS microbeads for 10 days reduced growth, reproduction, and clearance rates (Xu et al., 2017). Coral species such as *Acropora* presented a 41% decrease in photosynthetic activity after 12 days of fiber MP exposure, and high concentrations (66–300 mg/L) disrupted calcification (Mendrik et al., 2021). Another study revealed that the exposure of Jacopever fish to PS MPs at a concentration of 106 particles per liter led to a significant decline in physiological performance, with weight gain decreasing by 65.4%, the growth rate decreasing by 65.9%, and the overall energy content decreasing by 9.5% (Yin et al., 2018).

1.3.1.3 *MPs in the aquatic food chain*

The ingestion of MPs by a wide range of aquatic organisms may disrupt food web dynamics, suppress zooplankton grazing, reduce the top-down control of phytoplankton, and result in algal overgrowth (Malinowski et al., 2023). In addition to direct uptake from the aquatic environment, organisms may also ingest MPs indirectly through trophic transfer, which occurs when they feed on prey that have previously consumed these particles (Walkinshaw et al., 2020). This trophic transfer can result in the bioaccumulation of MPs and their associated toxicants along the food chain, but evidence for biomagnification across levels is limited (Miller et al., 2020). Over time, organisms at relatively high trophic levels, including fish, seabirds, marine mammals, and benthic species such as crabs and shellfish, can accumulate high concentrations of MPs and the pollutants they carry. These substances can cause serious biological effects, such as hormonal disruption, reproductive disorders, and immune system compromise (Liang et al., 2024). Furthermore, since many of these organisms are part of the human diet, MPs and associated toxins can enter the human food chain, raising concerns about potential health risks (Egbeocha et al., 2018; Walkinshaw et al., 2020).

1.3.2 Terrestrial organisms

While most studies have focused on aquatic ecosystems, emerging research highlights the adverse impacts of MPs on terrestrial organisms. For example, the combined presence of PE MPs and antibiotics such as tetracycline, especially under drought conditions, has been shown to significantly disrupt the growth and physiological functions of rice plants. These interactions not only impair plant development but also suggest that soil fauna and soil contamination can interact with climatic stressors to amplify adverse outcomes, even in drought-resistant cultivars (de Souza Machado et al., 2018; Khan et al., 2023; Pastorino and Barceló, 2023).

1.3.2.1 *Effects of MPs on soil animals*

MPs in soil ecosystems have significant impacts on soil fauna and ecosystem functioning. Studies have shown that MPs can disrupt the structure of soil microbial communities by promoting the growth of certain bacterial groups (Yi et al., 2021). These tiny particles can induce oxidative stress, DNA damage, and other forms of cellular toxicity in soil fauna, potentially impairing their survival, reproduction, and ecological functions (Su et al., 2022). In addition, earthworms may ingest MPs, potentially altering their digestion, fecal production, and soil bioturbation, thereby impairing nutrient cycling and soil structure (Zhang et al., 2023). MPs significantly affect the growth, feeding, burrowing, and avoidance behavior of earthworms. Studies have shown that high concentrations of MPs, especially smaller particles such as nanoplastics, can reduce growth, increase mortality, disrupt feeding behavior, and decrease energy reserves in earthworms. These impacts include the accumulation of plastics in the digestive tract, tissue damage, and interference with immune function.

Moreover, MPs can increase the toxicity of heavy metals and organic pollutants, increasing avoidance behavior, which may contribute to increased migration of contaminants in soil (Cui et al., 2022).

1.3.2.2 *Effects of MPs on birds and mammals*

Sea and coastal birds are integral to ecosystem functioning, notably by mediating nutrient transfer between marine and terrestrial systems. Nonetheless, MPs have emerged as critical environmental hazards, impacting >200 bird species across diverse habitats (Carrasco et al., 2025; Signa et al., 2021; Yaghmour et al., 2025). Owing to their global distribution, varied diets, diverse habitats, and position at high trophic levels, birds are widely recognized as effective bioindicators for detecting environmental changes and pollution impacts, including those from persistent organic pollutants (POPs), heavy metals, and plastics. Notably, plastic ingestion by seabirds has been documented since the 1960s (Zeng et al., 2025). Recent evidence suggests that MPs can enter bird bodies not only through ingestion but also via inhalation. These particles have been detected in both avian feces and lung tissues, indicating multiple pathways of exposure and raising concerns about their potential health impacts on birds (Jiang et al., 2024; Wang et al., 2025). MPs can impair feeding behavior in seabirds, potentially resulting in malnutrition and mortality (Susanti et al., 2020). These particles may also interfere with nutrient uptake, reproductive success, and chick development. To assess the extent of plastic pollution and its toxicological impacts, researchers often utilize long-lived bird species, human-associated birds, and established avian model organisms (Susanti et al., 2020). Toxicological impacts include gastrointestinal damage, oxidative stress, and bioaccumulation of toxic chemicals (Carrasco et al., 2025). For example, a study on *Coturnix japonica* (a Japanese bird) demonstrated that MP ingestion led to reduced body biomass, increased malondialdehyde production, suppressed hepatic nitric oxide production, and altered enzyme activities in various organs (de Souza et al., 2022).

Mammals represent another group of organisms at risk from MP exposure. Research on rodents has demonstrated that MPs can cross the intestinal barrier, accumulate in internal organs, and induce a range of adverse physiological effects, suggesting potential health threats to both terrestrial mammals and humans (Banerjee and Shelver, 2021). Fig. 1.2 shows the main exposure routes of MPs in birds and terrestrial mammals, along with their potential target organs. The primary pathways include ingestion of contaminated food, water, or prey; inhalation of airborne MPs; and, less commonly, dermal contact. Once absorbed, MPs can enter the GIT and bloodstream, eventually accumulating in vital organs such as the liver, kidneys, lungs, brain, and spleen, posing potential health risks. In toxicological research, in vitro studies often reveal acute cellular impacts of MPs, whereas in vivo studies tend to reveal subchronic or chronic effects (Tang, 2024). Caco-2 cells, which are derived from the human intestinal epithelium, are widely used to model gut barrier responses, whereas other human and animal cell lines are used to assess organ-specific toxicity. Rodents, especially mice, are frequently employed in in vivo studies because of their genetic uniformity and sensitivity, with ingestion being the predominant exposure pathway (Lehner et al., 2020; Palaniappan et al., 2022).

Numerous studies have demonstrated the adverse effects of MPs, particularly those composed of PS and PE, on both animal models and cell lines. In mice, oral exposure to PS

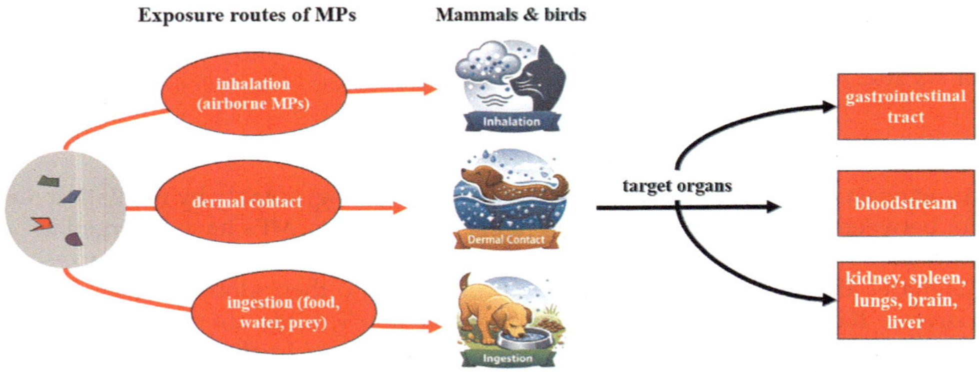

FIG. 1.2 Diagram of the main microplastic exposure routes in mammals and birds. Microplastic entry via inhalation of airborne particles; ingestion through food, water, or prey; dermal contact; and potential target organs, including the gastrointestinal tract, bloodstream, kidneys, spleen, lungs, brain, and liver.

MPs (0.01–0.5 mg/day, sizes 0.5–20 μm) for 28–90 days led to decreased testosterone levels, a reduced sperm count, impaired lipid metabolism, increased oxidative stress, liver inflammation, ovarian fibrosis, and reproductive dysfunction. In vitro experiments revealed that PS MPs at concentrations of 20–200 μg/mL induced reactive oxygen species generation and decreased cell viability, membrane integrity, and apoptosis in intestinal and hepatic cell lines (e.g., Caco-2 and HepG2). Exposure to PE MPs (0.125–2 mg/day, 10–150 μm) in mice resulted in testicular degeneration, ovarian cysts, kidney and lung damage, intestinal inflammation, and reduced birth rates and offspring body weights. Collectively, these findings highlight the potential toxicity of MPs to reproductive health, immune function, and gastrointestinal integrity (Tang, 2024).

1.3.2.3 MPs in the terrestrial food chain

Studies have demonstrated the transfer of MPs from soil to earthworms and then to livestock, such as chickens, cows, and goats, ultimately entering the human food chain through milk, eggs, and meat (Huerta Lwanga et al., 2017; Wahyudi et al., 2025). This pathway is driven mainly by agricultural practices that introduce MPs into soil, such as compost application, wastewater use, sewage sludge application, and plastic mulch films. These particles may then be taken up by crops or ingested by soil fauna, making terrestrial exposure a significant route for human intake through contaminated agricultural products and drinking water (Pathak et al., 2024). Therefore comprehensive studies are still needed to elucidate the mechanisms of MP uptake, internal distribution, and ecological effects across terrestrial species, especially regarding their movement along the soil–plant–animal continuum, which ultimately contributes to their entry into the food chain.

1.3.3 Impacts on specific species

In addition to environments heavily influenced by human activities, MPs have been detected in remote, seemingly pristine regions such as polar regions. Alarming evidence of their presence has been reported in the feces of apex predators, such as polar bears, as well as in the GIT and internal organs (lungs and trachea) of species such as the Adélie penguin (Bhattacharjee et al., 2024; Iyare et al., 2024). Moreover, MPs have been confirmed in terrestrial habitats, including tigers, leopard cats, and takins, further illustrating their global penetration into both aquatic and terrestrial ecosystems, regardless of their remoteness or ecological significance (Huang et al., 2025; Wu et al., 2024).

Given the detection of MPs in remote and ecologically sensitive habitats and their known ability to adsorb and transport POPs and other toxic substances, there is growing concern about their potential to act as vectors for harmful chemicals within living organisms. However, despite these risks, a comprehensive understanding of the physiological and ecological impacts of MPs on these species is lacking, underscoring the urgent need for further research to elucidate their effects on wildlife health and ecosystem integrity.

1.4 Global evidence and case studies

1.4.1 MP pollution around the world

MP pollution has emerged as a pervasive global concern, with particles being detected not only in surface waters and marine sediments but also across a broad spectrum of ecosystems, including farmlands, remote polar regions, and wildlife habitats (Bhattacharjee et al., 2024; Iyare et al., 2024; Kibria et al., 2022; Lackner and Branka, 2024). Studies have reported MPs in the bodies of various animals, ranging from farm livestock and fish to polar species such as penguins and polar bears. Moreover, MPs have been detected in animal tissues, including muscle and the GIT, and even in regions far from direct human activity, underscoring their extensive dispersion and environmental persistence (Bhattacharjee et al., 2024; Haave et al., 2021). Table 1.2 shows the widespread occurrence of MPs across a broad range of animal species from diverse ecosystems.

Ruminants such as cattle, sheep, and goats rely on gut microbes to digest plant polymers, such as cellulose. Ingesting large plastic items (e.g., bags, caps, and ropes) can lead to ruminal impaction, causing indigestion, bloating, and potentially death (Khan et al., 2024). Plastics can also contain toxic pollutants such as PAHs, PCBs, and pesticides, which may leach into the environment and bioaccumulate through the food chain. In poultry, larger plastic fragments are retained in crops and gizzards, whereas smaller MPs are excreted in feces, indicating partial breakdown during digestion. The accumulation of plastics in digestive organs may reduce food intake and organ size, with implications for both bird health and human food safety. In hares and rabbits, plastic particles have been linked to intestinal inflammation and exposure to agrochemical residues, suggesting similar risks to health and well-being in terrestrial mammals (Arena et al., 2017; Khan et al., 2024).

In the United States, cats and dogs are estimated to ingest 35 and 19 ng/kg polyethylene terephthalate (PET) daily, respectively, with fecal concentrations significantly higher than those in food (Zhang et al., 2019). In Indonesia, domesticated ducks from Salatiga, Semarang, and Magelang have 49, 39, and 27 MP filaments per individual, respectively (Susanti et al., 2021). In Portugal, MPs ranging in size from 1 to 10 μm were detected in 69.4% of ileum, 53.5% of lung, and 44.2% of kidney samples from 49 cats and dogs (Prata et al., 2020).

TABLE 1.2 Detection of MPs in various animal species across ecosystems.

Animal species	Habitat type	Detected location in organisms	References
Bivalves	Marine	Tissue	(Rabelo et al., 2025)
Fish	Marine/ freshwater	Gills, gastrointestinal tract, tissue	(Mukhopadhyay and Valsalan, 2025; Varol et al., 2025; Zhou et al., 2025)
Zooplankton	Marine	Whole organisms	(Corradini et al., 2025; Paternina-Ramos et al., 2025)
Sea lion	Marine	Scat	(Moreira-Mendieta et al., 2023)
Dolphin	Marine	Tissue	(Padula et al., 2025)
Earthworm	Terrestrial	Whole organisms	(Cui et al., 2022)
Ruminant animals	Terrestrial	Feces, tissue, gastrointestinal tract	(Khan et al., 2024)
Tiger	Terrestrial	Feces	(Huang et al., 2025)
Cat Leopard	Terrestrial	Scat	(Wu et al., 2024)
Polar Bear	Terrestrial	Feces	(Iyare et al., 2024)
Penguin	Terrestrial	Feces	(Bhattacharjee et al., 2024)

1.4.2 Important findings from recent studies

Recent studies have revealed that MPs pose a growing ecological threat to a wide range of animal species in both terrestrial and aquatic environments. Owing to their small size and persistence, these particles are readily ingested by organisms across various trophic levels and are often mistaken for food. In marine ecosystems, fish, bivalves, and crustaceans experience physiological stress, reduced reproductive success, and bioaccumulation of associated toxicants following MP exposure. Similarly, terrestrial species such as earthworms and birds experience adverse effects, including inflammation, oxidative damage, and disruption of feeding behavior. Moreover, MP pollution has become so pervasive that these contaminants have been detected in remote and pristine environments, including polar regions where direct human activity is minimal or absent. The presence of MPs in such isolated ecosystems raises significant concerns, as they can potentially disrupt fragile food webs and threaten the survival of cold-adapted species. These findings highlight the global scale of MP contamination and emphasize the urgent need for international efforts to mitigate its spread and ecological consequences.

1.5 Challenges and future outlook

1.5.1 Gaps in the knowledge of toxic effects

Despite growing concern about the toxicological effects of MPs, significant knowledge gaps remain regarding their long-term health impacts. While some studies have demonstrated adverse effects in specific animal models, the extent to which MPs pose a risk to human or other animal health remains unclear. Most existing research has focused on short-term exposure, leaving chronic and cumulative effects largely unexplored. Furthermore, the lack of standardized methods and quality assurance/quality control (QA/QC) protocols for detecting and quantifying MPs in biological tissues hinders consistent risk assessment. Systematic reviews by international authorities have confirmed the current insufficiency of evidence to draw definitive conclusions (Damaj et al., 2024). Addressing these gaps, alongside enhanced public education and the integration of citizen science into monitoring efforts, is essential to guide future policy and protect public health.

1.5.2 Challenges in detecting MPs

The detection of MPs in biological matrices, especially feces, poses significant analytical and methodological challenges. Fecal samples often contain low MP concentrations embedded in complex organic matter, necessitating meticulous sample collection, digestion (30% H_2O_2, 0.05 M Fe(II)), and contamination control to ensure reliable results. There is currently no universally accepted protocol, and inconsistencies in sample processing, polymer identification methods, and particle size thresholds limit comparability across studies. Techniques such as fourier-transform infrared spectroscopy (FTIR) and Raman spectroscopy are widely used. FTIR is practical for routine analysis but lacks resolution for particles < 20 μm, whereas Raman spectroscopy offers higher sensitivity but is costly and prone to fluorescence interference in organic-rich samples. Furthermore, standard density separation methods (e.g., NaCl or $ZnCl_2$ solutions) may fail to recover denser polymers such as PVC or polycarbonate (PC), potentially leading to an underestimation of MP loads. Additional concerns include interindividual variation, dietary influences, and contamination risks during handling. Overall, the lack of standardized methods and validated protocols hinders large-scale biomonitoring efforts and interpretations of MP exposure and risk in both humans and animals (Sangkham et al., 2025).

1.5.3 Importance of animal protection

Protecting animals from MP exposure is essential for safeguarding biodiversity and ecosystem stability. A wide range of species are vulnerable to MP contamination through direct ingestion or indirect transfer via the food chain. These exposures can result in digestive blockages, immune suppression, reproductive disorders, and behavioral changes. Integrating MP monitoring into conservation efforts is therefore critical for preventing long-term ecological damage.

Equally important is raising public awareness about the sources and impacts of plastic pollution. Educational campaigns, policy reforms, and behavioral shifts, such as reducing single-use plastic consumption, encouraging eco-friendly alternatives, and promoting responsible waste management, can significantly reduce plastic leakage into the environment. Protecting animals from MP threats requires both scientific intervention and widespread societal engagement.

1.5.4 Future research perspective

Future research on MPs must address several critical gaps to better understand their environmental and health implications:

- **Species-specific responses:** As noted by Jeong et al. (2024), more studies are needed on how different animal species, including wild and farm animals, mammals, and birds, especially terrestrial ones, respond to MP exposure.
- **Soil–plant–animal transfer:** Limited understanding of how MPs move through soil food webs and accumulate in animals is known.
- **Chemical interactions:** The role of MPs in transporting contaminants (e.g., heavy metals, antibiotics, PAHs, PCBs, and PFASs) needs further investigation.
- **Interdisciplinary collaboration:** Cross-disciplinary research is essential for creating risk assessment models and mitigation strategies. As noted by Fowzi et al. (2025), indicators such as the pollution hazard index, pollution load index, and potential ecological risk index are commonly used. However, their applicability and accuracy in real environmental settings require further validation and refinement.
- **Detection methods:** Submicron detection, automated imaging, and chemical analysis techniques should be further developed and standardized, especially for complex biological matrices such as animal tissues and feces, where MPs are often present at low concentrations and embedded within organic matter.
- **Toxicity mechanisms:** The long-term and chronic effects of MP exposure, including physiological adaptations, are poorly understood.

1.6 Conclusion

MP pollution has rapidly evolved from an emerging environmental concern into a pressing global threat, affecting ecosystems from densely populated urban coasts to remote polar regions. This chapter highlights the pervasive presence of MPs across aquatic and terrestrial habitats, demonstrating their ability to penetrate biological systems, disrupt physiological processes, and bioaccumulate across trophic levels. Research has shown that MPs impair feeding, reproduction, immunity, and behavior across fish, marine mammals, birds, ruminants, and domestic animals. For example, in the United States, cats and dogs are estimated to ingest up to 35 ng/kg of PET daily, whereas in Portugal, MPs are detected in the lungs, kidneys, and intestines of more than half of the sampled domestic pets. In Indonesia, domesticated ducks carry up to 49 MP filaments per individual. Laboratory studies in mice have shown that exposure to polystyrene MPs leads to reproductive dysfunction, oxidative stress, and liver inflammation.

In contrast, PE MPs have been linked to testicular degeneration, ovarian cysts, and multiorgan damage. In aquatic systems, exposure to MPs has led to reduced reproductive output in bivalves, disrupted calcification in corals (*Acropora*), and impaired growth and energy balance in fish such as Jacopever, with growth rate reductions of up to 65.9%. Despite the growing body of evidence, significant knowledge gaps remain, especially concerning chronic toxicity, MP interactions with co-contaminants, and transfer through terrestrial food webs. Moreover, the lack of standardized detection methods continues to hinder risk assessment. Addressing these issues requires species-specific toxicological studies, refined analytical methods, and improved ecological risk assessment frameworks. Public awareness and behavioral shifts are equally critical to limit plastic emissions. Ultimately, mitigating MP pollution is essential not only for protecting ecosystem integrity but also for safeguarding human health through shared exposure through environmental and food pathways.

References

Andrady, A.L., 2011. Microplastics in the marine environment. Mar. Pollut. Bull. 18793363. 62 (8), 1596–1605. doi:10.1016/j.marpolbul.2011.05.030.

Arena, M., Auteri, D., Barmaz, S., Bellisai, G., Brancato, A., Brocca, D., Bura, L., Byers, H., Chiusolo, A., Marques, D.C., Crivellente, F., De Lentdecker, C., De Maglie, M., Egsmose, M., Erdos, Z., Fait, G., Ferreira, L., Goumenou, M., Greco, L., Ippolito, A., Istace, F., Jarrah, S., Kardassi, D., Leuschner, R., Lythgo, C., Magrans, J. O., Medina, P., Miron, I., Molnar, T., Nougadere, A., Padovani, L., Morte, J.M.P., Pedersen, R., Reich, H., Sacchi, A., Santos, M., Serafimova, R., Sharp, R., Stanek, A., Streissl, F., Sturma, J., Szentes, C., Tarazona, J., Terron, A., Theobald, A., Vagenende, B., Verani, A., Villamar-Bouza, L., 2017. Peer review of the pesticide risk assessment of the active substance tribenuron-methyl. EFSA J. 18314732. 15 (7). Available from: efsa.onlinelibrary.wiley.com/journal/18314732/. doi:10.2903/j.efsa.2017.4912.

Arienzo, M., Ferrara, L., Trifuoggi, M., 2021. Research progress in transfer, accumulation and effects of microplastics in the oceans. J. Mar. Sci. Eng. 2077-1312. 9 (4), 433. doi:10.3390/jmse9040433.

Ashrafy, A., Liza, A.A., Islam, M.N., Billah, M.M., Arafat, S.T., Rahman, M.M., Rahman, S.M., 2023. Microplastics pollution: a brief review of its source and abundance in different aquatic ecosystems. J. Hazard. Mater. Adv. 27724166. 9. Available from: https://www.journals.elsevier.com/journal-of-hazardous-materials-advances. doi:10.1016/j.hazadv.2022.100215.

Avio, C.G., Gorbi, S., Regoli, F., 2015. Experimental development of a new protocol for extraction and characterization of microplastics in fish tissues: first observations in commercial species from Adriatic Sea. Mar. Environ. Res. 18790291. 111, 18–26. Available from: https://www.elsevier.com/locate/marenvrev. doi:10.1016/j.marenvres.2015.06.014.

Avio, C.G., Gorbi, S., Regoli, F., 2017. Plastics and microplastics in the oceans: from emerging pollutants to emerged threat. Mar. Environ. Res. 18790291. 128, 2–11. Available from: https://www.elsevier.com/locate/marenvrev. doi:10.1016/j.marenvres.2016.05.012.

Banerjee, A., Shelver, W.L., 2021. Micro- and nanoplastic-mediated pathophysiological changes in rodents, rabbits, and chickens: a review. J. Food Prot. 19449097. 84 (9), 1480–1495. Available from: https://watermark.silverchair.com/i0362-028x-84-9-1480.pdf. doi:10.4315/JFP-21-117.

Bank, M.S., 2022. Microplastic in the Environment: Pattern and Process Environmental Contamination Remediation and Management. Springer International Publishing. 10.1007/978-3-030-78627-4.

Bhattacharjee, S., Rathore, C., Naik, A., Saha, M., Tudu, P., Dastidar, P.G., Bhattacharyya, S., de Boer, J., Chaudhuri, P., 2024. Do microplastics accumulate in penguin internal organs? Evidence from Svenner island, Antarctica. Sci. Total Environ. 00489697. 951, 175361. doi:10.1016/j.scitotenv.2024.175361.

Browne, M.A., 2015. Sources and Pathways of Microplastics to Habitats Marine Anthropogenic Litter. Springer International Publishing, United States, pp. 229–244. Available from: http://dx.doi.org/10.1007/978-3-319-16510-3. doi:10.1007/978-3-319-16510-3_9.

Carrasco, L., Jiménez-Mora, E., Utrilla, M.J., Pizarro, I.T., Reglero, M.M., Román, L.R.-S., Martin-Maldonado, B., 2025. Birds as bioindicators: revealing the widespread impact of microplastics. Birds. 26736004. 6 (1). Available from: https://www.mdpi.com/journal/birds. doi:10.3390/birds6010010.

Chae, Y., An, Y.J., 2017. Effects of micro- and nanoplastics on aquatic ecosystems: current research trends and perspectives. Mar. Pollut. Bull. 18793363. 124 (2), 624–632. Available from: https://www.elsevier.com/locate/marpolbul. doi:10.1016/j.marpolbul.2017.01.070.

Chen, X., Zhou, S., Liu, Y., Feng, Z., Mu, C., Zhang, T., 2024. The combined effects of microplastics and bisphenol-A on the innate immune system response and intestinal microflora of the swimming crab *Portunus trituberculatus*. Aquat. Toxicol. 0166445X. 268, 106855. doi:10.1016/j.aquatox.2024.106855.

Corradini, R., Menéndez, M.C., Gonella, F.M., Baleani, C.A., Gomes Teixeira, V., Larrea Valdivia, A.E., Valenzuela Huillca, C., Oliva, A.L., Arias, A.H., 2025. Microplastic ingestion by zooplankton in surf zone waters of sandy beaches: are copepods potential reservoirs of these emerging pollutants? Environ. Pollut. 18736424. 383. Available from: https://www.sciencedirect.com/science/journal/02697491. doi:10.1016/j.envpol.2025.126772.

Cui, W., Gao, P., Zhang, M., Wang, L., Sun, H., Liu, C., 2022. Adverse effects of microplastics on earthworms: a critical review. Sci. Total Environ. 00489697. 850, 158041. doi:10.1016/j.scitotenv.2022.158041.

Damaj, S., Trad, F., Goevert, D., Wilkesmann, J., 2024. Bridging the gaps between microplastics and human health. Microplastics. 2673-8929. 3 (1), 46–66. doi:10.3390/microplastics3010004.

de Souza Machado, A.A., Kloas, W., Zarfl, C., Hempel, S., Rillig, M.C., 2018. Microplastics as an emerging threat to terrestrial ecosystems. Glob. Change Biol. 13652486. 24 (4), 1405–1416. Available from: https://www.blacksci.co.uk/~cgilib/jnlpage.bin?Journal=gcb&File=gcb&Page=aims. doi:10.1111/gcb.14020.

de Souza, S.S., Freitas, Í.N., de Oliveira Gonçalves, S., Luz, T.M.D., Araújo, A.P.D.C., Rajagopal, R., Balasubramani, G., Rahman, M.M., Malafaia, G., 2022. Toxicity induced via ingestion of naturally-aged polystyrene microplastics by a small-sized terrestrial bird and its potential role as vectors for the dispersion of these pollutants. SSRN. 15565068. Available from: https://www.ssrn.com/index.cfm/en/. doi:10.2139/ssrn.4056816.

Duis, K., Coors, A., 2016. Microplastics in the aquatic and terrestrial environment: sources (with a specific focus on personal care products), fate and effects. Environ. Sci. Eur. 2190-4707. 28 (1). doi:10.1186/s12302-015-0069-y.

Egbeocha, C.O., Malek, S., Emenike, C.U., Milow, P., 2018. Feasting on microplastics: ingestion by and effects on marine organisms. Aquat. Biol. 1864-7782. 27, 93–106. doi:10.3354/ab00701.

Eriksen, M., Mason, S., Wilson, S., Box, C., Zellers, A., Edwards, W., Farley, H., Amato, S., 2013. Microplastic pollution in the surface waters of the Laurentian Great Lakes. Mar. Pollut. Bull. 0025326X. 77 (1-2), 177–182. doi:10.1016/j.marpolbul.2013.10.007.

Fernández-Míguez, M., Puvanendran, V., Burgerhout, E., Presa, P., Tveiten, H., Vorkamp, K., Hansen, Ø.J., Johansson, G.S., Bogevik, A.S., 2023. Effects of weathered polyethylene microplastic ingestion on sexual maturation, fecundity and egg quality in maturing broodstock Atlantic cod *Gadus morhua*. Environ. Pollut. 02697491. 320, 121053. doi:10.1016/j.envpol.2023.121053.

Fonte, E., Ferreira, P., Guilhermino, L., 2016. Temperature rise and microplastics interact with the toxicity of the antibiotic cefalexin to juveniles of the common goby (*Pomatoschistus microps*): post-exposure predatory behaviour, acetylcholinesterase activity and lipid peroxidation. Aquat. Toxicol. 0166445X. 180, 173–185. doi:10.1016/j.aquatox.2016.09.015.

Fowzi, M.M., Ebrahimpour, K., Dehnavi, A., Jamshidi, S., Andaluri, G., 2025. Microplastics as emerging contaminants in municipal solid waste compost: distribution, characterization, and ecological risk. Environ. Res. 00139351. 284, 122248. doi:10.1016/j.envres.2025.122248.

Gall, S.C., Thompson, R.C., 2015. The impact of debris on marine life. Mar. Pollut. Bull. 0025326X. 92 (1-2), 170–179. doi:10.1016/j.marpolbul.2014.12.041.

Galloway, T.S., Cole, M., Lewis, C., 2017. Interactions of microplastic debris throughout the marine ecosystem. Nat. Ecol. Evol. 2397334X. 1 (5). Available from: https://www.nature.com/natecolevol/. doi:10.1038/s41559-017-0116.

Gheisari, L., Ebrahimpour, K., Fowzi, M.M., Pourzamani, H., 2025. Assessment of microplastic contamination in compost leachate: insights from a municipal compost plant in Isfahan, Iran. Environ. Monit. Assess. 1573-2959. 197 (6). doi:10.1007/s10661-025-14079-2.

Ghosh, T., 2025. Microplastics bioaccumulation in fish: its potential toxic effects on hematology, immune response, neurotoxicity, oxidative stress, growth, and reproductive dysfunction. Toxicol. Rep. 22147500. 14, 101854. doi:10.1016/j.toxrep.2024.101854.

Gunawardhana, G.M.S.S., Perera, U.L.H.P., Ratnayake, A.S., Weerasingha, W.A.D.B., Subasinghe, H.C.S., 2024. Formation of secondary microplastics during degradation of plastics originating from the MV X-Press Pearl maritime disaster. Discov. Environ. 27319431. 2 (1). Available from: https://link.springer.com/journal/44274. doi:10.1007/s44274-024-00044-2.

Haave, M., Gomiero, A., Schönheit, J., Nilsen, H., Olsen, A.B., 2021. Documentation of microplastics in tissues of wild coastal animals. Front. Environ. Sci. 2296665X. 9. Available from: https://journal.frontiersin.org/journal/environmental-science. doi:10.3389/fenvs.2021.575058.

Hamidian, A.H., Feizi, M., 2025. Investigating the behavioral implications of microplastic exposure in animal species. Ann. Anim. Sci. 23008733. Available from: https://sciendo.com/journal/aoas. doi:10.2478/aoas-2025-0028.

Hasan, A.K.M.M., Hamed, M., Hasan, J., Martyniuk, C.J., Niyogi, S., Chivers, D.P., 2024. A review of the neuro-behavioural, physiological, and reproductive toxicity of microplastics in fishes. Ecotoxicol. Environ. Saf. 10902414. 282. Available from: https://www.sciencedirect.com/science/journal/01476513. doi:10.1016/j.ecoenv.2024.116712.

Huang, Z., Liu, D., Cheng, W., Zhang, W., Zhijian, H., Huang, B., Guo, S., Zhao, B., Wang, Y., Zhang, Y., Jiang, G., 2025. Microplastics in the Amur tiger's habitat: occurrence, characteristics, and risk assessment. J. Hazard. Mater. 03043894. 493, 138380. doi:10.1016/j.jhazmat.2025.138380.

Huerta Lwanga, E., Mendoza Vega, J., Ku Quej, V., Chi, J.D.L.A., Sanchez Del Cid, L., Chi, C., Escalona Segura, G., Gertsen, H., Salánki, T., van der Ploeg, M., Koelmans, A.A., Geissen, V., 2017. Field evidence for transfer of plastic debris along a terrestrial food chain. Sci. Rep. 20452322. 7 (1). Available from: https://www.nature.com/srep/index.html. doi:10.1038/s41598-017-14588-2.

Ibrahim, Y.S., Razak, N.I.A., Roslan, N.S., Yusof, K.M.K.K., Ali, A.A.M., Omar, N.F., Chinglenthoiba, C., Mohamad, N.N., Anuar, S.T., 2025. Morphochemical information on microplastic fibers found in edible tissue of local commercial fishes from the South China Sea and the Straits of Malacca for potential human consumption. Environ. Sci. Adv. 27547000. 4 (6), 964–979. Available from: https://www.rsc.org/journals-books-databases/about-journals/environmental-science-advances. doi:10.1039/d4va00425f.

Islam, M.S., Karim, M.R., Islam, M.T., Oishi, H.T., Tasnim, Z., Das, H., Kabir, A.H.M.E., Sekine, M., 2023. Abundance, characteristics, and ecological risks of microplastics in the riverbed sediments around Dhaka city. Sci. Total Environ. 18791026. 877. Available from: https://www.elsevier.com/locate/scitotenv. doi:10.1016/j.scitotenv.2023.162866.

Islam, M.S., Kekre, K.M., Shah, T.A., Tsai, P.C., Ponnusamy, V.K., Andaluri, G., 2025. Unraveling the complexities of microplastics and PFAS synergy to foster sustainable environmental remediation and ecosystem protection: a critical review with novel insights. J. Hazard. Mater. Adv. 27724166. 17. Available from: https://www.sciencedirect.com/science/journal/27724166. doi:10.1016/j.hazadv.2025.100621.

Iyare, P.U., Vanderlip, H.L., Dias, M., Provencher, J.F., Zou, S., Lougheed, S.C., Van Coeverden de Groot, P.J., Whitelaw, G., Branigan, M., Dyck, M., Orihel, D.M., 2024. An assessment of microplastics in fecal samples from polar bears (*Ursus maritimus*) in Canada's North. Arctic Sci. 2368-7460. 10 (2), 409–423. doi:10.1139/as-2023-0060.

Jabeen, K., Li, B., Chen, Q., Su, L., Wu, C., Hollert, H., Shi, H., 2018. Effects of virgin microplastics on goldfish (*Carassius auratus*). Chemosphere. 00456535. 213, 323–332. doi:10.1016/j.chemosphere.2018.09.031.

Jeong, E., Lee, J.-Y., Redwan, M., 2024. Animal exposure to microplastics and health effects: a review. Emerg. Contam. 24056650. 10 (4), 100369. doi:10.1016/j.emcon.2024.100369.

Jiang, H., Cheng, H., Shiyue, W., Hongshan, L., Chen, H., Zhenghao, L., Yao, X., Zhang, Y., Chen, Y., Chen, S., Chen, S., Zheng, L., Sui, Y., Shao, R., 2024. Microplastics footprint in nature reserves-a case study on the microplastics in the guano from Yancheng Wetland Rare Birds National Nature Reserve, China. Environ. Res. 00139351. 256, 119252. doi:10.1016/j.envres.2024.119252.

Kabir, A.H.M.E., Sekine, M., Imai, T., Yamamoto, K., Kanno, A., Higuchi, T., 2021. Assessing small-scale freshwater microplastics pollution, land-use, source-to-sink conduits, and pollution risks: perspectives from Japanese rivers polluted with microplastics. Sci. Total Environ. 18791026. 768. Available from: https://www.elsevier.com/locate/scitotenv. doi:10.1016/j.scitotenv.2020.144655.

Kavery Chengappa, S., Rao, A., Aparna, K.S., Jodalli, P.S., Kudpi, R.S., 2023. Microplastic content of over-the-counter toothpastes - a systematic review. F1000Research. 2046-1402. 12, 390. doi:10.12688/f1000research.132035.1.

Khan, A., Qadeer, A., Wajid, A., Ullah, Q., Rahman, S.U., Ullah, K., Safi, S.Z., Ticha, L., Skalickova, S., Chilala, P., Bernatova, S., Samek, O., Horky, P., 2024. Microplastics in animal nutrition: occurrence, spread, and hazard in animals. J. Agric. Food Res. 26661543. 17. Available from: https://www.sciencedirect.com/science/journal/26661543. doi:10.1016/j.jafr.2024.101258.

Khan, K.Y., Ali, B., Ghani, H.U., Fu, L., Shohag, M.J.U.I., Zhang, S., Cui, X., Xia, Q., Tan, J., Ali, Z., Guo, Y., 2023. Single and combined effect of tetracycline and polyethylene microplastics on two drought contrasting cultivars of

Oryza sativa L. (Rice) under drought stress. Environ. Toxicol. Pharmacol. 18727077. 101. Available from: https://www.elsevier.com/locate/etap. doi:10.1016/j.etap.2023.104191.

Khanjani, M.H., Mohammadi, A., 2025. Microplastic pollution in aquatic ecosystems: a comprehensive review of impacts on aquatic animals. Ann. Anim. Sci. 23008733. Available from: https://sciendo.com/journal/aoas. doi:10.2478/aoas-2025-0051.

Kibria, G., Nugegoda, D., Yousuf Haroon, A.K., 2022. Microplastic (MP) Pollution in the Context of Occurrence, Distribution, Composition and Concentration in Surface Waters and Sediments: A Global Overview. Springer Science and Business Media LLC, pp. 133–166. doi:10.1007/978-3-030-89220-3_7.

Koelmans, A.A., Bakir, A., Burton, G.A., Janssen, C.R., 2016. Microplastic as a vector for chemicals in the aquatic environment: critical review and model-supported reinterpretation of empirical studies. Environ. Sci. Technol. 15205851. 50 (7), 3315–3326. Available from: http://pubs.acs.org/journal/esthag. doi:10.1021/acs.est.5b06069.

Lackner, M., Branka, M., 2024. Microplastics in farmed animals—a review. Microplastics. 2673-8929. 3 (4), 559–588. doi:10.3390/microplastics3040035.

Lee, H., Lee, H.J., Kwon, J.H., 2019. Estimating microplastic-bound intake of hydrophobic organic chemicals by fish using measured desorption rates to artificial gut fluid. Sci. Total Environ. 18791026. 651, 162–170. Available from: https://www.elsevier.com/locate/scitotenv. doi:10.1016/j.scitotenv.2018.09.068.

Lehner, R., Wohlleben, W., Septiadi, D., Landsiedel, R., Petri-Fink, A., Rothen-Rutishauser, B., 2020. A novel 3D intestine barrier model to study the immune response upon exposure to microplastics. Arch. Toxicol. 0340-5761. 94 (7), 2463–2479. doi:10.1007/s00204-020-02750-1.

Liang, J., Ji, F., Wang, H., Zhu, T., Rubinstein, J., Worthington, R., Abdullah, A.L.B., Tay, Y.J., Zhu, C., George, A., Li, Y., Han, M., 2024. Unraveling the threat: microplastics and nano-plastics' impact on reproductive viability across ecosystems. Sci. Total Environ. 18791026. 913. Available from: https://www.sciencedirect.com/science/journal/00489697. doi:10.1016/j.scitotenv.2023.169525.

Lu, Y., Zhang, Y., Deng, Y., Jiang, W., Zhao, Y., Geng, J., Ding, L., Ren, H., 2016. Uptake and accumulation of polystyrene microplastics in zebrafish (*Danio rerio*) and toxic effects in liver. Environ. Sci. Technol. 0013-936X. 50 (7), 4054–4060. doi:10.1021/acs.est.6b00183.

Lusher, A.L., Tirelli, V., O'Connor, I., Officer, R., 2015. Microplastics in Arctic polar waters: the first reported values of particles in surface and sub-surface samples. Sci. Rep. 20452322. 5. Available from: https://www.nature.com/srep/index.html. doi:10.1038/srep14947.

Malinowski, C.R., Searle, C.L., Schaber, J., Höök, T.O., 2023. Microplastics impact simple aquatic food web dynamics through reduced zooplankton feeding and potentially releasing algae from consumer control. Sci. Total Environ. 18791026. 904. Available from: https://www.elsevier.com/locate/scitotenv. doi:10.1016/j.scitotenv.2023.166691.

Massahi, T., Sharafi, M., Ahmadi, B., Parnoon, K., Hossini, H., 2024. Microplastic pollution in compost derived from mixed municipal waste in Kermanshah City: abundance, characteristics, and ecological risk evaluation. Water, Air, & Soil Pollution. 0049-6979. 235 (9). doi:10.1007/s11270-024-07357-1.

Mendrik, F.M., Henry, T.B., Burdett, H., Hackney, C.R., Waller, C., Parsons, D.R., Hennige, S.J., 2021. Species-specific impact of microplastics on coral physiology. Environ. Pollut. 02697491. 269, 116238. doi:10.1016/j.envpol.2020.116238.

Merrill, G.B., Hermabessiere, L., Rochman, C.M., Nowacek, D.P., 2023. Microplastics in marine mammal blubber, melon, & other tissues: evidence of translocation. Environ. Pollut. 18736424. 335. Available from: https://www.journals.elsevier.com/environmental-pollution. doi:10.1016/j.envpol.2023.122252.

Meyer, W.K., Jamison, J., Richter, R., Woods, S.E., Partha, R., Kowalczyk, A., Kronk, C., Chikina, M., Bonde, R.K., Crocker, D.E., Gaspard, J., Lanyon, J.M., Marsillach, J., Furlong, C.E., Clark, N.L., 2018. Ancient convergent losses of Paraoxonase 1 yield potential risks for modern marine mammals. Science. 10959203. 361 (6402), 591–594. Available from: http://science.sciencemag.org/content/361/6402/591/tab-pdf. doi:10.1126/science.aap7714.

Miller, M.E., Hamann, M., Kroon, F.J., Mukherjee, A., 2020. Bioaccumulation and biomagnification of microplastics in marine organisms: a review and meta-analysis of current data. PLOS ONE. 1932-6203. 15 (10), e0240792. doi:10.1371/journal.pone.0240792.

Moreira-Mendieta, A., Garcia-Garin, O., Muñoz-Pérez, J.P., Urquía, D.O., Drago, M., Borrell, A., Páez-Rosas, D., 2023. Detection and quantification of microplastic pollution in the endangered Galapagos sea lion. Sci. Total Environ. 18791026. 896. Available from: https://www.elsevier.com/locate/scitotenv. doi:10.1016/j.scitotenv.2023.166223.

Mukhopadhyay, P., Valsalan, S.A., 2025. Comparative evaluation of biodegradable microplastic presence in edible and non-edible tissues of cage-cultured and wild fishes of Periyar River. Environ. Pollut. 02697491. 373, 126073. doi:10.1016/j.envpol.2025.126073.

Nabi, G., Ahmad, S., Ullah, S., Zada, S., Sarfraz, M., Guo, X., Ismail, M., Wanghe, K., 2022. The adverse health effects of increasing microplastic pollution on aquatic mammals. J. King Saud Univ. Sci. 10183647. 34 (4), 102006. doi:10.1016/j.jksus.2022.102006.

Nguyen, M.K., Rakib, M.R.J., Lin, C., Hung, N.T.Q., Le, V.G., Nguyen, H.L., Malafaia, G., Idris, A.M., 2023. A comprehensive review on ecological effects of microplastic pollution: an interaction with pollutants in the ecosystems and future perspectives. TrAC Trends Anal. Chem. 18793142. 168. Available from: https://www.elsevier.com/locate/trac. doi:10.1016/j.trac.2023.117294.

Padula, A.D., Ronda, A.C., Rodríguez Pirani, L.S., Picone, A.L., Romano, R.M., Giardino, G.V., Seco Pon, J.P., Machado, R., Gerpe, M., Rodriguez, D., Denuncio, P., 2025. Microplastics in the digestive tract of an endangered cetacean of the Southwest Atlantic Ocean: the franciscana dolphin. Mar. Pollut. Bull. 18793363. 210. Available from: https://www.sciencedirect.com/science/journal/0025326X. doi:10.1016/j.marpolbul.2024.117348.

Palaniappan, S., Sadacharan, C.M., Rostama, B., 2022. Polystyrene and polyethylene microplastics decrease cell viability and dysregulate inflammatory and oxidative stress markers of MDCK and L929 cells in vitro. Expo. Health. 24519685. 14 (1), 75–85. Available from: http://link.springer.com/journal/volumesAndIssues/12403. doi:10.1007/s12403-021-00419-3.

Pastorino, P., Barceló, D., 2023. Microplastics and their environmental effects. Environ. Toxicol. Pharmacol. 13826689. 104, 104324. doi:10.1016/j.etap.2023.104324.

Paternina-Ramos, A., Adriana Gracia, C., Rangel-Buitrago, N., 2025. Zooplankton assemblage and microplastics associated with a coastal sandspit (Tubará, Atlántico). Mar. Pollut. Bull. 0025326X. 215, 117853. doi:10.1016/j.marpolbul.2025.117853.

Pathak, G., Mangla, S., Bhan, V., 2024. Microplastics in the Food Chain Microplastics: Origins, Risks, and Mitigation. Elsevier, India, pp. 31–57. Available from: https://www.sciencedirect.com/book/9780443298042. doi:10.1016/B978-0-443-29804-2.00002-0.

Porcino, N., Bottari, T., Mancuso, M., 2023. Is wild marine biota affected by microplastics? Animals. 2076-2615. 13 (1), 147. doi:10.3390/ani13010147.

Prata, J.C., da Costa, J.P., Lopes, I., Duarte, A.C., Rocha-Santos, T., 2020. Environmental exposure to microplastics: an overview on possible human health effects. Sci. Total Environ. 18791026. 702. Available from: https://www.elsevier.com/locate/scitotenv. doi:10.1016/j.scitotenv.2019.134455.

Prietl, B., Meindl, C., Roblegg, E., Pieber, T.R., Lanzer, G., Fröhlich, E., 2014. Nano-sized and micro-sized polystyrene particles affect phagocyte function. Cell Biol. Toxicol. 0742-2091. 30 (1), 1–16. doi:10.1007/s10565-013-9265-y.

Rabelo, E.F., Oliveira, R.C.M., Freitas, J.C.D.N., Silva, M.L.C.D., Brabo, L., Ziveri, P., Grelaud, M., Bessa, F., Sobral, P., Guilherme, A., Giarrizzo, T., Soares, M.O., 2025. Microplastic pollution in bivalve species from tropical mangrove habitats. Reg. Stud. Mar. Sci. 23524855. 87. Available from: https://www.sciencedirect.com/science/journal/23524855. doi:10.1016/j.rsma.2025.104227.

Sangkham, S., Phairuang, W., Sakunkoo, P., Ta, A.T., 2025. A review on microplastics in mammalian feces: monitoring techniques and associated challenges. Environ. Chall. 26670100. 20, 101217. doi:10.1016/j.envc.2025.101217.

Santos, R.G., Machovsky-Capuska, G.E., Andrades, R., 2021. Plastic ingestion as an evolutionary trap: toward a holistic understanding. Science. 10959203. 373 (6550), 56–60. Available from: https://science.sciencemag.org/content/373/6550/56.full. doi:10.1126/science.abh0945.

Sarkar, S., Diab, H., Thompson, J., 2023. Microplastic pollution: chemical characterization and impact on wildlife. Int. J. Environ. Res. Public Health. 1660-4601. 20 (3), 1745. doi:10.3390/ijerph20031745.

Savoca, M.S., McInturf, A.G., Hazen, E.L., 2021. Plastic ingestion by marine fish is widespread and increasing. Glob. Change Biol. 13652486. 27 (10), 2188–2199. Available from: http://onlinelibrary.wiley.com/journal/10.1111/(ISSN)1365-2486. doi:10.1111/gcb.15533.

Sharda, N., Antil, J., Yadav, S., Tripathi, P., Lamba, M., Kumar, N., 2024. Microplastics and biodiversity: a comprehensive analysis of their ecological impact. Int. J. Agric. Sci. 0973130X. 20 (RAAEALSES-2024), 78–94. doi:10.15740/has/ijas/20/raaealses-2024/78-94.

Signa, G., Mazzola, A., Vizzini, S., 2021. Seabird influence on ecological processes in coastal marine ecosystems: an overlooked role? A critical review. Estuar. Coast. Shelf Sci. 02727714. 250, 107164. doi:10.1016/j.ecss.2020.107164.

Singh, A., Mishra, B.K., 2023. Microbeads in personal care products: an overlooked environmental concern. J. Clean. Prod. 09596526. 427, 139082. doi:10.1016/j.jclepro.2023.139082.

Su, P., Bu, N., Li, Z., Wang, J., Zhang, D., Chu, K., Yao, Y., Sun, Q., Luo, Y., Zhang, R., Su, X., Wang, Z., 2022. Hierarchical and cascading changes in the functional traits of soil animals induced by microplastics: a meta-analysis. SSRN Electron. J. 1556–5068. doi:10.2139/ssrn.4156269.

Susanti, N.K.Y., Mardiastuti, A., Wardiatno, Y., 2020. Microplastics and the impact of plastic on wildlife: a literature review. IOP Conf. Ser. Earth Environ. Sci. 1755-1307. 528 (1), 012013. doi:10.1088/1755-1315/528/1/012013.

Susanti, R., Yuniastuti, A., Fibriana, F., 2021. The evidence of microplastic contamination in central javanese local ducks from intensive animal husbandry. Water, Air, & Soil Pollution. 0049-6979. 232 (5). doi:10.1007/s11270-021-05142-y.

Tang, K.H.D., 2024. A review of the toxic effects of microplastics based on studies on mammals and mammalian cell lines. Environ. Sci. Adv. 27547000. 3 (12), 1669–1678. Available from: https://www.rsc.org/journals-books-databases/about-journals/environmental-science-advances. doi:10.1039/d4va00227j.

Varol, M., Karakaya, G., Arısoy, G., Çelik, B., 2025. Comprehensive analysis of microplastics in water, sediment and fish from a large recreational lake. Environ. Res. 00139351. 279 (Pt 1), 121799. doi:10.1016/j.envres.2025.121799.

Wahyudi, A., Setyobudi, R.H., Iswahyudi, I., Garfansa, M.P., Savitri, E.S., Prahardika, B.A., Patmawati, P., Putri, R. D., Dodi, D.S., Putra, N.S., Yunita, E., Wahono, S.K., 2025. Determination and quantification of microplastics in goat farms. Bioresour. Technol. Rep. 2589014X. 31. Available from: https://www.sciencedirect.com/science/journal/2589014X. doi:10.1016/j.biteb.2025.102199.

Walkinshaw, C., Lindeque, P.K., Thompson, R., Tolhurst, T., Cole, M., 2020. Microplastics and seafood: lower trophic organisms at highest risk of contamination. Ecotoxicol. Environ. Saf. 01476513. 190, 110066. doi:10.1016/j.ecoenv.2019.110066.

Wang, M., Zhou, P., DuBay, S., Zhang, S., Yang, Z., Wang, Y., Zhang, J., Cao, Y., Hu, Z., He, X., Wang, S., Li, M., Fan, C., Zou, B., Zhou, C., Wu, Y., 2025. Assessing microplastic and nanoplastic contamination in bird lungs: evidence of ecological risks and bioindicator potential. J. Hazard. Mater. 03043894. 487, 137274. doi:10.1016/j.jhazmat.2025.137274.

Wu, T., Shu, X., Wang, C., Li, W., Zhu, D., Wang, J., Zhang, Y., Yang, X., Wang, X., 2024. Microplastic pollution of threatened terrestrial wildlife in nature reserves of Qinling Mts., China. Glob. Ecol. Conserv. 23519894. 51, e02865. doi:10.1016/j.gecco.2024.e02865.

Xu, X.-Y., Lee, W.T., Chan, A.K.Y., Lo, H.S., Shin, P.K.S., Cheung, S.G., 2017. Microplastic ingestion reduces energy intake in the clam *Atactodea striata*. Mar. Pollut. Bull. 0025326X. 124 (2), 798–802. doi:10.1016/j.marpolbul.2016.12.027.

Yaghmour, F., Naqbi, M.J.A., Kanan, S., Dronjak, L., Knuteson, S., Mohammed, A., Almazrouei, S., Alqaydi, M., Alzaabi, A., Al Bousi, M., Al Marashda, A., Al Naqbi, H., Al Naqbi, J., Samara, F., 2025. Seabirds in a flap: the ingestion of meso/macro marine debris, microplastics and oil/tar by marine and coastal birds in the United Arab Emirates. Mar. Pollut. Bull. 0025326X. 217, 118063. doi:10.1016/j.marpolbul.2025.118063.

Yi, M., Zhou, S., Zhang, L., Ding, S., 2021. The effects of three different microplastics on enzyme activities and microbial communities in soil. Water Environ. Res. 1061-4303. 93 (1), 24–32. doi:10.1002/wer.1327.

Yin, L., Chen, B., Xia, B., Shi, X., Qu, K., 2018. Polystyrene microplastics alter the behavior, energy reserve and nutritional composition of marine jacopever (*Sebastes schlegelii*). J. Hazard. Mater. 03043894. 360, 97–105. doi:10.1016/j.jhazmat.2018.07.110.

Zantis, L.J., Carroll, E.L., Nelms, S.E., Bosker, T., 2021. Marine mammals and microplastics: a systematic review and call for standardisation. Environ. Pollut. 18736424. 269. Available from: https://www.journals.elsevier.com/environmental-pollution. doi:10.1016/j.envpol.2020.116142.

Zeng, Y., Cai, J., Zhu, Y., Wang, J., Guo, R., Jian, L., Zheng, X., Mai, B.-X., 2025. Species-specific accumulation of microplastics in different bird species from South China: a comprehensive analysis. J. Hazard. Mater. 03043894. 483, 136607. doi:10.1016/j.jhazmat.2024.136607.

Zhang, J., Li, Z., Zhou, X., Ding, W., Wang, X., Zhao, M., Li, H., Zou, G., Chen, Y., 2023. Long-term application of organic compost is the primary contributor to microplastic pollution of soils in a wheat–maize rotation. Sci. Total Environ. 00489697. 866, 161123. doi:10.1016/j.scitotenv.2022.161123.

Zhang, J., Wang, L., Kannan, K., 2019. Polyethylene terephthalate and polycarbonate microplastics in pet food and feces from the United States. Environ. Sci. Technol. 0013-936X. 53 (20), 12035–12042. doi:10.1021/acs.est.9b03912.

Zhang, Y., Wang, C., Jia, R., Long, H., Zhou, J., Sun, G., Wang, Y., Zhang, Z., Rong, X., Jiang, Y., 2024. Transfer from ciliate to zebrafish: unveiling mechanisms and combined effects of microplastics and heavy metals. J. Hazard. Mater. 03043894. 479, 135645. doi:10.1016/j.jhazmat.2024.135645.

Zhou, C., Ran, M., Wang, Z., Li, M., Wang, X., Zhang, C., Song, Z., 2025. Baseline biomonitoring of microplastic pollution in freshwater fish from the Chishui River, China: insights into accumulation patterns and influencing factors. J. Hazard. Mater. 03043894. 495, 139055. doi:10.1016/j.jhazmat.2025.139055.

Zorin, I.M., Chelushkin, P.S., Senchukova, A.S., Fetin, P.A., Kenny, J.M., Lyulin, S.V., 2025. Study on primary microplastics in cosmetics: their isolation, spectral and thermal analysis. Anal. Methods. 17599679. 17 (22), 4599–4606. Available from: http://pubs.rsc.org/en/journals/journal/ay. doi:10.1039/d5ay00178a.

CHAPTER

2

Scientific advancements in microplastic pollution research

Shuvra Debnath[a], *Xinlei Huang*[a], *and Suvash C. Saha*[b]

[a]Faculty of Engineering and Information Technology, University of Technology Sydney, Ultimo, NSW, Australia

[b]School of Mechanical and Mechatronic Engineering, University of Technology Sydney, Sydney, NSW, Australia

2.1 Introduction

Plastics and their microscopic residues are now embedded in virtually every sector of modern society. Despite ongoing advances in recycling, mismanaged end-of-life handling remains a major driver of plastic pollution (Pabortsava and Lampitt, 2020), with only a small fraction of plastics recycled in some jurisdictions (Gu and Ozbakkaloglu, 2016). Annual plastic waste is on the order of hundreds of millions of tons (Murray, 2023), and environmental stockpiles are projected to grow substantially in the near term (Brahney et al., 2020). Over time, macroplastics fragment into microplastics (MPs) and, via coupled physical, chemical, and microbial weathering (e.g., UV irradiation, mechanical abrasion, biofilm activity), further into nanoscale particles (Song et al., 2017; Wei et al., 2021; Zhu et al., 2019).

Over the last two decades, particles spanning fragments, fibers, films, foams, and beads from consumer products, textiles, tires, paints, and industrial feedstocks have been reported in virtually all environmental compartments, from mountain air and agricultural soils to river mouths and the deep sea. The smallest residues, MPs and nanoplastics (NPs), have shifted from a peripheral curiosity to a central focus of monitoring, exposure assessment, and policy. MPs are commonly defined as plastic particles <5 mm with diverse morphologies (fibers, fragments, films, beads) and polymers (e.g., polyethylene (PE), polypropylene (PP), polystyrene (PS)). In contrast, NPs denote the colloidal fraction whose upper bound is operational rather than universal (≤ 100 nm, by analogy to engineered nanomaterials, or ≤ 1000 nm in colloid-informed frameworks) (Gigault et al., 2018; Hartmann et al., 2019).

This chapter highlights scientific advancements rather than the general background for two reasons. First, measurement science for micronanoplastics (MNPs) has progressed

Microplastics in Animal and Human Physiology
https://doi.org/10.1016/B978-0-443-45635-0.00006-9

from exploratory surveys to increasingly standardized, automated, and interlaboratory-validated workflows, directly improving data quality, comparability, and policy relevance. International standards now articulate cross-cutting principles for sampling, preparation, and analysis, and emerging water-focused standards (ISO 16094 series) codify spectroscopic and thermoanalytical approaches; summarizing these developments helps align method choice with evolving norms. Second, the policy landscape is rapidly growing as a global instrument to address plastic pollution is negotiated; regulators increasingly request technical evidence on sources, loads, and exposure pathways, rendering recent analytical improvements particularly salient.

MNPs originate from two major source classes: primary particles manufactured at small sizes (e.g., industrial powders, preproduction pellets, cosmetic beads, and synthetic microfibers) and secondary particles produced by fragmentation, tire and road wear, and paint degradation (Koelmans et al., 2022). Global mass balance and river emission models attribute the dominant ocean inputs to mismanaged waste and riverine transport, on the order of several megatons per year from land and ~1–2 Mt yr^{-1} *via* rivers, concentrated in a subset of catchments (Jambeck et al., 2015; Lebreton et al., 2017). Atmospheric transport and deposition disperse fibers and fragments to remote regions, indicating cross-media cycling beyond coastal sources (Allen et al., 2019). Accordingly, MNPs are now documented across marine, freshwater, and terrestrial systems; in indoor and outdoor air; and throughout food webs. Soils and agroecosystems accumulate particles *via* biosolids, irrigation, and deposition, whereas wastewater treatment plants act as both sinks (sludge) and sources (effluents).

Human exposure is plausible *via* ingestion, inhalation, and, under specific conditions, dermal contact. Diet, drinking water, and dust ingestion account for most oral intake, with bottled water potentially increasing exposure. The detection of MPs in human stool confirms gastrointestinal contact (Schwabl et al., 2019). Inhalation pathways are supported by indoor air measurements and micro-Fourier transform infrared spectroscopy (μFTIR) identification of MPs in human lung tissue (Jenner et al., 2022; Vianello et al., 2019). Dermal uptake appears to be limited for microsized particles, although NPs may penetrate under certain conditions; *in vivo* evidence remains sparse. Across models, toxicological studies frequently report oxidative stress, inflammation, and membrane damage, with hazard attribution complicated by particle size/surface area, as well as the presence of additives and sorbed contaminants (Prata et al., 2020).

No single analytical method answers all of these questions; modern workflows are complementary and orthogonal. Micro-FTIR and micro-Raman enable automated, polymer-type particle counts with size/area distributions but require weathering-aware spectral libraries and conservative automation to limit false positives. Pyrolysis–gas chromatography–mass spectrometry (Py-GC/MS) and thermal extraction–desorption–gas chromatography–mass spectrometry (TED-GC/MS) quantify polymer mass and additive fingerprints, supporting mass budgets and cross-validation against spectroscopic inventories. At the nanoscale, asymmetric flow field–flow fractionation coupled with multiangle light scattering (AF4-MALS) (often coupled offline to Py-GC/MS) provides size-resolved number and mass distributions; dynamic light scattering (DLS) and nanoparticle tracking analysis (NTA) yield hydrodynamic sizing but require rigorous controls; and transmission electron microscopy (TEM) provides high-resolution morphology. FTIR is used for polymer identification in the MP size range.

Standards and quality assurance are foundational. ISO 24187 sets cross-cutting principles for MP analysis, whereas the ISO 16094 series codifies spectroscopy-based (counts) and thermoanalysis-based (mass) approaches, encouraging dual reporting where feasible. Core QA/QC guidance emphasizes clean air practices, comprehensive blanks (field, laboratory, equipment), and transparent reporting of detection limits and blank corrections. Interlaboratory comparisons benchmark uncertainty, and community resources [e.g., OpenSpecy; National Oceanic and Atmospheric Administration's (NOAA's) NCEI portal] facilitate harmonized spectral identification and global synthesis (Cowger et al., 2021). Together, these advances are enabling comparable, decision-grade datasets that more reliably link environmental measurements to exposure and risk.

Accordingly, this chapter adopts a practice-oriented literature review, as illustrated in Fig. 2.1: analytical breakthroughs for detecting and characterizing MNPs are first synthesized; advances in environmental fate, transport, and transformation are then integrated; and biological effects and risk assessment are finally discussed, highlighting where methods and evidence are decision-ready and where critical gaps remain.

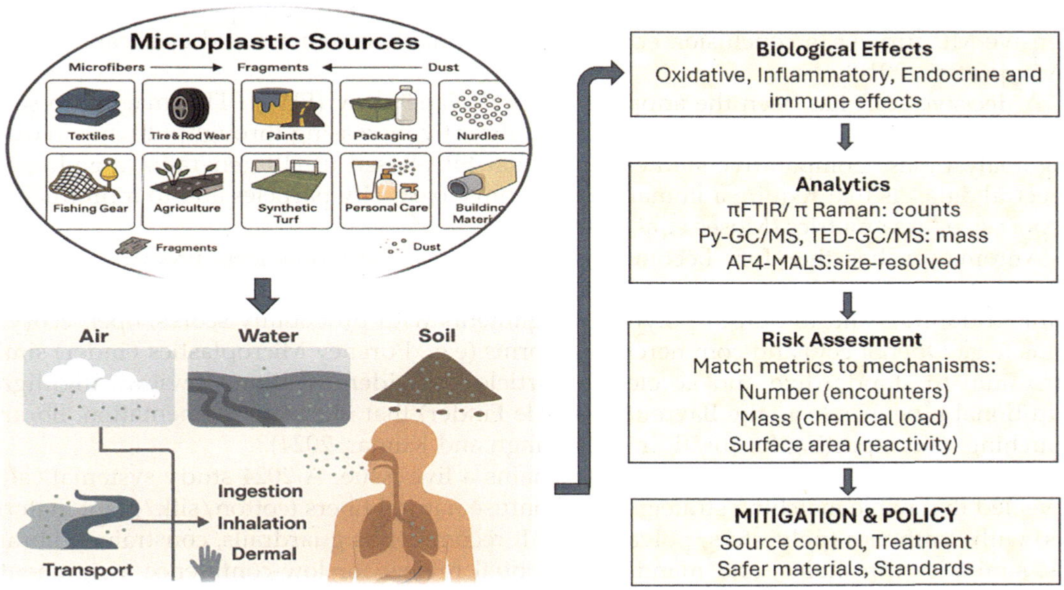

FIG. 2.1 Microplastics from common sources enter the air, water, and soil and are transported across environments. Exposure occurs *via* inhalation, ingestion, and dermal contact, contributing to oxidative, inflammatory, endocrine, and immune effects. Evidence is generated by micro-Fourier transform infrared spectroscopy (μFTIR) and micro-Raman spectroscopy (μRaman) for particle identification and counting; pyrolysis–gas chromatography–mass spectrometry (Py-GC/MS) and thermal extraction/desorption–gas chromatography–mass spectrometry (TED-GC/MS) for polymer mass and additive fingerprints; and asymmetric flow field–flow fractionation coupled with multiangle light scattering (AF4-MALS) for size-resolved distributions. These outputs support risk assessment *via* number, mass, and surface-area metrics and inform mitigation and policy (source control, treatment, safer materials, and standards).

2.2 Detecting and characterizing MNPs

Vibrational spectroscopy, thermoanalytical chemistry, and nanoscale separations/microscopy have rapidly matured into complementary pillars for MNP measurement. In parallel, international standards, interlaboratory trials, and expanded reference materials are improving data quality and comparability across studies and jurisdictions. This section synthesizes recent advances with an emphasis on (1) high-resolution spectroscopy and hyperspectral workflows, (2) thermoanalytical quantification via Py-GC/MS and TED-GC/MS, (3) the nanoscale frontier (operational definitions and AF4-MALS-centered pipelines), and (4) quality assurance and harmonized reporting.

2.2.1 High-resolution spectroscopy and imaging

Vibrational microspectroscopy remains the central strategy for polymer-specific identification at the particle level, linking spectra to chemical fingerprints with minimal sample destruction. Micro-FTIR and micro-Raman provide complementary strengths: the broad functional group sensitivity of FTIR and high spatial resolution (often down to ~1 μm, with narrower bands and reduced water interference) make them natural partners for comprehensive MP surveys, a conclusion echoed by recent method reviews and application notes (Araujo et al., 2018).

A decisive shift has been the adoption of focal-plane array (FPA) FTIR imaging to scan entire filters without manual preselection, significantly increasing throughput while reducing analyst bias. Comparative studies report that FPA-based μFTIR generates large hyperspectral datasets that require automated algorithms to classify particles and to reproducibly export size/area metrics (Moses et al., 2023).

Automation has therefore become inseparable from best practices. Freeware such as siMPle (Systematic Identification of MPs) was built to analyze full-spectral maps against curated libraries and to output polymer assignments with probability scores; open ecosystems (e.g., OpenSpecy) and commercial platforms (e.g., Purency Microplastics Finder) similarly aim to standardize and accelerate particle-level identification. Reviews highlight additional pipelines (e.g., the Bayreuth Particle Finder) that integrate segmentation, library matching, and reporting for μFTIR images (Singh and Kumar, 2024).

However, quality-assured automation remains a live issue. A 2024 study systematically revealed that naive matching strategies can confuse natural fibers (cotton/silk/plant materials) with synthetics (polyesters/polyamides). It recommends guardrails, constrained libraries, similarity thresholds, and mandatory manual review for low-confidence assignments (Kozloski et al., 2024). These concerns align with broader calls to expand libraries with spectra of weathered polymers and common biogenic interferents to reduce false positives in environmental matrices (Cowger et al., 2021).

Artificial Intelligence (AI)-assisted spectroscopy is moving quickly from proof of concept to usable tooling. Studies applying deep learning to μFTIR/Raman images have reported promising gains in classification, particularly for fibers, while also highlighting persistent weaknesses in beads and irregular fragments, underscoring the need for balanced training sets and explainable models (Devipriya et al., 2025). Broader surveys of AI-empowered workflows for collection/characterization similarly point to the value of algorithmic assistance under heavy sample throughput and mixed matrices.

2.2.2 Thermoanalytical chemistry

Where spectroscopy excels at counts and sizes, thermoanalytical methods provide the polymer mass critical for mass budgets, source apportionment, and regulatory metrics. Py-GC/MS thermally cracks polymers into diagnostic products that can be calibrated for the masses of polythylene (PE), polypropylene (PP), polystyrene (PS), polythylene terepthalate (PET), polyvinyl chloride (PVC), and other compounds in sediments, sludges, air particulates, and water. Recent reviews emphasize strengths (direct mass units, additive profiling) and pitfalls (matrix effects, marker specificity), with guidance for optimized digestion and pressurized liquid extraction before Py-GC/MS (Seeley and Lynch, 2023). Multiple 2024–2025 applications have demonstrated robust quantification across terrestrial and aquatic samples when polymer-specific markers and appropriate standards are used (Peel et al., 2025).

TED-GC/MS couples thermogravimetry with sorbent trapping of evolved volatiles before GC/MS, enabling larger sample masses and high throughput with minimal prehandling (e.g., analyzing a whole loaded filter), and has reported relative standard deviations on the order of 6%–12% in validated workflows. Agilent method papers and vendor application notes describe decoupled pyrolysis/analysis that reduces overloading and allows automation, positioning TED-GC/MS as a screening tool complementary to Py-GC/MS.

Critically, interlaboratory comparisons (ILCs) organized under versailles project on advanced materials and standards (VAMAS) have benchmarked spectroscopic (µFTIR/µRaman) and thermoanalytical (Py-GC/MS, TED-GC/MS) methods across 84 laboratories, providing unprecedented external validation of accuracy, comparability, and uncertainty. The 2025 *Analytical Chemistry* report summarizes method-specific performance and highlights areas for harmonization and improvement of reference materials (Ciornii et al., 2025).

The frontier now extends to size-resolved mass. Offline coupling of AF4-MALS fractionation with Py-GC/MS apportions the polymer mass to nano- vs. microsize classes, bridging spectroscopic counts with mass budgets and offering a route to "mass↔count" closure in complex waters. Proof-of-concept and follow-up studies have revealed detection down to tens of nanometers (by separation) with ng-level mass detection limits, depending on the polymer type (Hayder et al., 2025).

A recent perspective on standardizing Py-GC/MS underscores the need for matrix-aware markers, weathering-informed calibrations (to mitigate underestimation of oxidized polypropylene), and interlaboratory protocols for maintenance and QC, reinforcing that thermoanalysis, such as spectroscopy, benefits from shared reference materials and community guidance (Jing et al., 2025).

2.2.3 The nanoscale frontier

Consensus on what constitutes an "NP" remains unsettled and operational. Two frameworks dominate: (1) NPs defined by analogy to engineered nanomaterials (1–100 nm) and

(2) colloid-informed definitions extending up to 1–1000 nm (often for unintentionally produced particles), advocated by Gigault and coworkers and adopted in several recent assessments. The authors increasingly recommend stating the operational size window used in any study to improve comparability (Hartmann et al., 2019).

Batch-sizing methods (Dynamite Light Scattering (DLS), Nanoparticle Tracking Analysis (NTA), Tunable Resistive Pulse Sensing (TRPS)) are attractive for speed but must be interpreted cautiously for heterogeneous plastics: signals depend strongly on the refractive index, brightness, and aggregation, leading to bias toward larger/brighter scatterers and undercounting of dim, oxidized particles. Recent evaluations and methods have quantified these limitations and proposed reporting parameters (ionic strength, viscosity, surfactants, and RI assumptions) to improve reproducibility (Tian et al., 2024).

AF4-MALS has emerged as a workhorse for size-resolved separations of polydisperse MNP mixtures in water, with validated methodologies demonstrating good precision and low detection limits for standard polystyrene latexes and environmental samples. Hyphenated detectors (UV/MALS; increasingly ICP–MS for co-associated metals) expand compositional insight, and offline coupling to Py-GC–MS adds polymer fingerprinting and mass closure. Recent studies and interviews detail practical lower size bounds (influenced by the membrane cut-off and MALS sensitivity) and demonstrate fraction-wise mass quantification. A multitechnique platform combining enrichment, AF4-MALS, and Py-GC/MS has been shown to deliver size-resolved mass concentrations of NPs, along with particle number and size distributions (Li et al., 2024).

Electron microscopy including transmission electron microscopy (TEM) and scanning electron microscopy (SEM) remains essential for direct nanoscale visualization (morphology, roughness, and agglomeration). Nevertheless, it is susceptible to preparation- and beam-induced artifacts (e.g., contamination, stain residues, and beam damage), a long-standing problem across TEM domains. Recent tutorials and studies emphasize rigorous grid cleaning, low-dose imaging, and cautious interpretation, all of which are critical for differentiating carbonaceous debris from true NPs.

2.2.4 Quality assurance and comparability

Robust quality control and assurance depend on realistic reference materials (RMs) spanning sizes, shapes, and weathering states. The US National Institute of Standards and Technology (NIST) is actively expanding MP RM inventories, prioritizing the use of 1–500 μm materials to support extraction method development, additive identification, and in-house spectral library building. Related metrology programs are also developing protocols to reproducibly generate irregular NP test materials.

The VAMAS prestandardization platform has coordinated large ILCs (84 laboratories) that benchmark μFTIR/μRaman and Py-GC/MS/TED-GC/MS across shared test materials, producing the most comprehensive accuracy and uncertainty assessment to date and directly informing standard work and library content (Ciornii et al., 2025). Classical NOAA guidance (2015) remains foundational: clean-air workspaces (ideally laminar flow/HEPA), procedural blanks (field/lab/equipment), cotton laboratory coats and nonsynthetic

PPE, prerinsed glassware, and covered samples to prevent fiber fallout. Blank results should be reported separately and used to correct sample results. The ITRC's living guidance (2023–2024) consolidates current best practices and provides outreach/training materials for practitioners.

To enable cross-study comparability, guidance urges dual reporting whenever workflows permit: particle counts (e.g., items L^{-1}, items kg^{-1}, items m^{-2}) from spectroscopy and polymer mass (e.g., μg L^{-1}, mg kg^{-1}) from thermoanalysis, with explicit documentation of size-binning rules and, where segmentation is robust, surface area metrics that better correlate with sorption/reactivity. Recent ITRC fact sheets summarize unit choices and conversion pitfalls for decision-makers.

In addition to ISO 24187 and ISO 16094, ongoing regional and network efforts (e.g., NORMAN) are evaluating a dedicated database of (FT)IR spectra for different polymer types and differently aged variants to improve the interoperability of automated classifiers. In parallel, NOAA's NCEI Marine Microplastics database aggregates global records, applies quality control, and provides a public map portal, enabling metaanalyses and policy-relevant synthesis.

Reference materials, standardized methods, and transparent QA/QC (including blanks and unit choices) are now the principal levers for comparability; data infrastructure (NCEI) and spectral-library initiatives (NORMAN) are accelerating harmonization.

2.2.5 Practical synthesis

The study design determines the primary decision variable up front; if mass loading, polymer budgets, or additives are central, Py-GC/MS or TED-GC/MS can be used; if fate/transport or exposure by size class is key, μFTIR/μRaman hyperspectral imaging with robust segmentation and size binning can be prioritized (Seeley and Lynch, 2023). Automation with safeguards uses curated, context-appropriate libraries (including weathered spectra) and document classifier versions/thresholds. Probability-scored hierarchical classification and manual review of borderline spectra can be used to minimize false positives, especially for natural fibers (Kozloski et al., 2024). The operational transparency at the nanoscale is reported as the operational definition (100 vs. 1000 nm upper bound), the DLS/NTA acquisition parameters (ionic strength, assumed RI), and the AF4 method settings (membrane, channel flow, cross flow). For microscopy, grid preparation, cleaning, and dose conditions are recorded to contextualize artifacts (Hartmann et al., 2019).

To summarize, the field is actually converging on multimethod, QA/QC-driven workflows: (1) hyperspectral μFTIR/μRaman with library-aware automation for object-level counts and size distributions; (2) Py-GC/MS and TED-GC/MS for polymer mass and additive fingerprints; and (3) AF4-MALS-centered nanoscale pipelines with targeted TEM and offline Py-GC/MS for size-resolved mass. Interlaboratory trials (VAMAS), expanding RMs (NIST), and codified standards (ISO 24187; ISO 16094 series) now provide a stable foundation. Continued work on weathered polymer libraries, explainable AI, and fraction-wise mass closure across the nanophotocentric continuum will further reduce uncertainty and enable more defensible risk assessments (Ciornii et al., 2025).

2.3 Environmental fate, transport, and transformation

This section synthesizes recent advances in understanding how NMPs are generated, transformed, transported across environmental compartments, and transferred through food webs.

2.3.1 Source apportionment and flux

Recent studies have quantified the relative contributions of textiles, tire and road wear particles (TRWPs), and wastewater pathways to environmental MP fluxes across urban and remote settings. Urban source apportionment studies increasingly focus on traffic-derived particles and stormwater as decisive controls on MP emissions, with TRWPs emerging as a dominant, multicompartment signal. A city-scale survey across Beijing's ring road network illustrates how TRWPs pervade road dust, snowpacks, surface waters, and sediments, scale tightly with traffic intensity, and can be translated into emission factors and fate pathways, revealing substantial retention by street sweeping and treatment while still exporting material to freshwater and the atmosphere (Ren et al., 2024). In high-impervious catchments, stormwater infrastructure does not merely convey but can also intercept and archive MP assemblages; sediments in a São Paulo detention reservoir are tire wire particles (TWP) dominated, and the authors argue that imperviousness and street length are essential normalizers for urban flux comparisons (Goehler et al., 2022).

Risk-oriented analyses of European road runoff have shown that TRWPs co-occur with priority and traffic-related organics, with modeled and measured exposures indicating nontrivial risks in surface water and sediments and algal effects in whole effluent tests, underscoring the need to consider MPs and chemical stressors jointly (Tamis et al., 2021). Airborne pathways further broaden the source profile: PM10 road dust in Seoul contains substantial TRWMPs alongside phthalates, with polymer signatures implicating passenger car tires in the finer fractions and highlighting the coemission of plastic additives (Yoo et al., 2025). In addition to roads, domestic laundering contributes large fiber loads that mix synthetic fibers with abundant cellulosics, complicating polymer-resolved budgets and underscoring the need for source-specific metrics (Sudheshna et al., 2022). Event-driven runoff dynamics, metal mobilization, and persistent emissions from snow storage sites in northern cities reinforce the idea that urban MP fluxes are episodic, mixed, and infrastructure dependent, with management leverage at street sweeping, stormwater treatment, and road surface interfaces (Soininen and Koistinen, 2025). Taken together, the literature indicates a coherent trend: TRWPs and textile fibers dominate urban emissions; stormwater systems act as both conduits and partial sinks; and co-released additives increase composite risk, implicating traffic intensity, catchment imperviousness, and maintenance practices as first-order controls on source apportionment. Because environmental processing alters particle properties and detectability, we next consider weathering and the plastisphere.

2.3.2 Weathering and the plastisphere

After release, MP particles undergo physicochemical weathering and biofilm formation ("plastisphere"), processes that transform their size, surface chemistry, density, and mechanical integrity. Across polymers and settings, the literature converges on MPs as dynamic, long-lived sources of plasticizers, whose release is jointly governed by polymer chemistry, additive loading, boundary-layer mass transfer, and weathering. Environmentally relevant leaching protocols demonstrate that cumulative additive release varies by several orders of magnitude across common plastics, underscoring the need for standardized, sink-validated methods to support robust risk assessment (Bridson et al., 2023). Mechanistic modeling for PVC shows that aqueous boundary-layer diffusion commonly controls phthalate desorption because of strong polymer–water partitioning, implying multidecadal release and sustained background exposure (Henkel et al., 2022). Empirical work on dibutyl phthalate (DBP) from polyvinyl chloride (PVC) confirms that higher additive content, UV-induced oxidation, and hydrochemical conditions accelerate leaching, positioning MPs as persistent DBP hotspots in rivers and coastal waters and motivating targeted management of DBP-bearing waste in coastal zones (Li et al., 2024). Polymer-specific behaviors are evident: in polystyrene, phthalate plasticizer release is enhanced by hydrodynamic thinning of the boundary layer, whereas bisphenol release is controlled by temperature- and size-dependent intraparticle diffusion, highlighting how agitation, particle size, and additive chemistry decouple leaching pathways (Gulizia et al., 2023). Under natural UV, additive degradation interacts with MPs aging to reshape diffusion and partitioning parameters, yielding nonmonotonic release profiles that depend on the polymer type, particle size, and additive content (Xu et al., 2025). Finally, weathering at the material scale, especially UV-driven embrittlement and microcracking of recycled polyolefins, creates new pathways for additive egress and accelerates the transition from intact fragments to smaller particles, with greater variability linked to waste origin than to the nominal polymer class (Kuka et al., 2024). Overall, these studies provide a clear picture: weathering and biofouling convert MPs into sustained chemical sources, with release kinetics largely controlled by boundary-layer processes and polymer-specific aging, underscoring the need for harmonized leaching protocols and management focused on high-additive materials. Given that weathering modifies transport-relevant properties, we integrate these transformations into multimedia transport frameworks and predictive models.

2.3.3 Multimedia transport and modeling

Together, observations and models describe cross-media transport from rivers to coasts and the open ocean, exchanges with soil or sediments, and atmospheric pathways. Airborne pathways now link urban sources, remote deposition, and human exposure, with convergent evidence that fibers dominate atmospheric MPs and travel farther than spherical particles. Urban deposition surveys document ubiquitous fallout composed of multiple polymers and predominantly fibrous morphologies; rates scale with city size in Spain's nationwide network and were first demonstrated for London, with wind-dependent source sectors (Wright et al., 2020). Yearlong monitoring in subtropical Guangzhou further implicates textiles, chiefly polyester fibers, as major contributors and identifies rivers as secondary aerosol sources via reentrainment (Huang et al., 2021). Mechanistically, laboratory settling experiments and shape-aware modeling have shown that fiber geometry lowers effective settling velocities, enabling long-range and even stratospheric transport that current sphere-based models underpredict (Tatsii et al., 2024). At the receptor end, computational fluid–particle dynamics via full respiratory tract reconstruction indicate that particle size and shape, breathing regime, and airway anatomy jointly govern deposition hotspots from the nasal passages to the distal bronchi, providing a plausible physical basis for observed respiratory associations (Huang et al., 2024). At basin-to-global scales, interpretable machine learning syntheses suggest that biogeochemical and anthropogenic controls outweigh purely physical atmospheric predictors in shaping surface-ocean MP patterns, underscoring cross-domain coupling among atmospheric inputs, ocean processes, and source regions (Zhang et al., 2025). In summary, the transport literature points to fiber-dominated atmospheric delivery, strong sensitivity to particle shape, and cross-compartment coupling that links urban textile sources to remote marine sinks and human inhalation, highlighting the imperative for shape-aware models and harmonized air, land, and water monitoring. Transport pathways determine biological encounters, motivating a focus on exposure routes and trophic transfer.

2.3.4 Trophic transfer and exposure pathways

Convergent laboratory, field, and modeling evidence indicates widespread ingestion across taxa and context-dependent trophic transfer of particles and associated chemicals without uniform biomagnification. Table 2.1 presents a selection of MP discoveries in human and animal bodies. In a tiered estuarine chain, polystyrene particles are transferred from brine shrimp to shrimp and then to fish but largely remain confined to the gut, with limited tissue translocation yet measurable organismal stress, indicating exposure without consistent biomagnification (Saikumar et al., 2024). Experiments with a predatory crab likewise revealed comparable contributions from contaminated food and ambient water,

TABLE 2.1 Selection of discoveries of microplastics in human and animal bodies.

Human body			**Animal body**		
Organism	**MPs found within**	**References**	**Species**	**MPs found within**	**References**
Whole blood	22 healthy volunteers	(Leslie et al., 2022)	Marine fish	504 fishes (10 species)	(Lusher et al., 2013)
Lung tissue	5 females and 6 males	(Jenner et al., 2022)	Bivalves	36 *Mytilus edulis* and 10 *Crassostrea gigas*	(Van Cauwenberghe and Janssen, 2014)
Placenta	62 placenta samples	(Garcia et al., 2024)	Seabirds	1295 Northern Fulmar	(Van Franeker et al., 2011)
Breast milk	34 females	(Schwabl et al., 2019)	Sea turtles	102 marine turtles (7 species)	(Duncan et al., 2019)
Stool	8 healthy volunteers	(Schwabl et al., 2019)	Marine mammals	Seal scat samples	(Nelms et al., 2018)
Atherosclerotic plaques	304 carotid endarterectomy patients	(Marfella et al., 2024)	Companion animals	25 dogs and 24 cats	(Prata et al., 2022)

pronounced organ-specific burdens, and compensatory detoxification that failed beyond a threshold, emphasizing physiological defenses, egestion-driven loss, and dose-dependent collapse of homeostasis (Wang et al., 2021). Empirical observations of captive gray seals fed wild mackerel further corroborate the transfer to higher trophic levels under realistic diets, underscoring whole-prey consumption as an essential route for top predators (Nelms et al., 2018). Extending beyond marine systems, a lettuce snail chain has been characterized by NP passage, with evidence of biodilution yet growth impairment, illustrating that mass transfer may decline across levels even as sublethal effects persist (Li et al., 2023). The overarching pattern is consistent: trophic transfer is standard but typically constrained by egestion and limited assimilation, with effects manifesting as sublethal physiological stress rather than routine biomagnification.

Ecosystem-scale models help reconcile these heterogeneous outcomes: coastal Ecopath–ecosystem analysis projects trophic magnification across functional groups when environmental inflow increases, highlighting the primacy of external loading in setting long-term food web burdens (Gao et al., 2024). Finally, contaminant–particle interactions add further complexity: MPs alter the trophic transfer and effects of fluoxetine, increasing accumulation in prey but attenuating specific neurotoxicity in fish while exacerbating oxidative and immune responses, demonstrating endpoint-specific, bidirectional "carrier" and "cleaning" effects (Yan et al., 2024). Collectively, these studies suggest that trophic transfer is prevalent and often constrained by egestion and limited assimilation; however, it is ecologically consequential through sublethal stress, interactions with co-contaminants, and system-level responses modulated by external inputs and organismal energetics. Collectively, the modeling and mixture evidence indicate that system-level outcomes hinge on external inputs and organismal energetics. Moreover, co-contaminants can shift risk profiles, underscoring the need to integrate particle budgets, biological performance, and chemical interactions into trophic transfer assessments.

Together, source strengths, weathering transformations, cross-media transport, and trophic processes form a coupled system that sets environmental burdens and exposure, informing monitoring design and risk assessment. This integrated perspective suggests prioritizing (1) traffic and textile-focused source controls, (2) standardized leaching and aging protocols, (3) shape-aware transport modeling with coordinated atmospheric and aquatic observations, and (4) trophic-level studies that couple organismal energetics with co-contaminant dynamics to refine exposure and risk characterization.

2.4 Biological effects and risk assessment advances

Biological effects and risk assessments for MNPs have shifted from questions of presence to those of mechanism, dose, and consequences. Across cell systems, model organisms, and sentinel wildlife, this interplay repeatedly manifests as oxidative stress, low-grade inflammation, and endocrine perturbations that often intensify as particle size decreases and their interfacial area increases (Prata et al., 2020). Turning these signals into decision-grade risk requires matching dose metrics to mechanisms (particle number for encounter-driven effects, mass for chemical load, surface area for catalysis and sorption), building test systems that reflect real exposures (heterogeneous shapes, polydisperse sizes, environmentally weathered polymers), and tracing lines of evidence from controlled studies to humans and ecosystems. Early biomonitoring findings, including those concerning MPs identified in human stool, lungs, blood, and placenta, underscore plausible exposure and translocation routes, as well as the need for contamination-resistant analytics and robust dosimetry (Jenner et al., 2022; Ragusa et al., 2021).

2.4.1 Mechanistic insights

MNPs affect biology through three linked components: the particle (size, shape, surface), the chemistry it contains or carries, and the host response. Across many cell and animal studies, the typical patterns are oxidative stress (increased reactive oxygen species and lipid peroxidation), inflammation (NF-κB and cytokine signaling), and membrane/mitochondrial disturbances. Smaller particles with larger surface areas tend to produce stronger responses, which fits basic surface chemistry expectations. These general mechanisms are reflected in recent reviews that weigh both particle effects and chemical effects together (Koelmans et al., 2022). These plastics contain additives (plasticizers, monomers, stabilizers, and flame retardants) that can leach, and their surfaces sorb other pollutants from the environment. In the lipid-rich environment of the gut, some of these chemicals can desorb and activate endocrine pathways (e.g., steroid or thyroid signaling). Current health assessments, therefore, treat MPs as a dual exposure: (1) a particle that can irritate tissues and (2) a potential carrier for other chemicals that may increase the local dose beyond water-only exposure.

Accordingly, early human evidence strengthens plausibility. A recent study quantified MPs in human placentas (62 of 62 samples; polyethylene most common) via Py-GC/MS with careful tissue cleanup, demonstrating that polymer-level identification in human tissues is possible (Garcia et al., 2024). Another 2024 prospective study revealed MNPs in carotid atherosclerotic plaques and linked their presence to a higher rate of cardiovascular events during follow-up; microscopy and polymer chemistry were used together to confirm the material (Marfella et al., 2024). These studies do not prove causation, but they show tissue-level presence aligned with outcomes that fit the oxidative–inflammatory story and set a new bar for biomonitoring (Garcia et al., 2024). Recent risk frameworks recommend selecting or combining metrics by mechanism to avoid under- or overestimating hazards when translating between laboratory tests and real exposures (Koelmans et al., 2022). The test systems are improved in two practical ways. First, studies aim for environmental realism: they use heterogeneous particles (fibers, fragments, films, not only pristine polystyrene spheres), span field-relevant concentrations (background to hotspots), and use media with salts, natural organic matter, and proteins so that agglomeration and coronas resemble real conditions. Second, more studies now report the number and mass, and estimate the surface area where possible. This allows risk assessors to select the metric that matches the hypothesized mechanism without having to guess conversions.

For the nanoscale, where the most significant uncertainties lie, their separation and measurement are improving. In 2024 the platform combined AF4-MALS to size-resolve submicrometric fractions and then used Py-GC/MS offline to quantify the polymer mass in those same fractions. The result is a size-resolved number and mass for the nano–micro boundary, precisely what is needed to align number-based toxicology with mass-based exposure (Li et al., 2024). Risk analysis is also becoming more probabilistic. A recent study developed species sensitivity distributions from mass-normalized chronic effect data to estimate community-level thresholds. The authors note that better representation of fibers and weathered fragments is still needed. Nevertheless, the method shows that aligning metrics and quality filtering can deliver thresholds with transparent uncertainty rather than a single safety factor (Iwasaki et al., 2023).

2.4.2 From the laboratory to humans and ecosystems

To connect the controlled experiments to the real world, three steps are used. First, sentinel species that integrate exposure over time and offer tractable biomarkers should be used. Bivalves and benthic invertebrates, for example, exhibit oxidative–inflammatory responses at realistic concentrations and can be sampled across gradients (urban to remote). These field data complement laboratory mechanisms and are directly useful for local risk screening. Second, humans should be paired with exposure data via robust analytics. Indoor air studies with breathing manikins and μFTIR imaging have revealed airborne MPs in the breathing zone, especially indoors. Biomonitoring has detected MPs in blood via (double-shot) Py-GC/MS and in the lungs and placenta via spectroscopy plus targeted chemistry. The World Health Organization's technical review concludes that health risk cannot yet be quantified with confidence, but the biological plausibility and exposure potential, especially for particles <10 μm, justify better dosimetry and contamination-controlled methods (Vianello et al., 2019). Third, the mixtures should be treated as a rule. In water, sediment, soils, and indoor dust, MPs cooccur with metals, per and polyfluoroalkyl substances (PFAS), pharmaceuticals, and polycyclic aromatic hydrocarbons (PAHs). Because particles can carry chemicals and microbes, tests that combine particles with chemicals at field-realistic ratios are more informative than one-agent tests. The goal is to show not only co-occurrence but also a vector effect, for example, a higher internal dose of a chemical when particles are present. The WHO and recent reviews recommend incorporating this logic into the study design to prevent risk assessments from missing additive effects. On the human translation side, two recent studies illustrate the new standard. The placenta study developed a cleanup and Py-GC/MS pipeline that reports μg plastic per g tissue by polymer (with microscopy for >1-μm objects), answering "what and how much?" in the same tissue. The atheroma study reported MNPs in plaques and a higher rate of cardiovascular events over ~3 years among patients with plastics present. These findings do not indicate causality, but they reveal how to tie polymer-confirmed internal doses to clinically relevant outcomes, which is what risk frameworks need next (Garcia et al., 2024).

2.5 Conclusion

Research on MNPs has advanced rapidly, from simple detection to understanding their behavior and potential effects on living systems. Today, powerful tools such as hyperspectral micro-FTIR and micro-Raman can scan whole filters to reveal how many particles are present, how large they are, and which polymers they are made of. Thermoanalytical methods such as Py-GC/MS and TED-GC/MS fill in missing pieces by measuring the polymer mass and profiling additives. For most minor size fractions, nanoscale separations such as AF4-MALS, coupled with polymer-specific chemistry, now provide size-resolved numbers and masses, helping scientists align "how many particles" with "how much plastic" in the same samples.

Moreover, we have a clearer picture of where these particles come from and how they move. Textiles, tires and road wear, paints, and mismanaged waste are primary sources. Rivers, wind, and wastewater move particles across water, land, and air, whereas weathering and biofilms change particle surfaces, density, and shape. These transformations affect how long particles persist, where they end up, and how organisms encounter them.

Across cell studies, animal models, and sentinel species, a consistent biological pattern has emerged, revealing oxidative stress, inflammation, and membrane and mitochondrial disruption. Early human biomonitoring detected MPs in the stool, lung tissue, blood, and placenta. These findings do not prove disease, but they strengthen the case for improved dosimetry, rigorous contamination control, and longitudinal studies linking the internal dose to health outcomes. In the future, progress will depend on stronger standards, realistic tests with weathered and mixed particles, and routine consideration of mixtures and the microbiome. For NPs, more precise size definitions and broader use of size-resolved methods are currently priorities. With these steps, sharper measurements can drive sharper decisions targeting sources, improve treatment, guide safer material design, and inform policies that lower exposure while supporting a more sustainable plastics economy.

AI disclosure

During the preparation of this work, the author(s) used ChatGPT 4.0 to improve the readability and language of the work. After using this tool/service, the author(s) reviewed and edited the content as needed and take(s) full responsibility for the content of the publication.

References

Allen, S., Allen, D., Phoenix, V.R., Roux, G.L., Durántez Jiménez, P., Simonneau, A., Binet, S., Galop, D., 2019. Atmospheric transport and deposition of microplastics in a remote mountain catchment. Nat. Geosci. 17520908. 12 (5), 339–344. Available from: http://www.nature.com/ngeo/index.html. doi:10.1038/s41561-019-0335-5.

Araujo, C.F., Nolasco, M.M., Ribeiro, A.M.P., Ribeiro-Claro, P.J.A., 2018. Identification of microplastics using Raman spectroscopy: latest developments and future prospects. Water Res. 18792448. 142, 426–440. Available from: https://www.elsevier.com/locate/watres. doi:10.1016/j.watres.2018.05.060.

Brahney, J., Hallerud, M., Heim, E., Hahnenberger, M., Sukumaran, S., 2020. Plastic rain in protected areas of the United States. Science. 0036-8075. 368 (6496), 1257–1260. doi:10.1126/science.aaz5819.

Bridson, J.H., Abbel, R., Smith, D.A., Northcott, G.L., Gaw, S., 2023. Release of additives and non-intentionally added substances from microplastics under environmentally relevant conditions. Environ. Adv. 26667657. 12. Available from: https://www.journals.elsevier.com/environmental-advances. doi:10.1016/j.envadv.2023.100359.

Ciornii, D., Hodoroaba, V.D., Benismail, N., Maltseva, A., Ferrer, J.F., Wang, J., Parra, R., Jézéquel, R., Receveur, J., Gabriel, D., Scheitler, A., van Oversteeg, C., Roosma, J., van Renesse van Duivenbode, A., Bulters, T., Zanella, M., Perini, A., Benetti, F., Mehn, D., Dierkes, G., Soll, M., Ishimura, T., Bednarz, M., Peng, G., Hildebrandt, L., Peters, M., Kim, S.K., Türk, J., Steinfeld, F., Jung, J., Hong, S., Kim, E.J., Yu, H.W., Klockmann, S., Krafft, C., Süssmann, J., Zou, S., ter Halle, A., Giovannozzi, A.M., Sacco, A., Fadda, M., Putzu, M., Im, D.H., Nhlapo, N.,

Carrillo-Barragán, P., Schmidt, N., Herzke, D., Gomiero, A., Jaén-Gil, A., Cabanes, D.J.E., Doedt, M., Cardoso, V., Schmitz, A., Hawly, M., Mo, H., Jacquin, J., Mechlinski, A., Adediran, G.A., Andrade, J., Muniategui-Lorenzo, S., Ramsperger, A., Löder, M.G.J., Laforsch, C., Velickovic, T.C., Fabbri, D., Coralli, I., Federici, S., Scholz-Böttcher, B. M., la Nasa, J., Biale, G., Rauert, C., Okoffo, E.D., Undas, A., An, L., Wachtendorf, V., Fengler, P., Altmann, K., 2025. Interlaboratory comparison reveals state of the art in microplastic detection and quantification methods. Anal. Chem. 15206882. 97 (16), 8719–8728. Available from: http://pubs.acs.org/journal/ancham. doi:10.1021/acs.analchem.4c05403.

Cowger, W., Steinmetz, Z., Gray, A., Munno, K., Lynch, J., Hapich, H., Primpke, S., De Frond, H., Rochman, C., Herodotou, O., 2021. Microplastic spectral classification needs an open source community: open specy to the rescue! Anal. Chem. 0003-2700. 93 (21), 7543–7548. doi:10.1021/acs.analchem.1c00123.

Devipriya, K., Tlija, M., Kumar, C.K., Kumar, V.C., Jana, S., Jana, C., 2025. Automated micro-plastic detection and classification using deep convolution neural network pre-trained models and transfer learning. AIP Adv. 21583226. 15 (2). Available from: http://scitation.aip.org/content/aip/journal/adva. doi:10.1063/5.0243976.

Duncan, E.M., Broderick, A.C., Fuller, W.J., Galloway, T.S., Godfrey, M.H., Hamann, M., Limpus, C.J., Lindeque, P. K., Mayes, A.G., Omeyer, L.C.M., Santillo, D., Snape, R.T.E., Godley, B.J., 2019. Microplastic ingestion ubiquitous in marine turtles. Glob. Change Biol. 13652486. 25 (2), 744–752. Available from: http://onlinelibrary.wiley.com/journal/10.1111/(ISSN)1365-2486. doi:10.1111/gcb.14519.

Gao, S., Li, Z., Zhang, S., 2024. Trophic transfer and biomagnification of microplastics through food webs in coastal waters: A new perspective from a mass balance model. Mar. Pollut. Bull. 200, 116082. doi:10.1016/j.marpolbul.2024.116082.

Garcia, M.A., Liu, R., Nihart, A., Hayek, E.E., Castillo, E., Barrozo, E.R., Suter, M.A., Bleske, B., Scott, J., Forsythe, K., Gonzalez-Estrella, J., Aagaard, K.M., Campen, M.J., 2024. Quantitation and identification of microplastics accumulation in human placental specimens using pyrolysis gas chromatography mass spectrometry. Toxicol. Sci. 10960929. 199 (1), 81–88. Available from: http://toxsci.oxfordjournals.org/. doi:10.1093/toxsci/kfae021.

Gigault, J., Halle, A.T., Baudrimont, M., Pascal, P.Y., Gauffre, F., Phi, T.L., Hadri, H.E., Grassl, B., Reynaud, S., 2018. Current opinion: what is a nanoplastic? Environ. Pollut. 18736424. 235, 1030–1034. Available from: https://www.elsevier.com/inca/publications/store/4/0/5/8/5/6. doi:10.1016/j.envpol.2018.01.024.

Goehler, L.O., Moruzzi, R.B., Tomazini da Conceição, F., Júnior, A.A.C., Speranza, L.G., Busquets, R., Campos, L.C., 2022. Relevance of tyre wear particles to the total content of microplastics transported by runoff in a high-imperviousness and intense vehicle traffic urban area. Environ. Pollut. 18736424. 314. Available from: https://www.journals.elsevier.com/environmental-pollution. doi:10.1016/j.envpol.2022.120200.

Gulizia, A.M., Patel, K., Philippa, B., Motti, C.A., van Herwerden, L., Vamvounis, G., 2023. Understanding plasticiser leaching from polystyrene microplastics. Sci. Total Environ. 18791026. 857. Available from: https://www.elsevier.com/locate/scitotenv. doi:10.1016/j.scitotenv.2022.159099.

Hartmann, N.B., Hüffer, T., Thompson, R.C., Hassellöv, M., Verschoor, A., Daugaard, A.E., Rist, S., Karlsson, T., Brennholt, N., Cole, M., Herrling, M.P., Hess, M.C., Ivleva, N.P., Lusher, A.L., Wagner, M., 2019. Are we speaking the same language? recommendations for a definition and categorization framework for plastic debris. Environ. Sci. Technol. 15205851. 53 (3), 1039–1047. Available from: http://pubs.acs.org/journal/esthag. doi:10.1021/acs.est.8b05297.

Hayder, M., Veclin, C., Ahern, A., Chojnacka, A., Roex, E., Meier, F., Gruter, G.-J.M., van Wezel, A.P., Astefanei, A., 2025. Integrating AF4 and Py-GC-MS for combined size-resolved polymer-compositional analysis of nanoplastics with application to wastewater. Anal. Chem. 0003-2700. 97 (28), 15216–15224. doi:10.1021/acs.analchem.5c01766.

Henkel, C., Hüffer, T., Hofmann, T., 2022. Polyvinyl Chloride Microplastics Leach Phthalates into the Aquatic Environment over DecadesPolyvinyl Chloride Microplastics Leach Phthalates into the Aquatic Environment over Decades. Environ. Sci. Technol. 56 (20), 14507–14516. doi:10.1021/acs.est.2c0510810.1021/acs.est.2c05108.s001.

Huang, X., Saha, S.C., Saha, G., Francis, I., Luo, Z., 2024. Transport and deposition of microplastics and nanoplastics in the human respiratory tract. Environ. Adv. 26667657. 16, 100525. doi:10.1016/j.envadv.2024.100525.

Huang, Y., He, T., Yan, M., Yang, L., Gong, H., Wang, W., Qing, X., Wang, J., 2021. Atmospheric transport and deposition of microplastics in a subtropical urban environment. J. Hazard. Mater. 03043894. 416, 126168. doi:10.1016/j.jhazmat.2021.126168.

Iwasaki, Y., Takeshita, K.M., Ueda, K., Naito, W., 2023. Estimating species sensitivity distributions for microplastics by quantitatively considering particle characteristics using a recently created ecotoxicity database. Microplast.

Nanoplast. 26624966. 3 (1). Available from: https://www.springer.com/journal/43591. doi:10.1186/s43591-023-00070-6.

Jambeck, J.R., Geyer, R., Wilcox, C., Siegler, T.R., Perryman, M., Andrady, A., Narayan, R., Law, K.L., 2015. Plastic waste inputs from land into the ocean. Science. 10959203. 347 (6223), 768–771. Available from: http://www.sciencemag.org/content/347/6223/768.full.pdf. doi:10.1126/science.1260352.

Jenner, L.C., Rotchell, J.M., Bennett, R.T., Cowen, M., Tentzeris, V., Sadofsky, L.R., 2022. Detection of microplastics in human lung tissue using μFTIR spectroscopy. Sci. Total Environ. 18791026. 831. Available from: https://www.elsevier.com/locate/scitotenv. doi:10.1016/j.scitotenv.2022.154907.

Jing, S., Wang, Y., Chen, Y., Chen, Y., Gong, X., Wanger, T.C., 2025. Standardizing pyrolysis gas chromatography mass spectrometry for nanoplastics and microplastics detection to advance environmental research. Npj Emerg. Contam. 3059-3719. 1 (1). doi:10.1038/s44454-025-00001-5.

Koelmans, A.A., Redondo-Hasselerharm, P.E., Nor, N.H.M., de Ruijter, V.N., Mintenig, S.M., Kooi, M., 2022. Risk assessment of microplastic particles. Nat. Rev. Mater. 20588437. 7 (2), 138–152. Available from: https://www.nature.com/natrevmats/. doi:10.1038/s41578-021-00411-y.

Kozloski, R., Cowger, W., Arienzo, M.M., 2024. Moving toward automated μFTIR spectra matching for microplastic identification: addressing false identifications and improving accuracy. Micropl.&Nanopl. 4 (1), doi:10.1186/s43591-024-00106-5.

Kuka, E., Cirule, D., Andersone, I., Vasiljevs, L.O., Merna, J., Sarakovskis, A., Kurnosova, N., Sansonetti, E., Vevere, L., Andersons, B., 2024. A step to microplastic formation: microcracking and associated surface transformations of recycled LDPE, LLDPE, HDPE, and PP plastics exposed to UV radiation. Polym. Degrad. Stab. 01413910. 229, 110967. doi:10.1016/j.polymdegradstab.2024.110967.

Lebreton, L.C.M., Zwet, J.V.D., Damsteeg, J.W., Slat, B., Andrady, A., Reisser, J., 2017. River plastic emissions to the world's oceans. Nat. Commun. 20411723. 8. Available from: http://www.nature.com/ncomms/index.html. doi:10.1038/ncomms15611.

Lei, G., Ozbakkaloglu, T., 2016. Use of recycled plastics in concrete: a critical review. Waste Manag. 0956053X. 51, 19–42. doi:10.1016/j.wasman.2016.03.005.

Leslie, H.A., van Velzen, M.J.M., Brandsma, S.H., Vethaak, A.D., Garcia-Vallejo, J.J., Lamoree, M.H., 2022. Discovery and quantification of plastic particle pollution in human blood. Environ. Int. 18736750. 163. Available from: https://www.elsevier.com/locate/envint. doi:10.1016/j.envint.2022.107199.

Li, H., Lee, L.M., Yu, D., Chan, S.H., Li, A., 2024. An optimized multi-technique based analytical platform for identification, characterization and quantification of nanoplastics in water. Talanta. 00399140. 272. Available from: https://www.sciencedirect.com/science/journal/00399140. doi:10.1016/j.talanta.2024.125800.

Li, Y., Lin, X., Wang, J., Xu, G., Yu, Y., 2023. Mass-based trophic transfer of polystyrene nanoplastics in the lettuce-snail food chain. Sci. Total Environ. 00489697. 897, 165383. doi:10.1016/j.scitotenv.2023.165383.

Lusher, A.L., McHugh, M., Thompson, R.C., 2013. Occurrence of microplastics in the gastrointestinal tract of pelagic and demersal fish from the English Channel. Mar. Pollut. Bull. 0025326X. 67 (1-2), 94–99. doi:10.1016/j.marpolbul.2012.11.028.

Marfella, R., Prattichizzo, F., Sardu, C., Fulgenzi, G., Graciotti, L., Spadoni, T., D'Onofrio, N., Scisciola, L., Grotta, R.L., Frigé, C., Pellegrini, V., Municinò, M., Siniscalchi, M., Spinetti, F., Vigliotti, G., Vecchione, C., Carrizzo, A., Accarino, G., Squillante, A., Spaziano, G., Mirra, D., Esposito, R., Altieri, S., Falco, G., Fenti, A., Galoppo, S., Canzano, S., Sasso, F.C., Matacchione, G., Olivieri, F., Ferraraccio, F., Panarese, I., Paolisso, P., Barbato, E., Lubritto, C., Balestrieri, M.L., Mauro, C., Caballero, A.E., Rajagopalan, S., Ceriello, A., D'Agostino, B., Iovino, P., Paolisso, G., 2024. Microplastics and nanoplastics in atheromas and cardiovascular events. N. Engl. J. Med. 0028-4793. 390 (10), 900–910. doi:10.1056/nejmoa2309822.

Moses, S.R., Roscher, L., Primpke, S., Hufnagl, B., Löder, M.G.J., Gerdts, G., Laforsch, C., 2023. Comparison of two rapid automated analysis tools for large FTIR microplastic datasets. Anal. Bioanal. Chem. 16182650. 415 (15), 2975–2987. Available from: https://www.springer.com/journal/216. doi:10.1007/s00216-023-04630-w.

Murray, N.J., 2023. The extent and drivers of global wetland loss. Nature. 14764687. 614 (7947), 234–235. Available from: https://www.nature.com/nature/. doi:10.1038/d41586-023-00268-x.

Nelms, S.E., Galloway, T.S., Godley, B.J., Jarvis, D.S., Lindeque, P.K., 2018. Investigating microplastic trophic transfer in marine top predators. Environ. Pollut. 18736424. 238, 999–1007. Available from: https://www.elsevier.com/inca/publications/store/4/0/5/8/5/6. doi:10.1016/j.envpol.2018.02.016.

Pabortsava, K., Lampitt, R.S., 2020. High concentrations of plastic hidden beneath the surface of the Atlantic Ocean. Nat. Commun. 11 (1), doi10.1038/s41467-020-17932-9.

Peel, R.H., Lloyd, C.E., Roberts, S.J., Naafs, B.D., Bull, I.D., 2025. Quantification of microplastic targets in environmental matrices using pyrolysis-gas chromatography-mass spectrometry. Environ. Sci.: Adv. 4 (1), 159–171. doi:10.1039/D4VA00269E.

Prata, J.C., da Costa, J.P., Lopes, I., Duarte, A.C., Rocha-Santos, T., 2020. Environmental exposure to microplastics: an overview on possible human health effects. Sci. Total Environ. 18791026. 702. Available from: https://www.elsevier.com/locate/scitotenv. doi:10.1016/j.scitotenv.2019.134455.

Prata, J.C., Silva, A.L.P., da Costada Costa, J.P., Dias-Pereira, P., Carvalho, A., Fernandes, A.J.S., da Costada Costa, F.M., Duarte, A.C., Rocha-Santos, T., 2022. Microplastics in internal tissues of companion animals from urban environments. Animals. 20762615. 12 (15). Available from: http://www.mdpi.com/journal/animals. doi:10.3390/ani12151979.

Ragusa, A., Svelato, A., Santacroce, C., Catalano, P., Notarstefano, V., Carnevali, O., Papa, F., Rongioletti, M.C.A., Baiocco, F., Draghi, S., D'Amore, E., Rinaldo, D., Matta, M., Giorgini, E., 2021. Plasticenta: first evidence of microplastics in human placenta. Environ. Int. 18736750. 146. Available from: https://www.elsevier.com/locate/envint. doi:10.1016/j.envint.2020.106274.

Ren, Y., Li, W., Zhou, P., Wu, H., Yu, L., Wang, R., Qu, C., Zhao, Y., Liu, J., Wu, C., 2024. Occurrence, emission, and transport of tire and road wear particles across four environmental compartments along ring road networks in Beijing. Environ. Sci. Technol. 0013-936X. 58 (52), 23160–23168. doi:10.1021/acs.est.4c08466.

Saikumar, S., Mani, R., Ganesan, M., Dhinakarasamy, I., Palanisami, T., Gopal, D., 2024. Trophic transfer and their impact of microplastics on estuarine food chain model. J. Hazard. Mater. 03043894. 464, 132927. doi:10.1016/j.jhazmat.2023.132927.

Schwabl, P., Köppel, S., Königshofer, P., Bucsics, T., Trauner, M., Reiberger, T., Liebmann, B., 2019. Detection of various microplastics in human stool. Ann. Intern. Med. 0003-4819. 171 (7), 453–457. doi:10.7326/m19-0618.

Seeley, M.E., Lynch, J.M., 2023. Previous successes and untapped potential of pyrolysis–GC/MS for the analysis of plastic pollution. Anal. Bioanal. Chem. 16182650. 415 (15), 2873–2890. Available from: https://www.springer.com/journal/216. doi:10.1007/s00216-023-04671-1.

Singh, B., Kumar, A., 2024. Advances in microplastics detection: a comprehensive review of methodologies and their effectiveness. TrAC Trends Anal. Chem. 01659936. 170, 117440. doi:10.1016/j.trac.2023.117440.

Soininen, T., Koistinen, A., 2025. Microplastic discharge and other anthropogenic pollution in urban runoff. Sci. Total Environ. 00489697. 1001, 180572. doi:10.1016/j.scitotenv.2025.180572.

Song, Y.K., Hong, S.H., Jang, M., Han, G.M., Jung, S.W., Shim, W.J., 2017. Combined effects of UV exposure duration and mechanical abrasion on microplastic fragmentation by polymer type. Environ. Sci. Technol. 15205851. 51 (8), 4368–4376. Available from: http://pubs.acs.org/journal/esthag. doi:10.1021/acs.est.6b06155.

Sudheshna, A.A., Srivastava, M., Prakash, C., 2022. Characterization of microfibers emission from textile washing from a domestic environment. Sci. Total Environ. 852, 158511. doi:10.1016/j.scitotenv.2022.158511.

Tamis, J.E., Koelmans, A.A., Dröge, R., Kaag, N.H.B.M., Keur, M.C., Tromp, P.C., Jongbloed, R.H., 2021. Environmental risks of car tire microplastic particles and other road runoff pollutants. Microplast. Nanoplast. 26624966. 1 (1). Available from: https://www.springer.com/journal/43591. doi:10.1186/s43591-021-00008-w.

Tatsii, D., Bucci, S., Bhowmick, T., Guettler, J., Bakels, L., Bagheri, G., Stohl, A., 2024. Shape Matters: Long-Range Transport of Microplastic Fibers in the AtmosphereShape Matters: Long-Range Transport of Microplastic Fibers in the Atmosphere. Environ. Sci. Technol. 58 (1), 671–682. doi:10.1021/acs.est.3c0820910.1021/acs.est.3c08209.s001.

Tian, Y., Tian, D., Peng, X., Qiu, H., 2024. Critical parameters to standardize the size and concentration determination of nanomaterials by nanoparticle tracking analysis. Int. J. Pharm. 03785173. 656, 124097. doi:10.1016/j.ijpharm.2024.124097.

Van Cauwenberghe, L., Janssen, C.R., 2014. Microplastics in bivalves cultured for human consumption. Environ. Pollut. 18736424. 193, 65–70. Available from: https://www.journals.elsevier.com/environmental-pollution. doi:10.1016/j.envpol.2014.06.010.

Van Franeker, J.A., Blaize, C., Danielsen, J., Fairclough, K., Gollan, J., Guse, N., Hansen, P.L., Heubeck, M., Jensen, J. K., Le Guillou, G., Olsen, B., Olsen, K.O., Pedersen, J., Stienen, E.W.M., Turner, D.M., 2011. Monitoring plastic ingestion by the northern fulmar *Fulmarus glacialis* in the North Sea. Environ. Pollut. 18736424. 159 (10),

2609–2615. Available from: https://www.journals.elsevier.com/environmental-pollution. doi:10.1016/j.envpol.2011.06.008.

Vianello, A., Jensen, R.L., Liu, L., Vollertsen, J., 2019. Simulating human exposure to indoor airborne microplastics using a Breathing Thermal Manikin. Sci. Rep. 20452322. 9 (1). Available from: https://www.nature.com/srep/index.html. doi:10.1038/s41598-019-45054-w.

Wang, T., Hu, M., Xu, G., Shi, H., Leung, J.Y.S., Wang, Y., 2021. Microplastic accumulation via trophic transfer: can a predatory crab counter the adverse effects of microplastics by body defence? Sci. Total Environ. 00489697. 754, 142099. doi:10.1016/j.scitotenv.2020.142099.

Wei, X.-F., Bohlén, M., Lindblad, C., Hedenqvist, M., Hakonen, A., 2021. Microplastics generated from a biodegradable plastic in freshwater and seawater. Water Res. 00431354. 198, 117123. doi:10.1016/j.watres.2021.117123.

Wright, S.L., Ulke, J., Font, A., Chan, K.L.A., Kelly, F.J., 2020. Atmospheric microplastic deposition in an urban environment and an evaluation of transport. Environ. Int. 01604120. 136, 105411. doi:10.1016/j.envint.2019.105411.

Xu, Z., Zhang, J., Qi, R., Liu, Q., Cao, H., Wen, F., Liao, Y., Shih, K., Tang, Y., 2025. Complex release dynamics of microplastic additives: an interplay of additive degradation and microplastic aging. J. Hazard. Mater. 03043894. 490, 137711. doi:10.1016/j.jhazmat.2025.137711.

Yan, Z., Zhao, H., Zhu, P., Wang, Y., Hou, J., Lu, G., He, C., 2024. Polystyrene microplastics alter the trophic transfer and biotoxicity of fluoxetine in an aquatic food chain. J. Hazard. Mater. 03043894. 470, 134179. doi:10.1016/j.jhazmat.2024.134179.

Yoo, H., Park, P., Le, Y.T.H., Park, S., Jung, Y.W., Youn, J.S., Kim, Y.M., Jeon, K.J., 2025. Characterization of tire and road wear microplastics and phthalates in inhalable PM10 road Dust: implications for urban air pollution. Environ. Pollut. 18736424. 384. Available from: https://www.sciencedirect.com/science/journal/02697491. doi:10.1016/j.envpol.2025.126991.

Zhang, L., Wang, W., Wang, F., Wu, D., Su, Y., Zhan, M., Li, K., Shi, H., Xie, B., 2025. Decoding the plastic patch: exploring the global microplastic distribution in the surface layers of marine regions with interpretable machine learning. Environ. Sci. Technol. 0013-936X. 59 (15), 7656–7666. doi:10.1021/acs.est.4c12227.

Zhu, F., Zhu, C., Wang, C., Gu, C., 2019. Occurrence and ecological impacts of microplastics in soil systems: a review. Bull. Environ. Contam. Toxicol. 0007-4861. 102 (6), 741–749. doi:10.1007/s00128-019-02623-z.

CHAPTER

3

Microplastics in the food chain and related consequences

Ehsan Jafarpisheh[a] *and Karim Ebrahimpour*[b]

[a]Department of Civil Engineering, University of Isfahan, Isfahan, Iran

[b]Department of Environmental Health Engineering, University of Medical Sciences, Isfahan, Iran

3.1 Introduction

Plastics have become an essential part of modern life because of their versatility, durability, and low cost. However, the widespread use and improper disposal of plastic materials have led to their accumulation in various ecosystems. Estimates indicate that by 2050, global plastic waste will exceed 12 billion tons (Zhang et al., 2021). A significant share of plastic waste ends up in rivers and oceans, either through direct disposal or through wind- and runoff-driven transport from landfills (Schmidt et al., 2017; Wang et al., 2024).

Plastic items, including bottles, bags, and toys, can infiltrate ecosystems and slowly break down under sunlight, heat, and mechanical forces, turning into tiny particles. Owing to their resistance to complete breakdown, these particles can remain in the environment for extended periods (He et al., 2018; Margot et al., 2013). These small particles, known as microplastics, are defined as plastic particles <5 mm in diameter and are classified into two major categories: primary microplastics, which are intentionally manufactured into small sizes (e.g., microbeads in cosmetics and fibers from synthetic textiles), and secondary microplastics, which result from the breakdown of larger plastic debris under environmental conditions (Auta et al., 2017; Fowzi et al., 2025).

Microplastics have been detected in various environmental compartments, including air, water, and soil (Akbari and Jaafari, 2024; Kabir et al., 2023). Their small size makes them easily transportable across long distances, enabling them to reach even remote and pristine ecosystems (Allen et al., 2019). In aquatic environments, microplastics are particularly concerning because of their high persistence and ability to interact with organisms at multiple trophic levels.

One of the most critical concerns surrounding microplastics is their potential to be ingested by living organisms. Numerous studies have reported the ingestion of microplastic particles

Microplastics in Animal and Human Physiology
https://doi.org/10.1016/B978-0-443-45635-0.00008-2

by a wide range of aquatic species, including plankton, mollusks, crustaceans, and fish (Botterell et al., 2019; Cáceres-Farias et al., 2023; Kristanti et al., 2022; Marmara et al., 2023; Santillo et al., 2017). Microplastics often resemble natural food sources in size, shape, and texture, and their surfaces can become coated with biofilms that emit chemical cues similar to those of edible particles, which frequently leads organisms to mistakenly consume them (Xu et al., 2020). Upon ingestion, these particles may cause gastrointestinal blockage, disrupt nutrient assimilation, and lead to physiological stress and altered feeding behavior (Jeon et al., 2023; Jovanović, 2017).

The potential impacts of microplastics extend beyond physical effects. Their hydrophobic surfaces enable them to attract and bind a range of harmful contaminants, including persistent organic pollutants, heavy metals, and other toxic chemicals (Rafa et al., 2024; Fowzi et al., 2025). When ingested by organisms, these contaminated microplastics act as vectors, facilitating the transfer and accumulation of hazardous substances within biological tissues, thereby triggering toxicological responses (Weis and Alava, 2023). Over time, these microplastics and their absorbed pollutants accumulate through successive levels of the food chain, ultimately reaching top predators, including humans, and posing serious ecological and health risks.

This chapter provides an overview of the role of microplastics in the food chain. It begins with their ingestion by lower-trophic-level organisms and continues through trophic interactions. The discussion extends to the presence of microplastics in human-consumed food products and concludes with their ecological and health-related consequences.

3.2 Microplastics in lower-trophic-level organisms

Low-trophic-level organisms are the primary entry points for microplastics into aquatic food webs, given their close interaction with water and sediments. These species, including plankton, benthic invertebrates, and filter-feeding organisms, are particularly vulnerable due to their feeding strategies and ecological functions. Understanding microplastic interactions at these foundational levels is essential for revealing how contamination moves through higher-trophic-level consumers and affects overall ecosystem integrity.

3.2.1 Ingestion by plankton, benthic invertebrates, and filter-feeding species

Microplastics have become widespread pollutants in aquatic ecosystems and are frequently ingested by various organisms, particularly those occupying lower trophic levels. Among these, plankton, benthic invertebrates, and filter-feeding species are particularly vulnerable due to their feeding mechanisms and continuous exposure to suspended particles. They serve as fundamental links in food webs and play essential roles in maintaining ecosystem health and productivity.

Planktonic organisms, including copepods, rotifers, and the larval stages of larger aquatic animals, function as essential primary consumers, forming a vital trophic link between primary producers and organisms at higher trophic levels. Owing to their small size and generally nonselective feeding habits, these organisms are particularly vulnerable to ingesting microplastics that resemble their natural food sources in size and appearance

(Cáceres-Farias et al., 2023; Valdez-Cibrián et al., 2024). Evidence from field observations and laboratory experiments indicates that various zooplankton species consume microplastics, potentially leading to negative impacts on biological processes. To date, the ingestion of microplastics has been reported in 39 zooplankton species spanning 28 taxonomic groups (Botterell et al., 2019). In copepods, exposure to microplastics has been associated with decreased ingestion of algal cells and lower overall carbon uptake (Bai et al., 2021). Nearly 45% of studies investigating these interactions have reported adverse effects on key biological traits of zooplankton, such as growth, feeding efficiency, reproductive success, and longevity (Botterell et al., 2019).

Furthermore, microplastics can attach to the outer body surfaces of zooplankton, including their appendages and exoskeletons, potentially interfering with their normal functions (Cole et al., 2013). The extent to which zooplankton are exposed to and affected by microplastics is influenced by several factors, including particle size, shape, age, and environmental abundance (Botterell et al., 2019). Importantly, microplastics may serve as carriers for hazardous chemicals, thereby introducing secondary exposure to pollutants when ingested. This can intensify physiological stress and toxic effects in planktonic organisms (Pestana et al., 2021). Such impacts at the base of the food web may propagate through trophic levels, potentially affecting entire aquatic ecosystems.

Similarly, microplastic ingestion has been widely reported among benthic invertebrates, particularly in sediment-rich environments where plastic particles tend to accumulate. Species such as polychaete annelids, amphipods, isopods, and chironomid larvae are particularly prone to ingesting microplastics because of their sediment-feeding behaviors, which involve the processing of large volumes of sediment to obtain organic matter (Sandgaard et al., 2023). Empirical evidence revealed that microplastics were detected in approximately 42% of *the Sabella spallanzanii* samples and 93% of the *Hermodice carunculata* samples examined (Vecchi et al., 2021), with a 48% prevalence observed among the deep-sea invertebrates analyzed (Courtene-Jones et al., 2019). Furthermore, microplastic contamination extends to deep-sea waters and benthic macroinvertebrates inhabiting the North Atlantic Ocean (Courtene-Jones et al., 2019), as well as benthic fauna in the Antarctic region (Sfriso et al., 2020), underscoring the extensive and global reach of microplastic pollution in marine benthic ecosystems. Although specific investigations suggest that ambient environmental concentrations of microplastics may exert limited effects on some taxa, sensitive species such as *Gammarus pulex* exhibit inhibited growth at low exposure levels, indicating potential adverse impacts on biodiversity and ecosystem functionality within aquatic communities (Redondo-Hasselerharm et al., 2018).

In contrast, filter-feeding organisms, including bivalve mollusks (e.g., mussels and oysters), are affected primarily by microplastic exposure through water filtration. The ecological role and commercial value of microplastics have positioned them at the center of research on microplastics. By filtering large volumes of water to extract plankton and suspended particulate matter, these organisms inevitably ingest microplastics present in their surrounding aquatic environment (Ding et al., 2022). Despite their capacity to selectively retain particles on the basis of their physicochemical properties, such as size, morphology, and surface texture (Ward et al., 2019), microplastic debris consistently accumulates within their tissues (Tibbetts, 2015). This accumulation results from both direct uptake and trophic transfer mechanisms, which can be intensified in regions with strong water currents or during plankton blooms.

This bioaccumulation phenomenon is of growing concern given its potential human health implications, as the consumption of contaminated bivalves may result in the annual ingestion of thousands of microplastic particles (Tibbetts, 2015). Empirical evidence from numerous investigations has confirmed the ubiquity of microplastic contamination across both wild and aquaculture-derived bivalve populations. A systematic review of 33 studies revealed that marine bivalves accumulate microplastics from diverse environmental compartments, with fibers and fragment particles as the predominant ingested forms (Saraswati, 2023). Notably, research has indicated that farmed specimens of *Mytilus edulis* and *Crassostrea gigas* present higher levels of microplastic contamination than their wild counterparts do (Mathalon and Hill, 2014).

Overall, microplastic contamination is pervasive across multiple trophic levels in aquatic ecosystems, from plankton to benthic invertebrates and filter-feeding bivalves. Their diverse feeding strategies and ecological roles make them vulnerable to ingesting and accumulating microplastics, with potential consequences that cascade through food webs and may ultimately impact ecosystem health and human consumption. Understanding the extent and effects of microplastic exposure in these foundational organisms is, therefore, critical for assessing broader environmental risks and guiding effective mitigation strategies.

3.2.2 Key factors influencing uptake at lower trophic levels

The uptake of microplastics by organisms at lower trophic levels of the food chain is influenced by a multifaceted set of factors, including physiological traits, feeding behaviors, and the physical and chemical characteristics of both the particles and the surrounding environment. Gaining insight into these interrelated variables is essential for accurately assessing ecological exposure and forecasting the likelihood of microplastic movement through trophic networks.

Building on this understanding, the physical form of microplastic particles—particularly their size and shape—plays a critical role in determining their likelihood of ingestion by lower-trophic-level organisms. Many species at these levels have feeding mechanisms adapted to specific particle size ranges, which often coincide with the dimensions of microplastics. In aquatic systems, some zooplankton exhibit selective feeding behavior on the basis of particle geometry. For example, *Calanus helgolandicus* tends to ingest fragment-shaped microplastics, *Acartia tonsa* prefers fibers, and *Homarus gammarus* larvae are more inclined to consume bead-shaped particles (Botterell et al., 2020). Similarly, filter-feeding bivalves such as mussels and oysters exhibit particle-size selectivity, favoring shorter microplastic fibers over longer fibers (Mladinich et al., 2022). Moreover, studies indicate that particles smaller than 300 micrometers in aquatic environments and <1 micrometer in soil systems can be ingested by microfauna, expanding the scope of microplastic bioavailability across ecosystems (Kim et al., 2024).

Considering the significant impact of microplastic particle traits on ingestion by lower-trophic-level organisms, it is equally crucial to examine how feeding strategies influence microplastic uptake in aquatic biota. Filter feeders continuously process large quantities of water, increasing their risk of ingesting suspended microplastics. In contrast, deposit feeders ingest microplastics directly from contaminated sediments, leading to distinct exposure patterns. Studies have demonstrated that filter-feeding polychaetes consume more microfibers than deposit feeders do (Porter et al., 2023). Similarly, amphipod microplastic uptake varies with feeding mode, underscoring the complex link between feeding behavior and contaminant exposure (Hiki and Nakajima, 2022).

In addition to feeding modes, microplastic exposure is further modulated by environmental conditions that govern both particle availability and ingestion dynamics. Hydrodynamic forces, particularly current velocity, play a central role in shaping the distribution and accumulation of microplastics in marine environments (Liu et al., 2022). Under shear flow, zooplankton such as *Daphnia magna* exhibit enhanced filtration performance and increased survival despite exposure to microplastics (Colomer et al., 2019). Temperature and salinity variations affect seawater density and circulation, indirectly influencing the dispersal of microplastics and their local concentrations (Liu et al., 2022). Notably, elevated temperatures have been associated with greater physiological stress from microplastic ingestion in filter-feeding zooplankton (Hoffschröer et al., 2021). In addition, higher particle concentrations lead to greater uptake, particularly among bivalves and pelagic crustaceans, than they do among benthic species (Setälä et al., 2016).

Finally, in addition to environmental factors and feeding behaviors, the chemical properties of microplastics play crucial roles in determining their uptake and potential effects on aquatic organisms. The polymer composition, particularly types such as polyethylene and polyvinyl chloride, has been identified as having a pronounced capacity to adsorb heavy metals, including lead, chromium, and zinc, from surrounding waters, thereby increasing toxicological risk upon ingestion (Godoy et al., 2019). Moreover, dissolved organic matter in natural aquatic systems further enhances the adsorption of these metals onto microplastic surfaces, thereby altering their environmental interactions (Godoy et al., 2019). In addition to these chemical characteristics, the colonization of microplastic particles by microbial communities and the subsequent formation of heteroaggregates further complicate their bioavailability and ingestion dynamics within freshwater ecosystems (Scherer et al., 2018).

3.2.3 Role of lower-trophic-level organisms in the transfer of microplastics to higher consumers

Organisms at lower trophic levels in food webs play a pivotal role in facilitating the movement of microplastics through the food web. The ecological ubiquity of these contaminants, combined with their frequent exposure to and ingestion of microplastic particles, positions them as key intermediaries in the transfer of these contaminants to higher-trophic-level predators.

These lower-trophic-level organisms serve as primary prey for various higher-trophic-level consumers, thereby facilitating indirect microplastic uptake through feeding interactions. For example, copepods that ingest microplastic particles have been reported to pass these contaminants to fish larvae, thereby negatively impacting their growth rates and survival probabilities (Uy and Johnson, 2022). Further evidence indicates that the uptake of microplastics through prey consumption can exceed direct absorption from the surrounding aquatic environment (Hasegawa and Nakaoka, 2021). In a similar vein, mysid shrimp exposed to polyethylene microplastics facilitated the transfer of these particles to European sea bass (*Dicentrarchus labrax*), leading to measurable accumulation of microplastics in the fish gastrointestinal tract (Hasegawa and Nakaoka, 2021). Prolonged exposure in juvenile sea bass additionally resulted in the translocation of microplastic particles into muscle tissues, with an estimated ratio of approximately one particle detected in the fillet for every 1.87×10^7 particles ingested (Zeytin et al., 2020).

In conclusion, lower-trophic-level aquatic organisms serve both as accumulators of microplastics and as vectors enabling their transmission to higher trophic levels within the food web. Gaining insight into the mechanisms and pathways of such trophic transfer is critical for assessing the potential long-term impacts of microplastic contamination on ecosystem stability.

3.3 Trophic transfer and bioaccumulation

The movement of microplastics through food webs represents a critical pathway by which these particles extend beyond the initial points of environmental exposure. Through successive feeding interactions and organismal retention processes, microplastics can persist within biological systems and progressively increase in abundance at relatively high trophic levels. This section examines the ecological mechanisms governing trophic transfer and bioaccumulation, highlighting their roles in shaping exposure risks for wildlife and, ultimately, human consumers.

3.3.1 Movement of microplastics through feeding interactions

Microplastics can enter food webs through trophic transfer, in which particles are passed from one organism to another, typically from prey to predator. This typically begins with lower-trophic aquatic organisms that directly ingest microplastics from water or sediment. These organisms are then consumed by small fish and crustaceans, which, in turn, become prey for larger predatory fish, seabirds, and marine mammals. The trophic transfer of microplastics from lower to higher organisms in aquatic ecosystems is summarized in Fig. 3.1.

At each level of the food chain, microplastics can be transferred from one organism to the next, where they can accumulate in the digestive tract or even in the tissues of higher-trophic -level consumers. Studies have shown that microplastics are present in fish-eating killer whales (Harlacher et al., 2023), tuna, large pelagic fish (Justino et al., 2023), gray seals, and Atlantic mackerel (Nelms et al., 2018). The presence of microplastics has been documented in >560 marine species, including fish, crustaceans, mammals, and sea turtles (Ozturk and Altinok, 2020).

The efficiency of the trophic transfer of microplastics is governed by a range of interacting variables, including the size relationship between prey and predator, the duration of particle retention in the digestive tract, and the inherent physicochemical characteristics of the plastic particles (Au et al., 2017). Research has indicated that smaller, more flexible forms, particularly fibers, tend to resist digestive breakdown more effectively and are therefore more likely to be passed on to higher trophic levels (Santonicola et al., 2023). Interestingly, findings by Bour et al. (2020) demonstrated that, in both brine shrimp and fish, microbeads remained

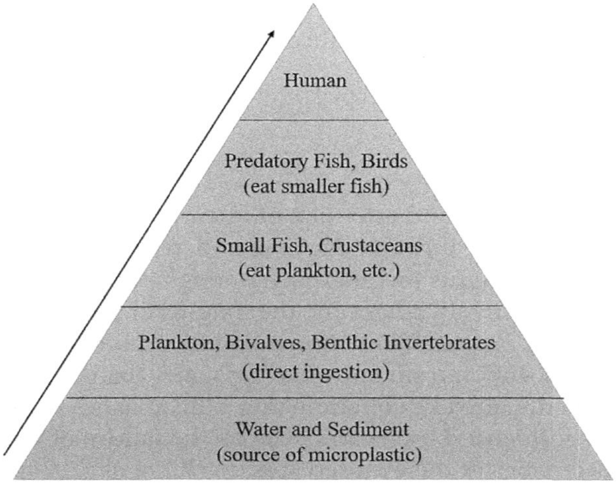

FIG. 3.1 Schematic pyramid illustrating the trophic transfer of microplastics through the aquatic food web, from water and sediment sources to plankton, small fish, predatory fish, and finally humans. The trophic transfer of microplastics from lower to higher organisms in aquatic ecosystems.

within the organisms longer than fibers did (Bour et al., 2020). In addition, as microplastics are transferred through multiple trophic levels, they may experience physical fragmentation or alterations in surface chemistry, which could influence their chemical reactivity and toxicological profile (Benson et al., 2022).

Although the ecological implications of microplastic transfer across trophic levels remain incompletely characterized, increasing scientific evidence suggests the potential for sublethal impacts on predators. Documented effects in species such as fish and crustaceans that consume prey contaminated with microplastics include impaired growth, disruptions in feeding behavior, neurotoxic responses, and elevated oxidative stress (Dioses-Salinas et al., 2019). Of particular concern is the potential for microplastic bioaccumulation to propagate through aquatic food webs, thereby increasing the likelihood of human exposure (Parmar et al., 2025).

Trophic transfer extends beyond aquatic environments, encompassing interactions between aquatic and terrestrial ecosystems. One pathway for this cross-ecosystem movement of microplastics involves terrestrial animals and birds that consume aquatic organisms. Research has documented the transfer of microplastics from aquatic invertebrates to aerial insectivores such as birds and bats (Arnold et al., 2022). In addition, the presence of microplastics in the digestive tracts of both nestling and adult migratory birds indicates that they are likely transferred through aquatic insects (Hoang and Mitten, 2022). In addition, microplastics have been detected in the gastrointestinal systems of terrestrial raptors in central California, providing clear evidence of microplastic movement from aquatic to terrestrial food webs (Carlin et al., 2020). Notably, 55% of freshwater-associated birds and 56% of seabirds were reported to have microplastics in their stomachs. Plastic ingestion in aquatic bird species has been linked to a range of adverse physiological outcomes, such as pronounced inflammatory responses, diminished feeding behavior, loss of energy reserves, and disruptions in reproductive timing, including delayed ovulation, reduced reproductive success, and, in some instances, death (Carlin et al., 2020).

Ultimately, as microplastics ascend through food webs, they may reach species consumed by humans, raising concerns about potential dietary exposure through seafood and other products.

3.3.2 Bioaccumulation across the food chain

Bioaccumulation refers to the gradual accumulation and retention of microplastics within living organisms, resulting from ongoing exposure through feeding and respiration. In contrast to simple trophic transfer, the concept highlights the interplay among uptake, tissue storage, and elimination mechanisms, which collectively determine the long-term burden of these particles in organisms (Ibrahim et al., 2021). Research has shown that bioaccumulation occurs primarily through two routes: direct absorption from surrounding water (bioconcentration) and the ingestion of prey or food items already contaminated with plastics (biomagnification). The level of accumulation is influenced by factors such as the nature of the polymer, particle size, and biological characteristics of the exposed species (Xu et al., 2020).

Empirical evidence illustrates how plastic loads increase across trophic levels. For example, copepods consume an average of 2.4 particles per individual, and when eaten by fish larvae, the number increases to 9–15 particles per larva (Desforges et al., 2015). Similarly, fish that fed on mysids previously exposed to polyethylene beads accumulated three to eleven times more microplastics than fish directly exposed to the same contamination in water did, demonstrating the amplifying role of trophic transfer (Hasegawa and Nakaoka, 2021). Feeding strategies can further intensify this process. In the amberstripe scad, 80% of individuals contained blue polyethylene fragments resembling the appearance of their copepod prey, indicating that visual predation can accelerate uptake (Ory et al., 2017). In freshwater systems, red tilapia exposed to contaminated prey exhibit bioaccumulation in internal organs, with particle loads following the order of gut > gills > liver ≈ brain, indicating not only retention but also translocation into sensitive tissues (Ding et al., 2018). These patterns of increasing loads provide strong evidence of the cumulative nature of bioaccumulation throughout food webs.

Importantly, bioaccumulation is not confined to digestive organs. Microplastics have also been detected in edible tissues, which is particularly relevant to human health. For example, commercial fish from Lagos Lagoon carry 7.2–9.5 particles per liver sample, whereas mussels and oysters produce 0.36–2.46 particles per individual (Akinhanmi et al., 2024). Laboratory studies have confirmed that once ingested, microplastics can persist in organisms for extended periods, resisting excretion and progressively accumulating in organs such as the liver, gonads, and even muscle tissue (Fernández and Albentosa, 2019; Garcia et al., 2023).

3.4 Microplastics in food

The movement of microplastics through food webs and their accumulation in living organisms has shifted scientific attention to their presence in foods directly consumed by humans. Evidence from numerous investigations reveals their presence in a variety of food products, with seafood emerging as a significant route of human exposure. Microplastics have been detected within the human body itself, appearing in organs and bodily fluids, including the liver, blood, heart, placenta, breast milk, sputum, semen, and urine, and affecting 8 of the 12 human organ systems, with reported concentrations ranging from negligible levels to over 1200 particles per gram (Enyoh et al., 2023; Roslan et al., 2024). Exposure to microplastics is associated with a range of adverse health outcomes, such as cytotoxicity, DNA damage, inflammation, oxidative stress, and cellular injury, and may further contribute to metabolic disorders, neurotoxicity, and increased cancer risk (Bao et al., 2024; Kibria, 2024; Rahman et al., 2021). Moreover, microplastics can act as carriers for hazardous chemicals and pathogenic microorganisms, potentially amplifying toxicity and compromising immune function, thereby promoting chronic inflammation and tumor development (Bao et al., 2024; Prata et al., 2020). Importantly, their contamination is not confined to marine-based foods; they have also been identified in drinking water, table salt, processed products, and agricultural products such as fruits and vegetables. The occurrence and concentration of microplastics in various food sources reported in recent studies are summarized in Table 3.1. The widespread distribution of microplastics across diverse dietary sources underscores growing concerns about food safety, cumulative exposure, and long-term implications for human health.

TABLE 3.1 Summary of microplastic characteristics and concentrations reported in various food sources.

Food source	Dominant particle type	Particle size	Microplastic concentration	Reference
Bivalves (mussels, oysters)	Fibers and fragments	50–100 μm	0.36–2.46 particles/individual	(Van Cauwenberghe and Janssen, 2014)
Fish (cod, sardines, anchovies)	Fragments and fibers	Variable	9–15 particles/larva (from copepod prey), 3–11 × higher in fish via trophic transfer	(Desforges et al., 2015)
Drinking water (tap)	Fragments and fibers	<300 μm	Mean 62.38 particles/L	(Zhu et al., 2024)
Drinking water (bottled)	Fragments and fibers	<300 μm	Mean 38.45 particles/L	(Zhu et al., 2024)
Table salt	Fibers and fragments	<500 μm	50–6430 particles/kg (varies by country)	(Lee et al., 2019)
Processed foods (honey, milk, beverages)	Fibers and fragments	0.1–5000 μm	Up to 100 particles/L (beverages)	(Diaz-Basantes et al., 2020)
Fruits and vegetables	Small fragments	2–100 μm	Highest in apples and carrots (exact numbers not specified)	(Aydın et al., 2023)

3.4.1 Occurrence of microplastics in seafood and aquatic products

Among the food categories examined for microplastic contamination, seafood—particularly fish, shellfish, and crustaceans—has received the most research attention. Global surveys of commercially available aquatic species have consistently confirmed the presence of microplastics. A metaanalysis estimated that the maximum annual human intake of microplastics from seafood is approximately 55,000 particles (Danopoulos et al., 2020). In commonly consumed mussels and oysters, the reported microplastic burdens range from 0.36 to 2.46 particles per bivalve (Van Cauwenberghe and Janssen, 2014). Studies have shown that European shellfish consumers potentially ingest up to 11,000 microplastic particles annually (Van Cauwenberghe and Janssen, 2014). Laboratory studies investigating microplastic uptake in bivalves often apply concentrations higher than those found in natural environments, although a range of 100–1000 particles/L has been recommended to ensure ecological relevance (Baroja et al., 2021). Analyses have revealed the presence of various polymers, with microplastics most frequently occurring as gray fragments measuring 50–100 μm in diameter (Phuong et al., 2018).

In addition to bivalves, numerous studies have documented microplastic contamination in a variety of coastal and commercial fish species, including Atlantic cod, sardines, anchovies, and shrimp. Investigations in the Northeast Atlantic Ocean revealed the presence of microplastics in multiple tissues of European sardine, European anchovy, and horse mackerel, with the latter exhibiting the highest contamination levels (Lopes et al., 2023). A global survey reported the presence of microplastics in 926 seafood species from 57 countries, with Asian seafood showing the highest prevalence (Kibria, 2023). Microplastics have been detected in diverse fish tissues, including the skin, gills, stomach, liver, intestine, and muscle, as well as in processed products such as dried and canned fish (Kibria et al., 2022). Notably, (Silva et al., 2024) reported microplastics in all the cannabis seafood varieties examined, averaging 3.5 per can.

In contrast, Gündoğdu and Köşker (2023) reported similar findings in Turkish canned fish products, with fragments as the dominant shape. Methodological advances by (Duman et al., 2023) have improved the extraction of microplastics from packaged fish, revealing concentrations of 0.05–0.33 microplastics per gram. These findings underscore the potential for microplastic bioaccumulation in fish tissues and raise concerns over their transfer to humans through seafood consumption (Kibria, 2023).

3.4.2 Presence of microplastics in drinking water, table salt, and processed foods

Drinking water, whether distributed through municipal supply systems or sold commercially in bottles, constitutes a significant and consistent route of human exposure to microplastics. Multiple studies have confirmed the presence of these particles in both sources, with mean concentrations reported at approximately 62.38 particles/L in tap water and 38.45 particles/L in bottled water (Zhu et al., 2024). The detected microplastics are predominantly small fragments or fibers measuring <300 μm and are primarily composed of polyethylene terephthalate, polyethylene, and polypropylene. Globally, concentrations in drinking water have shown substantial variability, ranging from 0.0007 particles/L to 6292 particles/L, corresponding to an estimated daily ingestion of approximately 175 particles per person

(Zhu et al., 2024). This variation is likely attributable to differences in analytical methodologies and contamination during sampling and laboratory procedures (Kirstein et al., 2021).

In addition to drinking water, table salt has emerged as a significant dietary route of human exposure to microplastics. Comprehensive surveys of commercially available salt products from various regions have revealed that microplastic particles are present in the overwhelming majority of samples, with detection rates reaching 94% globally (Lee et al., 2019). The reported concentrations exhibit wide geographic variability, ranging from 50–280 particles/kg in Spain (Iñiguez et al., 2017) to 595–6430 particles/kg in Portugal (Sofia Soares et al., 2021). The predominant polymer types identified include polyethylene, polypropylene, and polyethylene terephthalate, which most commonly occur as fragments and fibers smaller than 500 μm (Maharjan and Dhungel, 2024; Suteja et al., 2025). Contamination levels are influenced primarily by the type and geographic origin of the salt, with sea salt generally exhibiting higher microplastic loads than lake or rock salt (Gündoğdu, 2018). On the basis of average consumption rates, it has been estimated that an individual may ingest several hundred to >1000 microplastic particles annually from saltwater alone (Gündoğdu, 2018; Lee et al., 2019).

In addition to drinking water and table salt, a range of processed foods have been recognized as significant contributors to dietary microplastic exposure. Empirical evidence has demonstrated the occurrence of microplastics in items such as honey, sugar, milk, beer, soft drinks, and other commercially processed commodities (Diaz-Basantes et al., 2020; Giri et al., 2024; Kadac-Czapska et al., 2024). These synthetic particles, which typically range from 0.1 to 5000 μm in size, have been detected at concentrations that vary considerably across products, with some beverages containing up to 100 particles/L (Diaz-Basantes et al., 2020). Polyethylene, polypropylene, and polyacrylamide are the most frequently identified polymers in these matrices (Diaz-Basantes et al., 2020). The ingestion of microplastics through processed foods raises concerns due to potential adverse health outcomes, including oxidative stress, inflammatory responses, and disruptions to the gut microbiome (Giri et al., 2024). Sources of contamination include not only the environmental pollution of raw materials but also manufacturing, processing, and packaging practices (Diaz-Basantes et al., 2020).

3.4.3 Microplastics in agricultural products: Fruits, vegetables, and grains

Recent studies have shown that microplastics are not limited to marine and freshwater environments but are also present in agricultural systems, including fruits, vegetables, and grains. These particles can enter farmland through various pathways, including irrigation with contaminated water, atmospheric deposition, and the use of plastic mulch or fertilizers that contain plastic residues (Qi et al., 2020). Once in the soil, microplastics can change their structure and chemical properties, reduce water retention, disrupt nutrient cycling, and affect microbial communities (Vignesh et al., 2025). They can also carry harmful chemicals and heavy metals, which may be taken up by plants (Vignesh et al., 2025).

Microplastics found in soils and on plant surfaces can be eaten by animals, including livestock and wild herbivores, as part of their regular diet (Prata and Dias-Pereira, 2023). Over time, these particles can accumulate in animal tissues and may be transmitted to humans through the consumption of meat, milk, and eggs (Prata and Dias-Pereira, 2023).

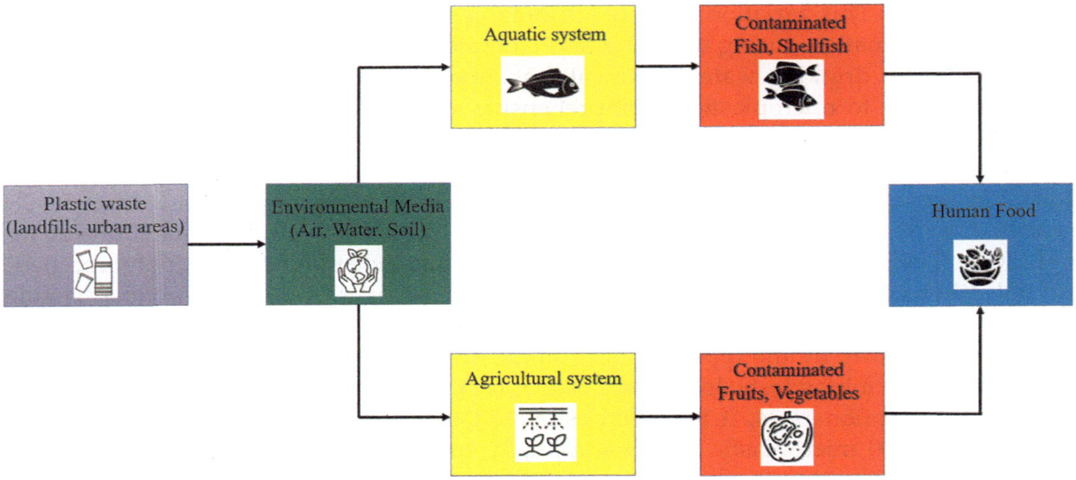

FIG. 3.2 Flowchart illustrating the main pathways of microplastic entry into the human food chain. Plastic waste from urban and landfill sources disperses through environmental media (air, water, and soil), contaminating aquatic and agricultural systems. Microplastics then accumulate in seafood (fish and shellfish) and agricultural products (fruits and vegetables), ultimately entering the human food supply.

This process represents an indirect pathway of human exposure through the terrestrial food chain.

There is also a direct route of exposure, as microplastics have been detected on and inside fresh produce such as apples, tomatoes, carrots, and potatoes. The levels of contamination differ by crop, with apples showing the highest concentrations among fruits and carrots among vegetables (Aydın et al., 2023; Conti et al., 2020). The detected particles are generally tiny, ranging from 0.002 mm to 0.01 mm (Rajendran et al., 2022). While the health impacts of consuming microplastic-contaminated produce are not yet fully understood, their widespread presence in agricultural products underscores the need for further research to assess risks and identify effective ways to reduce contamination. The major pathways of microplastic entry into the human food chain are summarized in Fig. 3.2.

3.5 Conclusion

Microplastics have become pervasive pollutants across aquatic, terrestrial, and atmospheric environments, integrating into food chains through ingestion by lower-trophic-level organisms such as plankton, benthic invertebrates, and filter-feeding bivalves. These particles are transferred and sometimes bioaccumulate at relatively high trophic levels, crossing ecosystem boundaries and ultimately reaching humans through seafood, drinking water, salt, processed foods, and agricultural products. Their persistence, mobility, and surface properties enable them to adsorb toxic chemicals and microorganisms, creating combined hazards that may alter their toxicological effects as they move through food webs. In multiple human organs and biological fluids, microplastics are linked to inflammation, oxidative stress, metabolic disruption, neurotoxicity, and potentially cancer. Addressing this growing problem demands coordinated strategies, including reducing plastic release at the source, improving waste management, advancing detection and monitoring methods, and standardizing analytical approaches. Without comprehensive intervention, microplastics continue to accumulate in ecosystems and food systems, posing escalating risks to biodiversity, food safety, and public health.

References

Akbari, A., Jaafari, J., 2024. Entry pathway and potential impacts of microplastics in air, water, soil and human health: a review. Caspian J. Health Res. 2423-8171. 9 (4), 181–198. doi:10.32598/cjhr.9.4.254.2.

Akinhanmi, F.O., Ayanda, O.I., Dedeke, G.A., 2024. Assessment of microplastic-contaminated liver through gene expression profiling of four commercial fish species in the Lagos Lagoon, Nigeria. Sci. Afr. 24682276. 24. Available from: https://www.sciencedirect.com/science/journal/24682276. doi:10.1016/j.sciaf.2024.e02242.

Allen, S., Allen, D., Phoenix, V.R., Roux, G.L., Durántez Jiménez, P., Simonneau, A., Binet, S., Galop, D., 2019. Atmospheric transport and deposition of microplastics in a remote mountain catchment. Nat. Geosci. 17520908. 12 (5), 339–344. Available from: http://www.nature.com/ngeo/index.html. doi:10.1038/s41561-019-0335-5.

Arnold, G., Muller, G., Ormerod, S., 2022. Field data support the transfer of microplastics to aerial insectivores via flying insects. SSRN Electron. J. 1556–5068. doi:10.2139/ssrn.4250871.

Au, S.Y., Lee, C.M., Weinstein, J.E., van den Hurk, P., Klaine, S.J., 2017. Trophic transfer of microplastics in aquatic ecosystems: identifying critical research needs. Integr. Environ. Assess. Manag. 15513793. 13 (3), 505–509. Available from: http://www.interscience.wiley.com/jpages/1551-3777. doi:10.1002/ieam.1907.

Auta, H.S., Emenike, C.U., Fauziah, S.H., 2017. Distribution and importance of microplastics in the marine environment: a review of the sources, fate, effects, and potential solutions. Environ. Int. 01604120. 102, 165–176. doi:10.1016/j.envint.2017.02.013.

Aydın, R.B., Yozukmaz, A., Şener, İ., Temiz, F., Giannetto, D., 2023. Occurrence of microplastics in most consumed fruits and vegetables from Turkey and public risk assessment for consumers. Life. 20751729. 13 (8). Available from: www.mdpi.com/journal/life/. doi:10.3390/life13081686.

Bai, Z., Wang, N., Wang, M., 2021. Effects of microplastics on marine copepods. Ecotoxicol. Environ. Saf. 01476513. 217, 112243. doi:10.1016/j.ecoenv.2021.112243.

Bao, Y.B., Wang, C.C., Peng, W.G., Nong, D.Q., Xiang, P., 2024. Human accumulation and toxic effects of microplastics: a critical review. Huanjing Kexue/Environmental Science. 02503301. 45 (2), 1173–1184. Available from: https://www.hjkx.ac.cn/hjkx/ch/html/20240253.htm. doi:10.13227/j.hjkx.202303260.

Baroja, E., Christoforou, E., Lindström, J., Spatharis, S., 2021. Effects of microplastics on bivalves: are experimental settings reflecting conditions in the field? Mar. Pollut. Bull. 0025326X. 171, 112696. doi:10.1016/j.marpolbul.2021.112696.

Benson, N.U., Agboola, O.D., Fred-Ahmadu, O.H., De-la-Torre, G.E., Oluwalana, A., Williams, A.B., 2022. Micro(nano)plastics prevalence, food web interactions, and toxicity assessment in aquatic organisms: a review. Front. Mar. Sci. 22967745. 9. Available from: https://www.frontiersin.org/journals/marine-science#. doi:10.3389/fmars.2022.851281.

Botterell, Z.L.R., Beaumont, N., Cole, M., Hopkins, F.E., Steinke, M., Thompson, R.C., Lindeque, P.K., 2020. Bioavailability of microplastics to marine zooplankton: effect of shape and infochemicals. Environ. Sci. Technol. 15205851. 54 (19), 12024–12033. Available from: http://pubs.acs.org/journal/esthag. doi:10.1021/acs.est.0c02715.

Botterell, Z.L.R., Beaumont, N., Dorrington, T., Steinke, M., Thompson, R.C., Lindeque, P.K., 2019. Bioavailability and effects of microplastics on marine zooplankton: a review. Environ. Pollut. 18736424. 245, 98–110. Available from: https://www.elsevier.com/inca/publications/store/4/0/5/8/5/6. doi:10.1016/j.envpol.2018.10.065.

Bour, A., Hossain, S., Taylor, M., Sumner, M., Almroth, B.C., 2020. Synthetic microfiber and microbead exposure and retention time in model aquatic species under different exposure scenarios. Front. Environ. Sci. 2296-665X. 8. doi:10.3389/fenvs.2020.00083.

Cáceres-Farias, L., Espinoza-Vera, M.M., Orós, J., Garcia-Bereguiain, M.A., Alfaro-Núñez, A., 2023. Macro and microplastic intake in seafood variates by the marine organism's feeding behaviour: is it a concern to human health? Heliyon. 24058440. 9 (5). Available from: http://www.journals.elsevier.com/heliyon/. doi:10.1016/j.heliyon.2023.e16452.

Carlin, J., Craig, C., Little, S., Donnelly, M., Fox, D., Zhai, L., Walters, L., 2020. Microplastic accumulation in the gastrointestinal tracts in birds of prey in central Florida, USA. Environ. Pollut. 02697491. 264, 114633. doi:10.1016/j.envpol.2020.114633.

Cole, M., Lindeque, P., Fileman, E., Halsband, C., Goodhead, R., Moger, J., Galloway, T.S., 2013. Microplastic ingestion by zooplankton. Environ. Sci. Technol. 15205851. 47 (12), 6646–6655. doi:10.1021/es400663f.

Colomer, J., Müller, M.F., Barcelona, A., Serra, T., 2019. Mediated food and hydrodynamics on the ingestion of microplastics by *Daphnia magna*. Environ. Pollut. 18736424. 251, 434–441. Available from: https://www.journals.elsevier.com/environmental-pollution. doi:10.1016/j.envpol.2019.05.034.

Conti, G.O., Ferrante, M., Banni, M., Favara, C., Nicolosi, I., Cristaldi, A., Fiore, M., Zuccarello, P., 2020. Micro- and nano-plastics in edible fruit and vegetables. The first diet risks assessment for the general population. Environ. Res. 00139351. 187, 109677. doi:10.1016/j.envres.2020.109677.

Courtene-Jones, W., Quinn, B., Ewins, C., Gary, S.F., Narayanaswamy, B.E., 2019. Consistent microplastic ingestion by deep-sea invertebrates over the last four decades (1976–2015), a study from the North East Atlantic. Environ. Pollut. 18736424. 244, 503–512. Available from: https://www.elsevier.com/inca/publications/store/4/0/5/8/5/6. doi:10.1016/j.envpol.2018.10.090.

Danopoulos, E., Jenner, L.C., Twiddy, M., Rotchell, J.M., 2020. Microplastic contamination of seafood intended for human consumption: a systematic review and meta-analysis. Environ. Health Perspect. 15529924. 128 (12). Available from: https://ehp.niehs.nih.gov/doi/pdf/10.1289/EHP7171. doi:10.1289/EHP7171.

Desforges, J.P.W., Galbraith, M., Ross, P.S., 2015. Ingestion of microplastics by zooplankton in the Northeast Pacific Ocean. Arch. Environ. Contam. Toxicol. 14320703. 69 (3), 320–330. Available from: https://link.springer.com/article/10.1007/s00244-015-0172-5. doi:10.1007/s00244-015-0172-5.

Diaz-Basantes, M.F., Conesa, J.A., Fullana, A., 2020. Microplastics in honey, beer, milk and refreshments in Ecuador as emerging contaminants. Sustainability. 2071-1050. 12 (14), 5514. doi:10.3390/su12145514.

Ding, J., Sun, C., Li, J., Shi, H., Xu, X., Ju, P., Jiang, F., Li, F., 2022. Microplastics in global bivalve mollusks: a call for protocol standardization. J. Hazard. Mater. 03043894. 438, 129490. doi:10.1016/j.jhazmat.2022.129490.

Ding, J., Zhang, S., Razanajatovo, R.M., Zou, H., Zhu, W., 2018. Accumulation, tissue distribution, and biochemical effects of polystyrene microplastics in the freshwater fish red tilapia (*Oreochromis niloticus*). Environ. Pollut. 18736424. 238, 1–9. Available from: https://www.elsevier.com/inca/publications/store/4/0/5/8/5/6. doi:10.1016/j.envpol.2018.03.001.

Dioses-Salinas, D., Pérez-Baca, B., De-la-Torre, G., 2019. Ecotoxicological effects of microplastics and adsorbed contaminants on aquatic organisms. Manglar. 24141046. 16 (2), 173–182. doi:10.17268/manglar.2019.024.

Duman, S., Doyen, P., Merveillie, P., Andersson, N., Bayeuil, R., Grard, T., Dehaut, A., Duflos, G., 2023. Optimization of a method designed to extract and characterize microplastics in different packaged fish products. Food Control. 09567135. 154, 110029. doi:10.1016/j.foodcont.2023.110029.

Enyoh, C.E., Devi, A., Kadono, H., Wang, Q., Rabin, M.H., 2023. The plastic within: microplastics invading human organs and bodily fluids systems. Environments. 2076-3298. 10 (11), 194. doi:10.3390/environments10110194.

Fernández, B., Albentosa, M., 2019. Insights into the uptake, elimination and accumulation of microplastics in mussel. Environ. Pollut. 02697491. 249, 321–329. doi:10.1016/j.envpol.2019.03.037.

Fowzi, M., Ebrahimpour, K., Dehnavi, A., Jamshidi, S., Andaluri, G., 2025. Microplastics as emerging contaminants in municipal solid waste compost: distribution, characterization, and ecological risk. Environ. Res. 00139351. 284, 122248. doi:10.1016/j.envres.2025.122248.

Garcia, M.M., Romero, A.S., Merkley, S.D., Meyer-Hagen, J.L., Forbes, C., Hayek, E.E., Sciezka, D.P., Templeton, R., Gonzalez-Estrella, J., Jin, Y., Gu, H., Benavidez, A., Hunter, R.P., Lucas, S., Herbert, G., Kim, K.J., Cui, J.Y., Gullapalli, R., In, J.G., Campen, M.J., Castillo, E.F., 2023. In Vivo tissue distribution of microplastics and systemic metabolomic alterations after gastrointestinal exposure. bioRxiv. 26928205. Available from: https://www.biorxiv.org. doi:10.1101/2023.06.02.542598.

Giri, S., Lamichhane, G., Khadka, D., Devkota, H.P., 2024. Microplastics contamination in food products: occurrence, analytical techniques and potential impacts on human health. Curr. Res. Biotechnol. 25902628. 7, 100190. doi:10.1016/j.crbiot.2024.100190.

Godoy, V., Blázquez, G., Calero, M., Quesada, L., Martín-Lara, M.A., 2019. The potential of microplastics as carriers of metals. Environ. Pollut. 02697491. 255 (Pt 3), 113363. doi:10.1016/j.envpol.2019.113363.

Gündoğdu, S., 2018. Contamination of table salts from Turkey with microplastics. Food Addit. Contam. Part A. 1944-0049. 35 (5), 1006–1014. doi:10.1080/19440049.2018.1447694.

Gündoğdu, S., Köşker, A.R., 2023. Microplastic contamination in canned fish sold in Türkiye. PeerJ. 2167-8359. 11, e14627. doi:10.7717/peerj.14627.

Harlacher, J., Emmons, C., Hanson, B., Olsen, D., Matkin, C., Phan, S., Luscombe, C.K., daSilva, D., Park, L., Parsons, K.M., 2023. Evidence of microplastics in marine top predators: microplastic particles isolated from the feces of fish-eating killer whales. SSRN, United States. 15565068. Available from: https://www.ssrn.com/index.cfm/en/. doi:10.2139/ssrn.4364184.

Hasegawa, T., Nakaoka, M., 2021. Trophic transfer of microplastics from mysids to fish greatly exceeds direct ingestion from the water column. Environ. Pollut. 02697491. 273, 116468. doi:10.1016/j.envpol.2021.116468.

He, D., Luo, Y., Lu, S., Liu, M., Song, Y., Lei, L., 2018. Microplastics in soils: analytical methods, pollution characteristics and ecological risks. TrAC. Trends. Anal. Chem. 01659936. 109, 163–172. doi:10.1016/j.trac.2018.10.006.

Hiki, K., Nakajima, F., 2022. Microplastic ingestion by a benthic amphipod in different feeding modes. J. Water Environ. Technol. 1348-2165. 20 (5), 137–144. doi:10.2965/jwet.22-057.

Hoang, T.C., Mitten, S., 2022. Microplastic accumulation in the gastrointestinal tracts of nestling and adult migratory birds. Sci. Total Environ. 00489697. 838 (Pt 1), 155827. doi:10.1016/j.scitotenv.2022.155827.

Hoffschröer, N., Grassl, N., Steinmetz, A., Sziegoleit, L., Koch, M., Zeis, B., 2021. Microplastic burden in Daphnia is aggravated by elevated temperatures. Zoology. 09442006. 144, 125881. doi:10.1016/j.zool.2020.125881.

Ibrahim, A.A., Yusuf, A.G., Ismail, G., Ibrahim, M.A., Musa, A.R., Sulaiman, M.S., 2021. Conceptual background of bioaccumulation in environmental science. World J. Adv. Pharm. Life Sci. 27990222. 1 (1), 035–041. doi:10.53346/wjapls.2021.1.1.0015.

Iñiguez, M.E., Conesa, J.A., Fullana, A., 2017. Microplastics in Spanish table salt. Sci. Rep. 20452322. 7 (1). Available from: https://www.nature.com/srep/index.html. doi:10.1038/s41598-017-09128-x.

Jeon, M.J., Yoo, J.W., Lee, K.W., Won, E.J., Lee, Y.M., 2023. Microplastics disrupt energy metabolism in the brackish water Flea Diaphanosoma Celebensis. SSRN. 15565068. Available from: https://www.ssrn.com/index.cfm/en/. doi:10.2139/ssrn.4365533.

Jovanović, B., 2017. Ingestion of microplastics by fish and its potential consequences from a physical perspective. Integr. Environ. Assess. Manag. 1551-3777. 13 (3), 510–515. doi:10.1002/ieam.1913.

Justino, A.K.S., Ferreira, G.V.B., Fauvelle, V., Schmidt, N., Lenoble, V., Pelage, L., Martins, K., Travassos, P., Lucena-Frédou, F., 2023. From prey to predators: evidence of microplastic trophic transfer in tuna and large pelagic species in the southwestern Tropical Atlantic. Environ. Pollut. 18736424. 327. Available from: https://www.journals.elsevier.com/environmental-pollution. doi:10.1016/j.envpol.2023.121532.

Kabir, M.S., Wang, H., Luster-Teasley, S., Zhang, L., Zhao, R., 2023. Microplastics in landfill leachate: sources, detection, occurrence, and removal. Environ. Sci. Ecotechnol. 26664984. 16, 100256. doi:10.1016/j.ese.2023.100256.

Kadac-Czapska, K., Knez, E., Grembecka, M., 2024. Food and human safety: the impact of microplastics. Crit. Rev. Food Sci. Nutr. 1040-8398. 64 (11), 3502–3521. doi:10.1080/10408398.2022.2132212.

Kibria, G., 2023. Impacts of microplastic on fisheries and seafood security — global analysis and synthesis. Sci. Total Environ. 00489697. 904, 166652. doi:10.1016/j.scitotenv.2023.166652.

Kibria, G., 2024. Exposure routes of microplastics (MPs) to humans and possible risks of MPs to human health from food and the environment: a short review. J. Food Saf. Hyg. 2476–3241. doi:10.18502/jfsh.v10i1.16440.

Kibria, G., Nugegoda, D., Yousuf Haroon, A.K., 2022. Microplastic Pollution and Contamination of Seafood (Including Fish, Sharks, Mussels, Oysters, Shrimps and Seaweeds): A Global Overview. Springer Science and Business Media LLC, pp. 277–322. doi:10.1007/978-3-030-89220-3_14.

Kim, D., Kim, S.A., Nam, S.H., Kwak, J.I., Kim, L., Lee, T.Y., Kim, H., An, S., An, Y.J., 2024. Microplastic ingestion in aquatic and soil biota: a comprehensive review of laboratory studies on edible size and intake pattern. Mar. Pollut. Bull. 18793363. 200. Available from: https://www.sciencedirect.com/science/journal/0025326X. doi:10.1016/j.marpolbul.2024.116056.

Kirstein, I.V., Gomiero, A., Vollertsen, J., 2021. Microplastic pollution in drinking water. Curr. Opin. Toxicol. 24682020. 28, 70–75. Available from: https://www.journals.elsevier.com/current-opinion-in-toxicology/. doi:10.1016/j.cotox.2021.09.003.

Kristanti, R.A., Wong, W.L., Darmayati, Y., Hatmanti, A., Wulandari, N.F., Sibero, M.T., Afianti, N.F., Hernandes, E., Lopez-Martinez, F., 2022. Characteristics of Microplastic in Commercial Aquatic Organisms. Trop. Aquat. Soil Pollut. 2798-3056. 2 (2), 134–158. doi:10.53623/tasp.v2i2.134.

Lee, H., Kunz, A., Shim, W.J., Walther, B.A., 2019. Microplastic contamination of table salts from Taiwan, including a global review. Sci. Rep. 20452322. 9 (1). Available from: https://www.nature.com/srep/index.html. doi:10.1038/s41598-019-46417-z.

Liu, J., Liu, H., Dongyang, H., Zhang, T., Jiao, Q., Yihan, L., Zhang, Y.-N., 2022. Comprehensive effects of temperature, salinity, and current velocity on the microplastic abundance in offshore area. Pol. J. Environ. Stud. 1230-1485. 31 (2), 1727–1736. doi:10.15244/pjoes/142389.

Lopes, C., Ambrosino, A.C., Figueiredo, C., Caetano, M., Santos, M.M., Garrido, S., Raimundo, J., 2023. Microplastic distribution in different tissues of small pelagic fish of the Northeast Atlantic Ocean. Sci. Total Environ. 18791026. 901. Available from: https://www.elsevier.com/locate/scitotenv. doi:10.1016/j.scitotenv.2023.166050.

Maharjan, K.K., Dhungel, R.P., 2024. First-ever study uncovers microplastic contamination in Nepalese table salt. Heliyon. 24058440. 10 (14). Available from: https://www.sciencedirect.com/science/journal/24058440. doi:10.1016/j.heliyon.2024.e34621.

Margot, J., Kienle, C., Magnet, A., Weil, M., Rossi, L., de Alencastro, L.F., Abegglen, C., Thonney, D., Chèvre, N., Schärer, M., Barry, D.A., 2013. Treatment of micropollutants in municipal wastewater: ozone or powdered activated carbon? Sci. Total Environ. 18791026. 461-462, 480–498. doi:10.1016/j.scitotenv.2013.05.034.

Marmara, D., Katsanevakis, S., Brundo, M.V., Tiralongo, F., Ignoto, S., Krasakopoulou, E., 2023. Microplastics ingestion by marine fauna with a particular focus on commercial species: a systematic review. Front. Mar. Sci. 22967745. 10. Available from: https://www.frontiersin.org/journals/marine-science#. doi:10.3389/fmars.2023.1240969.

Mathalon, A., Hill, P., 2014. Microplastic fibers in the intertidal ecosystem surrounding Halifax Harbor, Nova Scotia. Mar. Pollut. Bull. 0025326X. 81 (1), 69–79. doi:10.1016/j.marpolbul.2014.02.018.

Mladinich, K., Holohan, B.A., Shumway, S.E., Brown, K., Ward, J.E., 2022. Determining the properties that govern selective ingestion and egestion of microplastics by the blue mussel (*Mytilus edulis*) and Eastern Oyster (*Crassostrea virginica*). Environ. Sci. Technol. 15205851. 56 (22), 15770–15779. Available from: http://pubs.acs.org/journal/esthag. doi:10.1021/acs.est.2c06402.

Nelms, S.E., Galloway, T.S., Godley, B.J., Jarvis, D.S., Lindeque, P.K., 2018. Investigating microplastic trophic transfer in marine top predators. Environ. Pollut. 18736424. 238, 999–1007. Available from: https://www.elsevier.com/inca/publications/store/4/0/5/8/5/6. doi:10.1016/j.envpol.2018.02.016.

Ory, N.C., Sobral, P., Ferreira, J.L., Thiel, M., 2017. Amberstripe scad Decapterus muroadsi (Carangidae) fish ingest blue microplastics resembling their copepod prey along the coast of Rapa Nui (Easter Island) in the South Pacific subtropical gyre. Sci. Total Environ. 18791026. 586, 430–437. Available from: https://www.elsevier.com/locate/scitotenv. doi:10.1016/j.scitotenv.2017.01.175.

Ozturk, R.C., Altinok, I., 2020. Interaction of plastics with marine species. Turk. J. Fish. Aquat. Sci. 13032712. 20 (8), 647–658. Available from: http://www.trjfas.org/pdf.php?&id=1472. doi:10.4194/1303-2712-v20_8_07.

Parmar, R., Thakur, S., Singh, A., Rajesh, P., Singh, A.K., Rajpoot, S., 2025. Microplastics: understanding the interaction with the food web and potential health hazards. J. Environ. Eng. Sci. 1496256X. Available from: http://www.icevirtuallibrary.com/toc/jenes/current. doi:10.1680/jenes.24.00085.

Pestana, C.J., Moura, D.S., Capelo-Neto, J., Edwards, C., Dreisbach, D., Spengler, B., Lawton, L.A., 2021. Potentially poisonous plastic particles: microplastics as a vector for cyanobacterial toxins microcystin-LR and microcystin-LF. Environ. Sci. Technol. 15205851. 55 (23), 15940–15949. Available from: http://pubs.acs.org/journal/esthag. doi:10.1021/acs.est.1c05796.

Phuong, N.N., Poirier, L., Pham, Q.T., Lagarde, F., Zalouk-Vergnoux, A., 2018. Factors influencing the microplastic contamination of bivalves from the French Atlantic coast: location, season and/or mode of life? Mar. Pollut. Bull. 18793363. 129 (2), 664–674. Available from: https://www.elsevier.com/locate/marpolbul. doi:10.1016/j.marpolbul.2017.10.054.

Porter, A., Barber, D., Hobbs, C., Love, J., Power, A.L., Bakir, A., Galloway, T.S., Lewis, C., 2023. Uptake of microplastics by marine worms depends on feeding mode and particle shape but not exposure time. Sci. Total Environ. 18791026. 857. Available from: https://www.elsevier.com/locate/scitotenv. doi:10.1016/j.scitotenv.2022.159287.

Prata, J.C., da Costa, J.P., Lopes, I., Duarte, A.C., Rocha-Santos, T., 2020. Environmental exposure to microplastics: an overview on possible human health effects. Sci. Total Environ. 18791026. 702. Available from: https://www.elsevier.com/locate/scitotenv. doi:10.1016/j.scitotenv.2019.134455.

Prata, J.C., Dias-Pereira, P., 2023. Microplastics in terrestrial domestic animals and human health: implications for food security and food safety and their role as sentinels. Animals. 20762615. 13 (4). Available from: http://www.mdpi.com/journal/animals. doi:10.3390/ani13040661.

Qi, R., Jones, D.L., Li, Z., Liu, Q., Yan, C., 2020. Behavior of microplastics and plastic film residues in the soil environment: a critical review. Sci. Total Environ. 00489697. 703, 134722. doi:10.1016/j.scitotenv.2019.134722.

Rafa, N., Ahmed, B., Zohora, F., Bakya, J., Ahmed, S., Ahmed, S.F., Mofijur, M., Chowdhury, A.A., Almomani, F., 2024. Microplastics as carriers of toxic pollutants: source, transport, and toxicological effects. Environ. Pollut. 18736424. 343. Available from: https://www.sciencedirect.com/science/journal/02697491. doi:10.1016/j.envpol.2023.123190.

Rahman, A., Sarkar, A., Yadav, O.P., Achari, G., Slobodnik, J., 2021. Potential human health risks due to environmental exposure to nano- and microplastics and knowledge gaps: a scoping review. Sci. Total Environ. 00489697. 757, 143872. doi:10.1016/j.scitotenv.2020.143872.

Rajendran, K., Rajendiran, R., Pasupathi, M.S., Ahamed, S.B.N., Kalyanasundaram, P., Velu, R.K., 2022. Authentication of Microplastic Accumulation in Customary Fruits and Vegetables. Research Square, India, Available from https://www.researchsquare.com/browse. doi:10.21203/rs.3.rs-1314420/v1.

Redondo-Hasselerharm, P.E., Falahudin, D., Peeters, E.T.H.M., Koelmans, A.A., 2018. Microplastic effect thresholds for freshwater benthic macroinvertebrates. Environ. Sci. Technol. 15205851. 52 (4), 2278–2286. Available from: http://pubs.acs.org/journal/esthag. doi:10.1021/acs.est.7b05367.

Roslan, N.S., Lee, Y.Y., Ibrahim, Y.S., Anuar, S.T., Yusof, K.M.K.K., Lai, L.A., Brentnall, T., 2024. Detection of microplastics in human tissues and organs: a scoping review. J. Glob. Health. 20472986. 14. Available from: http://jogh.org/. doi:10.7189/jogh.14.04179.

Sandgaard, M.H., Palmqvist, A., Bour, A., Grønlund, S.N., Hooge, A., Selck, H., Thit, A., Syberg, K., 2023. Sediment matters as a route of microplastic exposure: a call for more research on the benthic compartment. Front. Mar. Sci. 22967745. 9. Available from: https://www.frontiersin.org/journals/marine-science#. doi:10.3389/fmars.2022.1100567.

Santillo, D., Miller, K., Johnston, P., 2017. Microplastics as contaminants in commercially important seafood species. Integr. Environ. Assess. Manag. 1551-3777. 13 (3), 516–521. doi:10.1002/ieam.1909.

Santonicola, S., Volgare, M., Cocca, M., Dorigato, G., Giaccone, V., Colavita, G., 2023. Impact of fibrous microplastic pollution on commercial seafood and consumer health: a review. Animals. 2076-2615. 13 (11), 1736. doi:10.3390/ani13111736.

Saraswati, N.L.G.R.A., 2023. Microplastics ingestions by wild and aquaculture marine bivalves: a systematic review on field investigation study. Sustinere: J. Environ. Sustain. 2549-1253. 7 (1), 15–26. doi:10.22515/sustinerejes.v7i1.294.

Scherer, C., Weber, A., Lambert, S., Wagner, M., 2018. Interactions of microplastics with freshwater biota. Handb. Environ. Chem. 58, 153–180. Available from http://www.springer.com/series/698. doi:10.1007/978-3-319-61615-5_8.

Schmidt, C., Krauth, T., Wagner, S., 2017. Export of plastic debris by rivers into the sea. Environ. Sci. Technol. 0013-936X. 51 (21), 12246–12253. doi:10.1021/acs.est.7b02368.

Setälä, O., Norkko, J., Lehtiniemi, M., 2016. Feeding type affects microplastic ingestion in a coastal invertebrate community. Mar. Pollut. Bull. 0025326X. 102 (1), 95–101. doi:10.1016/j.marpolbul.2015.11.053.

Sfriso, A.A., Tomio, Y., Rosso, B., Gambaro, A., Sfriso, A., Corami, F., Rastelli, E., Corinaldesi, C., Mistri, M., Munari, C., 2020. Microplastic accumulation in benthic invertebrates in Terra Nova Bay (Ross Sea, Antarctica). Environ. Int. 01604120. 137, 105587. doi:10.1016/j.envint.2020.105587.

Shen, X., Jie, M., Rong, J., Pan, K., Miao, A.-J., 2020. Microplastics in aquatic environments: occurrence, accumulation, and biological effects. Sci. Total Environ. 00489697. 703, 134699. doi:10.1016/j.scitotenv.2019.134699.

Silva, D.M., Almeida, C.M.R., Guardiola, F.A., Pereira, R., Rodrigues, S.M., Ramos, S., 2024. Uncovering microplastics contamination in canned seafood. Food Chem. 18737072. 448. Available from: https://www.sciencedirect.com/science/journal/03088146. doi:10.1016/j.foodchem.2024.139049.

Sofia Soares, A., Pinheiro, C., Oliveira, U., Vieira, M.N., 2021. Microplastic pollution in Portuguese saltworks. Inland Waters - Dynamics and Ecology. IntechOpen. doi: 10.5772/intechopen.91476.

Suteja, Y., Dirgayusa, I.G.N.P., Purnama, S.G., Purwiyanto, A.I.S., 2025. From sea to table: assessing microplastic contamination in local and non-local salt in Bali, Indonesia. Chemosphere. 18791298. 374. Available from: https://www.sciencedirect.com/science/journal/00456535. doi:10.1016/j.chemosphere.2025.144192.

Tibbetts, J.H., 2015. Plastics on the half shell. BioScience. 15253244. 65 (8), 836. Available from: http://bioscience.oxfordjournals.org/. doi:10.1093/biosci/biv089.

Uy, C.A., Johnson, D.W., 2022. Effects of microplastics on the feeding rates of larvae of a coastal fish: direct consumption, trophic transfer, and effects on growth and survival. Mar. Biol. 14321793. 169 (2). Available from: https://www.link.springer.de/link/service/journals/00227/index.htm. doi:10.1007/s00227-021-04010-x.

Valdez-Cibrián, A., Kozak, E.R., Franco-Gordo, C., 2024. Microplastic ingestion in marine mesozooplankton species associated with functional feeding traits. Mar. Environ. Res. 01411136. 200, 106650. doi:10.1016/j.marenvres.2024.106650.

Van Cauwenberghe, L., Janssen, C.R., 2014. Microplastics in bivalves cultured for human consumption. Environ. Pollut. 18736424. 193, 65–70. Available from: https://www.journals.elsevier.com/environmental-pollution. doi:10.1016/j.envpol.2014.06.010.

Vecchi, S., Bianchi, J., Scalici, M., Fabroni, F., Tomassetti, P., 2021. Field evidence for microplastic interactions in marine benthic invertebrates. Sci. Rep. 2045-2322. 11 (1). doi:10.1038/s41598-021-00292-9.

Vignesh, M., Alagesan, A., Ragavan, T., Gurusamy, A., Kannan, P., Sabarinathan, K.G., 2025. Tiny toxins, big problems: the hidden threat of microplastic in agroecosystems. Plant Sci. Today. 23481900. 12 (3). Available from: https://horizonepublishing.com/journals/index.php/PST/article/view/8337. doi:10.14719/pst.8337.

Wang, T., Baojie, L., Shi, H., Ding, Y., Chen, H., Yuan, F., Liu, R., Zou, X., 2024. The processes and transport fluxes of land-based macroplastics and microplastics entering the ocean via rivers. J. Hazard. Mater. 03043894. 466, 133623. doi:10.1016/j.jhazmat.2024.133623.

Ward, J.E., Rosa, M., Shumway, S.E., 2019. Capture, ingestion, and egestion of microplastics by suspension-feeding bivalves: a 40-year history. Anthropocene Coasts. 25614150. 2 (1), 39–49. Available from: https://www.nrcresearchpress.com/journal/anc. doi:10.1139/anc-2018-0027.

Weis, J.S., Alava, J.J., 2023. (Micro)Plastics are toxic pollutants. Toxics. 23056304. 11 (11). Available from: http://www.mdpi.com/journal/toxics. doi:10.3390/toxics11110935.

Zeytin, S., Wagner, G., Mackay-Roberts, N., Gerdts, G., Schuirmann, E., Klockmann, S., Slater, M., 2020. Quantifying microplastic translocation from feed to the fillet in European sea bass *Dicentrarchus labrax*. Mar. Pollut. Bull. 0025326X. 156, 111210. doi:10.1016/j.marpolbul.2020.111210.

Zhang, Z., Su, Y., Zhu, J., Shi, J., Huang, H., Xie, B., 2021. Distribution and removal characteristics of microplastics in different processes of the leachate treatment system. Waste Manag. 18792456. 120, 240–247. Available from: https://www.elsevier.com/locate/wasman. doi:10.1016/j.wasman.2020.11.025.

Zhu, L., Pan, W., Zhao, X., Wang, F., Wang, C., Kang, Y., Dong, Z., Shao, B., Wu, F., An, L., 2024. Insights into human exposure to microplastics through drinking water: current state of the science. Crit. Rev. Environ. Sci. Technol. 1064-3389. 54 (24), 1875–1901. doi:10.1080/10643389.2024.2371622.

CHAPTER

4

Primary dermatological impacts of microplastics

Nahid Khoshnamvand[a] *and Adel Mokammel*[b]

[a]**Department of Environmental Health Engineering, Lorestan University of Medical Sciences, Khorramabad, Iran**

[b]**Ardabil University of Medical Sciences, Social Determinants of Health Research Center, Ardabil, Iran**

4.1 Introduction

Microplastics (MPs) were initially described as plastic debris approximately 20 μm in size, but the current definition includes particles ≤5 mm in size. Plastic pollutants are classified by size into macroplastics (>5 mm), mesoplastics (≤5 mm to ≤1 mm), MPs (≤1 mm to >0.1 μm), and nanoplastics (NPs) (≤0.1 μm). Common plastic types found as MPs include acrylic, nylon (polyamide), polypropylene (PP), polyester, polyethene (PE), and polystyrene. MPs have become a significant global health and environmental concern owing to their widespread presence in diverse environments—from ocean surfaces to deep sediments, farmlands to mountain peaks, and snow and ice—where they enter the human food chain. Terrestrial sources contribute 80%–90% of MPs, mainly from plastic bags, bottles, and personal care products, whereas 10%–20% come from aquatic sources such as fishing lines and nets. MPs originate either as primary MPs manufactured for commercial use in products such as cosmetics, detergents, paints, and insecticides or as secondary MPs produced by the degradation of larger plastic materials. Owing to their extensive distribution and bioavailability, MPs are pervasive pollutants that affect ecosystems and human health (Thompson et al., 2024). Research on the local effects of MPs on human skin, the body's first physical barrier, is limited compared with that on systemic impacts. However, continuous and direct skin contact with MPs, which are common in cosmetic and health products such as exfoliants, creams, cleansers, and sunscreens, has emerged as a public health concern. Studies using 3D skin models have shown that MPs can penetrate healthy skin on their own and that chemical and toxic pollutants can attach to MPs and be absorbed through the skin into the bloodstream. Absorption is greater in hydrated skin. Dermal exposure to MPs may result in skin

Microplastics in Animal and Human Physiology
https://doi.org/10.1016/B978-0-443-45635-0.00014-8

inflammation and toxicity, as these chemicals accumulate over time. MPs can also penetrate skin pores, especially very small synthetic fibers (<25 μm), potentially bypassing the outer barrier and interacting with deeper skin layers. Damaged skin increases the permeability to MPs. This exposure may contribute to various inflammatory skin diseases, which are among the leading nonfatal health burdens worldwide. An increased understanding of MP dermal absorption, molecular mechanisms of toxicity, and associated skin diseases is essential for guiding public health policies and mitigation strategies (Liu et al., 2025).

4.2 MP characterization and dermal exposure pathways

The skin is directly exposed to MPs throughout the day via various routes, including inhalation of airborne particles, contact with water, and contact with soil. Exposure to higher concentrations of MPs, especially in related industries such as hairdressing or beauty care products, and contact with contaminated water or MP fibers from washing synthetic textiles, are also problems of modern life (Nawab et al., 2026).

4.2.1 Physical and chemical properties of dermatologically relevant MPs

The physical and chemical properties of these particles, including their high stability, high specific surface area, and ability to absorb and transport various chemical compounds, have made them among the most challenging emerging pollutants and have attracted the attention of researchers. The use of cosmetic and health products such as creams, lotions, shampoos, exfoliating scrubs, or personal care products that contain polymer MPs is another route of direct skin exposure to MPs.

4.2.1.1 Particle size and morphological characteristics

The physical and chemical properties of MPs, such as their size and shape, chemical composition, ability to absorb toxic substances, ability to impact the skin's defense barrier, high resistance and stability against biodegradation, and long persistence in the surrounding environment and the human body, lead to their accumulation and harmful effects. These characteristics make MPs a serious threat to skin health, and their interaction with the skin can lead to inflammatory damage, sensitivity, and chronic toxic effects (Kumari and Arora, 2025).

4.2.1.2 Surface chemistry and polymer type variations

There are various types of plastic polymers, such as PE, PP, polyvinyl chloride, polystyrene, polyamide, and polyethene terephthalate. These polymers vary in properties such as hydrophobicity, crystallinity, and chemical stability, which influence their ability to adhere to and interact with the outer layer of the skin. The surface charge, hydrophilicity, and reactivity of functional groups attached to MPs, such as hydroxyl, carbonyl, or carboxyl groups, can be modified. These changes affect how MPs bind to skin lipids and proteins, potentially disrupting interactions among skin cells, enzymes, and microbial biofilms. For example, PE MPs tend to adsorb more sulfamethazine than PP MPs or PVC MPs do, whereas PP MPs have a greater capacity for adsorbing polycyclic aromatic hydrocarbons (Das et al., 2025).

4.2.1.3 *Fragment vs. spherical particle behavior*

Spherical particles have smooth, symmetrical shapes. This makes it harder for them to penetrate the lipid layers between skin cells. Instead, they tend to enter the skin primarily through appendages, such as hair follicles, where they accumulate uniformly and cause minimal mechanical damage. These particles can act as long-term reservoirs in the follicles without severely damaging the skin barrier. In contrast, irregular and broken particles with uneven edges have a greater potential to disrupt the skin barrier mechanically by creating cracks in the stratum corneum.

4.2.2 Mechanisms of dermal penetration and bioavailability

The main routes of penetration of MPs into the skin are as follows:

- Intercellular route: MPs pass between keratinocytes of the stratum corneum and reach deeper epidermal layers. Smaller, hydrophilic particles prefer this path.
- In the intracellular route, NPs enter keratinocytes and fibroblasts via endocytosis, leading to accumulation and potential inflammation.
- Follicular route: Hair follicles and glands act as reservoirs, facilitating accumulation and penetration, especially for larger MPs.
- Damaged skin: Skin injuries or disorders weaken the barrier, increasing the likelihood of penetration.

The bioavailability of MPs in the human body depends on their size, surface chemistry, and exposure route. After crossing the skin layers, MPs can accumulate in organs such as the liver, kidneys, and intestines through mechanisms such as endocytosis, posing potential health risks, including inflammatory responses and plastic absorption.

4.2.2.1 *Passive diffusion and active transport pathways*

MPs penetrate the skin through two main mechanisms:

- Passive diffusion: Small particles move from areas of high concentration to areas of low concentration through spaces between skin cells without requiring energy.
- Active transport: Cells use energy and carrier proteins to internalize particles that cannot diffuse freely.

These pathways involve passage through intercellular spaces, skin cells (keratinocytes, fibroblasts), sweat glands, and capillaries. MPs can enter the skin via sweat glands and subsequently reach the bloodstream, potentially causing inflammatory and biological damage.

4.2.2.2 *Role of skin hydration and barrier function*

The outermost layer of the skin, called the stratum corneum, acts as the primary defense barrier. It consists mainly of dead skin cells surrounded by natural fats, such as ceramides and fatty acids, which reduce excessive moisture loss. Proper skin hydration is crucial for maintaining the effectiveness of this barrier. Well-moisturized skin remains soft and flexible, effectively blocking MPs and other particles. In contrast, dry, cracked skin loses this protective ability, allowing easier penetration of MPs, which can cause inflammation and damage.

4.2.2.3 *Appendageal penetration routes*

The appendageal route (via hair follicles, sebaceous glands, and sweat glands) is an important pathway through which MPs penetrate deep into the skin. These structures act like tunnels and, because of their direct connection to deeper tissues, allow particles smaller than a few microns to accumulate beneath the surface and trigger local inflammation, even though they cover only a small portion of the skin (Menichetti et al., 2025).

4.3 Cellular and molecular mechanisms of MP-induced dermatotoxicity

When MPs penetrate the skin, immune cells, including macrophages and dendritic cells, sense danger and become activated, leading to the secretion of inflammatory cytokines and ultimately inflammation, redness, swelling, and tissue damage (Mahmud et al., 2024). MPs produce reactive oxygen species (ROS), which generate free radicals, leading to oxidative stress and damage to DNA, proteins, and cellular lipids. MPs also disrupt the lipid and protein structure of the epidermal layers and increase skin permeability.

4.3.1 Keratinocyte responses and barrier dysfunction

The epidermis is mainly formed by keratinocytes, which build the stratum corneum as a protective barrier. When exposed to MPs, keratinocytes trigger inflammation, release cytokines and growth factors, and produce free radicals that damage lipids and proteins, thereby weakening the barrier. This damage increases skin permeability and facilitates MP penetration. In addition, keratinocytes support skin immunity by signaling to immune cells and releasing antimicrobial peptides to maintain homeostasis and fight infections.

4.3.1.1 *Cellular internalization mechanisms*

The internalization of MPs in the skin is a complex process. In aqueous environments, MPs adsorb biological molecules (proteins, lipids, carbohydrates) from their surroundings, forming an "eco-corona" that enhances their recognition and uptake by immune cells, such as macrophages, monocytes, neutrophils, and dendritic cells. Internalization begins with attachment to the cell membrane and occurs either passively (without energy) or actively via energy-dependent endocytosis, during which the cell engulfs MPs (Liu et al., 2021).

4.3.1.2 *Cytotoxic and cytoprotective effects*

MPs have both harmful and protective effects on skin cells, such as cytotoxic and cytoprotective effects. Their cytotoxic effects include chronic skin inflammation, dryness, redness, and irritation, which disrupt normal skin and immune functions. MPs carrying toxic chemicals can penetrate the skin, causing oxidative stress, inflammation, cell death, and impaired skin function. Some cellular defense and antioxidant mechanisms may be activated in response to MPs but are generally insufficient to prevent damage. Cytoprotective effects have been less well reported, and there is insufficient evidence to support them. Overall, the evidence suggests that MPs are harmful, with continued caution advised because of their negative impact on the skin and overall health (Aristizabal et al., 2024).

4.3.1.3 *Epidermal homeostasis disruption*

Disruption of epidermal homeostasis is a response triggered when the immune system encounters MPs in the skin. This occurs when specific receptors on skin cells, such as keratinocytes, Langerhans cells, dendritic cells, melanocytes, macrophages, and T cells, sense a threat of cellular or molecular damage from the entry of MPs, leading to the disruption of the skin's defense barrier. Disruption of homeostasis leads to symptoms such as dryness, inflammation, destruction of skin tissues, accelerated aging, and increased skin sensitivity (Aristizabal et al., 2024).

4.3.2 Inflammasome activation and immune dysfunction

Inflammasomes play dual roles in the skin. They are crucial for defending against pathogens and supporting tissue repair after skin damage by releasing cytokines such as IL-1β and IL-18. However, exposure to MPs can cause aberrant activation of inflammasomes, leading to chronic inflammation and impaired keratinocyte maturation. This results in a compromised skin barrier and the development of inflammatory and autoimmune skin diseases. Understanding inflammasome activation in response to MPs is important for reducing skin-related health risks from MP exposure.

4.3.2.1 *NLRP3 inflammasome signaling pathways*

The NLRP3 inflammasome, a key protein complex in skin tissues, regulates inflammatory responses triggered by cellular stress, injury, and MP infiltration. Its activation occurs in two phases: initially, pattern recognition receptors, such as Toll-like receptors (TLRs), detect inflammatory signals and increase NLRP3 gene expression via the NF-κB pathway. Subsequently, stimuli such as ROS and ionic changes induce structural changes in $NLRP_3$, leading to the assembly of the inflammasome complex with apoptosis-associated speck-like protein containing a CAR (ASC) proteins. This complex activates inflammatory cytokines, causing inflammation (kodi et al., 2024). Fig. 4.1 shows that the penetration of MPs into the skin layers activates inflammatory signaling pathways, leading to skin inflammation. In addition, MPs alter the structural proteins of the skin by producing ROS and inducing oxidative stress, thereby leading to the gradual accumulation of MPs in skin tissues and progressive, chronic damage.

4.3.2.2 *Cytokine production and inflammatory cascades*

Cytokines regulate immune and inflammatory responses in the skin. Produced by cells such as keratinocytes, macrophages, and dendritic cells, they initiate cascades that attract

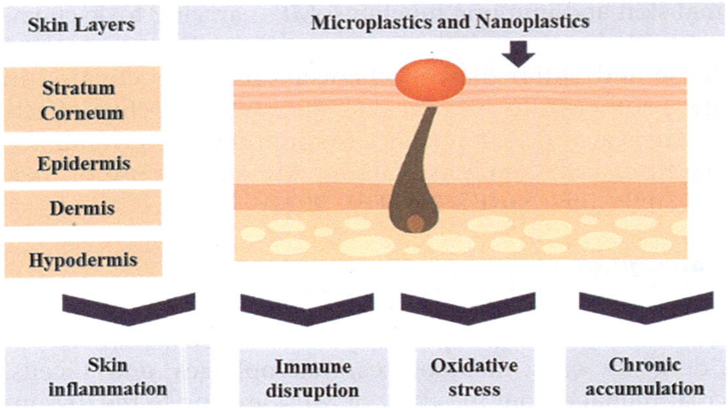

FIG. 4.1 Microplastic penetration into skin layers and its potential effects. The penetration of MPs into the skin layers activates inflammatory signaling pathways, leading to skin inflammation. In addition, MPs alter the structural proteins of the skin by producing ROS and inducing oxidative stress, thereby leading to the gradual accumulation of MPs in skin tissues and progressive, chronic damage.

immune cells to sites of inflammation. Proinflammatory cytokines (e.g., IL-1 and TNF-α) promote inflammation, whereas antiinflammatory cytokines (e.g., IL-10) maintain balance. Their dysregulation or overproduction leads to skin disorders such as psoriasis and atopic dermatitis through excessive immune activity and chronic inflammation.

4.3.2.3 *Autoimmune response modulation*

MPs disrupt intracellular signaling pathways, causing immune imbalance and tissue damage. They induce ROS production, trigger danger-associated molecular patterns, and impair TLR function, thereby promoting cytokine production and inflammatory responses. Immune responses vary by MP type, size, charge, and concentration, increasing the levels of proinflammatory cytokines such as IL-6, TNF-α, and IL-1β while also inducing antiinflammatory signals to counterbalance them. These complex immune disruptions contribute to chronic inflammation and tissue damage. MPs can also carry external pollutants and pathogens, amplifying their toxic effects.

4.3.3 Mitochondrial dysfunction and oxidative stress

Mitochondria, the energy-producing organelles in cells, are targeted by MPs, which cause mitochondrial oxidative stress, DNA damage, and dysfunction.

MPs induce overproduction of mitochondrial ROS, damage mitochondrial membranes, and impair ATP production.

This mitochondrial damage activates proinflammatory cytokines, leading to inflammation, disrupting skin homeostasis, and promoting skin aging, inflammatory diseases, and, potentially, cancer cell proliferation.

Experimental studies have shown that NPs and MPs can penetrate skin cells, disrupt the mitochondrial membrane potential, trigger mitophagy, and cause cellular aging and inflammation (han et al., 2024).

4.3.3.1 *ROS production and membrane potential alterations*

MPs trigger excessive mitochondrial ROS production (superoxide, hydroxyl radical, and hydrogen peroxide), leading to oxidative stress that damages DNA, proteins, and lipids in skin cells. They disrupt mitochondrial membranes, alter mitochondrial morphology, and cause dysfunction, driving inflammation, apoptosis, and the progression of skin diseases. Oxidative damage is size dependent and affects both extracellular and intracellular pathways, weakening cell membranes and impairing lysosomal function.

4.3.3.2 *NRF2/KEAP1 pathway modulation*

The KEAP1–NRF2 pathway is a crucial thiol-dependent sensor-effector system that maintains skin and epidermal homeostasis by regulating the oxidative stress response. NRF2, a transcription factor, activates protective genes to shield skin cells from oxidative damage and environmental toxins such as MPs. This pathway controls the redox balance by maintaining the thiol gradient and enhances the expression of antioxidants and detoxifying proteins, such as loricrin, aiding in skin barrier repair and differentiation. NRF2 limits inflammation by reducing free radical production, protects against DNA damage, and modulates the skin's innate immune response. While protective, overactivation of NRF2 can also be harmful, underscoring the need for balanced regulation.

4.3.3.3 *Cell death mechanisms and DNA damage*

MPs induce both accidental cell death (ACD) and programmed cell death (PCD), including autophagy, apoptosis, and ferroptosis. ACD involves hemolysis and necrosis due to external damage, whereas PCD maintains cellular balance and immunity. MPs trigger ROS production, oxidative stress, and mitochondrial dysfunction, with smaller MPs causing greater membrane damage. They also increase the levels of proinflammatory cytokines (IL-6, TNF-α, and IL-1β), activate autophagy and inflammation, and lead to cellular senescence and dysfunction (han et al., 2024).

4.4 Dermatological health consequences and disease associations

This section explores how MPs can affect skin health, ranging from acute inflammatory responses to long-term impacts on skin structure and function. These findings highlight associations with inflammatory conditions, autoimmune reactions, and potential carcinogenic effects. The discussion emphasizes the clinical relevance of these findings and their implications for human dermatological well-being.

4.4.1 Inflammatory skin conditions and autoimmune disease

The skin is affected by various pathological changes due to exposure to environmental hazards, including MPs. Continuous stimulation of the skin's immune system by MPs creates conditions that promote the development or worsening of inflammatory and autoimmune skin diseases. In autoimmune skin disorders, the immune system mistakenly attacks healthy skin cells, worsening inflammation, promoting carcinogenesis, accelerating skin aging, and leading to diverse skin symptoms.

4.4.1.1 *Psoriasis exacerbation mechanisms*

Psoriasis is a chronic inflammatory skin disorder characterized by disrupted communication between keratinocytes and immune cells, including antigen-presenting cells and T lymphocytes. MPs act as triggers by inducing oxidative stress and inflammatory pathways, leading to the release of cytokines (TNF-α, IL-17, and IL-23) that promote disease progression. Barrier damage caused by MPs allows antigens to penetrate, further activating the immune response. This creates a feedback loop that amplifies cytokine production and psoriasis inflammation (Diotallevi et al., 2023).

4.4.1.2 *Eczema and atopic dermatitis associations*

Atopic dermatitis is a chronic inflammatory skin disease characterized by dryness, itching, and inflammation. Environmental factors, such as MPs, exacerbate this condition by disrupting the skin microbiome and reducing the levels of protective lipids, such as ceramides. MPs can penetrate skin tissues, damage the epithelial barrier, and interact with structural proteins. They regulate immune cells, particularly TH2 lymphocytes, and increase the levels of proinflammatory cytokines. MPs also influence cell cycle protein expression, induce inflammatory gene transcription, and promote apoptosis, contributing to skin inflammation and worsening atopic dermatitis (Han and Kim, 2025).

4.4.1.3 *Contact dermatitis and sensitivity reactions*

Contact dermatitis is an eczematous skin condition caused by direct contact with allergens or irritants such as MPs. It commonly occurs due to occupational exposure and affects the hands and face. There are two types of contact dermatitis: irritant contact dermatitis, which is caused by direct immune irritation from MPs, and allergic contact dermatitis, which is a delayed hypersensitivity reaction triggered when MPs bind to skin proteins to form antigen complexes. This activates immune cells, including Langerhans and dendritic cells, leading to T-cell-mediated inflammation (Celebi Sozener et al., 2022).

4.4.2 Carcinogenic potential and cell transformation

Emerging evidence suggests that MPs may influence cellular processes in ways that could promote cancer development. This section explores how DNA damage, genotoxic stress, and dysregulated cell proliferation contribute to malignant transformation, emphasizing the potential long-term risks of MP exposure to skin health.

4.4.2.1 *Squamous cell carcinoma proliferation enhancement*

Squamous cell carcinoma of the skin is a malignant tumor derived from keratinizing cells. MPs increase the proliferation of squamous cell carcinoma cells in a dose- and time-dependent manner. MPs, as external pollutants, exert their biological influence by regulating intracellular molecules and, by inducing an inflammatory response, play a role in processes such as tumor cell proliferation and migration. The mechanism of MPs involves the proliferation of skin cancer cells through the activation of $NLRP_3$, which promotes the maturation and secretion of inflammatory factors and ultimately leads to skin cancer.

4.4.2.2 *Genotoxic effects and DNA damage*

The genotoxic effects of MPs are a particular concern for particles and depend on the particle dose and size. Genotoxicity generally refers to a pollutant's ability to damage a cell's genetic material. Direct or indirect exposure often results in DNA damage, altered gene expression, and chromosomal abnormalities. MPs can cause chromosome damage by inducing clastogenesis, disrupt chromosome segregation by inducing aneugenesis, and alter gene expression in other biological processes (Table 4.1).

4.4.2.3 *Tumor-promoting mechanisms*

MPs interact with tumor cells within the tumor microenvironment, thereby influencing tumor growth and behavior. They can increase tumor cell proliferation, alter metabolism, and promote metastasis. In skin squamous cell carcinoma, MPs increase mitochondrial ROS and inflammation in a time- and dose-dependent manner, leading to cancer cell proliferation. Thus MPs act as a double-edged sword, stimulating tumor growth while damaging normal skin cells.

TABLE 4.1 Particle sizes of micro- and nanoplastics, their potential dermal penetration pathways, and the associated cellular and tissue-level effects observed upon skin exposure.

Relationship between particle size and dermal toxicity of micro/nanoplastics

Particle size	Scientific classification	Potential skin penetration pathways	Penetration capacity	Toxicological effects
>100 μm	MPs	Low penetration into the stratum corneum; accumulation on the skin surface	Very low	Mechanical irritation; superficial contact dermatitis
10–100 μm	Large MPs	Limited penetration through hair follicles and sweat gland ducts	Low	Local mechanical irritation; disruption of skin barrier integrity
1–10 μm	Small MPs	Paracellular transport through disrupted tight junctions in the epidermis	Moderate	Activation of inflammatory pathways (NF-κB, MAPK), cytokine release (TNF-α, IL6)
100 nm–1 μm	Transitional MPs/NPs	Cellular uptake by endocytosis	High	Oxidative stress, mitochondrial dysfunction, apoptosis
<100 nm	NPs	Direct intercellular penetration and potential systemic transport	Very high	Genotoxicity, DNA damage, chronic inflammatory responses

Notes: *MPs*, microplastics; *NPs*, nanoplastics; *NF-κB*, nuclear factor kappa-light-chain-enhancer of activated B cells; *MAPK*, mitogen-activated protein kinase; *TNF-α*, tumor necrosis factor alpha; *IL-6*, interleukin 6.

4.4.3 Skin aging and senescence

This section explores how MPs and NPs can accelerate skin aging and cellular senescence. This highlights mechanisms such as oxidative stress, disruption of collagen and elastin fibers, and impaired regenerative capacity. Understanding these effects helps clarify the long-term impact of MP exposure on skin health and appearance.

4.4.3.1 *Premature aging acceleration*

Intrinsic skin aging occurs naturally over time, owing to genetics and hormones, and is marked by telomere shortening. Premature aging, however, results from persistent DNA damage, oxidative stress, inactivation of tumor suppressor genes, mitochondrial dysfunction, and exposure to environmental pollutants such as MPs. MPs impair the skin barrier by inhibiting keratinocyte differentiation and disrupting the extracellular matrix, leading to the senescence of keratinocytes and fibroblasts and the release of aging-related factors.

4.4.3.2 *Disruption of collagen synthesis*

MPs damage fibroblasts by increasing the production of ROS and inflammatory molecules and disrupting collagen synthesis.

The presence of MPs can increase the activity of collagen-degrading enzymes, such as matrix metalloproteinases, thereby disrupting the balance between collagen synthesis and degradation. Impairment of the synthesis and regeneration of interstitial collagen in the skin leads to reduced skin elasticity, increased wrinkling and thinning of skin layers, and exacerbated skin aging.

4.4.3.3 *Regenerative capacity impairment*

MPs impair skin regeneration by triggering mitochondrial DNA release, activating inflammatory pathways, such as the AIM2 pathway, and stimulating the immune system. This inflammation damages the skin structure, reduces collagen production, weakens the protective barrier, and increases permeability to harmful substances, ultimately decreasing self-repair capacity and promoting premature aging (Wang et al., 2023).

4.5 Chemical co-contaminants and enhanced toxicity

The skin can absorb both cosmetics and their chemical additives, creating a combined exposure that may increase toxicity. Flame retardants, often added to plastics, include brominated compounds such as polybrominated diphenyl ethers (PBDEs), hexabromocyclododecane (HBCD), tetrabromobisphenol A, and polybrominated biphenyls, which are chemically bonded into the polymer structure (De Wit, 2002). Plasticizers enhance plastic flexibility and reduce viscosity, with approximately 50 types being used commercially. Additives such as flame retardants (PBDEs) and plasticizers (HBCDs) can leach from MPs and be absorbed through the skin. Dermal models estimate adult and toddler absorption at 0.01–22 ng/kg/day, indicating that the skin is an important route of exposure (Abafe et al., 2023).

4.5.1 Flame retardants and plasticizer absorption

The dermal absorption of flame retardants and plasticizers can be influenced by the polymer type, particle size, and chemical composition of cosmetic formulations. This finding indicates that the skin is exposed not only to the MP particles themselves but also to the chemical additives they release.

4.5.1.1 *PBDE and HBCDD dermal bioavailability*

The skin can serve as a route of combined exposure, increasing the potential for toxicity from both microbeads in cosmetics and chemical additives in their formulations. Additives such as flame retardants (PBDEs) and plasticizers (HBCDDs) can leach from MPs and enter the body through skin contact. Factors such as particle size and product formulation—smaller particles and oily creams or lotions—increase skin absorption. Overall, contaminants in MPs, including PBDEs and HBCDDs, can penetrate skin layers, potentially increasing the dose beyond that of the particles alone (Abafe et al., 2023). Fig. 4.2 illustrates the mechanism of dermal absorption of these compounds and their entry into the bloodstream.

4.5.1.2 *Particle type and chemical release kinetics*

The type of MP and its size affect how easily the associated pollutants are released. For example, rigid polymers such as PE and PP differ in their ability to absorb semi-volatile organic compounds. Abafe et al. (2023) reported that PBDEs leach more readily from PE beads than from PP beads under similar skin conditions. Another study reported a strong negative correlation ($r^2 = -0.98$ for PE and -0.90 for PP MPs) between PBDE accumulation in skin tissue and log K_ow, suggesting that polymer–chemical compatibility governs release rates. Smaller MPs, with higher surface-area-to-volume ratios, release adsorbed chemicals

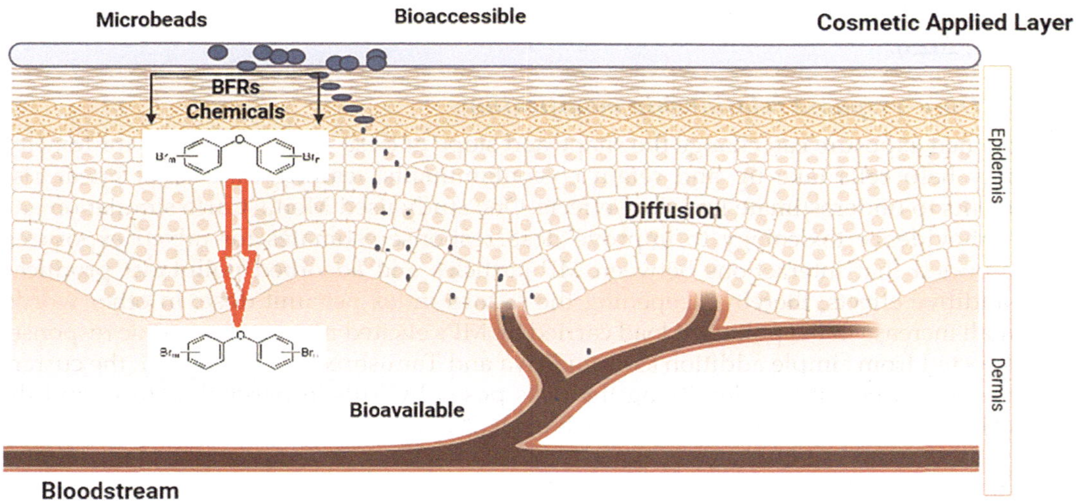

FIG. 4.2 Dermal absorption process of brominated flame retardants (BFRs) from applied cosmetics. The mechanism of dermal absorption of these compounds and their entry into the bloodstream.

more quickly. A previous study revealed that plastic nanoparticles penetrate the skin barrier more effectively than do MPs. In addition, aging or surface roughness from weathering or oxidation can accelerate the release of chemical additives from MP surfaces (Kissel et al., 2025).

4.5.1.3 *Cosmetic formulation interactions*

Cosmetic ingredients can affect how chemicals are released and absorbed from MPs. Studies on simulated skin have shown that oily and emollient components in creams and sunscreens promote the detachment of PBDEs and increase their absorption. For example, adding products such as antiperspirants, foundations, moisturizers, or sunscreens to extraction media increases PBDE bioaccessibility, suggesting that consumers may absorb MPs and their additives at rates different from those observed under simple laboratory conditions, such as water exposure alone (Abafe et al., 2023).

4.5.2 Synergistic toxicological effects

MPs can penetrate the skin and transport toxic chemicals to deeper layers, increasing the risk of combined harmful effects. Studies have shown strong synergy, such as in drinking water contaminated with MPs carrying heavy metals, PAHs, and additives, where the cumulative hazard for children is far greater than that for individual pollutants, highlighting the role of MPs as carriers (Swathi Priya and Tanushree, 2025).

4.5.2.1 *Combined particle and chemical toxicity*

MPs often contain additives and impurities that can increase chemical toxicity. Studies have shown that when MPs are combined with pollutants such as PAHs or heavy metals, they cause greater oxidative stress, inflammation, and macrophage activation than do MPs alone, disrupt detoxification pathways, lead to DNA damage, and weaken protective barriers (Lin et al., 2025).

4.5.2.2 *Dose–response relationship modifications*

Combined exposure can change the shape and slope of dose–response curves for toxicity. MPs may help cells take up and retain co-contaminants, leading traditional linear risk assessments to underestimate low-dose effects. One study used Monte Carlo simulation and weight-of-evidence toxicology to improve dose–response estimates for mixed exposures. They reported that interactions between MP particles and adsorbed pollutants produced supra-additive effects, thereby steepening the hazard index per unit dose. In other words, each small increase in the pollutant load carried by MPs elicited much greater toxic responses than expected from simple addition (Swathi Priya and Tanushree, 2025). Overall, the current evidence is insufficient for identifying the risks posed by MPs in particulate form and the chemical contaminants added to them (Fitch et al., 2025).

4.5.2.3 *Effects of chronic exposure accumulation*

Chronic, low-level exposure to NPs and MPs increases the likelihood that these particles will accumulate in body tissues. When both organic and inorganic additives are used, the particles persist in the body because of their resistance to breakdown. Over time, they can slowly release these chemicals, maintaining continuous internal exposure to toxic substances. According to Lin et al. (2025), MPs are poorly metabolized and cleared, leading to long biological half-lives and accumulation in the body. Experimental studies have shown that repeated exposure increases inflammation and oxidative DNA damage across multiple organs, with elevated levels of markers such as C-reactive protein, IL-6, and 8-OHdG associated with long-term MP exposure.

4.6 Detection, characterization, and analytical methods

The detection and measurement of MPs and associated chemicals in cosmetics and the environment require advanced analytical methods. Techniques such as microscopy, spectroscopy, and chromatography are used to identify particles, polymer types, and co-contaminants. Modern approaches favor nondestructive vibrational spectroscopy [Fourier transform infrared spectroscopy (FTIR), Raman] with imaging, complemented by mass spectrometry for confirmation. The sensitivity of submicron plastics and complex samples continues to increase (Kissel et al., 2025).

4.6.1 Advanced microscopic and spectroscopic techniques

Advanced microscopic and spectroscopic techniques enable reliable identification, characterization, and quantification of MPs and associated additives in complex matrices. The integration of imaging with vibrational spectroscopy and mass spectrometry improves analytical sensitivity and reduces uncertainty, particularly for small and irregular particles.

4.6.1.1 *Raman spectroscopy applications*

Raman spectroscopy and FTIR, often combined with microscopy, are widely used to identify MPs without damaging samples, providing unique molecular fingerprints. Advanced Raman techniques, such as stimulated Raman spectroscopy, coherent anti-Stokes Raman spectroscopy, and surface-enhanced Raman spectroscopy, can nondestructively detect particles as small as 1 μm. A recent study introduced flow-through Raman, which can detect MPs of approximately 4 μm directly in liquids, eliminating the need for filtration and improving throughput. This method can identify polymer types and even detect certain additives (Kissel et al., 2025).

4.6.1.2 *Fluorescence and electron microscopy*

Fluorescence microscopy using lipophilic dyes, such as Nile red, is a fast, cost-effective method for detecting MPs. These dyes bind to plastic particles, causing them to fluoresce under specific light. Lee and Lee (2021) reported that Nile red staining enables the visualization of MPs smaller than 100 μm in cosmetic extracts without damaging the particles or filters. This technique allows rapid screening of plastic debris; however, it can produce false positives because dyes may also stain organic materials. Therefore fluorescence results are typically confirmed via spectroscopic techniques, such as FTIR or SEM, which provide complementary chemical and morphological information.

4.6.1.3 *Multimodal analytical approaches*

No single method can fully identify and analyze MPs, so multimodal approaches are increasingly used. For example, combining Raman and infrared imaging allows rapid particle mapping with laser direct infrared imaging, followed by targeted Raman spectroscopy for precise polymer identification. Mass spectrometry techniques, such as pyrolysis-GC/MS, thermally decompose particles to detect polymers via monomer fragments while also revealing additives or cocontaminants. In addition, new software tools are being developed to automate particle counting and polymer classification in large datasets (Zhang et al., 2025).

4.6.2 Biomarker development and exposure assessment

Given the challenges of directly measuring particles in biological samples, researchers have sought biological markers of MP exposure and effects. Biomonitoring methods include measuring indicators of inflammation and oxidative stress, as well as developing exposure models. The goal is to connect external exposure to internal doses and biological responses.

4.6.2.1 *Inflammatory and oxidative stress markers*

MPs can induce inflammation and oxidative stress in cells, increasing the levels of ROS, antioxidant enzyme activity, and the levels of DNA damage markers such as 8-OHdG. Polystyrene and PE MPs also increase the levels of proinflammatory cytokines (IL-1β, IL-6, and TNF-α) via NLRP3 inflammasome activation. Lin et al. (2025) reported that MPs and nanoparticles impair mitochondrial function and increase biomarkers of DNA damage, such as 8-OHdG and γ-H2AX, in both in vitro and in vivo studies. While researchers seek MP-specific signatures, the consistent elevation of ROS and cytokines (e.g., IL-6 and TNF-α) suggests that these markers could serve as early indicators of plastic-induced cellular stress.

4.6.2.2 *Exposure quantification methods*

Modeling is an effective method for quantifying human exposure to MPs and is used following their initial identification. Physiologically based toxicokinetic (PBK) models play an important role; they can convert external exposure—whether through ingestion, inhalation, or dermal absorption—into a corresponding internal dose, providing a more accurate picture of the actual body burden. This approach uses probability density functions to model the distribution of MPs of varying types, sizes, and porosities within organs under different conditions. Moreover, direct exposure assessment methods are emerging, in which researchers measure plastic-related additives, such as plasticizers and flame retardants, in blood or urine as indirect indicators of plastic exposure (Abafe et al., 2023).

4.6.2.3 *Individual susceptibility factors*

Not everyone reacts in the same way to exposure to MPs. Genetic differences, age, health status, and preexisting conditions can influence vulnerability. For example, people with compromised skin barriers (e.g., dermatitis) or respiratory problems (e.g., asthma) may ingest more particles or experience more inflammation (Vethaak and Legler, 2021). A previous study revealed that these plants are more susceptible to negative effects than other plants because of their high growth rate and smaller body mass (Swathi Priya and Tanushree, 2025). Antioxidant status and nutrition influence oxidative stress levels. Given the limited data on human sensitivity to various types of MPs, it is prudent to apply precautionary measures, especially for vulnerable population groups (Thompson et al., 2024).

4.7 Regulatory frameworks and risk management

Governments and international organizations are beginning to address MPs in products, but their regulatory approaches differ significantly. This section outlines current laws and highlights gaps in aligning controls on MPs and related chemicals. It then discusses how risk assessment practices could be improved to better protect health.

4.7.1 Current regulatory approaches and limitations

Although several regions have introduced restrictions on MPs in cosmetic products, existing regulations often cover only limited product categories and specific particle sizes. The lack of harmonized definitions and standardized monitoring methods remains a major barrier to effective enforcement.

4.7.1.1 EU MP restrictions

The European Union has implemented one of the most comprehensive restrictions on intentional MPs under the REACH 2023/2055 regulation. Under this law, synthetic polymer particles are prohibited if they are intentionally added to a product and represent 0.01% or more of the total weight of the product. This ban applies to all product types, including cosmetics and detergents. By extending its scope beyond cosmetics to include polishes, sealants, and even fertilizers, the EU approach addresses a wide variety of plastic particle sources. However, the 0.01% threshold and specific exclusions (for example, MPs in agricultural seed coatings) have raised questions about enforcement. Implementation depends on manufacturers' self-reporting, and many consumer products contain plastic derivatives that are not explicitly covered. Some of the restrictions imposed by countries and relevant organizations are summarized in Table 4.2.

4.7.1.2 Cosmetic product regulations

In other places, regulations have focused on "microbeads" in rinse-off products. In the United States, the Microbead-Free Waters Act of 2015 amended the Federal Food, Drug, and Cosmetic Act to ban the ability of cosmetic manufacturers to make rinse-off personal care products that contain plastic microbeads (defined as intentionally added solid plastic particles 5 mm or smaller for exfoliation). It set deadlines of July 1, 2017, to halt manufacturing and July 1, 2018, to stop sales of such products. Similar bans were enacted at the state level before 2015. Other regions have taken similar actions; for example, South Korea banned plastic microbeads in cosmetic cleansing products starting in July 2017. Taiwan

TABLE 4.2 Details of restrictions on the use of MPs in cosmetic and personal care products.

Region/ organization	Regulatory approach	Key measures	Limitations
European Union (ECHA/ Commission)	In January 2019, ECHA proposed a wide-ranging restriction on MPs in products placed on the EU/EEA market to avoid or reduce their release into the environment.	In 2023 first measures, e.g., bans on loose glitter and microbeads in many products	Restriction does not include secondary MPs
Canada (Health Canada)	In 2017 a ban on microbeads in toiletries and cosmetics was implemented under the mandatory notification system.	As of July 1, 2018, the manufacture and import of all toiletries that contain plastic microbeads are prohibited	Restriction does not include secondary MPs
United Kingdom	A ban on tiny plastic pieces is often added to products such as face scrubs, soaps, toothpaste, and shower gels.	2018 ban on the manufacture/ sale of rinse-off cosmetics with microbeads; cosmetic safety assessment and responsible person requirements continue	Scope limited to targeted microbeads; secondary MPs not regulated
United States	Microbead-free Waters Act of 2015	Microbead-specific ban enacted via federal law	Broader microplastic pathways (fibers shed from textiles and wear) are outside the scope of standard cosmetic controls

and New Zealand also enacted bans on microbead cosmetics in 2018. China's government announced that it would phase out rinse-off cosmetics containing plastic beads by the end of 2020 and stop sales by the end of 2022. However, many leave-on cosmetics (such as foundations and lipsticks) were not fully covered by early laws, nor were pollutants that stick to nonpolymeric particles. As a result, many product types remain unregulated (Thompson et al., 2024).

4.7.1.3 *International harmonization challenges*

Various countries have attempted to impose legal restrictions to prevent the use of microbeads in multiple products. Eight countries have officially banned the use of microbeads. However, the types of products being monitored, imported, exported, and manufactured with microbeads differ. Some laws or regulations include exemptions. For example, Canada does not regulate goods in transit that contain microbeads in toiletries, Sweden does not include cosmetics with natural plastic polymers, and the United States excludes medical drugs.

4.7.2 Risk assessment methodologies and future directions

Risk assessment approaches are increasingly shifting toward the consideration of combined exposures and synergistic effects, as traditional methods may underestimate real-world risks. Further development of predictive models and validation of exposure data through experiments are essential to improve assessment accuracy.

4.7.2.1 *Science-based regulatory development*

Decision-making based on science is one of the main criteria for regulation when dealing with emerging pollutants. Some environmental agencies have published guidelines on scientific evidence and have sought to use standard, reliable studies to include data on MPs in chemical risk assessments. The recent EU MP restriction, which explicitly addresses scientific concerns about the persistence of the particles and the associated increased risks, is an example of such a decision on the basis of science. In the future, regulatory frameworks should also be adjusted to consider the effects of mixtures and vulnerable groups, as recent risk assessments suggest (Swathi Priya and Tanushree, 2025). International organizations (such as the World Health Organization—WHO and European Food Safety Authority—EFSA) are reviewing guidelines for safe levels of MP exposure, although no formal health limits have yet been established.

4.7.2.2 *Predictive modeling approaches*

New tools, such as computational modeling, are increasingly playing a role in exposure and risk assessment. Physiologically based kinetic (PBK) modeling frameworks simulate how particles of various sizes and types move throughout the body after exposure, linking external intake to internal tissue doses. By including probability distributions for particle size, age, and polymer type, PBK models can account for variability within human populations. This technique helps predict which exposure scenarios (for example, high versus low consumers) might exceed safety limits. Similarly, other researchers are investigating quantitative structure–activity relationship models and machine learning to predict polymer toxicity on the basis of chemical descriptors, although data on this topic remain sparse. In summary, predictive modeling has become a vital complement to experimental data in regulatory decision-making.

4.7.2.3 *International collaboration initiatives*

Plastic pollution crosses borders, making international collaboration essential. The United Nations Environment Programme (UNEP) and WHO have warned about MP risks and urged joint action, whereas the Organisation for Economic Co-operation and Development (OECD) emphasizes its connection to the Sustainable Development Goals and encourages the sharing of best practices. Effective management of this crisis will depend on global cooperation and coordination.

4.8 Mitigation strategies and alternative approaches

Growing evidence of MP and pollutant hazards has increased the need for solutions. Strategies include removing harmful compounds from cosmetics, educating consumers and professionals on their safe use, redesigning products through green chemistry, and empowering individuals to minimize personal chemical exposure.

4.8.1 Cosmetic reformulation and sustainable alternatives

Cosmetic reformulation focuses on eliminating intentionally added MPs and hazardous additives by using safer, sustainable materials and green chemistry principles. These efforts aim to maintain product performance while reducing environmental persistence and potential human health risks.

4.8.1.1 Natural and biodegradable alternatives

One primary strategy to reduce environmental impacts is replacing synthetic plastic exfoliants with natural or biodegradable materials. These alternatives include nut shells, sugar or salt granules, cellulose beads, and biodegradable polymers such as polylactic acid (PLA), polycaprolactone (PLC), and polyhydroxyalkanoates (PHA). These materials exhibit similar exfoliation performance but decompose faster in the environment, causing less ecological harm. Bikiaris et al. (2024) noted that aliphatic biopolyesters (such as PLA and PHA) are promising because they degrade through hydrolysis and are made from renewable sources. Natural materials such as ground coffee, bamboo powder, or jojoba esters (wax beads) are used in clean-label cosmetics (Liu et al., 2025). For example, Badische Anilin- und Soda-Fabrik (BAFS) has marketed a "safe synthetic" jojoba-oil-based wax bead that is fully biodegradable. Ongoing research will be necessary to enhance their performance and cost-effectiveness.

4.8.1.2 Industry innovation and best practices

Industry initiatives have moved beyond simply swapping ingredients. Regulatory pressure and consumer demand drive innovation in product design. According to Bikiaris et al. (2024), a global trend is emerging in which companies voluntarily eliminate MPs and advertise plastic-free formulas. Innovations include the use of microfluidic techniques to produce uniform biodegradable microspheres, enzyme-responsive polyesters that degrade during wastewater treatment, and the repurposing of agricultural by-products as exfoliants (Krzyżostan et al., 2024). Improving processes to limit particle release includes encapsulating active ingredients in dissolving capsules rather than in solid beads. Overall, the industry's best practices involve both innovation (such as plant-based polymers and biotech fermentation of PHA) and proactive removal of known harmful additives (Bikiaris et al., 2024).

4.8.1.3 Performance and safety evaluation

Any new ingredient or process must comply with safety and efficacy standards. Performance evaluations compare innovative exfoliants with traditional microbeads. For example, a study showed that PHA copolymer beads removed sunscreen from pig skin as effectively as PE beads did. Notably, these biodegradable beads absorb much lower levels of persistent organic pollutants than conventional plastic beads do, indicating a reduced risk of pollutant transfer and release (Choi et al., 2023). Similarly, natural exfoliants are tested to ensure that they are gentle on the skin (avoiding microabrasions) and do not trigger allergic reactions. However, regulators and industry groups suggest ongoing monitoring (such as postmarket surveillance) to address any unexpected problems with new materials (Bikiaris et al., 2024).

4.8.2 Personal protection and exposure reduction

Individual exposure to MPs can be reduced through informed product selection, improved personal hygiene practices, and minimal contact with contaminated sources. These personal-level actions complement regulatory and industrial measures in overall risk management.

4.8.2.1 Consumer education and product selection

In addition to product changes, informing consumers is essential. Experts from Stanford Medicine stress that while avoiding MPs is not feasible, people can lower their exposure by choosing cosmetics labeled "plastic-free" or those with only natural exfoliants. Reading labels for ingredients such as "PE" or "PP" helps consumers avoid microbead-containing products. People should also be aware of other sources of exposure, such as MP-contaminated dust and water. Governments and advocacy groups encourage sustainable choices and plastic reduction, whereas education and labeling programs guide safer consumer decisions. Reducing the use of high-plastic products can significantly lower personal MP exposure.

4.8.2.2 Professional dermatological guidance

Dermatologists and healthcare professionals can reduce MP exposure by recommending safer practices. These findings suggest that people with sensitive skin avoid microbead exfoliants and opt for chemical or physical alternatives. In clinical settings, the use of vacuum systems and barrier creams during procedures with plastic materials, such as polymer fillers, is advised. Educating patients about potential hormonal or inflammatory effects and promoting "clean" products further minimizes exposure. Overall, a cautious approach to MP-containing products is recommended (Han and Kim, 2025).

4.8.2.3 Individual risk management strategies

Individuals can reduce MP exposure by maintaining skin and digestive health, regularly cleansing with products free of plastic exfoliating beads, and minimizing the inhalation of airborne microfibers through proper home ventilation and effective dust control, including vacuuming and air filtration (Chen et al., 2022). On the contrary, diet may also play a supporting role; some research has shown that high dietary fiber intake can bind to ingested MPs, preventing their absorption in the gut, although human evidence is still inconclusive (Han and Kim, 2025).

4.9 Conclusion

MPs have attracted significant attention because of their potential effects on skin health. Most evidence from laboratory and animal studies shows that MPs can weaken the skin barrier, trigger inflammatory responses, disrupt the skin microbiome, accelerate aging, and possibly contribute to carcinogenesis. However, these findings are primarily derived from controlled experiments and may not fully reflect the complexities of human exposure. Therefore interpretations should be cautious, and further studies using more reliable biological models are needed to clarify the actual impact of MPs on human skin. Upon contact with the skin, MPs activate immune receptors in cutaneous cells, initiating inflammatory responses. This triggers the release of proinflammatory and antimicrobial mediators, including interleukins and tumor necrosis factor, which can weaken the skin barrier, induce cellular damage, and trigger apoptosis. MPs may also interact with proteins, altering their structure and function. Proteins adsorbed on MPs form a "protein corona," changing how particles interact with cells. The accumulation of MPs in cells induces oxidative stress and inflammation, leading to cellular injury. Repeated exposure can limit cell proliferation, activate autophagy pathways, and accelerate premature skin aging. Cells respond by increasing the levels of antioxidant enzymes to counter oxidative stress. Future studies should focus on human-relevant skin models and long-term, low-dose exposure scenarios to better reflect real-life conditions. Particular attention to NPs, chronic exposure, and interactions with co-contaminants is essential for understanding their actual impact on the dermis and supporting evidence-based risk assessment.

References

Abafe, O.A., Harrad, S., Abdallah, M.A.E., 2023. Novel insights into the dermal bioaccessibility and human exposure to brominated flame retardant additives in microplastics. Environ. Sci. Technol. 15205851. 57 (29), 10554–10562. Available from: http://pubs.acs.org/journal/esthag. doi:10.1021/acs.est.3c01894.

Aristizabal, M., Jiménez-Orrego, K.V., Caicedo-León, M.D., Páez-Cárdenas, L.S., Castellanos-García, I., Villalba-Moreno, D.L., Ramírez-Zuluaga, L.V., Hsu, J.T.S., Jaller, J., Gold, M., 2024. Microplastics in dermatology: potential effects on skin homeostasis. J. Cosmet. Dermatol. 14732165. 23 (3), 766–772. Available from: http://onlinelibrary.wiley.com/journal/10.1111/(ISSN)1473-2165. doi:10.1111/jocd.16167.

Bikiaris, N., Nikolaidis, N.F., Barmpalexis, P., 2024. Microplastics (MPs) in cosmetics: a review on their presence in personal-care, cosmetic, and cleaning products (PCCPs) and sustainable alternatives from biobased and biodegradable polymers. Cosmetics. 20799284. 11 (5). Available from: www.mdpi.com/journal/cosmetics. doi:10.3390/cosmetics11050145.

Chen, Y., Li, X., Zhang, X., Zhang, Y., Gao, W., Wang, R., He, D., 2022. Air conditioner filters become sinks and sources of indoor microplastics fibers. Environ. Pollut. 18736424. 292. Available from: https://www.journals.elsevier.com/environmental-pollution. doi:10.1016/j.envpol.2021.118465.

Choi, Y.H., Park, J.J., Sim, E.J., Lee, E., Yoon, K.C., Park, W.H., 2023. Ecofriendly poly(3-hydroxybutyrate-co-4-hydroxybutyrate) microbeads for sanitary products. Int. J. Biol. Macromol. 18790003. 224, 1487–1495. Available from: www.elsevier.com/locate/ijbiomac. doi:10.1016/j.ijbiomac.2022.10.237.

Das, B.K., Das, S., Kumar, V., Roy, S., Mitra, A., Mandal, B., 2025. Microplastics in ecosystems: ecotoxicological threats and strategies for mitigation and governance. Front. Mar. Sci. 22967745. 12. Available from: https://www.frontiersin.org/journals/marine-science#. doi:10.3389/fmars.2025.1672484.

De Wit, C.A., 2002. An overview of brominated flame retardants in the environment. Chemosphere. 00456535. 46 (5), 583–624. Available from: www.elsevier.com/locate/chemosphere. doi:10.1016/S0045-6535(01)00225-9.

Diotallevi, F., Offidani, A., 2023. Skin, Autoimmunity and inflammation: a comprehensive exploration through scientific research. Int. J. Mol. Sci. 1422-0067. 24 (21), 15857. doi:10.3390/ijms242115857.

Fitch, S., Rogers, J., Marty, S., Norman, J., Schneider, S., Rushton, E., Wikoff, D., Ellis-Hutchings, R., 2025. Systematic review of potential developmental and reproductive toxicity of microplastics. Toxicol. Sci. 10960929. 207 (2), 289–305. Available from: http://toxsci.oxfordjournals.org/. doi:10.1093/toxsci/kfaf108.

Han, J.H., Kim, H.S., 2025. Microplastics in cosmetics: emerging risks for skin health and the environment. Cosmetics. 20799284. 12 (4). Available from: www.mdpi.com/journal/cosmetics. doi:10.3390/cosmetics12040171.

han, W., Cui, J., Sun, G., Miao, X., Pufang, Z., nannan, L., 2024. Nano-sized microplastics exposure induces skin cell senescence via triggering the mitochondrial localization of GSDMD. Environ. Pollut. 18736424. 349. Available from: https://www.sciencedirect.com/science/journal/02697491. doi:10.1016/j.envpol.2024.123874.

Kissel, A., Nogowski, A., Kienle, A., Foschum, F., 2025. Flow Raman spectroscopy for the detection and identification of small microplastics. Sensors. 14248220. 25 (5). Available from: http://www.mdpi.com/journal/sensors. doi:10.3390/s25051390.

kodi, T., Sankhe, R., Gopinathan, A., Nandakumar, K., Kishore, A., 2024. New insights on NLRP3 inflammasome: mechanisms of activation, inhibition, and epigenetic regulation. J. Neuroimmune Pharmacol. 15571904. 19 (1). Available from: https://www.springer.com/journal/11481. doi:10.1007/s11481-024-10101-5.

Krzyżostan, M., Wawrzyńczak, A., Nowak, I., 2024. Use of waste from the food industry and applications of the fermentation process to create sustainable cosmetic products: a review. Sustainability. 2071-1050. 16 (7), 2757. doi:10.3390/su16072757.

Kumari, D., Arora, S., 2025. Multi-dimensional view of microplastic toxicity: environmental and health impacts. Water Air Soil Pollut. 15732932. 236 (9). Available from: https://www.springer.com/journal/11270. doi:10.1007/s11270-025-08243-0.

Lee, S., Lee, T.G., 2021. A novel method for extraction, quantification, and identification of microplastics in CreamType of cosmetic products. Sci. Rep. 20452322. 11 (1). Available from: www.nature.com/srep/index.html. doi:10.1038/s41598-021-97557-0.

Lin, R.H., Chen, H.T., Lee, I.T., Vo, T.T.T., Wang, Y.L., 2025. Microplastics and nanoplastics in health concerning cellular toxicity mechanisms, exposure pathways, and global mitigation strategies. Life. 20751729. 15 (9). Available from: www.mdpi.com/journal/life/. doi:10.3390/life15091449.

Liu, H.C., Hsieh, C.Y., Tsai, P.W., Chou, T.Y., Yang, S.C., Chang, C.H., Huang, Y.P., Chien, C.C., Lee, S.C., Shih, H.D., Liu, Y.C., 2025. Development and applications of Bambusa stenostachya leaf extract in personal care products. Processes. 22279717. 13 (1). Available from: http://www.mdpi.com/journal/processes. doi:10.3390/pr13010233.

Liu, L., Xu, K., Zhang, B., Ye, Y., Zhang, Q., Jiang, W., 2021. Cellular internalization and release of polystyrene microplastics and nanoplastics. Sci. Total Environ. 18791026. 779. Available from: www.elsevier.com/locate/scitotenv. doi:10.1016/j.scitotenv.2021.146523.

Mahmud, F., Sarker, D.B., Jocelyn, J.A., Sang, Q.X.A., 2024. Molecular and cellular effects of microplastics and nanoplastics: focus on inflammation and senescence. Cells. 20734409. 13 (21). Available from: https://www.mdpi.com/journal/cells. doi:10.3390/cells13211788.

Menichetti, A., Mordini, D., Montalti, M., 2025. Penetration of microplastics and nanoparticles through skin: effects of size, shape, and surface chemistry. J. Xenobiotics. 20394713. 15 (1). Available from: https://www.mdpi.com/2039-4713/11. doi:10.3390/jox15010006.

Nawab, A., Khan, M.T., Ihsanullah, I., Nafees, M., Shah, A.M., 2026. From pollution to ocean warming: the climate impacts of marine microplastics. J. Hazard. Mater. Plast. 30510600. 2, 100032. doi:10.1016/j.hazmp.2025.100032.

Sozener, Z.C., Özbey Yücel, Ü., Altiner, S., Ozdel Oztürk, B., Cerci, P., Türk, M., Gorgülü Akin, B., Akdis, M., Yilmaz, I., Ozdemir, C., Mungan, D., Akdis, C.A., 2022. The external exposome and allergies: from the perspective of the epithelial barrier hypothesis. Front. Allergy. 2673-6101. 3. doi:10.3389/falgy.2022.887672.

Swathi Priya, P., Tanushree, P., 2025. Synergistic human health risks of microplastics and co-contaminants: a quantitative risk assessment in water. J. Hazard. Mater. 03043894. 491, 137809. doi:10.1016/j.jhazmat.2025.137809.

Thompson, R.C., Courtene-Jones, W., Boucher, J., Pahl, S., Raubenheimer, K., Koelmans, A.A., 2024. Twenty years of microplastic pollution research—what have we learned? Science. 10959203. 386 (6720). Available from: https://www.science.org/doi/10.1126/science.adl2746. doi:10.1126/science.adl2746.

Vethaak, A.D., Legler, J., 2021. Microplastics and human health. Science. 0036-8075. 371 (6530), 672–674. doi:10.1126/science.abe5041.

Wang, Y., Xu, X., Jiang, G., 2023. Microplastics exposure promotes the proliferation of skin cancer cells but inhibits the growth of normal skin cells by regulating the inflammatory process. Ecotoxicol. Environ. Saf. 10902414. 267. Available from: http://www.elsevier.com/inca/publications/store/6/2/2/8/1/9/index.htt. doi:10.1016/j.ecoenv.2023.115636.

Zhang, M., Pan, J., Song, D., Jiang, J., Jiang, Y., 2025. Recent applications in analytical techniques of microplastics. Crit. Rev. Anal. Chem. 15476510. Available from: www.tandf.co.uk/journals/titles/10408347.asp. doi:10.1080/10408347.2025.2518328.

CHAPTER

5

Systemic toxicity of microplastics

Pinar Arslan Yüce[a], *Aysel Caglan Gunal*[b], *and Caterina Faggio*[c,d]

[a]Department of Biology, Faculty of Science, Çankırı Karatekin University, Çankırı, Turkey
[b]Department of Biology Education, Faculty of Gazi Education, Gazi University, Ankara, Turkey
[c]Department of Chemical, Biological, Pharmaceutical and Environmental Sciences, University of Messina Viale Ferdinando Stagno d'Alcontres, Messina, Italy
[d]Department of Eco-sustainable Marine Biotechnology, Stazione Zoologica Anton Dohrn, Naples, Italy

5.1 Introduction

Plastics are polymeric substances produced primarily by combining monomers derived from petrochemicals, which are moldable under pressure (Cho et al., 2024). Owing to their low cost, light weight, and durability, plastics are used in a wide range of applications, including packaging materials, containers, toys, and medical supplies. In addition to these advantages, their greatest disadvantage is that they are persistent and have very slow degradation rates, which leads to significant problems when they are discarded into the environment (Donkers et al., 2022). Therefore, the plastic age, one of the most important periods of our current anthropogenic era, represents the most significant and concerning timeline during which aquatic and terrestrial ecosystems face abiotic and biotic interactions (Chen et al., 2023).

Regardless of their field of use, plastic materials that enter the environment as waste undergo physical and chemical weathering, breaking down into smaller sizes. Microplastics (MPs) are small plastic particles ranging in size from 1 μm to 5 mm (Shan et al., 2022; Cho et al., 2024; Forutan et al., 2025). MPs are classified as primary or secondary MPs. Primary MPs include microbeads found in personal care products such as facial cleansers and toothpaste. Secondary MPs form when large plastic pieces break down during disposal due to artificial processes or natural weathering.

Furthermore, the chemical structures of MPs are diverse. Polystyrene (PS), polyethylene (PE), and polyethylene terephthalate are among the most common chemical structures (Cho et al., 2024). These different chemical structures also affect MP density. For example, while

Microplastics in Animal and Human Physiology
https://doi.org/10.1016/B978-0-443-45635-0.00005-7

high-density MPs derived from polyvinyl chloride (PVC) tend to sink when they enter aquatic ecosystems, low-density PE MPs remain on the water surface, increasing their potential to absorb other toxic substances present in the water (Forutan et al., 2025). Owing to their environmental persistence and intensive use, MPs pose a potential threat to ecosystems and human health (Shan et al., 2022).

MPs can be ingested by organisms through ingestion (via food and water), inhalation, and dermal contact. Owing to their small size, high biological permeability, and large specific surface area, these particles can easily penetrate the tissues and cells of an organism. Thus, they accumulate in different tissues. Through accumulation along the food chain, they are detected at very high concentrations in higher organisms. This accumulation and increasing concentration can impair the systemic functions of organisms (Cho et al., 2024; Li et al., 2024; Forutan et al., 2025).

This review provides a systematic review, starting from the absorption and bioavailability of MPs, through detailed mechanisms of toxicity, to their role as vectors for associated chemicals and their impact on bioaccumulation. Findings from various model organisms are synthesized to establish a framework for human health risk assessment. Despite the growing body of evidence, significant knowledge gaps remain, particularly regarding the long-term effects of chronic low-dose exposure, the behavior of MPs, and the toxicology of MP mixtures (cocktail effect). Future research must prioritize addressing these gaps to enable more robust risk assessments and inform effective public health strategies.

5.2 Absorption and bioavailability

Owing to their small physical size, MPs can translocate across biological barriers, enter the circulatory system, and reach internal organs. The process of cellular uptake and distribution occurs via various mechanisms, such as endocytosis, which leads to particle interactions with intracellular organelles.

5.2.1 Translocation across biological barriers

MP enters the body by mixing with air, water, soil, and food through processes such as inhalation, ingestion, dermal penetration, and even placental transfer. Their detection in various tissues of organisms, such as blood, lungs, brain, and placenta, indicates that they can cross physical barriers.

Epithelial tissue is a type of tissue that covers the internal and external surfaces of organisms and acts as a barrier to their structural and functional integrity. Therefore, it constitutes the first line of defense against environmental threats at the interface between the body and its surroundings. The barrier function in the external region of the body, the skin, is the stratum corneum, which is formed by the differentiation of keratinocytes. This multilayered structure acts as a shield, preventing the entry of foreign substances, including pollutants, from the external environment. The skin also serves as a chemical barrier, secreting antimicrobial peptides (Pat et al., 2024).

The gastrointestinal system has a similar structure to that of the skin. It is selectively permeable, allowing the passage of specific nutrients while maintaining microbial integrity and preventing the entry of pathogens (Pat et al., 2024). Organisms cause these particles to undergo gastrointestinal processes by ingesting MPs, either directly or indirectly. The primary region of the gastrointestinal system, the small intestine, absorbs nutrients, water, and other compounds while also playing a protective role. Like the skin, the small intestine also contains different cell types. Enterocytes form the epithelial cell layer that creates a physical barrier via tight junctions, while goblet cells secrete mucus into the lumen. Microfold cells (or M cells) transport antigens from the intestinal lumen to the basolateral region and deliver them to the mucosa-associated lymphoid tissue. Furthermore, small particles are taken up by M cells, cross the intestinal barrier, and are transported to the lymphatic system, liver, and gallbladder. These particles can be released back into the intestine via bile without being excreted in feces (Stock et al., 2019; Yang et al., 2022).

In the lungs, while larger MPs that enter through inhalation are cleared by mucociliary clearance, smaller MPs pass through bronchial epithelial cells, contact endothelial cells, and enter the circulatory system (Yang et al., 2022).

The blood–brain barrier forms a selective diffusion barrier that protects the homeostasis of the central nervous system. Composed of brain microvascular endothelial cells connected by tight junctions and adherens junctions, the blood–brain barrier provides a strong defense against foreign substances (Cho et al., 2024; Li et al., 2024). However, MPs can cross the blood–brain barrier and accumulate in the central nervous system. MPs can enter the central nervous system directly via the systemic circulation and through olfactory nerve fibers.

In mammals, the placenta is a temporary barrier with a complex structure that is vital to the embryo/fetus during rapid growth. Following maternal exposure to MPs, they reach the placenta via the bloodstream and subsequently enter the amniotic fluid and the fetus's body (Sharma et al., 2024; Haddadi et al., 2025).

5.2.2 Cellular uptake and distribution of MPs

After entering an organism through various routes, such as ingestion and inhalation, MPs reach organs, tissues, and finally, cells. The entry of these particles into the cell is first achieved by their adhesion to the cell membrane. These adhering particles are then internalized into the cell via passive penetration or active endocytosis. Active endocytosis occurs in two ways: pinocytosis and phagocytosis. Pinocytosis occurs via mechanisms such as clathrin-mediated endocytosis and macropinocytosis, whereas phagocytosis occurs in specialized phagocytic cells, such as macrophages, monocytes, and neutrophils. As the size of MPs increases (in other words, larger MPs), they can be taken up by macrophages via phagocytosis and macropinocytosis. Therefore, the size of MPs determines their cellular entry pathway (Hua and Wang, 2022).

MPs that enter the cell progress within the cell and between organelles. MPs encapsulated in an endosome formed by the cell membrane are transported within the cell via different pathways. One of these pathways involves the maturation and differentiation of an early endosome into a late endosome, which engulfs MPs and facilitates their transfer to lysosomes containing hydrolytic enzymes. Another pathway involves the transport of MPs via autophagy. Particles that enter the lysosome can overflow into the lysosome, forming

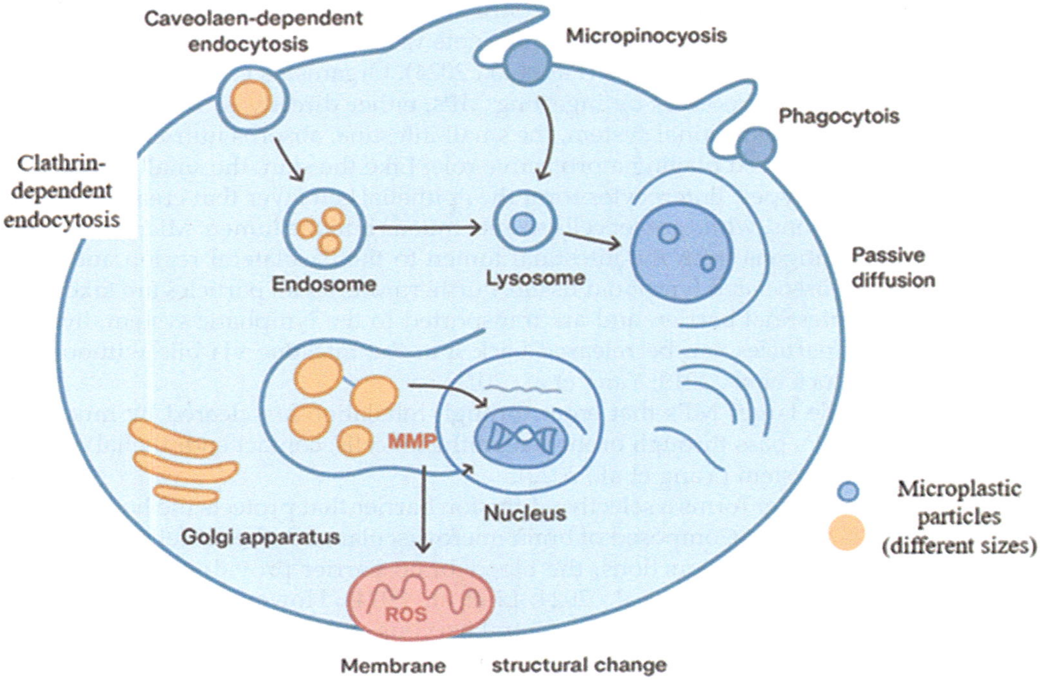

FIG. 5.1 The entry mechanisms of MPs into the cell are endocytosis and micropinocytosis. The exit mechanism from the cell is phagocytosis, and MPs initiate oxidative stress mechanisms within the cell. *MPs*, microplastics; *MMP*, mitochondrial membrane potential; *ROS*, reactive oxygen species. How MPs enter cells via various pathways, including clathrin- and caveolin-mediated endocytosis, macropinocytosis, phagocytosis, and passive diffusion, and their subsequent trafficking through the endosome–lysosome system. This highlights the potential of MPs to induce structural and functional damage at the membrane, mitochondrial, and nuclear levels by increasing intracellular ROS production. Adapted from Hua, X., Wang, D., 2022. Cellular uptake, transport, and organelle response after exposure to microplastics and nanoplastics: current knowledge and perspectives for environmental and health risks. Rev. Environ. Contam. Toxicol. 260 (1). https://doi.org/10.1007/s44169-022-00013-x.

autophagosomes. Consequently, lysosomal damage leads to the release of enzymes into the cytoplasm, which can harm other organelles. Simultaneously, the particles leak into the cell. They are transported to other organelles, such as the mitochondria and endoplasmic reticulum, or they are exocytosed by recycling endosomes and expelled from the cell (Hua and Wang, 2022). If the particles within the lysosome are to be excreted via a normal exocytosis process, they are transported to the cell membrane for release (Fig. 5.1).

5.3 Mechanisms of toxicity

When ingested by organisms, MPs cause a range of harmful effects, primarily oxidative stress and inflammation. As a result, the immune system can be impaired, the hormonal balance can be disrupted, and genetic damage may occur in organisms.

5.3.1 Oxidative stress and inflammation

The cell is the smallest unit of a living organism and is a structure where countless reactions occur to sustain life. Among these reactions, redox reactions facilitate electron transfer between molecules. Redox reactions maintain or alter the balance between electron donors and acceptors. To maintain this balance, or homeostasis, the cell utilizes a buffer system composed of small molecules, such as glutathione–glutathione disulfide, and proteins. To maintain the equilibrium of this buffer system, it uses antioxidant enzymes [such as superoxide dismutase (SOD), catalase, and glutathione peroxidase] and nonenzymatic antioxidants (such as α-tocopherol and ascorbate) (Lugrin et al., 2014). Despite these protective systems, cells exposed to xenobiotics develop functional impairments when molecules such as lipids, carbohydrates, proteins, and DNA are targeted by oxidative stress (Hussain et al., 2016).

Inflammation, on the contrary, is a natural defense system developed by an organism against pathogens. It is closely related to an organism's exposure to allergens or chemicals (Hussain et al., 2016). It is a process that demonstrates the ability to cope with acute and chronic pathological conditions through a biological process involving different cell types and humoral components. These pathological conditions trigger various repair mechanisms to preserve tissue integrity and restore tissue homeostasis following events caused by oxidative stress in the cell (Lugrin et al., 2014; Hsieh et al., 2021; Subaramaniyam et al., 2023; Kadac-Czapska et al., 2024).

Studies suggest that the cellular events described above are interconnected (Gill et al., 2010; Bondia-Pons et al., 2012; McGarry et al., 2018). While the redox reactions occurring within the cell can cause oxidative stress, they can also be precursors to inflammation (Lugrin et al., 2014).

Numerous studies have shown that MPs cause oxidative stress. In ecotoxicology, oxidative stress encompasses not only the oxidative effects of these MPs but also the reactive oxygen species (ROS) produced in the cells or tissues exposed to them. This is because MPs taken up by cells not only disrupt cell membrane integrity but also increase intracellular ROS production, leading to impaired mitochondrial function, the release of proinflammatory cytokines, and cell damage (Kadac-Czapska et al., 2024) (Fig. 5.2).

5.3.2 Immune system modulation

There is limited information on the effects of MPs on the immune system or immune response (Segovia-Mendoza et al., 2020). Following their uptake into the organism, particularly through ingestion, the immune system is activated, which affects defense mechanisms. The immune system works to regulate the disruption of energy metabolism caused by both the chemical structure of MPs (and any adsorbed chemicals) and reduced nutrient absorption resulting from the physical blockage of digestive organs (Espinosa et al., 2018). After these particles cross the intestinal epithelium and enter the circulatory system, they interact directly with immune cells (Schwabl et al., 2019).

The immune system is the first line of defense developed by organisms against threats in the external environment. Innate immunity, on the contrary, is also referred to as nonspecific immunity, the system in which an organism develops against foreign agents. Leukocytes

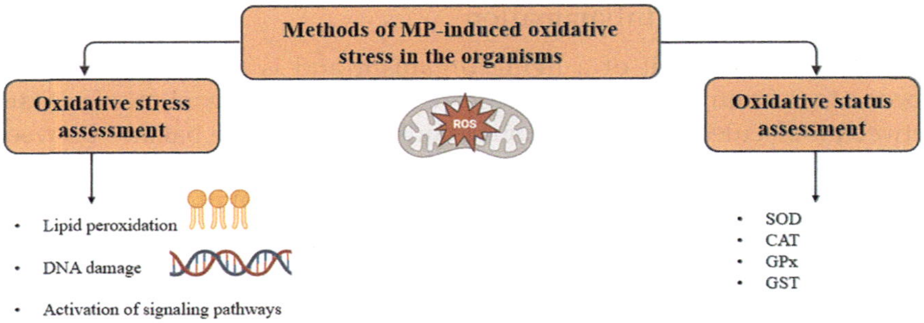

FIG. 5.2 Effects of MPs on oxidative stress: oxidative stress assessments, including lipid peroxidation, DNA damage, and signal transduction, and oxidative status assessments, encompassing enzymatic and nonenzymatic antioxidant parameters. *MPs*, microplastics; *SOD*, superoxide dismutase; *CAT*, catalase; *GPx*, glutathione peroxidase; *GST*, glutathione-S-transferase. Key methodologies used to evaluate oxidative stress triggered by MPs. Indicators such as lipid peroxidation, DNA damage, and the activation of specific signaling pathways are employed to detect oxidative stress. Moreover, the levels of antioxidant enzymes, including SOD, CAT, GPx, and GST, serve as fundamental parameters for determining the oxidative status. Adapted from Kadac-Czapska, K., Ośko, J., Knez, E., Grembecka, M., 2024. Microplastics and oxidative stress—current problems and prospects. Antioxidants 13 (5). https://doi.org/10.3390/antiox13050579.

(white blood cells), which are among the cells of the immune system, engulf and digest antigens presented at the cell surface, thereby forming the adaptive immune system. Interleukins, another element of the immune system, also play a role in this process (Yang et al., 2022). Depending on the disruption they cause to cellular homeostasis, MPs can trigger the release of proinflammatory cytokines (interleukin 1β, IL-6, and TNF-α) (Wu et al., 2019). Studies using mouse models have shown that PS MPs increase inflammatory infiltration in the spleen and alter the distribution of immune cells (Deng et al., 2017; Del Piano et al., 2023).

5.3.3 Endocrine disruption

Endocrine-disrupting chemicals are substances that interfere with the synthesis, secretion, transport, binding, or excretion of hormones responsible for the regulation of homeostasis, reproduction, development, and/or behavior in the body. Pesticides, including Endocrine Disrupting Chemicals (EDCs), are among the most studied and well-known chemicals. With the increasing use of plastics, the effects of their chemical structures on the endocrine functions of living organisms are being investigated, leading to the study of particles such as MPs (Aloisi and Poma, 2025). Plastic additives appear to be quite important in examining the endocrine-disrupting effects associated with MPs (Domenech et al., 2023).

MPs do not always exist in the environment as a single chemical entity. They can absorb various chemicals. Common among these chemical substances are bisphenol A and phthalates. These substances are chemicals widely used in industrial applications, particularly as plasticizers. The most significant effect of bisphenol A and phthalates on organisms is their impact on the endocrine system, which occurs after they bind to estrogen receptors. Therefore, these substances are endocrine-disrupting compounds, and their effects on living organisms involve the endocrine system (Segovia-Mendoza et al., 2020).

5.3.4 Genotoxicity and epigenetic alterations

Genotoxicity is the damage caused by a chemical substance or physical agent to a cell's genetic material. This interaction results in DNA damage, changes in gene expression, and chromosomal damage during cell division (Tang, 2025).

MPs exert their effects on chromosomes directly or indirectly through physical or chemical damage to chromosomes and chromosome segregation during cell division (Tang, 2025). In their study using human peripheral lymphocytes, Çobanoğlu et al. (2021) reported that PE MPs induced micronucleus formation. In a study with onion, Kaur et al. (2022) reported that PS MPs induced cytogenotoxicity in onion root cells. Various biomarkers, such as mitotic, chromosomal, and nuclear abnormalities, have been detected in cells. Furthermore, a decrease in the mitotic index with increasing particle size was observed.

Epigenetics is a branch of science that studies hereditary effects on DNA sequences that do not involve mutations. Therefore it is closely related to developmental processes. Processes such as DNA methylation and histone modifications are highly influential. Environmental epigenetics, on the contrary, examines the epigenetic mechanisms induced by environmental factors in organisms (Poma et al., 2023). Following the oral gavage of rats with different concentrations of PE MPs, Farag et al. (2023) reported an increase in DNA methylation and, through a comet assay, increased DNA damage. After exposure to PE and PS MPs, the DNA methylation patterns of water fleas (*Daphnia magna*) changed, and some genes were rapidly upregulated, whereas others were downregulated (Im et al., 2025).

5.4 MP-associated chemicals and additives

MPs act as vectors for environmental pollutants (such as heavy metals and persistent organic pollutants) by adsorbing them onto their surfaces, facilitating the accumulation of these toxic substances in living tissues. This significantly enhances the transport and concentration of pollutants through the food chain, leading to increased bioaccumulation and biomagnification at each trophic level.

5.4.1 Role of adsorbed environmental pollutants

Owing to their high specific surface area and hydrophobic nature, MPs are valuable carriers of environmental pollutants. Because of their persistence in the environment, MPs actively adsorb persistent and toxic chemicals from the surrounding water or soil. These include heavy metals, persistent organic pollutants, industrial chemicals, and endocrine-disrupting compounds that can leach from plastics. This process converts MPs into concentrated vectors of contamination, effectively functioning as "Trojan horses" that introduce these harmful substances into biological systems (Godoy et al., 2020; Fabra et al., 2021; Hildebrandt et al., 2021; Zhang and Xu, 2022; Samaei et al., 2025).

Primary ecological and health risks arise when organisms ingest these contaminated MPs. Upon ingestion, the complex and dynamic conditions within an organism's digestive tract (such as the presence of stomach acids and digestive enzymes) can facilitate the release of these adsorbed pollutants. This process, known as the "vector effect" or "Trojan horse effect," leads to the bioaccumulation of toxins in an organism's tissues. As predators consume these contaminated organisms, toxins biomagnify through the food web, reach potentially dangerous concentrations in top predators, including humans, and pose significant risks, such as oxidative stress, neurotoxicity, immune suppression, and reproductive disorders (Trevisan et al., 2020; Fabra et al., 2021; He et al., 2022; Menéndez-Pedriza et al., 2024).

5.4.2 Impact on bioaccumulation and biomagnification

The pollution of ecosystems by contaminants due to anthropogenic activities has significant consequences for the environment and human health. Industrial activities, agricultural practices, and increasing urbanization have led to the continuous release of various chemical substances into the environment. Some of these substances not only have toxic effects on organisms but also tend to accumulate in both biotic and abiotic components of the environment due to their specific physicochemical properties. The concepts of bioaccumulation and biomagnification describe the different dimensions of these accumulation processes (Bryan and Darracott, 1979; Streit, 1992; Franke et al., 1994).

The term "bioaccumulation" refers to the net accumulation of a pollutant in an organism's body over time, resulting from its uptake from the environment (water, air, soil, sediment, or food). This process occurs regardless of the organism's position in the food chain. Uptake occurs through respiration, dermal absorption, and, most commonly, dietary intake. The rate of accumulation depends on factors such as the chemical's lipophilicity, persistence (resistance to degradation), and metabolic rate. "Biomagnification" is the increase in the concentration of a pollutant at successive trophic levels (feeding stages) of a food chain. For example, a pollutant present at a low concentration in phytoplankton can reach its highest concentrations in zooplankton that consume it, then in small fish, predatory fish, and finally in birds or mammals that feed on fish. This phenomenon results from a pollutant's inability to be metabolized and excreted by organisms, leading to its storage in body fat (Suedel et al., 1994; Kelly et al., 2007).

Biomagnification is a consequence of food chain dynamics. An organism at a relatively high trophic level consumes a large amount of biomass to meet its energy needs. If the prey it consumes has accumulated persistent pollutants in its tissues, the predator ingests nearly the entire contaminant burden of its prey. Since predators cannot metabolize or excrete these chemicals, the pollutants accumulate in their body fat. As this process repeats at each trophic level, the concentration of the pollutant increases multiplicatively (Suedel et al., 1994; Kelly et al., 2007; Burkhard et al., 2013).

The bioaccumulation and biomagnification of MPs in the food or food web chain are similar to those of other xenobiotics. Particularly in aquatic ecosystems, the accumulation of MPs in organisms such as aquatic invertebrates and fish makes them a significant food source for the transfer of these particles to mammals. The ingestion of MPs at relatively high trophic levels leads to their accumulation at high concentrations. It causes toxicity in organisms, not only from their own chemical structure but also from any other chemicals they may carry. Consequently, ecosystem health is affected (Dey et al., 2025) (Fig. 5.3).

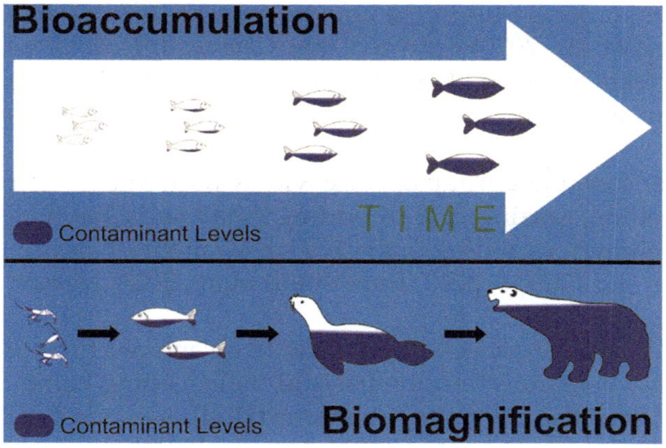

FIG. 5.3 Bioaccumulation of environmental MPs in the food web or through chronic exposure throughout the life of an organism and biomagnification of increasing concentrations through food web chains. The ingestion of MPs at relatively high trophic levels leads to their accumulation at high concentrations. It causes toxicity in organisms, not only from their own chemical structure but also from any other chemicals they may carry.

5.5 Systemic effects in model organisms

A wide range of model organisms, from aquatic species (e.g., fish, crustaceans, and algae) to mammalian models (e.g., rats and mice), have been used in numerous experimental studies in the scientific literature to understand the environmental impacts and biological consequences of MPs. These studies have provided crucial insights into the physical, chemical, and toxicological effects of MPs.

MPs are among the pollutants present in the biosphere and accumulate in abiotic and biotic compartments. The widespread use of these particles, especially in terrestrial areas, causes these particles to mix into aquatic ecosystems and even reach regions without human habitation, such as Antarctica, via atmospheric transport (Cunningham et al., 2020; Sfriso et al., 2020; Wright et al., 2020; Pakhomova et al., 2022; Fox et al., 2024). After entering aquatic ecosystems, aquatic organisms can take up MPs through respiration, dermal contact, or consumption (Rochman et al., 2017; D'Avignon et al., 2023; Dziobak et al., 2024; Zink et al., 2025).

In aquatic ecosystems, primary producers are the energy producers in the food web and/or food chain. The energy produced by this group is transferred to the next trophic level when other organisms consume it. The effects of MPs on these producer organisms have been studied to a limited extent (Yokota et al., 2017; Larue et al., 2021; Nava and Leoni, 2021). The photosynthetic systems of the algae *Chlorella pyrenoidosa* and *Microcystis flos-aquae* were damaged by exposure to polypropylene and PVC. Under short-term exposure, the chlorophyll content in *C. pyrenoidosa* remained unchanged, but after 5 days of exposure, the chlorophyll *a* content decreased. As the PVC concentration increased, the inhibition rate of the chlorophyll *a* content also increased. The inhibition rate of chlorophyll content in *M. flos-aquae* was lower with PP MPs (Wu et al., 2019). After PS MP exposure, the macrophyte *Utricularia vulgaris* presented a reduced plant growth rate, shoot length, chlorophyll *a* content, and ascorbate peroxidase enzyme activity. In contrast, the activities of SOD and peroxidase enzymes are increased (Yu et al., 2022).

Zooplankton constitute the primary consumer group in the aquatic food chain/web (He et al., 2022; Alfonso et al., 2023). Botterell et al. (2020) reported that *Calanus helgolandicus* preferentially ingested fragments, *Acartia tonsa* ingested fibers, and *Homarus gammarus* larvae ingested beads at a relatively high rate during feeding, indicating that these shapes affect the bioavailability of these species.

Aquatic invertebrates constitute another consumer group and are widely used due to their diverse feeding strategies, such as filter feeding, grazing, and shredding (Bhatt and Chauhan, 2023). MPs have various effects on invertebrates, including oxidative stress, mobility, reproduction, and genotoxicity (Burgos-Aceves and Faggio, 2017; Impellitteri et al., 2023; Hamed et al., 2025; Zeng et al., 2025).

Fish are a consumer group within the aquatic ecosystem food chain/web and are further classified as primary, secondary, or tertiary consumers on the basis of their feeding habits. Studies on the detection and toxicity of MPs in fish are quite numerous (Hodkovicova et al., 2022; Gokul et al., 2023; Banaee et al., 2025; Faramazinia et al., 2025; Ghosh, 2025).

MPs pose a serious threat to aquatic amphibians, especially their sensitive larval stages. These particles accumulate in the digestive system, leading to physical blockage, reduced nutrient uptake, and stunted growth. More importantly, MPs act as "Trojan horses," carrying pollutants used in their production or adsorbed from the water into the organisms' bodies. These toxic substances can disrupt the hormonal balance, weaken the amphibian immune system, and further endanger the lives of this already vulnerable group, many of which are facing the threat of extinction (Szkudlarek et al., 2023, 2024; Park et al., 2025; Tatlı et al., 2025).

Aquatic mammals, such as whales, dolphins, and seals, are typically exposed to MPs through the food chain, as particles accumulate in the bodies of their prey. Consequently, these animals are at risk of exposure to high levels of toxic substances due to bioaccumulation. While plastics that accumulate in the stomach and intestines can cause digestive tract blockage and injury, the chemicals they contain can lead to reproductive problems, immune system collapse, and chronic diseases. This situation directly threatens the long-term health and sustainability of their populations (Philipp et al., 2022; Onay et al., 2025).

5.5.1 Findings from rodent studies

Although many studies have investigated the effects of MPs on organisms in aquatic ecosystems, their impact on mammalian model organisms remains poorly understood. Therefore research using in vivo models, such as rats and mice, as well as their in vitro counterparts, aims to clarify the systemic toxicology of MPs (Shan et al., 2022).

The potential of MPs of different sizes to cross the blood–brain barrier was evaluated in mice. In mice orally exposed to PS MPs of varying sizes, as the particle size decreased and the particles approached nanodimensions, the particles crossed the blood–brain barrier in a very short time (e.g., 2 h). Furthermore, in silico analysis revealed that by adding and removing various coronas around plastic particles, these particles can easily cross the blood–brain barrier with the help of cholesterol molecules.

In a study by Forutan et al. (2025), research was conducted on the ability of male rats to cross the blood–brain barrier via low-density PE MPs, resulting in oxidative stress and neurotoxicity. MPs were administered orally to the rats over a chronic period (3 and

6 weeks). While no change was observed in the brain water content, the integrity of the blood–brain barrier was compromised after both exposure durations. Owing to increased oxidative stress, SOD levels decreased, and malondialdehyde (MDA) levels increased. With long-term exposure, there was also a significant decrease in brain-derived neurotrophic factor levels. Histological examination of brain tissue revealed neuronal damage and necrosis. Thus, this study demonstrated that low-density PE MPs have significant effects on the blood–brain barrier and are neurotoxic.

Haddadi et al. (2025) investigated the placental permeability of MPs in rats. For this purpose, in females exposed to spherical PS MPs with a 5 μm diameter during pregnancy, histopathological findings, including reduced tissue thickness in the labyrinth regions, oxidative stress, and induction of apoptosis, were observed. In addition, placental barrier permeability leads to changes in placental ion transport, which is associated with decreases in the expression of gap junction (CX43) and tight junction (ZO1, claudin 3, and claudin 4) proteins. Thus exposure to MPs during pregnancy adversely affects placental barrier permeability.

5.5.2 Human health risk assessment

MPs are widely distributed in the biosphere. Humans are constantly exposed to these particles, which have raised concerns about their potentially harmful effects on human health (Donkers et al., 2022; Xie et al., 2024). Although few studies have confirmed the presence of MPs in human tissues or body fluids, research on their effects on human health remains scarce. Therefore, emphasis should be placed on risk assessments of MPs for human health.

In a study by Ragusa et al. (2021), the authors collected placental samples from six pregnant women and examined them for MP residues via Raman microspectroscopy. A total of 12 MP particles of different shapes were detected in the four placentas. Their locations in these placental tissues were in the fetal (five particles), maternal (four particles), and chorioamniotic membrane (three particles) areas. All the MPs were pigmented, and of these, three were identified as polypropylene, one as a thermoplastic polymer, and nine as being used in artificial coatings, paints, personal care products, etc.

In a study by Donkers et al. (2022), MPs were assessed for membrane permeability, cytotoxicity, and proinflammatory cell activation via in vitro human lung and intestinal cell models. The MPs used in the study were selected from different materials (such as PS, PE, nylon, and rubber) and particles collected from the ocean, which had different shapes and sizes and were fluorescently labeled. In lung cell models, nylon fibers and high-density PE MPs were found to disrupt the membrane structure. In the intestinal cell model, PS and high-density PE MPs impair lumen function and reduce tissue viability. In both cell culture models, acute exposure led to the detection of interleukin-6 release, indicating proinflammatory cell activity.

In a study by Cho et al. (2024), the effects of PS MPs were investigated using human astrocytes, brain vascular pericytes, and brain microvascular endothelial cell models as a blood–brain barrier model. The smaller size of the MPs was shown to increase the membrane permeability of these particles, thereby leading to cellular damage. By creating a 3D blood–brain barrier model with these cell cultures, it was determined that the MPs exhibited highly toxic effects in the 3D cell model.

In a study by Xie et al. (2024), cerebrospinal fluid (CSF) was collected from 28 patients with and without central nervous system infections, and MP analysis was performed. A positive correlation was found between the CSF albumin index and PP and PE concentrations, and interleukin-6 and interleukin-8 levels were significantly increased in samples from patients with infections.

5.6 Conclusion

MPs are among the most significant and concerning causes of environmental pollution today. Their transport from their original area of use to other regions and their persistent presence in a continuous cycle led to the constant exposure of aquatic and terrestrial ecosystems to these particles. Regardless of the route of entry into a living organism, MPs cause systemic harm and adverse effects due to their size, chemical structure, and interactions with other chemical pollutants. In this chapter, the effects of MPs at the cellular and tissue levels after they cross biological barriers via exposure routes are described. Furthermore, the bioaccumulation and biomagnification of MPs in these ecosystems have been reported. Their effects on organisms living in these ecosystems have been explained with various examples. By reviewing the limited number of studies on human health, it has been emphasized that risk assessment is necessary. In conclusion, these microsized particles affect organisms at both the micro- and macrolevels.

References

Alfonso, M.B., Lindsay, D.J., Arias, A.H., Nakano, H., Jandang, S., Isobe, A., 2023. Zooplankton as a suitable tool for microplastic research. Sci. Total Environ. 18791026. 905. Available from: www.elsevier.com/locate/scitotenv. doi:10.1016/j.scitotenv.2023.167329.

Aloisi, M., Poma, A.M.G., 2025. Nanoplastics as gene and epigenetic modulators of endocrine functions: a perspective. Int. J. Mol. Sci. 14220067. 26 (5). Available from: http://www.mdpi.com/journal/ijms. doi:10.3390/ijms26052071.

Banaee, M., Multisanti, C.R., Impellitteri, F., Piccione, G., Faggio, C., 2025. Environmental toxicology of microplastic particles on fish: a review. Comp. Biochem. Physiol. C Toxicol. Pharmacol. 15320456. 287, 110042. doi:10.1016/j.cbpc.2024.110042.

Bhatt, V., Chauhan, J.S., 2023. Microplastic in freshwater ecosystem: bioaccumulation, trophic transfer, and biomagnification. Environ. Sci. Pollut. Res. 16147499. 30 (4), 9389–9400. Available from: https://www.springer.com/journal/11356. doi:10.1007/s11356-022-24529-w.

Bondia-Pons, I., Ryan, L., Martinez, J.A., 2012. Oxidative stress and inflammation interactions in human obesity. J. Physiol. Biochem. 18778755. 68 (4), 701–711. doi:10.1007/s13105-012-0154-2.

Botterell, Z.L.R., Beaumont, N., Cole, M., Hopkins, F.E., Steinke, M., Thompson, R.C., Lindeque, P.K., 2020. Bioavailability of microplastics to marine zooplankton: effect of shape and infochemicals. Environ. Sci. Technol. 15205851. 54 (19), 12024–12033. Available from: http://pubs.acs.org/journal/esthag. doi:10.1021/acs.est.0c02715.

Bryan, G.W., Darracott, A., 1979. Bioaccumulation of marine pollutants. Philos. Trans. R. Soc. London B Biol. Sci. 0080-4622. 286 (1015), 483–505. doi:10.1098/rstb.1979.0042.

Burgos-Aceves, M.A., Faggio, C., 2017. An approach to the study of the immunity functions of bivalve haemocytes: physiology and molecular aspects. Fish Shellfish Immunol. 10504648. 67, 513–517. doi:10.1016/j.fsi.2017.06.042.

Burkhard, L.P., Borgaȉš, K., Powell, D.E., Leonards, P., Muir, D.C.G., Parkerton, T.F., Woodburn, K.B., 2013. Improving the quality and scientific understanding of trophic magnification factors (TMFs). Environ. Sci. Technol. 15205851. 47 (3), 1186–1187. doi:10.1021/es305253r.

Chen, S., Li, S.W., Gu, X.Y., Ma, L.Q., Zhou, D.M., Li, H.B., 2023. Reduced dietary Ca, Cu, Zn, Mn, and Mg bioavailability but increased Fe bioavailability with polyethylene microplastic ingestion in a mouse model: changes in intestinal permeability and gut metabolites. Sci. Total Environ. 18791026. 885. Available from: www.elsevier.com/locate/scitotenv. doi:10.1016/j.scitotenv.2023.163853.

Cho, Y., Seo, E.U., Hwang, K.S., Kim, H., Choi, J., Kim, H.N., 2024. Evaluation of size-dependent uptake, transport and cytotoxicity of polystyrene microplastic in a blood-brain barrier (BBB) model. Nano Converg. 21965404. 11 (1). Available from: https://nanoconvergencejournal.springeropen.com. doi:10.1186/s40580-024-00448-z.

Çobanoğlu, H., Belivermiş, M., Sıkdokur, E., Kılıç, Ö., Çayır, A., 2021. Genotoxic and cytotoxic effects of polyethylene microplastics on human peripheral blood lymphocytes. Chemosphere. 00456535. 272, 129805. doi:10.1016/j.chemosphere.2021.129805.

Cunningham, E.M., Ehlers, S.M., Dick, J.T.A., Sigwart, J.D., Linse, K., Dick, J.J., Kiriakoulakis, K., 2020. High abundances of microplastic pollution in deep-sea sediments: evidence from antarctica and the Southern Ocean. Environ. Sci. Technol. 15205851. 54 (21), 13661–13671. Available from: http://pubs.acs.org/journal/esthag. doi:10.1021/acs.est.0c03441.

D'Avignon, G., Hsu, S.S.H., Gregory-Eaves, I., Ricciardi, A., 2023. Feeding behavior and species interactions increase the bioavailability of microplastics to benthic food webs. Sci. Total Environ. 00489697. 896, 165261. doi:10.1016/j.scitotenv.2023.165261.

Del Piano, F., Lama, A., Monnolo, A., Pirozzi, C., Piccolo, G., Vozzo, S., De Biase, D., Riccio, L., Fusco, G., Mercogliano, R., Meli, R., Ferrante, M.C., 2023. Subchronic exposure to polystyrene microplastic differently affects redox balance in the anterior and posterior intestine of Sparus aurata. Animals. 20762615. 13 (4). Available from: http://www.mdpi.com/journal/animals. doi:10.3390/ani13040606.

Deng, Y., Zhang, Y., Lemos, B., Ren, H., 2017. Tissue accumulation of microplastics in mice and biomarker responses suggest widespread health risks of exposure. Sci. Rep. 2045-2322. 7 (1). doi:10.1038/srep46687.

Dey, S., Sen, K., Saha, N.C., Saha, S., 2025. Analytical approaches for quantifying and characterizing microplastics: environmental impacts and bioaccumulation in aquatic Systems. Green Anal. Chem. 27725774. 12. Available from: https://www.sciencedirect.com/science/journal/27725774. doi:10.1016/j.greeac.2024.100191.

Domenech, J., Annangi, B., Marcos, R., Hernández, A., Catalán, J., 2023. Insights into the potential carcinogenicity of micro- and nano-plastics. Mutat. Res. Rev. Mutat. Res. 13835742. 791, 108453. doi:10.1016/j.mrrev.2023.108453.

Donkers, J.M., Höppener, E.M., Grigoriev, I., Will, L., Melgert, B.N., van der Zaan, B., van de Steeg, E., Kooter, I.M., 2022. Advanced epithelial lung and gut barrier models demonstrate passage of microplastic particles. Microplast. Nanoplast. 26624966. 2 (1). Available from: https://www.springer.com/journal/43591. doi:10.1186/s43591-021-00024-w.

Dziobak, M.K., Fahlman, A., Wells, R.S., Takeshita, R., Smith, C., Gray, A., Weinstein, J., Hart, L.B., Raverty, S., 2024. First evidence of microplastic inhalation among free-ranging small cetaceans. PLOS ONE. 1932-6203. 19 (10), e0309377. doi:10.1371/journal.pone.0309377.

Espinosa, C., García Beltrán, J.M., Esteban, M.A., Cuesta, A., 2018. In vitro effects of virgin microplastics on fish head-kidney leucocyte activities. Environ. Pollut. 18736424. 235, 30–38. Available from: www.elsevier.com/inca/publications/store/4/0/5/8/5/6. doi:10.1016/j.envpol.2017.12.054.

Fabra, M., Williams, L., Watts, J.E.M., Hale, M.S., Couceiro, F., Preston, J., 2021. The plastic Trojan horse: biofilms increase microplastic uptake in marine filter feeders impacting microbial transfer and organism health. Sci. Total Environ. 00489697. 797, 149217. doi:10.1016/j.scitotenv.2021.149217.

Farag, A.A., Youssef, H.S., Sliem, R.E., Gazzar, W.B.E., Nabil, N., Mokhtar, M.M., Marei, Y.M., Ismail, N.S., Radwaan, S.E., Badr, A.M., Sayed, A.E.D.H., 2023. Hematological consequences of polyethylene microplastics toxicity in male rats: oxidative stress, genetic, and epigenetic links. Toxicology. 18793185. 492. Available from: www.elsevier.com/locate/toxicol. doi:10.1016/j.tox.2023.153545.

Faramazinia, M., Sabzghabaei, G.R., Multisanti, C.R., Banaee, M., Piccione, G., Trivedi, A., Faggio, C., 2025. Individual and combined effects of microplastics and diphenyl phthalate as plastic additives on male goldfish:

a biochemical and physiological investigation. Comp. Biochem. Physiol. C Toxicol. Pharmacol. 15320456. 290, 110144. doi:10.1016/j.cbpc.2025.110144.

Forutan, G., Sarkaki, A., Dehbandi, R., Ghafouri, S., Hajipour, S., Farbood, Y., 2025. Chronic exposure to microplastics induces blood–brain barrier impairment, oxidative stress, and neuronal damage in rats. Mol. Neurobiol. 0893-7648. 62 (10), 13777–13785. doi:10.1007/s12035-025-05157-0.

Fox, S., Stefánsson, H., Peternell, M., Zlotskiy, E., Jón Ásbjörnsson, E., Sturkell, E., Wanner, P., Konrad-Schmolke, M., 2024. Physical characteristics of microplastic particles and potential for global atmospheric transport: a meta-analysis. Environ. Pollut. 02697491. 342, 122938. doi:10.1016/j.envpol.2023.122938.

Franke, C., Studinger, G., Berger, G., Böhling, S., Bruckmann, U., Cohors-Fresenborg, D., Jöhncke, U., 1994. The assessment of bioaccumulation. Chemosphere. 00456535. 29 (7), 1501–1514. doi:10.1016/0045-6535(94)90281-x.

Ghosh, T., 2025. Microplastics bioaccumulation in fish: its potential toxic effects on hematology, immune response, neurotoxicity, oxidative stress, growth, and reproductive dysfunction. Toxicol. Rep. 22147500. 14, 101854. doi:10.1016/j.toxrep.2024.101854.

Gill, R., Tsung, A., Billiar, T., 2010. Linking oxidative stress to inflammation: toll-like receptors. Free Radic. Biol. Med. 08915849. 48 (9), 1121–1132. doi:10.1016/j.freeradbiomed.2010.01.006.

Godoy, V., Martínez-Férez, A., Ángeles Martín-Lara, M., Antonio Vellido-Pérez, J., Calero, M., Blázquez, G., 2020. Microplastics as vectors of chromium and lead during dynamic simulation of the human gastrointestinal tract. Sustainability. 2071-1050. 12 (11), 4792. doi:10.3390/su12114792.

Gokul, T., Kumar, K.R., Prema, P., Arun, A., Balaji, P., Faggio, C., 2023. Particulate pollution and its toxicity to fish: an overview. Comp. Biochem. Physiol. C Toxicol. Pharmacol. 15320456. 270, 109646. doi:10.1016/j.cbpc.2023.109646.

Haddadi, A., Venditti, M., Kessabi, K., Messaoudi, I., 2025. Adverse effects of a realistic concentration of human exposure to microplastics on markers of placental barrier permeability in pregnant rats. Environ. Sci. Pollut. Res. 1614-7499. 32 (24), 14782–14795. doi:10.1007/s11356-025-36474-5.

Hamed, M., Said, R.E.M., Shaalan, W.M., Elbaghdady, H.A.M., Sayed, A.E.D.H., 2025. Immunological, neurological, and intestinal changes in red swamp crayfish (Procambarus clarkii) exposed to the combined toxicity of Pyrogallol and microplastics. Mar. Pollut. Bull. 18793363. 213. Available from: https://www.sciencedirect.com/science/journal/0025326X. doi:10.1016/j.marpolbul.2025.117641.

He, M., Yan, M., Chen, X., Wang, X., Gong, H., Wang, W., Wang, J., 2022. Bioavailability and toxicity of microplastics to zooplankton. Gondwana Res. 1342937X. 108, 120–126. doi:10.1016/j.gr.2021.07.021.

Hildebrandt, L., Nack, F.L., Zimmermann, T., Pröfrock, D., 2021. Microplastics as a Trojan horse for trace metals. J. Hazard. Mater. Lett. 26669110. 2, 100035. doi:10.1016/j.hazl.2021.100035.

Hodkovicova, N., Hollerova, A., Svobodova, Z., Faldyna, M., Faggio, C., 2022. Effects of plastic particles on aquatic invertebrates and fish – a review. Environ. Toxicol. Pharmacol. 13826689. 96, 104013. doi:10.1016/j.etap.2022.104013.

Hsieh, S.L., Wu, Y.C., Xu, R.Q., Chen, Y.T., Chen, C.W., Singhania, R.R., Dong, C.D., 2021. Effect of polyethylene microplastics on oxidative stress and histopathology damages in Litopenaeus vannamei. Environ. Pollut. 18736424. 288. Available from: https://www.journals.elsevier.com/environmental-pollution. doi:10.1016/j.envpol.2021.117800.

Hua, X., Wang, D., 2022. Cellular uptake, transport, and organelle response after exposure to microplastics and nanoplastics: current knowledge and perspectives for environmental and health risks. Rev. Environ. Contam. Toxicol. 0179-5953. 260 (1). doi:10.1007/s44169-022-00013-x.

Hussain, T., Tan, B., Yin, Y., Blachier, F., Tossou, M.C.B., Rahu, N., Rupasinghe, V., 2016. Oxidative stress and inflammation: what polyphenols can do for us? Oxid. Med. Cell. Longev. 1942-0900. 2016 (1). doi:10.1155/2016/7432797.

Im, H., Lee, J., Oh, J.-E., Song, J., Jeong, S., 2025. Epigenetic and gene expression responses of daphnia magna to polyethylene and polystyrene microplastics. Molecules. 1420-3049. 30 (7), 1608. doi:10.3390/molecules30071608.

Impellitteri, F., Yunko, K., Martyniuk, V., Matskiv, T., Lechachenko, S., Khoma, V., Mudra, A., Piccione, G., Stoliar, O., Faggio, C., 2023. Physiological and biochemical responses to caffeine and microplastics in Mytilus galloprovincialis. Sci. Total Environ. 00489697. 890, 164075. doi:10.1016/j.scitotenv.2023.164075.

Kadac-Czapska, K., Ośko, J., Knez, E., Grembecka, M., 2024. Microplastics and oxidative stress—current problems and prospects. Antioxidants. 2076-3921. 13 (5), 579. doi:10.3390/antiox13050579.

Kaur, M., Xu, M., Wang, L., 2022. Cyto–genotoxic effect causing potential of polystyrene micro-plastics in terrestrial plants. Nanomaterials. 2079-4991. 12 (12), 2024. doi:10.3390/nano12122024.

Kelly, B.C., Ikonomou, M.G., Blair, J.D., Morin, A.E., Gobas, F.A.P.C., 2007. Food web-specific biomagnification of persistent organic pollutants. Science. 10959203. 317 (5835), 236–239. doi:10.1126/science.1138275.

Larue, C., Sarret, G., Castillo-Michel, H., Pradas Del Real, A.E., 2021. A critical review on the impacts of nanoplastics and microplastics on aquatic and terrestrial photosynthetic organisms. Small. 16136829. 17 (20). Available from: http://onlinelibrary.wiley.com/journal/10.1002/(ISSN)1613-6829. doi:10.1002/smll.202005834.

Li, C., Chen, X., Du, Z., Geng, X., Li, M., Yang, X., Bo, C., Jia, Q., Gongchang, Y., Shi, L., 2024. Inhibiting ferroptosis in brain microvascular endothelial cells: a potential strategy to mitigate polystyrene nanoplastics–induced blood–brain barrier dysfunction. Environ. Res. 00139351. 250, 118506. doi:10.1016/j.envres.2024.118506.

Lugrin, J., Rosenblatt-Velin, N., Parapanov, R., Liaudet, L., 2014. The role of oxidative stress during inflammatory processes. Biol. Chem. 1437-4315. 395 (2), 203–230. doi:10.1515/hsz-2013-0241.

McGarry, T., Biniecka, M., Veale, D.J., Fearon, U., 2018. Hypoxia, oxidative stress and inflammation. Free Radic. Biol. Med. 18734596. 125, 15–24. Available from: www.elsevier.com/locate/freeradbiomed. doi:10.1016/j.freeradbiomed.2018.03.042.

Menéndez-Pedriza, A., Gual, M., Molina-Millán, L., Heeren, R.M.A., Barata, C., Navarro-Martín, L., 2024. Facilitating microplastic ingestion in aquatic models: a verified protocol for daphnia magna as a trojan horse vector. MethodsX. 22150161. 13, 102973. doi:10.1016/j.mex.2024.102973.

Nava, V., Leoni, B., 2021. A critical review of interactions between microplastics, microalgae and aquatic ecosystem function. Water Res. 00431354. 188, 116476. doi:10.1016/j.watres.2020.116476.

Onay, H., Er, A., Minaz, M., Emanet, M., Ceylan, Y., Akdemir, T., Karslı, B., Dalgıç, G., 2025. Assessing microplastic pollution in marine mammals: evidence from three cetacean species in the southeastern Black Sea. Front. Mar. Sci. 2296-7745. 12. doi:10.3389/fmars.2025.1657648.

Pakhomova, S., Berezina, A., Lusher, A.L., Zhdanov, I., Silvestrova, K., Zavialov, P., van Bavel, B., Yakushev, E., 2022. Microplastic variability in subsurface water from the Arctic to Antarctica. Environ. Pollut. 02697491. 298, 118808. doi:10.1016/j.envpol.2022.118808.

Park, J.K., Lee, J.E., Do, Y., 2025. Impacts and transport of microplastics: population dynamics in frogs and the transfer between aquatic and terrestrial ecosystems. J. Hazard. Mater. 18733336. 492. Available from: https://www.sciencedirect.com/science/journal/03043894. doi:10.1016/j.jhazmat.2025.138212.

Pat, Y., Yazici, D., D'Avino, P., Li, M., Ardicli, S., Ardicli, O., Mitamura, Y., Akdis, M., Dhir, R., Nadeau, K., Agache, I., Ogulur, I., Akdis, C.A., 2024. Recent advances in the epithelial barrier theory. Int. Immunol. 1460-2377. 36 (5), 211–222. doi:10.1093/intimm/dxae002.

Philipp, C., Unger, B., Siebert, U., 2022. Occurrence of microplastics in harbour seals (Phoca vitulina) and grey seals (Halichoerus grypus) from german waters. Animals. 2076-2615. 12 (5), 551. doi:10.3390/ani12050551.

Poma, A.M.G., Morciano, P., Aloisi, M., 2023. Beyond genetics: can micro and nanoplastics induce epigenetic and gene-expression modifications? Front. Epigenet. Epigenomics. 2813-706X. 1. doi:10.3389/freae.2023.1241583.

Ragusa, A., Svelato, A., Santacroce, C., Catalano, P., Notarstefano, V., Carnevali, O., Papa, F., Rongioletti, M.C.A., Baiocco, F., Draghi, S., D'Amore, E., Rinaldo, D., Matta, M., Giorgini, E., 2021. Plasticenta: first evidence of microplastics in human placenta. Environ. Int. 01604120. 146, 106274. doi:10.1016/j.envint.2020.106274.

Rochman, C.M., Parnis, J.M., Browne, M.A., Serrato, S., Reiner, E.J., Robson, M., Young, T., Diamond, M.L., Teh, S.J., 2017. Direct and indirect effects of different types of microplastics on freshwater prey (Corbicula fluminea) and their predator (Acipenser transmontanus). PLoS ONE. 19326203. 12 (11). Available from: http://journals.plos.org/plosone/article/file?id=10.1371/journal.pone.0187664&type=printable. doi:10.1371/journal.pone.0187664.

Samaei, S.H.A., Mojahednia, P., Chen, J., Li, Z., Jaszczyszyn, K., Kiedrzyńska, E., Xue, J., 2025. What does the "Trojan Horse" Carry? The pollutants associated with microplastics/nanoplastics in water environments. ACS ES T Water. 26900637. 5 (4), 1530–1545. Available from: https://pubs.acs.org/page/aewcaa/about.html. doi:10.1021/acsestwater.4c01089.

Schwabl, P., Köppel, S., Königshofer, P., Bucsics, T., Trauner, M., Reiberger, T., Liebmann, B., 2019. Detection of various microplastics in human stool. Ann. Intern. Med. 0003-4819. 171 (7), 453–457. doi:10.7326/m19-0618.

Segovia-Mendoza, M., Nava-Castro, K.E., Palacios-Arreola, M.I., Garay-Canales, C., Morales-Montor, J., 2020. How microplastic components influence the immune system and impact on children health: focus on cancer. Birth Defects Res. 24721727. 112 (17), 1341–1361. Available from: http://onlinelibrary.wiley.com/journal/10.1002/%28ISSN%292472-1727. doi:10.1002/bdr2.1779.

Sfriso, A.A., Tomio, Y., Rosso, B., Gambaro, A., Sfriso, A., Corami, F., Rastelli, E., Corinaldesi, C., Mistri, M., Munari, C., 2020. Microplastic accumulation in benthic invertebrates in Terra Nova Bay (Ross Sea, Antarctica). Environ. Int. 01604120. 137, 105587. doi:10.1016/j.envint.2020.105587.

Shan, S., Zhang, Y., Zhao, H., Zeng, T., Zhao, X., 2022. Polystyrene nanoplastics penetrate across the blood-brain barrier and induce activation of microglia in the brain of mice. Chemosphere. 18791298. 298. Available from: www.elsevier.com/locate/chemosphere. doi:10.1016/j.chemosphere.2022.134261.

Sharma, R.K., Kumari, U., Kumar, S., 2024. Impact of microplastics on pregnancy and fetal development: a systematic review. Cureus. 2168-8184. 16 (5). doi:10.7759/cureus.60712.

Stock, V., Böhmert, L., Lisicki, E., Block, R., Cara-Carmona, J., Pack, L.K., Selb, R., Lichtenstein, D., Voss, L., Henderson, C.J., Zabinsky, E., Sieg, H., Braeuning, A., Lampen, A., 2019. Uptake and effects of orally ingested polystyrene microplastic particles in vitro and in vivo. Arch. Toxicol. 14320738. 93 (7), 1817–1833. Available from: https://link.springer.de/link/service/journals/00204/index.htm. doi:10.1007/s00204-019-02478-7.

Streit, B., 1992. Bioaccumulation processes in ecosystems. Experientia. 0014-4754. 48 (10), 955–970. doi:10.1007/bf01919142.

Subaramaniyam, U., Allimuthu, R.S., Vappu, S., Ramalingam, D., Balan, R., Paital, B., Panda, N., Rath, P.K., Ramalingam, N., Sahoo, D.K., 2023. Effects of microplastics, pesticides and nano-materials on fish health, oxidative stress and antioxidant defense mechanism. Front. Physiol. 1664042X. 14. Available from: http://www.frontiersin.org/Physiology/archive/. doi:10.3389/fphys.2023.1217666.

Suedel, B.C., Boraczek, J.A., Peddicord, R.K., Clifford, P.A., Dillon, T.M., 1994. Trophic transfer and biomagnification potential of contaminants in aquatic ecosystems. Rev. Environ. Contam. Toxicol. 01795953. 136, 21–89. doi:10.1007/978-1-4612-2656-7_2.

Szkudlarek, M., Najbar, B., Jankowiak, Ł., 2023. Microplastics pollution in larvae of toads, frogs and newts in anthropopressure gradient. Ecol. Indic. 1470160X. 155, 110971. doi:10.1016/j.ecolind.2023.110971.

Szkudlarek, M., Najbar, B., Jankowiak, Ł., 2024. Similarity of microplastic characteristics between Amphibian Larvae and their aquatic environment. Animals. 2076-2615. 14 (5), 717. doi:10.3390/ani14050717.

Tang, K.H.D., 2025. Genotoxicity of microplastics on living organisms: effects on chromosomes, DNA and gene expression. Environments. 2076-3298. 12 (1), 10. doi:10.3390/environments12010010.

Tatlı, H., Gedik, K., Altunışık, A., 2025. Distribution of microplastics in tadpoles, adults, and habitats of three water frogs of Pelophylax spp. Environ. Sci. Eur. 2190-4715. 37 (1). doi:10.1186/s12302-025-01065-1.

Trevisan, R., Uzochukwu, D., Di Giulio, R.T., 2020. PAH sorption to nanoplastics and the Trojan Horse effect as drivers of mitochondrial toxicity and PAH localization in Zebrafish. Front. Environ. Sci. 2296665X. 8. Available from: https://journal.frontiersin.org/journal/environmental-science. doi:10.3389/fenvs.2020.00078.

Wright, S.L., Ulke, J., Font, A., Chan, K.L.A., Kelly, F.J., 2020. Atmospheric microplastic deposition in an urban environment and an evaluation of transport. Environ. Int. 01604120. 136, 105411. doi:10.1016/j.envint.2019.105411.

Wu, Y., Guo, P., Zhang, X., Zhang, Y., Xie, S., Deng, J., 2019. Effect of microplastics exposure on the photosynthesis system of freshwater algae. J. Hazard. Mater. 03043894. 374, 219–227. doi:10.1016/j.jhazmat.2019.04.039.

Xie, J., ji, J., Sun, Y., Ma, Y., Wu, D., Zhang, Z., 2024. Blood-brain barrier damage accelerates the accumulation of micro- and nanoplastics in the human central nervous system. J. Hazard. Mater. 03043894. 480, 136028. doi:10.1016/j.jhazmat.2024.136028.

Yang, W., Jannatun, N., Zeng, Y., Liu, T., Zhang, G., Chen, C., Yang, L., 2022. Impacts of microplastics on immunity. Front. Toxicol. 2673-3080. 4. doi:10.3389/ftox.2022.956885.

Yokota, K., Waterfield, H., Hastings, C., Davidson, E., Kwietniewski, E., Wells, B., 2017. Finding the missing piece of the aquatic plastic pollution puzzle: interaction between primary producers and microplastics. Limnol. Oceanogr. Lett. 2378-2242. 2 (4), 91–104. doi:10.1002/lol2.10040.

Yu, H., Qi, W., Cao, X., Wang, Y., Yang, L., Yan, X., Zhang, X., Peng, J., Jiuhui, Q., 2022. Impact of microplastics on the foraging, photosynthesis and digestive systems of submerged carnivorous macrophytes under low and high nutrient concentrations. Environ. Pollut. 02697491. 292 (Pt A), 118220. doi:10.1016/j.envpol.2021.118220.

Zeng, H., Zhong, Y., Wei, W., Luo, M., Xu, X., 2025. Combined exposure to microplastics and copper elicited size-dependent uptake and toxicity responses in red swamp crayfish (Procambarus clarkia). J. Hazard. Mater. 18733336. 487. Available from: https://www.sciencedirect.com/science/journal/03043894. doi:10.1016/j.jhazmat.2025.137263.

Zhang, M., Xu, L., 2022. Transport of micro- and nanoplastics in the environment: trojan-Horse effect for organic contaminants. Crit. Rev. Environ. Sci. Technol. 1064-3389. 52 (5), 810–846. doi:10.1080/10643389.2020.1845531.

Zink, L., Val, A.L., Wood, C.M., 2025. Perspectives on the impact of microplastics (MPs) on fish of the Amazon that exhibit air-breathing and aquatic surface respiration. Anais da Academia Brasileira de Ciencias. 16782690. 97. Available from: https://www.scielo.br/j/aabc/a/n9HTYRkjNTcCrtbdK9qT3Bc/?lang=en&format=pdf. doi:10.1590/0001-3765202520250254.

CHAPTER

6

Microplastic pollution and consequences on immune systems

M. Amin Mir[a], *Noushi Zaidi*[b], *Kim Andrews*[c], *Nouara Abdelli*[d], *and Mohammad Ayaz Ahmad*[e]

[a]Department of Mechanical Engineering, Prince Mohammad Bin Fahd University, Dammam, Saudi Arabia

[b]Department of Chemistry, College of Science, King Faisal University, Al-Ahsa, Saudi Arabia

[c]Department of Mathematics & Natural Sciences, Prince Mohammad Bin Fahd University, AL Khobar, Saudi Arabia

[d]Department of Biological Sciences, College of Science, King Faisal University Al-Ahsa, Saudi Arabia

[e]Department of Mathematics, Physics & Statistics Faculty of Natural Sciences, Turkeyen Campus University of Guyana, Greater Georgetown, Guyana

6.1 Introduction

Plastic pollution has become a serious environmental problem because of rapidly increasing plastic consumption worldwide. Recently, microplastics (MPs) have emerged as a new and persistent class of pollutants and have received widespread attention because of their chemical stability and long-term persistence in the environment (Moore, 2008). MPs can accumulate in living organisms (Sun et al., 2020), posing potential risks to ecological and human health. Owing to their small size and widespread occurrence in the environment, MPs are inadvertently incorporated into the diets of microorganisms, such as algae and lower trophic-level species (Steer et al., 2017), and after tissue deposition, they work their way up the food web. Measurable amounts of MPs (most commonly <3 mm) have been reported in the hemolymph of marine crustaceans (Von Moos et al., 2012) and in the digestive tracts of fish. More MPs, which may indicate MP exposure, have been reported in greater numbers in seabirds (Zhao et al., 2016). MPs have been found in human drinking water, seafood, and several human tissues, including the placenta. Therefore there is a high probability of exposure and bioaccumulation in humans and wildlife.

A growing body of literature confirms that MPs have negative effects on the health of organisms (Breschi et al., 2017). In an experimental mouse model, zebrafish have been

Microplastics in Animal and Human Physiology
https://doi.org/10.1016/B978-0-443-45635-0.00015-X

shown to deposit MPs in several tissues. Micron-sized MPs mainly enter the body via gut-associated lymphoid tissues (GALTs), whereas nanosized particles are absorbed into gut epithelial villus cells via pinocytosis. These MPs can affect gut barrier architecture, disrupt the internal gut microbiome, alter metabolic status, and translocate to other organs. Many assays are performed with fluorescent labeling to track absorption and distribution. MPs observed in the liver are implicated in various aspects of bile acid metabolism and oxidative stress (Wang et al., 2020). In addition, transcriptomics or functional genomics revealed metabolic alarms and immune metadata DSP constellations induced by the MPs consumed by the animals. Specifically, when distributed throughout the vertebrate body, the immune system serves as a complex defense mechanism against pathogenic invaders. Innate immunity serves as the first line of nonspecific host defense. The spleen acts as an immune organ, and the liver serves both as a metabolizing organ and as an immune organ (Kubes and Jenne, 2018). Sedimentation of antigens (matter) in the spleen and liver triggers rapid host intercellular responses that limit microbial growth (Rawat et al., 2025).

For any host, for example, eliminating an antigen during early infection is paramount. Clearing pathogenic infection typically requires fast engagement of innate immunity at the level of recognition, with the release of cytokines, activation of immune cells, and short-term clearance of pathogens (Heymann and Tacke, 2016). The extracellular matrix (ECM) is a complex structure composed of diverse macromolecules, and its architecture depends on the tissue type and its configuration (Humphries et al., 2019). Cell–ECM interactions involve a linkage between the ECM and the cell, providing a physical connection between the cytoskeleton and intracellular signaling pathways, and are facilitated by specialized adhesion complexes. These interactions impact cellular shape, motility, mechanical sensitivity, development, and immune reactions. Therefore ECM-mediated signaling plays a critical role in regulating effective immune defenses.

The anthropogenic burden on humans worldwide has increased MP pollution, resulting in chronic exposure to MPs and nanoplastics (NPs) via ingestion, inhalation, and dermal contact (Domenech and Marcos, 2021). Particle mass is an important determinant of risk assessment. A metaanalysis has quantified typical human exposure loads from exogenous sources, such as eating and breathing, as shown in Table 6.1, on the basis of rigorous study evaluation and standardized measurement protocols. The threshold exposure for disease has not yet been defined (Domenech and Marcos, 2021).

MPs are emerging contaminants that are intentionally inhaled by living organisms. As a consequence of continuous exposure to an open-air environment, particles are fragmented and discharged into the surrounding environment, likely increasing adverse effects on biological systems because of their small size, expansive surface area, and ability to penetrate (Xu et al., 2019). The adult human lung, which includes proteinaceous surfactant, contains approximately 150 m^2 of alveolar surface area and is approximately 1 μm (or less) away from the blood capillaries. Thus at least in theory, MPs can gain access to the circulation through the lungs, where they can be transported to different organs, tissues, and cells in the body if they cross the alveolar barrier (Lehner et al., 2019). Table 6.2 summarizes the immune-based effects following exposure to MPs. The skin, respiratory tract, and gastrointestinal tract (GIT) epithelial barriers function properly either physically, chemically, or immunologically to defend against biological or chemical agents.

TABLE 6.1 Approximate estimation of human intake of micro- and nanoplastics.

Exposure route	Source	Typical/recommended consumption	MP/NP Co (mean)	Daily intake	Yearly intake	Annual load
Ingestion	Fresh products	420 g/day	128,200 p/g	53.84×10^6	19.65×10^9	2.94×10^{10}
	Seafood	23.0 kg/year	1.05 p/g	66.30	24.20×10^3	–
	Water	1.9 L/day	12.90×10^6 p/L	24.51×10^6	8.94×10^9	–
	Table salt	4.8 g/day	150.25 p/kg	0.72	263.41	–
	Alcohols	6.2 L/year	4.20 p/L	0.07	26.04	–
Inhalation	Ambient air	8.50 m^3/day	0.70 p/m^3	5.95	2.17×10^3	2.17×10^3

Note: *MP*, microplastics; *NP*, nanoplastics; *p/g*, particles per gram; *p/L*, particles per liter; *p/m^3*, particles per cubic meter.

TABLE 6.2 Main exposure routes for micro/nanoplastics, their impacts on the immune system, and the resulting consequences for cellular and overall human health.

Exposure route	Immune system response	Implications for cellular and human health*
Gastrointestinal: predominantly from consuming food and beverages containing plastic contaminants (e.g., bottled water, table salt, bivalves, crustaceans, and various fish species).	Elevated IL-8 secretion by activated macrophages and monocytes	**Digestive system:** • Cellular toxicity linked to oxidative stress • Higher lipid peroxidation rates • Excess generation of reactive oxygen species (ROS) • Enhanced secretion of gastric mucus • Linked to gut microbiota imbalance and dysbiosis
Respiratory: primarily from airborne microfibers and particles shed from vehicle tires, fabrics, and household dust accumulation.	Increased IL-6 release from monocytes and macrophages	**Nervous system:** • Neurotoxic effects mediated by proinflammatory cytokines impacting neurons • Potential contribution to neurodegenerative processes suggested in some studies
Dermal: direct skin contact with microscopic plastic particles, disrupting epithelial integrity.	Elevated TNF-α production	**Respiratory system:** • Autophagy activation in response to oxidative stress and cytokine-induced inflammation • Worsening of chronic respiratory conditions • Barrier disruption linked to a higher incidence and severity of allergy and inflammation

Note: *Health effects vary depending on concentration, particle size, and duration of exposure.
Lehner, R., Weder, C., Petri-Fink, A., Rothen-Rutishauser, B., 2019. Emergence of nanoplastic in the environment and possible impact on human health. Environ. Sci. Technol. 53 (4), 1748–1765. https://doi.org/10.1021/acs.est.8b05512.

Damage to these epithelial barriers (i.e., various proteins in tight junctions that can be degraded) is connected to epithelial cell signaling and the release of cytokines such as IL-25, IL-33, and thymic stromal lymphopoietin (TSLP), which increase epithelial barrier permeability and allow entry of allergens, irritants, and potentially pathogens (Celebi Sözener et al., 2020). Similarly, disruption of epithelial barriers can alter the airway microbiota.

The principal ways in which MPs or NPs affect the respiratory system include epithelial barrier dysfunction, cytotoxicity, potential synergistic effects with allergens, the inflammatory response, and imbalances in oxidative stress. In the alveoli, MPs and NPs can reach the air–liquid interface, and fusing vesicles can transform the lung surfactant structure, mobility, and surface tension, which may lead to collapse (dependent on the particle type) (Shi et al., 2022). MPs can also alter alveolar structure, decrease barrier integrity, and cause a loss of cell viability with increasing exposure. Pathobiology is often characterized by alterations in morphology, oxidative stress, inflammation, and significant disruption of the epithelial barrier, leading to chronic respiratory disease (Mir and Banik, 2025) (*Sustainable healing*). Barrier disruptions caused by MPs and NPs can also increase the penetrability of allergens and pathogens, initiating inflammatory responses that can alternatively trigger type 2 immune responses that may promote asthma, rhinitis, or chronic rhinosinusitis (Zhao et al., 2020). The nasal microbial ecosystem may also be altered; specifically, high levels of exposure to MPs have previously been linked with increased relative abundances of respiratory disease-associated bacteria such as *Klebsiella and Helicobacter*. The Bacteroides were reduced.

Children have been identified as particularly vulnerable to exposure owing to their behaviors and developmental stages with respect to the neurobehavioral, immune, metabolic, and cardiovascular systems. In experimental work involving pregnant rats, NPs were transferred to the maternal lungs, heart, and spleen; placental tissue; and fetal lungs, hearts, livers, kidneys, and brains (Fournier et al., 2020). MPs, ranging from 5 to 10 μm in size, have been identified from human placental samples, including the fetal side, placental side, and chorioamniotic membrane (Ragusa et al., 2021). In trophoblasts, MPs are related to decreased viability, cell cycle arrest, diminished migration and invasion, increased reactive oxygen species (ROS), a dose–response relationship for proinflammatory cytokines (TNF-α and IFN-γ), and epithelial barrier disruption. Potential exposure in children can come from airborne MPs in nursery and schooling environments, which originate in uniforms, toys, furniture, packaging, and indoor/outdoor environments related to multiple sources, all of which are influenced by location and the ventilation system, as well as how close one is to traffic or industry (Abbasi et al., 2022). Inflamed airway mouse models have shown that MP exposure leads to the influx of inflammatory cells and the aggregation of macrophages and, perhaps, to a significant increase in the TNF-α concentration in bronchoalveolar lavage fluid, which is also associated with the severity of asthma; mice free of asthma exhibit an increase in IgG1. Transcriptomic analysis revealed altered regulation of genes involved in immunity, stress-related responses, and programmed cell death regulators (Mir and Banik, 2025). Together, long-term environmental exposure facilitates mixing mechanisms in the environment and atmosphere for MPs (Shi et al., 2022). Human lungs contain polyethene terephthalate (PET) and polypropylene (PP) MPs (<5.5 μm) and fibers (8.12–16.8 μm). Sputum studies involving the identification of 21 MPs <500 μm have largely identified the MPs as polyurethane, polyethene (PE), Procalcitonin (PCT), chlorinated PET, and alkyl groups, all of which match those associated with smoking. Bronchoalveolar lavage revealed that smokers have higher concentrations of MPs than nonsmokers do. A tumor lesion study revealed that patients with high-risk exposures had more microfibers and that microfiber accumulation increased with age. Although cigarette filters are marketed to be protective, they do not provide any health benefits to the user. Cigarette filters containing >15,000 fibers can release microfibers/MPs or break down into even smaller particles, providing another pathway for environmental and human exposure. Improper disposal of cigarette butts continues to introduce microfibers/MPs into ecosystems, representing an urgent, yet unresolved, area of waste management (Chen et al., 2022).

6.2 Digestive

MPs are produced through various environmental processes and are constantly moving across ecosystems. MPs also enter the food web and eventually the human diet. The health implications of ingesting MPs and nanoparticles are an increasing area of concern. The presence of MPs in human feces provides evidence of their ingestion and movement through the GIT. In vitro and in vivo studies in mammals and humans reported that MPs could alter gut homeostasis, affecting the microbiota, mucus integrity, and epithelial barrier function (Belzagui et al., 2021).

6.2.1 Inhalation

MPs are emerging contaminants that living organisms can inhale. The open-air (outdoor) environment facilitates the potential inhalation of MPs, and certain continuous exposures increase the fragmentation of MPs and their dispersal into the atmosphere, with all their subsequent negative impacts on biological systems due to their small size, wide-ranging potential to penetrate, and large surface area (Xu et al., 2019). The alveolar surface area of adult lungs is approximately $150\,m^2$, and the space that separates the alveolar air and blood capillaries is <1 μm thick, which allows easy passage of MPs to enter the bloodstream and circulate throughout the body (Lehner et al., 2019). Table 6.2 summarizes the immune health-related effects of exposure to MPs. The epithelial barriers of the skin, respiratory tract, and GIT are the first line of defense, physically, chemically, and immunologically, against biological and chemical threats from the external environment. Damage to these barriers, such as the degradation of proteins in the tight junctions of epithelial cells, initiates epithelial signaling and promotes the release of cytokines such as IL-25, IL-33, and TSLP, increasing epithelial barrier permeability and enabling increased access to allergens, irritants, and pathogens (Celebi Sözener et al., 2020). Similarly, damage to epithelial barriers may directly alter the composition of the airway microbiota.

6.2.2 Effects of MPs/NPs on the respiratory pathway

The main mechanisms by which MPs or NPs affect respiratory health are epithelial barrier destruction, cytotoxicity, synergistic effects with allergens, inflammation, and oxidative stress imbalance. MPs and NPs can reach the air–liquid interface in the alveoli and form vesicles, compromising the structure, mobility, and surface tension of the pulmonary surfactant, which, in turn, may lead to the collapse of the pulmonary surfactant (depending on the type of nanoparticle) (Shi et al., 2022). MPs may also change alveolar structure, reduce barrier function, and decrease cell viability in a dose-dependent manner. Damage is often associated with morphological disorganization, oxidative stress, inflammation, and eventual compromise of the epithelial barrier, which may lead to chronic lung diseases (Mir and Banik, 2025). Barrier dysfunction via MPs/NPs also increases allergen and pathogen penetration, leading to an inflammatory response and subsequent type 2 immune responses that promote conditions such as asthma, rhinitis, and chronic rhinosinusitis (Zhao et al., 2020). Nasal microbial communities could also be affected; for example, high levels of MPs have been associated with a greater abundance of respiratory disease-associated bacteria (e.g., *Klebsiella* and *Helicobacter*) and a lower abundance of beneficial microbes (e.g., Bacteroides). Children are thought to be particularly affected by complex behavioral patterns and developmental stages that can influence neurobehavioral function and the immune, metabolic, and cardiovascular systems. Experimental studies using pregnant rats demonstrated that NPs may be detected in maternal lungs, hearts, and spleens, as well as in placental and fetal tissues (liver, lungs, heart, kidneys, and brain). It has also been shown that MPs exist in human placental samples (fetal side, maternal side, and chorioamniotic membrane) at sizes ranging from 5 to 10 μm (Ragusa et al., 2021). In trophoblastic cytoplasm, MPs are associated with reduced cell viability, cell cycle arrest, impaired migration and invasion capacity, increased ROS, dose-dependent proinflammatory cytokine production (TNF-α and IFN-γ), and epithelial barrier disruption. Potential childhood exposure may occur through airborne MPs in day nurseries and nursery schools and from adults' uniforms, children's toys, furniture, packaging, and environmental sources both indoors and outdoors, as expected levels depend on location, ventilation, and proximity to traffic or industrial sources (Abbasi et al., 2022). In murine asthma models, MP exposure promoted inflammatory cell infiltration and macrophage clustering, increased TNF-α levels in bronchoalveolar lavage fluid, and increased asthma symptoms. In contrast, mild disease and nonasthmatic murine models are dependent on IgG1. Transcriptional profiles coding for B-cell activity, regulatory pathways, homeostasis, immune response, defense and stress responses, cell morphogenesis, and programmed cell death are affected by exposure (Mir and Banik, 2025). Aggregating long-term exposure from environmental contact results in the atmospheric mixing of MPs (Shi et al., 2022). In sputum studies, 21 types of MPs <500 μm in diameter were identified, with increasing litter durability; e.g., polyurethane was the most common, followed by PE, PCT, and chlorinated PET/alkyd varnish, with smoking linked the most to pets. Bronchoalveolar lavage fluid contained higher concentrations of MPs in smokers than in nonsmokers. Higher numbers of microfibers are found in the tumor lesions of patients at high risk of exposure, and the number of microfibers progressively increases with age (Chen et al., 2022). Cigarette filters marketed for health protection are not expected to convey any health benefits. Cigarette filters can be made from >15,000 fibers, and these fibers can be released as microfibers or MPs or fragmented into particles smaller than the MPs. There is now another conduit for environmental exposure and inadvertent exposure for human beings. The improper disposal of cigarette butts introduces MPs into ecosystems, making them an urgent but potentially persistent source of waste (Chen et al., 2022).

6.2.3 Effects of MPs or NPs on the digestive system

MPs originate from numerous environmental processes, colonize ecosystems, enter the food web, and are eventually consumed by humans. Ingested MPs and NPs represent an emerging health concern for consumers. Human feces containing MPs can be ingested and travel through the GIT. Studies, both in vitro and in vivo, conducted with mammals and humans, suggest that MPs may alter gut homeostasis, including the microbiota, mucosal integrity, and epithelial barrier function (Belzagui et al., 2021). It is unlikely that MPs will pass through the lining of the intestines via paracellular pathways, as the maximum functional diameter of the gaps is approximately 1.5 nm between the tight junctions of enterocytes. More likely, Peyer's patches are taken up via lymphatic tissues, phagocytosed, endocytosed, and then transported through the intestinal mucosa. Once ingested, MPs may undergo structural or surface modifications that can affect their absorption potential and absorption rate. Within the GIT, MPs can bind to biomolecules such as proteins, lipids, carbohydrates, nucleic acids, ions, and water, often encircled by a protein matrix known as the "corona." Although plastics are chemically inert, their surfaces can adsorb hazardous agents, including additives, heavy metals, proteins, and microorganisms, thereby increasing their toxicity. This allows MPs to act as "Trojan horses," transporting environmental degraders into contact with intestinal mucus, epithelial cells, and the microbiota, thereby triggering biological events associated with cellular responses and systemic dysregulation (Fournier et al., 2021). MP ingestion can affect gut microbial communities, leading to dysbiosis, reduced resilience, recurrent pathogen blooms, metabolic dysfunction, and increased levels of disease-associated bacteria (*Bifidobacterium, Streptococcus,* and *Sphingomonas*) while decreasing populations of beneficial taxa (*Ruminococcus, Fusobacterium,* and *Coprococcus*). Moreover, contaminants may be transferred from mother to infant through breastfeeding, and the infant's body burdens often overlap with their mother's cumulative exposures. Meconium analyses have indicated that newborns exhibit higher concentrations of MPs than adults do. Potential sources of MPs for infants include baby food packaging and baby feeding bottles. Bottles release 53–393 microparticles/mL depending on the type (plastic vs. glass) as well as the brand, which illustrates the uptake and transfer of microparticles in infancy (Xu et al., 2019). Exposure to bisphenols and phthalates in early life can negatively affect immune system development and has been associated with diseases such as cancer. Globally, human exposure through diet is estimated at approximately 0.1–5 g of MPs per week, and this contamination can result from environmental sources (air, water, and soil), contaminated organisms, food production processes, or leaching from packaging. MPs have been detected in seafood, drinking water, salt, honey, sugar, milk, beer, soft drinks, fruit, meat (chicken, beef, pork), and takeaway containers. Oliveri Conti et al. reported that apples and carrots were the most contaminated fruits and vegetables. In Mexico City, 100% of the commercial ice cube brands tested contained MPs, most notably PP, PET, polyvinyl chloride (PVC), polyamide, and cellophane (Shruti et al., 2023). MPs and NPs may also contribute to food allergies. Their small size allows them to penetrate or modify the biology of epithelial cells, compromise barrier integrity, alter protein digestibility, and induce inflammatory conditions conducive to sensitization. NPs form a protein coating rapidly upon contact with biological fluids, and in so doing, they may create what is called "an allergenic protein corona" when that protein is an allergen, increasing the potential for heightened allergic responses when ingested, for example, in the form of food processing that includes the NPs or during digestion (Zhdanov and Cho, 2016).

6.2.4 Impact of MPs/NPs on the skin

Human epithelial cells can be subjected to oxidative stress upon exposure to MPs and NPs, thereby disrupting the skin and mucosal barriers. Disruptions to the skin and mucosal barriers have been associated with an increased incidence and severity of allergic and inflammatory skin diseases, including atopic dermatitis. Lee et al. reported an association between perinatal exposure to bisphenol A and phthalates and the incidence of atopic dermatitis in six-month-old infants (Lee et al., 2021). Skin exposure can result from the use of personal care products that penetrate the skin or from contact with MP-contaminated water. Campbell et al. demonstrated that particles 20–200 nm in size can penetrate only the outer layers of the skin (~2 μm) and that particles <100 nm may breach the dermal barrier. Larger MPs can still penetrate the skin, but they normally do so through hair follicles, sweat glands, and lesions. Among the available products, it is expected that microbeads in exfoliating products, toothpaste, and dental materials can easily be degraded into MPs when they collide with other surfaces. Ingredients in personal care products, such as urea, glycerol, and α-hydroxy acids, have the potential to drive more MPs into the skin barrier (Llorca and Farré, 2021).

6.3 MPs and the immune system

6.3.1 MP internalization

Immune cells are constantly defending against foreign particles. For example, the lungs and intestines are the two main locations where our bodies come into contact with environmental particles, making them the most likely entry points. In the lungs, larger MPs are usually expelled via mucociliary clearance. In contrast, smaller particles (<1 μM) can traverse the bronchial epithelial layer and may enter endothelial cells before they enter the bloodstream (Wright and Kelly, 2017). As shown in Fig. 6.1, MPs are taken up into human alveolar epithelial (i.e., A549) cells after inhalation. Researchers have reported that MPs occupy areas of the nucleus typically occupied by DNA, induce cell cycle arrest and apoptosis in monocytes and macrophages, and upregulate the expression of inflammation-related genes. In the intestines, M cells are specialized epithelial cells in Peyer's patches that can take up MPs via endocytosis and transport them to associated mucosal lymphoid tissues (Mowat, 2003) where MPs can perturb cellular signaling pathways and gene expression (Yong et al., 2020). MPs may also stimulate autophagy, overcome barriers to access organs such as the liver, kidney, and brain, activate the innate immune system, and cause dysfunction (Mir and Banik, 2025).

The cellular uptake of MPs, in terms of size, surface charge, and functionalization properties, is also mediated by interactions with biomolecules, including proteins, carbohydrates, and phospholipids (Yee et al., 2021). Some MPs can readily cross biological membranes (Mani and Pandey, 2019). The different mechanisms of uptake include phagocytosis, macropinocytosis, and clathrin- and caveolae-mediated endocytosis. Interestingly, negatively charged

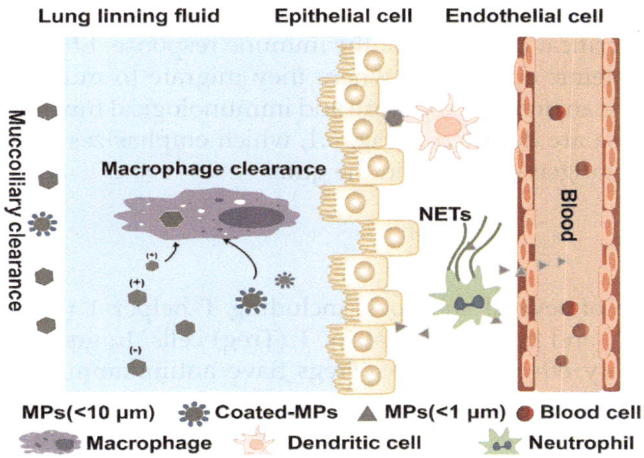

FIG. 6.1 Schematic illustration of how inhaled microplastics are deposited in the lungs, internalized by epithelial and immune cells through endocytic and phagocytic pathways, and ultimately removed via mucociliary transport or immune-mediated mechanisms. Microplastics occupy areas of the nucleus typically occupied by DNA, induce cell cycle arrest and apoptosis in monocytes and macrophages, and upregulate the expression of inflammation-related genes.

MPs can, at times, induce phagocytosis by mimicking bacterial pathogens and entering targeted phagocytes (Popa and Tabaran, 2025). Carboxylated polystyrene (PS) MPs (40 and 200 nm) were shown to enter HeLa, A549, and 1321N1 cells via energy-dependent mechanisms. These mechanisms vary across cell types, as shown in Fig. 6.1. For example, actin depolymerization reduces uptake into HeLa and 1321N1 cells, and disrupting microtubules (with genistein) has a greater effect on A549 cells (Dos Santos et al., 2011). The phagocytic efficiency of immune cells also varies depending on the tissue of origin and the type of MP. For example, in one study, PE MPs were taken up more than PS MPs (Abihssira-García et al., 2020). PS MPs (200 nm) present in the cytoplasm of RAW264.7 macrophages accumulate over time, impairing lysosomal acidification and function (Florance et al., 2021). Compared with neutrally or negatively charged MPs, positively charged MPs gain easier access through the cell membrane (Panariti et al., 2012). Neutrophil extracellular traps can trap nanoparticles, forming macroaggregates that facilitate phagocytosis by neutrophils or macrophages; similar mechanisms may apply to MPs; however, this has not been extensively studied (Bilyy et al., 2020). The immune response begins with innate immune system antigen-presenting cells (APCs) that capture and process pathogens, presenting antigens to adaptive immune cells. T cells differentiate into effector T cells to orchestrate immune responses, and B cells produce antibodies specific to antigens. This process can apply to the gut; however, the intestinal immune system has unique mechanisms for generating responses to ingested antigens.

6.3.2 Immunological response generation

In the GIT, almost all immune reactions occur in the GALT. Among its unique structures, Peyer's patches in the ileum play a key role in initiating an immune response. These patches, as depicted in Fig. 6.1, serve as the center points for immune activation. Microfold (M) cells continuously monitor the luminal contents, collect them, and transfer them to Peyer's patches, where dendritic cells take up the antigens and can present them, after processing, to naïve T and B lymphocytes. Once a naïve T cell recognizes its target antigen, it differentiates into effector T cells that serve as critical mediators of the immune response. Effector cells are then released back into the systemic circulation, where they migrate to mucosal tissues to serve their immune function. The anatomical structure and immunological function of Peyer's patches in the intestinal mucosa are depicted in Fig. 6.1, which emphasizes their importance in immune surveillance and antigen sampling in the gut.

6.3.3 T-cell polarization

Effector T cells differentiate into one of several subtypes, including T helper 1 (Th1), T helper 2 (Th2), and T helper 17 (Th17), and induce regulatory T (Treg) cells. In general, Th1 and Th17 cells have proinflammatory effects, whereas Tregs have antiinflammatory effects. The pathway for the polarization of a naïve T cell is governed by the costimulatory signals (often cytokines) present in its microenvironment. Naïve T cells that encounter high levels of proinflammatory cytokines tend to differentiate into proinflammatory Th subsets, with IL-6 and TGF-β inducing their differentiation into Th17 cells. Conversely, naïve T cells are more likely to differentiate into Tregs in the presence of low levels of proinflammatory cytokines, along with all-trans retinoic acid and TGF-β. The phenotypic plasticity of T cells allows switching between phenotypes; for example, induced Tregs can be switched to Th17 cells by IL-1 and IL-6 or to Th1 cells by IL-12. Similarly, Th17 cells can switch to a Th1 phenotype in the presence of IL-12. For proper immune system function, maintaining the balance of these Th subsets is essential, as states of imbalance are associated with autoimmune and inflammatory disease processes (Noack and Miossec, 2014).

6.3.4 Cytokines and immune signaling

Cytokines regulate T-cell polarization, but they also regulate broader immune signaling. While many stressed cell types can produce cytokines, the major source of many of the more common cytokines is lymphocytes, specifically T cells. The following are the major cytokines, according to their characteristics: TNF-α is a strong proinflammatory mediator that can exist in either membrane-bound or soluble forms (Neurath, 2014) and is certainly the key activator of effector T cells and macrophages in the adaptive immune response. IL-1 promotes the activation of T cells and macrophages. IL-6 promotes the proinflammatory activity of APCs and T cells and increases cytokine production (Neurath, 2014). IL-8 (CXCL8) has an extremely chemotactic effect on neutrophils.

6.4 Effects of plastic particles on immunity

Whether ingested MPs and NPs can stimulate immune responses in humans remains unclear, as there has been limited direct research. However, studies on animals and cell lines can provide some context. In blood clots, PS particles (500 nm and 30 μm) reduce the expression of the proinflammatory markers Tumor Necrosis Factor Receptor-Associated Factor 6 (TRAF6), Stands for Inhibitory-κB Kinase Alpha (IKKα), and Nuclear Factor kappa-light-chain-enhancer of activated B cells (NF-κB) (Tang et al., 2020). These findings suggest that the immune response was blunted. However, Huang et al. reported that guppies exposed for 28 days to PS MPs of 32–40 μm presented increases in TNF-α, IFN-γ, TLR4, and IL-6 at both 100 μg/L and 1000 μg/L, with a dose-dependent increase in IFN-γ, TLR4, and IL-6 (Huang et al., 2020). In zebrafish exposed to PS particles (100, 5, or 200 μm), TNF-α was not significantly altered. Li et al. utilized a murine model and administered doses of 10–150 μm PE particles at 6, 60, or 600 μg for up to 12 weeks; all the groups presented a twofold increase in IL-1α. In the 6 μg group at the end of 12 weeks, the level of IL-6 increased 3.2-fold after dosing, and the level of IL-2—which is associated with Treg induction—was reduced by half. Lehner et al., using a 3D intestinal barrier model (Caco-2, HT29-MTX, and cocultures with macrophages and dendritic cells), reported that exposure to tire-derived MPs (50–500 μm) did not result in significant differences in TNF-α, IL-8, or IL-1β levels. However, IL-8 levels increased to a nonsignificant extent (Lehner et al., 2019). Approximately 30% of airborne exposure to MPs originates from tire particles (Moore, 2008). Han et al. reported that smaller PVC particles (25–75 μm) resulted in greater secretion of IL-6 and TNF-α than did larger particles (75–200 μm), with TNF-α reaching significance. They also noted significant differences in IL-6 responses to smaller acrylonitrile–butadiene–styrene MPs; however, TNF-α responses did not differ with respect to particle size or dose (Han et al., 2020). Overall, most studies have shown that MPs induce a proinflammatory response, although variability in particle size, type, and dose has been observed (10, 18–20). Smaller particles appear to induce a stronger proinflammatory response characterized by cytokine release, suggesting that particle size may play a role in immune activation.

6.4.1 Size matters: NPs and immune modulation

Unlike MPs, NPs, which are defined as plastic particles <1 μm in size, can modulate immune responses. Forte et al. reported that particle size is an important determinant of immune activation. The authors examined PS NP samples of 44 nm (NP44) and 100 nm (NP100) in gastric adenocarcinoma cells (Forte et al., 2016). While NP100 exposure did not result in apparent changes in pro- or anti-inflammatory gene expression, NP44 caused an approximately sevenfold increase in IL-6 and IL-8 mRNA expression (Forte et al., 2016). Similarly, Brown et al. reported greatly increased neutrophil influx in rat lungs in response to 64-nm PS NPs compared with that in response to larger particles of 202 and 535 nm (Brown et al., 2001). A549 cells exposed to 64-nm particles presented significant IL-8 mRNA upregulation and oxidative stress in the form of ROS production, neither of which was observed in A549 cells exposed to larger particles (202 or 535 nm). Although the gastric adenocarcinoma cell line and A549 cells do not constitute the intestinal lining, these results demonstrate the high reactivity of human cells to NPs, with NP particles as small as 44 nm.

6.4.2 Effects of digestive processing on NPs

Most experimental work investigating NPs has examined pristine (virgin) plastic particles; however, human digestion can alter their physicochemical properties (Liu et al., 2020). Liu et al. reported that NPs that underwent an in vitro digestion process developed a corona layer composed of digestive enzymes and complement proteins on their surface. Critically, after digestion, 100-nm spherical PS nanoparticles formed larger particles averaging 440 nm due to reduced electrostatic repulsion (Liu et al., 2020). Caco-2 cell models revealed that the digestion of 100-nm (now 440-nm) particles led to significant increases in IL-8 and CCL2 levels, similar to those observed with undigested 100-nm particles, and demonstrated the need to evaluate not only the virgin forms of plastics but also their altered forms.

6.4.3 Immunotoxicity of MPs

The cellular uptake of MPs is immunotoxic, mainly through oxidative stress via ROS production and the activation of danger-associated molecular patterns (DAMPs) (AshaRani et al., 2009). ROS are thought to arise from two principal sources: (1) damage to mitochondrial electron transport chains and (2) oxidative bursts that originate from NADPH oxidases as the cell internalizes the MP (Bedard and Krause, 2007). DAMPs are well known to activate toll-like receptors (TLRs), which initiate inflammatory cascades in immune cells, as shown in Fig. 6.2. Studies examining NH_2-functionalized PS microplastics (NH_2-PS MPs) have shown that they can disrupt metabolic pathways in aquatic organisms. For example, in the freshwater cyanobacteria examined in this study, 48 h of exposure led to changes in arginine

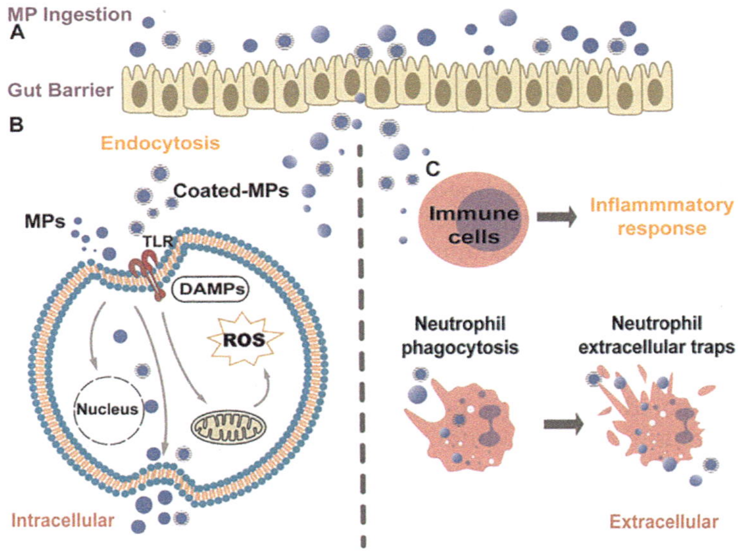

FIG. 6.2 Schematic of potential biological pathways associated with the cytotoxicity and immunotoxicity of microplastics (MPs). (A) After ingestion, MPs may attach to the plasma membrane and penetrate the intestinal barrier. (B) After endocytosis, MPs can affect intracellular signaling cascades, cause mitochondrial dysfunction, and induce elevated levels of reactive oxygen species (ROS), leading to oxidative stress and apoptosis. After being lysed, MPs may be recovered from the extracellular space. When MPs coated with biomolecules enter cells, they can induce immune responses by interacting with pathogen recognition receptors (PRRs), including toll-like receptors (TLRs). (C) MPs taken up by macrophages and neutrophils can induce inflammation. In addition, excessive neutrophil uptake can lead to neutrophil death and the release of neutrophil extracellular traps (NETs), which may entangle MPs.

biosynthesis and glutathione metabolism by downregulating L-glutamic acid, L-aspartic acid, and L-arginine while simultaneously increasing L-glutamine. Metabolism resulting from these changes triggers the release of further O-phosphoethanolamine, thereby inducing oxidative stress and membrane injury (Feng et al., 2019). Even in the short term, this study revealed that metabolism is potentially relevant to human health.

6.5 Organ storage and metabolic effects

Experimental evidence (from animal studies) suggests that the metabolism of MPs is not uniform in terms of size. For example, 20-μm PS MPs tend to accumulate in the liver, whereas 5-μm PS MPs seem to exist more in the gut than in the kidneys (Deng et al., 2017). Mice exposed to MPs presented decreased ATP levels in liver mitochondria and abnormal lipid metabolism (Mowat, 2003). Exposure to 1000 μg/L of 0.5- and 50-μm MPs is associated with altered triglyceride levels, reduced gut microbiota diversity, decreased mucus secretion, and weight loss (Liu et al., 2019), all of which weaken mucosal immunity and increase the risk of infection.

6.5.1 Size preferences in phagocytosis and blood interaction

Rat peritoneal macrophages showed greater uptake of the 1100-nm particles than the 100-nm particles did, suggesting that phagocytes may prefer to evaluate particle sizes close to those of bacteria. In addition, the small limits of latex spheres (~300 nm) caused platelet aggregation, and this aggregation was inhibited via chelation with EDTA (Movat et al., 1965). These findings further suggest that the entry of MPs into open skin wounds or damaged blood vessels from sewing could accelerate tissue injury.

6.5.2 Gene regulation and carcinogenesis relationships

MP exposure upregulated the transcription of the immune response regulator gene, thereby increasing cell proliferation and decreasing apoptosis. In addition, MPs have been associated with stages of human thyroid and lung cancers by impacting the expression of oncogenic genes (Limonta et al., 2019). Overall, the results illustrate that MPs can elicit immunotoxicity. Nevertheless, the immune system is also responsive to their presence, and additional in vitro and in vivo evaluations of MP-associated immunotoxicity are important.

6.6 Conclusion

MPs and NPs represent pervasive and insidious environmental threats with increasingly evident implications for human health, particularly through their interaction with the immune system. The evidence presented in this review demonstrates that MPs can enter the human body via ingestion, inhalation, and dermal contact; accumulate in critical organs; and disrupt immune homeostasis through oxidative stress, inflammatory signaling, and impairment of the epithelial barrier. Their capacity to adsorb and transport other toxicants amplifies their immunotoxic potential, whereas their small size—especially at the nanoscale—enables deeper tissue penetration and greater biological reactivity. The resulting immune dysregulation ranges from altered cytokine profiles and T-cell polarization shifts to microbiota imbalance and chronic inflammatory states, with emerging links to carcinogenesis. Despite mounting experimental data from animal and in vitro models, human-specific mechanistic pathways remain underexplored. Addressing these knowledge gaps will require integrative, multidisciplinary research to quantify real-world exposure, assess long-term effects, and develop effective mitigation strategies. Ultimately, understanding and curbing MP-related immune impacts is essential not only for safeguarding individual health but also for preserving ecosystem integrity in an era of escalating plastic pollution.

Acknowledgement

This work was supported by the Deanship of Scientific Research, Vice Presidency for Graduate Studies and Scientific Research, King Faisal University, Saudi Arabia **KFU263610.**

References

Abbasi, S., Turner, A., Sharifi, R., Nematollahi, M.J., Keshavarzifard, M., Moghtaderi, T., 2022. Microplastics in the school classrooms of Shiraz, Iran. Build. Environ. 03601323. 207, 108562. doi:10.1016/j.buildenv.2021.108562.

Abihssira-García, I.S., Park, Y., Kiron, V., Olsvik, P.A., 2020. Fluorescent microplastic uptake by immune cells of atlantic Salmon (Salmo salar L.). Front. Environ. Sci. 2296665X. 8. Available from: https://journal.frontiersin.org/journal/environmental-science. doi:10.3389/fenvs.2020.560206.

AshaRani, P.V., Mun, G.L.K., Hande, M.P., Valiyaveettil, S., 2009. Cytotoxicity and genotoxicity of silver nanoparticles in human cells. ACS Nano. 1936-0851. 3 (2), 279–290. doi:10.1021/nn800596w.

Bedard, K., Krause, K.H., 2007. The NOX family of ROS-generating NADPH oxidases: physiology and pathophysiology. Physiol. Rev. 15221210. 87 (1), 245–313. Available from: http://physrev.physiology.org/cgi/reprint/87/1/245.pdf. doi:10.1152/physrev.00044.2005.

Belzagui, F., Buscio, V., Gutiérrez-Bouzán, C., Vilaseca, M., 2021. Cigarette butts as a microfiber source with a microplastic level of concern. Sci. Total Environ. 00489697. 762, 144165. doi:10.1016/j.scitotenv.2020.144165.

Bilyy, R., Bila, G., Vishchur, O., Vovk, V., Herrmann, M., 2020. Neutrophils as main players of immune response towards nondegradable nanoparticles. Nanomaterials. 2079-4991. 10 (7), 1273. doi:10.3390/nano10071273.

Breschi, A., Gingeras, T.R., Guigó, R., 2017. Comparative transcriptomics in human and mouse. Nat. Rev. Genet. 14710064. 18 (7), 425–440. Available from: http://www.nature.com/reviews/genetics. doi:10.1038/nrg.2017.19.

Brown, D.M., Wilson, M.R., MacNee, W., Stone, V., Donaldson, K., 2001. Size-dependent proinflammatory effects of ultrafine polystyrene particles: a role for surface area and oxidative stress in the enhanced activity of ultrafines. Toxicol. Appl. Pharmacol. 0041008X. 175 (3), 191–199. doi:10.1006/taap.2001.9240.

Celebi Sözener, Z., Cevhertas, L., Nadeau, K., Akdis, M., Akdis, C.A., 2020. Environmental factors in epithelial barrier dysfunction. J. Allergy Clin. Immunol. 10976825. 145 (6), 1517–1528. Available from: http://www.elsevier.com/inca/publications/store/6/2/3/3/6/8/index.htt. doi:10.1016/j.jaci.2020.04.024.

Chen, Q., Gao, J., Yu, H., Su, H., Yang, Y., Cao, Y., Zhang, Q., Ren, Y., Hollert, H., Shi, H., Chen, C., Liu, H., 2022. An emerging role of microplastics in the etiology of lung ground glass nodules. Environ. Sci. Eur. 2190-4707. 34 (1). doi:10.1186/s12302-022-00605-3.

Deng, Y., Zhang, Y., Lemos, B., Ren, H., 2017. Tissue accumulation of microplastics in mice and biomarker responses suggest widespread health risks of exposure. Sci. Rep. 20452322. 7. Available from: www.nature.com/srep/index.html. doi:10.1038/srep46687.

Domenech, J., Marcos, R., 2021. Pathways of human exposure to microplastics, and estimation of the total burden. Curr. Opin. Food Sci. 22147993. 39, 144–151. doi:10.1016/j.cofs.2021.01.004.

Dos Santos, T., Varela, J., Lynch, I., Salvati, A., Dawson, K.A., Schnur, J.M., 2011. Effects of transport inhibitors on the cellular uptake of carboxylated polystyrene nanoparticles in different cell lines. PLoS ONE. 1932-6203. 6 (9), e24438. doi:10.1371/journal.pone.0024438.

Feng, L.J., Li, J.W., Xu, E.G., Sun, X.D., Zhu, F.P., Ding, Z., Tian, H., Dong, S.S., Xia, P.F., Yuan, X.Z., 2019. Short-term exposure to positively charged polystyrene nanoparticles causes oxidative stress and membrane destruction in cyanobacteria. Environ. Sci. Nano. 20518161. 6 (10), 3072–3079. Available from: http://pubs.rsc.org/en/journals/journal/en. doi:10.1039/c9en00807a.

Florance, I., Ramasubbu, S., Mukherjee, A., Chandrasekaran, N., 2021. Polystyrene nanoplastics dysregulate lipid metabolism in murine macrophages in vitro. Toxicology. 0300483X. 458, 152850. doi:10.1016/j.tox.2021.152850.

Forte, M., Iachetta, G., Tussellino, M., Carotenuto, R., Prisco, M., De Falco, M., Laforgia, V., Valiante, S., 2016. Polystyrene nanoparticles internalization in human gastric adenocarcinoma cells. Toxicol. Vitro. 08872333. 31, 126–136. doi:10.1016/j.tiv.2015.11.006.

Fournier, E., Etienne-Mesmin, L., Grootaert, C., Jelsbak, L., Syberg, K., Blanquet-Diot, S., Mercier-Bonin, M., 2021. Microplastics in the human digestive environment: a focus on the potential and challenges facing in vitro gut model development. J. Hazard. Mater. 03043894. 415, 125632. doi:10.1016/j.jhazmat.2021.125632.

Fournier, S.B., D'Errico, J.N., Adler, D.S., Kollontzi, S., Goedken, M.J., Fabris, L., Yurkow, E.J., Stapleton, P.A., 2020. Nanopolystyrene translocation and fetal deposition after acute lung exposure during late-stage pregnancy. Part. Fibre Toxicol. 17438977. 17 (1). Available from: http://www.particleandfibretoxicology.com/. doi:10.1186/s12989-020-00385-9.

Han, S., Bang, J., Choi, D., Hwang, J., Kim, T., Yoogyeong, O., Hwang, Y., Choi, J., Hong, J., 2020. Surface pattern analysis of microplastics and their impact on human-derived cells. ACS Appl. Polym. Mater. 2637-6105. 2 (11), 4541–4550. doi:10.1021/acsapm.0c00645.

Heymann, F., Tacke, F., 2016. Immunology in the liver-from homeostasis to disease. Nat. Rev. Gastroenterol. Hepatol. 17595053. 13 (2), 88–110. Available from: http://www.nature.com/nrgastro/index.html. doi:10.1038/nrgastro.2015.200.

Huang, J.N., Wen, B., Zhu, J.G., Zhang, Y.S., Gao, J.Z., Chen, Z.Z., 2020. Exposure to microplastics impairs digestive performance, stimulates immune response and induces microbiota dysbiosis in the gut of juvenile guppy (Poecilia reticulata). Sci. Total Environ. 18791026. 733. Available from: www.elsevier.com/locate/scitotenv. doi:10.1016/j.scitotenv.2020.138929.

Humphries, J.D., Chastney, M.R., Askari, J.A., Humphries, M.J., 2019. Signal transduction via integrin adhesion complexes. Curr. Opin. Cell Biol. 18790410. 56, 14–21. Available from: http://www.elsevier.com/locate/ceb. doi:10.1016/j.ceb.2018.08.004.

Kubes, P., Jenne, C., 2018. Immune responses in the liver. Annu. Rev. Immunol. 15453278. 36, 247–277. Available from: http://www.annualreviews.org/journal/immunol. doi:10.1146/annurev-immunol-051116-052415.

Lee, S., Park, S.K., Park, H., Lee, W., Lee, J.H., Hong, Y.C., Ha, M., Kim, Y., Lee, B.E., Ha, E., 2021. Joint association of prenatal bisphenol-A and phthalates exposure with risk of atopic dermatitis in 6-month-old infants. Sci. Total Environ. 18791026. 789. Available from: www.elsevier.com/locate/scitotenv. doi:10.1016/j.scitotenv.2021.147953.

Lehner, R., Weder, C., Petri-Fink, A., Rothen-Rutishauser, B., 2019. Emergence of nanoplastic in the environment and possible impact on human health. Environ. Sci. Technol. 0013-936X. 53 (4), 1748–1765. doi:10.1021/acs.est.8b05512.

Limonta, G., Mancia, A., Benkhalqui, A., Bertolucci, C., Abelli, L., Fossi, M.C., Panti, C., 2019. Microplastics induce transcriptional changes, immune response and behavioral alterations in adult zebrafish. Sci. Rep. 20452322. 9 (1). Available from: www.nature.com/srep/index.html. doi:10.1038/s41598-019-52292-5.

Liu, S., Wu, X., Gu, W., Yu, J., Wu, B., 2020. Influence of the digestive process on intestinal toxicity of polystyrene microplastics as determined by in vitro Caco-2 models. Chemosphere. 00456535. 256, 127204. doi:10.1016/j.chemosphere.2020.127204.

Liu, Z., Yu, P., Cai, M., Wu, D., Zhang, M., Chen, M., Zhao, Y., 2019. Effects of microplastics on the innate immunity and intestinal microflora of juvenile Eriocheir sinensis. Sci. Total Environ. 18791026. 685, 836–846. Available from: www.elsevier.com/locate/scitotenv. doi:10.1016/j.scitotenv.2019.06.265.

Llorca, M., Farré, M., 2021. Current insights into potential effects of micro-nanoplastics on human health by in-vitro tests. Front. Toxicol. 2673-3080. 3. doi:10.3389/ftox.2021.752140.

Mani, I., Pandey, K.N., 2019. Emerging concepts of receptor endocytosis and concurrent intracellular signaling: mechanisms of guanylyl cyclase/natriuretic peptide receptor-A activation and trafficking. Cell. Signal. 18733913. 60, 17–30. Available from: www.elsevier.com/locate/cellsig. doi:10.1016/j.cellsig.2019.03.022.

Mir, M.A., Banik, B.K., 2025. Heterocyclic phytochemicals as anticancer agents. Curr. Top. Med. Chem. 18734294. 25 (5), 533–553. Available from: http://www.eurekaselect.com/615/journal/current-topics-medicinal-chemistry. doi:10.2174/0115680266314693240914070250.

Moore, C.J., 2008. Synthetic polymers in the marine environment: a rapidly increasing, long-term threat. Environ. Res. 10960953. 108 (2), 131–139. doi:10.1016/j.envres.2008.07.025.

Movat, H.Z., Weiser, W.J., Glynn, M.F., Mustard, J.F., 1965. Platelet phagocytosis and aggregation. J. Cell Biol. 00219525. 27 (3), 531–543. doi:10.1083/jcb.27.3.531.

Mowat, A.M., 2003. Anatomical basis of tolerance and immunity to intestinal antigens. Nat. Rev. Immunol. 1474-1733. 3 (4), 331–341. doi:10.1038/nri1057.

Neurath, M.F., 2014. Cytokines in inflammatory bowel disease. Nat. Rev. Immunol. 14741741. 14 (5), 329–342. doi:10.1038/nri3661.

Noack, M., Miossec, P., 2014. Th17 and regulatory T cell balance in autoimmune and inflammatory diseases. Autoimmun. Rev. 15689972. 13 (6), 668–677. doi:10.1016/j.autrev.2013.12.004.

Panariti, A., Miserocchi, G., Rivolta, I., 2012. The effect of nanoparticle uptake on cellular behavior: disrupting or enabling functions? Nanotechnol. Sci. Appl. 11778903. 5 (1), 87–100. Available from: http://www.dovepress.com/getfile.php?fileID=13894. doi:10.2147/NSA.S25515Italy.

Popa, R.P., Tabaran, A.F., 2025. A systematic review of the toxicokinetics of micro- and nanoplastics in mammals following digestive exposure. Appl. Sci. 2076-3417. 15 (11), 6135. doi:10.3390/app15116135.

Ragusa, A., Svelato, A., Santacroce, C., Catalano, P., Notarstefano, V., Carnevali, O., Papa, F., Rongioletti, M.C.A., Baiocco, F., Draghi, S., D'Amore, E., Rinaldo, D., Matta, M., Giorgini, E., 2021. Plasticenta: first evidence of microplastics in human placenta. Environ. Int. 01604120. 146, 106274. doi:10.1016/j.envint.2020.106274.

Rawat, P., Chauhan, V., Chaudhary, J., Singh, C., Rawat, P., Yashaswi Singh, M.A.M., Chauhan, N., 2025. A novel approach to electroactivation: designing an electro activated reactor to constrain the growth of MDR Salmonella strains by sodium caprylate solutions. The Microbe. 29501946. 7, 100303. doi:10.1016/j.microb.2025.100303.

Shi, W., Cao, Y., Chai, X., Zhao, Q., Geng, Y., Liu, D., Tian, S., 2022. Potential health risks of the interaction of microplastics and lung surfactant. J. Hazard. Mater. 03043894. 429, 128109. doi:10.1016/j.jhazmat.2021.128109.

Shruti, V.C., Kutralam-Muniasamy, G., Pérez-Guevara, F., Priyadarsi, D.R., Elizalde-Martínez, I., 2023. First evidence of microplastic contamination in ready-to-use packaged food ice cubes. Environ. Pollut. 02697491. 318, 120905. doi:10.1016/j.envpol.2022.120905.

Steer, M., Cole, M., Thompson, R.C., Lindeque, P.K., 2017. Microplastic ingestion in fish larvae in the western English Channel. Environ. Pollut. 18736424. 226, 250–259. Available from: www.elsevier.com/inca/publications/store/4/0/5/8/5/6. doi:10.1016/j.envpol.2017.03.062.

Sun, X.D., Yuan, X.Z., Jia, Y., Feng, L.J., Zhu, F.P., Dong, S.S., Liu, J., Kong, X., Tian, H., Duan, J.L., Ding, Z., Wang, S. G., Xing, B., 2020. Differentially charged nanoplastics demonstrate distinct accumulation in Arabidopsis thaliana. Nat. Nanotechnol. 17483395. 15 (9), 755–760. Available from: http://www.nature.com/nnano/index.html. doi:10.1038/s41565-020-0707-4.

Tang, Y., Rong, J., Guan, X., Zha, S., Shi, W., Han, Y., Xueying, D., Fangzhu, W., Huang, W., Liu, G., 2020. Immunotoxicity of microplastics and two persistent organic pollutants alone or in combination to a bivalve species. Environ. Pollut. 02697491. 258, 113845. doi:10.1016/j.envpol.2019.113845.

Von Moos, N., Burkhardt-Holm, P., Köhler, A., 2012. Uptake and effects of microplastics on cells and tissue of the blue mussel Mytilus edulis L. after an experimental exposure. Environ. Sci. Technol. 15205851. 46 (20), 11327–11335. doi:10.1021/es302332w.

Wang, X., Zheng, H., Zhao, J., Luo, X., Wang, Z., Xing, B., 2020. Photodegradation elevated the toxicity of polystyrene microplastics to grouper (Epinephelus moara) through disrupting hepatic lipid homeostasis. Environ. Sci. Technol. 15205851. 54 (10), 6202–6212. Available from: http://pubs.acs.org/journal/esthag. doi:10.1021/acs.est.9b07016.

Wright, S.L., Kelly, F.J., 2017. Plastic and human health: a micro issue? Environ. Sci. Technol. 15205851. 51 (12), 6634–6647. Available from: http://pubs.acs.org/journal/esthag. doi:10.1021/acs.est.7b00423.

Xu, M., Halimu, G., Zhang, Q., Song, Y., Xuanhe, F., Yongqiang, L., Yansheng, L., Zhang, H., 2019. Internalization and toxicity: a preliminary study of effects of nanoplastic particles on human lung epithelial cell. Sci. Total Environ. 00489697. 694, 133794. doi:10.1016/j.scitotenv.2019.133794.

Yee, M.S.L., Hii, L.W., Looi, C.K., Lim, W.M., Wong, S.F., Kok, Y.Y., Tan, B.K., Wong, C.Y., Leong, C.O., 2021. Impact of microplastics and nanoplastics on human health. Nanomaterials. 20794991. 11 (2), 1–23. Available from: https://www.mdpi.com/2079-4991/11/2/496/pdf. doi:10.3390/nano11020496.

Yong, C.Q.Y., Valiyaveettil, S., Tang, B.L., 2020. Toxicity of microplastics and nanoplastics in Mammalian systems. Int. J. Environ. Res. Public Health. 16604601. 17 (5). Available from: https://www.mdpi.com/1660-4601/17/5/1509/pdf. doi:10.3390/ijerph17051509.

Zhao, C., Wang, Y., Su, Z., Pu, W., Niu, M., Song, S., Wei, L., Ding, Y., Lizhi, X., Tian, M., Wang, H., 2020. Respiratory exposure to PM2.5 soluble extract disrupts mucosal barrier function and promotes the development of experimental asthma. Sci. Total Environ. 00489697. 730, 139145. doi:10.1016/j.scitotenv.2020.139145.

Zhao, S., Zhu, L., Li, D., 2016. Microscopic anthropogenic litter in terrestrial birds from Shanghai, China: not only plastics but also natural fibers. Sci. Total Environ. 18791026. 550, 1110–1115. Available from: www.elsevier.com/locate/scitotenv. doi:10.1016/j.scitotenv.2016.01.112.

Zhdanov, V.P., Cho, N.J., 2016. Kinetics of the formation of a protein corona around nanoparticles. Math. Biosci. 18793134. 282, 82–90. Available from: www.elsevier.com/locate/mbs. doi:10.1016/j.mbs.2016.09.018.

CHAPTER

7

Microplastics and neural impacts

Valentina Iezzi[a], *Roberta Maria Ferri*[b], *and Elisabetta Benedetti*[b]

[a]Department of Clinical Medicine, Public Health, Life and Environmental Sciences, University of L'Aquila, L'Aquila, Italy
[b]Department of Life, Health and Environmental Sciences, University of L'Aquila, L'Aquila, Italy

7.1 Introduction

The ubiquitous distribution of microplastics and nanoplastics (MNPs) in every ecological niche has evolved from an environmental concern to a critical challenge for human health. While their presence in marine and terrestrial ecosystems is welldocumented, emerging evidence highlights a far more critical concern: their ability to cross biological barriers in living organisms, including humans. Among these, the central nervous system (CNS) is now recognized as a primary target for MNPsinduced toxicity. This chapter provides a comprehensive evaluation of the complex relationship between plastic particulate exposure and neural dysfunction. This chapter begins by dissecting the physicochemical properties of MNPs, including their polymer composition, size, and surface charge, which determine their bioavailability and capacity to act as vectors for organic pollutants. The focus then turns to the routes of neural exposure, exploring how these particles translocate across the gut–brain axis, penetrate the blood–brain barrier (BBB), or utilize direct nasal pathways via olfactory transport. At the cellular level, we elucidate the neuroinflammatory cascade initiated by MNPs interacting with microglia and astrocytes, which drives a positive feedback loop of oxidative stress and cellular damage.

Furthermore, this chapter investigates the molecular and functional implications of exposure, specifically the disruption of essential neurotransmitter systems and the impairment of synaptic plasticity. Combining insights from high resolution model organisms with emerging epidemiological evidence, we ultimately address the potential health implications of accelerating the progression of neurodevelopmental and neurodegenerative disorders, such as Alzheimer's disease (AD) and Parkinson's disease (PD). Through this analysis, this chapter

Microplastics in Animal and Human Physiology
https://doi.org/10.1016/B978-0-443-45635-0.00004-5

aims to underscore the urgent need for a deeper understanding of the neurotoxicological risks posed by our global plastic legacy.

7.2 Physicochemical properties of plastic particles

7.2.1 Polymer composition

Microplastics (MPs) are a heterogeneous set of particles derived from the fragmentation of various polymers with variable compositions and morphologies. The most widespread materials are polyethylene (PE), polypropylene (PP), polyethylene (PE), polystyrene (PS), and polyvinyl chloride (PVC), which together account for more than half of the global plastic production (Pelegrini et al., 2023; Salthammer, 2022). Scientific attention has focused mainly on PE, the most common polymer, which is characterized by linear $–CH_2–CH_2–$ chains, and PS, a vinyl polymer derived from styrene, which is hydrophobic and of high molecular weight. It is brittle, fragments easily into MNPs, yet persists in the environment for long periods due to its resistance to degradation (Cai et al., 2023; Singh et al., 2025).

7.2.2 Size, shape, and surface characteristics

MPs exhibit high variability in size, shape, and surface properties, which influence their environmental behavior and biological interactions. They are defined as synthetic polymer particles smaller than 5 mm (fibers, granules, fragments). Nanoplastics (NPs) (<1 μm), particularly those ~50 nm in size, have attracted increasing attention because of their high bioavailability, as they can readily cross biological barriers. Owing to their small size, large specific surface area, and hydrophobic nature, MPs can adsorb organic contaminants and act as vectors for pollutant transport.

PE typically occurs as fibers or films and is hydrophobic and smooth. However, under environmental aging processes (UV radiation, oxygen exposure, and physical abrasion), it undergoes photoinduced oxidation, leading to increased surface roughness and greater adsorption capacity for contaminants. PS is more brittle, initially forming spherical particles that fragment into irregular shapes. The presence of the aromatic phenyl group renders its surface more reactive, increasing its tendency to adsorb contaminants and interact with biomolecules, thus promoting the formation of a biomolecular corona in contact with biological fluids (Kihara et al., 2021).

7.2.3 Influence on biological interactions and toxicity

MNP pollution represents a global threat to ecosystems and their inhabitants, including humans, through exposure via food, water, air, and skin contact. Among the main toxicological processes are oxidative stress and inflammatory responses, which compromise cellular homeostasis and lead to tissue damage. These tiny particulates can be absorbed by various organisms, transported into the food chain, and consequently trigger adverse effects at the cellular and individual levels. Moreover, NPs are more likely to penetrate biological membranes (e.g., blood brain (Prüst et al., 2020), blood testis (Wei et al., 2022), and blood–placenta), directly affecting cellular function and consequently leading to systemic consequences (Jeong et al., 2022). Compelling evidence has also demonstrated increased toxicity of NPs, including increased mucus secretion and diverse perturbations in cell populations. In addition to their effects on human health, MNPs represent a serious threat to marine ecosystems (Guzzetti et al., 2018). They can be ingested by a wide range of organisms, from plankton and fish to large marine mammals, causing obstructions, mechanical tissue damage, and reduced feeding capacity. Once ingested, MNPs can accumulate in tissues and release toxic substances, including chemical additives and persistent organic pollutants, with adverse effects on the growth, reproduction, and survival of marine species (Alimba and Faggio, 2019).

7.3 Routes of CNS exposure

Prolonged exposure to MNPs can lead to a range of health issues in humans, particularly neurotoxicity, with consequences that include neurological developmental abnormalities, neuronal death, neuroinflammation, and neurodegenerative diseases. Model organisms and cell lines provide powerful tools for investigating these effects and uncovering the underlying mechanisms. Evidence from animal studies, including rodent and fish models, indicates that both NPs and MPs can accumulate in the brain, demonstrating their ability to cross the BBB (Shan et al., 2022; Yang et al., 2022). Compared with 100 nm particles, 50 nm particles cross the BBB more efficiently, an effect that has been observed in zebrafish models (Hwang et al., 2022). Once in the brain, MNPs can exert neurotoxic effects via various cellular and molecular mechanisms. These include dysregulation of neurotransmitter systems, oxidative stress, activation of microglia and astrocytes, and induction of neuroinflammatory pathways. Collectively, these mechanisms compromise neuronal function, disrupt synaptic plasticity, and contribute to the cognitive deficits and behavioral abnormalities observed in exposed models, highlighting the significant risk posed by environmental plastic pollution to the CNS, potentially affecting neurological development and thereby increasing the risk of neurological disorders (Chen et al., 2023; Zhao et al., 2023; Li et al., 2024).

As ubiquitous environmental pollutants, MNPs can enter the human body via ingestion, inhalation, and dermal contact (Yadav et al., 2022), as shown in Fig. 7.1. Among these routes, ingestion is considered the primary pathway of exposure, although inhalation also contributes significantly, accounting for nearly half of the total human exposure (Wang et al., 2022).

7.3.1 Gastrointestinal absorption and the gut–brain axis

The intestine and the gut microbiota communicate bidirectionally with the brain, forming the gut–brain axis, which is essential for nutrition, immunity, and hormonal regulation. The intestinal barrier, which is composed of a single layer of epithelial cells and tight junctions (TJs), is dynamic and selectively permeable. Similarly, the brain is protected by the BBB. Environmental factors and stress can impair gut function, leading to systemic inflammation and increased BBB permeability, with adverse effects on the brain. This mechanism is implicated in neurodegenerative diseases, in which environmental pollutants (e.g., pesticides) act as crucial triggers. MNPs are emerging disruptors of the gut environment: they

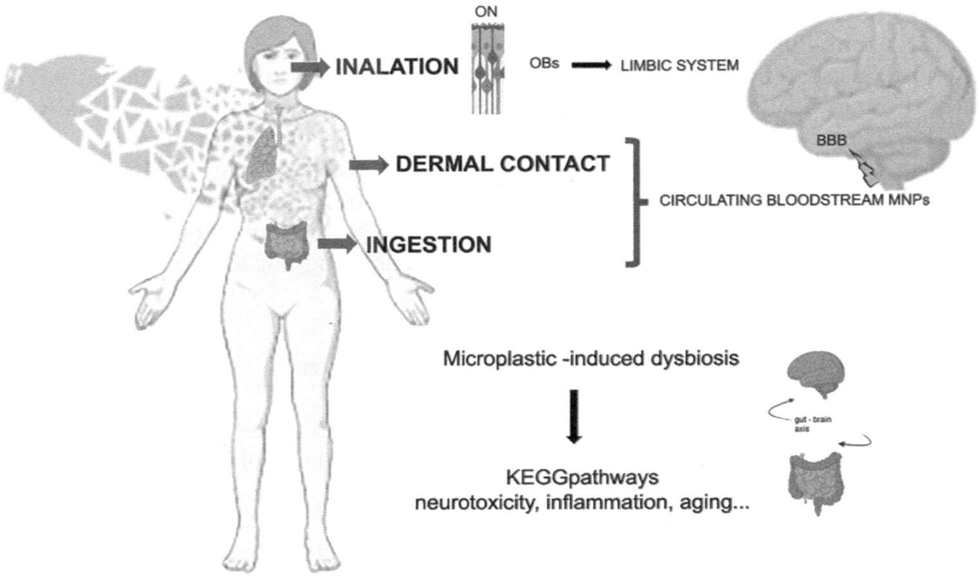

FIG. 7.1 Routes of CNS exposure to MNPs. Schematic of the three main exposure pathways: ingestion, inhalation, and dermal uptake, and their subsequent transport via olfactory neurons (ONs) or systemic circulation, allowing MNPs to cross the blood–brain barrier (BBB) and accumulate in the brain.

alter the microbiota (dysbiosis), activate the enteric immune system, and compromise metabolic, immune, and nervous system functions. MP induced dysbiosis modifies genes and metabolites, leading to alterations in Kyoto Encyclopedia of Genes and Genomes pathways related to metabolism, neurotoxicity, inflammation, aging, metabolic diseases, and cancer. Dysbiosis also increases intestinal permeability.

Studies using fluorescent MPs have shown that they can cross the intestinal barrier and accumulate in tissues in a size and shape dependent manner. Transport appears to occur both paracellularly and via epithelial cell internalization (Wu et al., 2019). The presence of MPs induces structural changes in the intestinal epithelium and mucosa, with inflammation characterized by edema, vacuolization, villus damage, enterocyte disruption, increased crypt depth, ciliary defects, and vascular proliferation from the duodenum to the colon.

7.3.2 BBB disruption and penetration

Owing to their charge, size, and physicochemical properties, MNPs can interact with biomolecules such that they can cross the BBB. Clinical evidence has shown that MNPs enter the circulatory system, deposit in the heart and blood vessels (Yang et al., 2022), and accumulate in the liver, kidneys, spleen, joints, and reproductive organs. While circulating in the bloodstream, MNPs reach the BBB, which is already vulnerable to pollutants and chronic inflammation. In vitro models demonstrated that MNP exposure reduced the expression of TJ proteins, including zonulin and occludin (Shan et al., 2022). In vivo, exposure increased

BBB permeability in the hippocampus, hypothalamus, and cortex of mice (Jin et al., 2022; Shan et al., 2022). MNPs up to 10 μm in size can penetrate the cortex, hippocampus, and cerebellum in a dose dependent manner after only 24 h of exposure. The transport of MNPs across the BBB, therefore, represents a significant risk for neuroinflammation and brain cellular damage. The BBB serves as the primary protective interface for neurons, safeguarding the brain from systemic toxins and maintaining the homeostasis necessary for proper neuronal function. By preventing circulating contaminants from entering the brain, the BBB represents a critical line of defense against neural damage. However, evidence suggests that the BBB is vulnerable to high particulate matter loads, including MNPs. PS NPs can cross the BBB and induce neurotoxic effects via multiple mechanisms. In vitro studies using human cerebral microvascular endothelial cells (hCMEC/D3), a widely accepted BBB model, have shown that PS NPs are internalized by endothelial cells, leading to the following:

- Increased production of reactive oxygen species (ROS)
- Activation of NF-κB (nuclear factor kappa-light-chain-enhancer of activated B cells) signaling
- Secretion of tumor necrosis factor alpha (TNF-α)

Moreover, PSNPs compromised TJ integrity, as evidenced by reduced transendothelial electrical resistance and downregulation of critical TJ proteins, including occludin and zonula occludens-1. Prolonged exposure induces necrotic cell death and direct endothelial damage, highlighting a strong link between oxidative stress, TJ disruption, and neurotoxicity (Teng et al., 2022; Shan et al., 2022). In particular, ROS production triggers necroptotic pathways mediated by receptor-interacting serine/threonine-protein kinase 1, receptor-interacting serine/threonine-protein kinase 3, and mixed lineage kinase-like protein, resulting in programmed necrotic cell death in endothelial cells and further impairment of BBB function. In developing zebrafish, PSNP exposure led to behavioral abnormalities and locomotor deficits that were not observed after MP exposure. Similarly, PS NPs have been detected in human dopaminergic neurons and developing neurospheres, where they induce gene expression changes and decrease cell viability at high concentrations, suggesting selective neuronal toxicity (Shan et al., 2022). Overall, current evidence demonstrates that PS nanoparticles pose a significant neurotoxic risk. The ability of MPs to disrupt the BBB, induce oxidative stress, activate microglia, and trigger necroptosis underscores the heightened vulnerability of the brain to nanoparticulate pollution, with considerably more pronounced effects than those observed with MPs (Shan et al., 2022).

7.3.3 Olfactory transport and direct nasal access

Inhalation represents one of the main routes of exposure to MNPs, along with ingestion and dermal contact (Prüst et al., 2020; Wright and Kelly, 2017). Once deposited in the respiratory tract, particles can interact with the nasal and pulmonary epithelium, promoting systemic translocation (Campanale et al., 2020). In particular, the nasal cavity constitutes a critical interface with the external environment and is already known as an entry point for airborne contaminants and environmental nanoparticles (Oberdörster et al., 2004; Maher et al., 2016). Olfactory neurons, located in the nasal mucosa, project their axons through the cribriform plate of the ethmoid bone to reach the olfactory bulbs (OBs), which are closely connected to olfactory cortex and several limbic structures, including the amygdala, hippocampus, and entorhinal cortex. This pathway, which has already been shown to facilitate the translocation of exogenous particles, has been associated with the accumulation of carbonaceous particulate matter at high concentrations within the OB (Maher et al., 2016; Calderón-Garcidueñas et al., 2008).

Given the ubiquity of MPs, it is plausible that they use the exact entry mechanism. In support of this hypothesis, a study conducted on 15 deceased individuals analyzed OB samples collected during autopsy and detected polymeric particles in eight cases. These findings suggest that the olfactory pathway may represent a relevant mechanism for the entry of MNPs into the human brain.

7.4 Cellular and immune responses in the brain

MNPs accumulate in the CNS, where they interact with key cells such as microglia and astrocytes. These interactions trigger immune and inflammatory responses, promote oxidative stress, and impair neuronal function. Understanding these mechanisms is crucial for assessing the impact of MNPs on brain health and cognitive processes.

7.4.1 Interaction with microglia and astrocytes

The internalization of NPs by microglia triggers the activation of the NF-κB transcription factor pathway. Usually, NF-κB is inactive by inhibitor of nuclear factor kappa B (IκB), but exposure to PS NP leads to IκB degradation, allowing NF-κB to translocate into the nucleus and activate proinflammatory genes (Oeckinghaus and Ghosh, 2009; Parkhurst et al., 2013). This creates a vicious cycle in which NPs activate microglia, promoting inflammation, and inflamed microglia, in turn, take up even more particles (Fig. 7.2).

This sustained activation induces inflammatory and phagocytic transcriptional programs characteristic of disease-associated microglia, with altered expression of genes such as *Spp1* (secreted phosphoprotein 1), *Cd9* (cluster of differentiation 9), and *lpl* (lipoprotein lipase), which are implicated in neuroinflammatory processes and neurodegenerative conditions. Activated microglia also release proinflammatory mediators, including TNF-α, interleukin-1 beta (IL-1β), interleukin-6 (IL-6), C3 (complement component 3), and lipocalin 2, which contribute to synaptic loss, neuronal death, and worsening neuroinflammatory stress.

This shift toward a proinflammatory phenotype is accompanied by elevated levels of proinflammatory cytokines (TNF-α, IL-1β, and IL-6) and chemokines such as C–X–C motif chemokine ligand 10 and monocyte chemoattractant protein-1. Studies have shown that microglia mount a robust functional response against MNPs in the brain; however, prolonged exposure ultimately leads to microglial cell death via pyroptosis and apoptosis (Jin et al., 2022). Importantly, microglial activation and subsequent immune alterations also occur secondary to MP induced intestinal dysbiosis (Fig. 7.2). Intestinal disruption triggers peripheral inflammation and elevated serum LPS levels, both of which compromise BBB integrity and induce neuroinflammation. Neuroinflammation represents a concerning consequence of

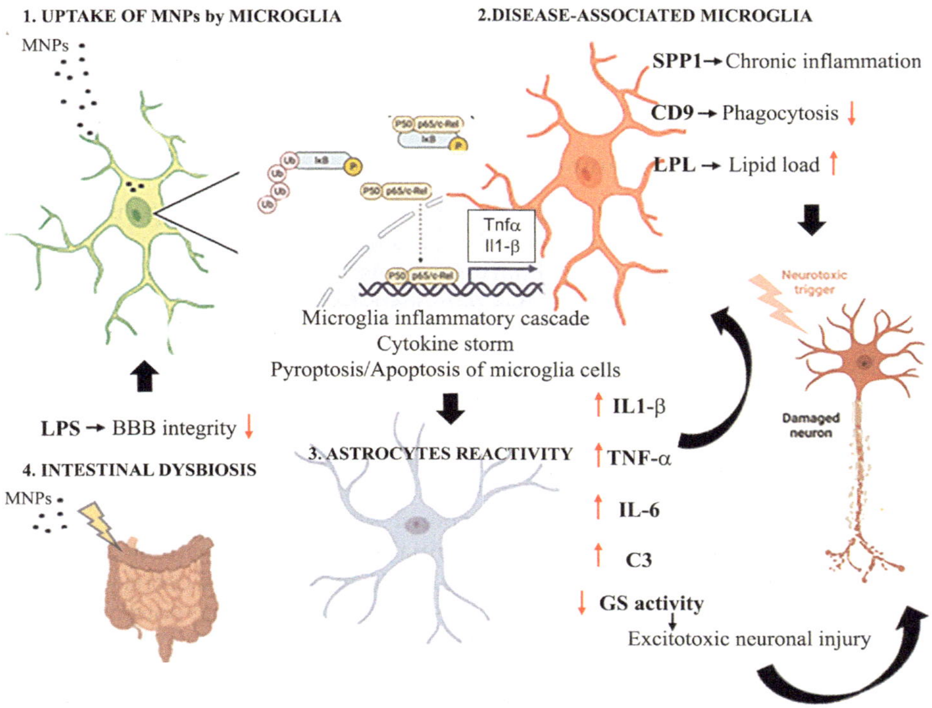

FIG. 7.2 Microglia–astrocyte crosstalk in MNP-induced neurotoxicity. Schematic representation of how MNPs activate microglia and astrocytes directly and indirectly, triggering nuclear factor kappa-light-chain-enhancer of activated B cells (NF-κB)-driven inflammatory signaling, cytokine storms, and glutamate toxicity, ultimately promoting synaptic impairment, neuronal damage, and neurodegenerative processes. Abbreviations: SPP1 (secreted phosphoprotein 1); CD9 (cluster of differentiation 9); LPL (lipoprotein lipase); GSH (glutathione); IL-1β (interleukin-1 beta); TNF-α (tumor necrosis factor alpha); IL-6 (interleukin-6); GS (glutamine synthetase); LPS (lipopolysaccharide); BBB (blood–brain barrier).

MP ingestion, as prolonged neuroinflammatory responses can lead to widespread alterations in brain structure and function.

Under physiological conditions, they maintain synaptic homeostasis by regulating glutamate uptake and recycling through the glutamate–glutamine cycle and by releasing trophic factors that support neuronal survival. However, exposure to NPs and the proinflammatory milieu generated by activated microglia drives astrocytes into a reactive state. Reactive astrocytes upregulate glial fibrillary acidic protein and secrete inflammatory mediators such as IL-1β, TNF-α, IL-6, and complement proteins (notably C3), which further amplify microglial activation and exacerbate synaptic damage (Liddelow et al., 2017; Guttenplan et al., 2021). In addition, oxidative stress induced by NPs can impair astrocytic glutamine synthetase (GS) activity, leading to inefficient detoxification of extracellular glutamate and promoting excitotoxic neuronal injury. Thus MP-induced neuroinflammation involves dynamic crosstalk between microglia and astrocytes: microglial activation triggers astrocytic reactivity, whereas reactive astrocytes release proinflammatory and neurotoxic signals that potentiate microglial dysfunction. This bidirectional interaction establishes a self-perpetuating cycle of inflammation, oxidative stress, and excitotoxicity, ultimately accelerating neuronal death and contributing to neurodegenerative processes.

7.4.2 Activation of neuroinflammatory pathways

The accumulation of MNPs within tissues promotes inflammatory responses and recruits immune cells, and the brain appears to exhibit a similar response. Microglia and brain-resident macrophages represent the primary sentinels against foreign particles that may bypass the BBB. Experimental evidence has demonstrated that microglia can phagocytose MPs and, upon their presence in brain tissue, initiate neuroinflammatory processes (Shan et al., 2022; Kwon et al., 2022). Moreover, oral exposure to MPs has been shown to induce microglial activation and polarization, characterized by a hypertrophic morphology (enlarged soma with reduced branching complexity) and by the upregulation of both M1- and M2-associated marker proteins (Kwon et al., 2022; Shan et al., 2022).

Moreover, PS NP exposure downregulates key neurotrophic genes, such as *Bdnf* (brain-derived neurotrophic factor), *Creb1* (cAMP response element-binding protein 1), and *Atf4* (activating transcription factor 4), which are essential for synaptic plasticity, neuronal activity, and hippocampus-dependent learning (Monteggia et al., 2004; Corona et al., 2018). In particular, reduced microglial BDNF may impair synaptic remodeling, potentially contributing to the cognitive deficits observed in exposed models (Parkhurst et al., 2013). Furthermore, additional inflammatory pathways are activated, as evidenced by increased iNOS (inducible nitric oxide synthase) mRNA and IL-1 levels, highlighting the engagement of multiple neuroinflammatory mechanisms.

7.4.3 Oxidative stress and cellular damage

Microglial inflammation is closely linked to oxidative stress. During immune activation, hydrogen peroxide and ROS, such as superoxide radicals and nitric oxide, are released, which amplify the inflammatory cascade and induce neuronal damage and demyelination. In murine models exposed to PS MPs (5 μm; 0.01–1 mg/day), histological alterations in the hippocampus, including neuronal disorganization and reduced Nissl bodies, were observed. Concurrently, increased markers of oxidative stress (ROS and malondialdehyde) and a dose-dependent decrease in glutathione (GSH)—the primary antioxidant—have been reported. MNPs cause cellular stress by disrupting mitochondrial function and altering energy metabolism, leading to excessive ROS production and mitochondrial structural damage (Gecegelen et al., 2025). ROS generation in response to plastic exposure has been observed in several cell types, including intestinal, hepatic, and neuronal cells, and is closely correlated with both dose and particle size (Das, 2023). In vitro exposure to MNPs induces mitochondrial damage and oxidative stress-mediated inflammatory responses. The antioxidant Nrf2 (nuclear factor erythroid 2-related factor 2) pathway is upregulated, which is likely to counteract damage, although it generally remains insufficient to restore mitochondrial integrity. These findings support the hypothesis that exposure to MNPs may contribute to neuronal damage through a combination of neuroinflammation and oxidative stress, highlighting a potential environmental factor involved in cognitive deficits and neurodegenerative processes (Li et al., 2024).

7.5 Molecular and functional effects

7.5.1 Disruption of neurotransmitter systems

The neurotoxic mechanisms of MNPs involve several molecular pathways, including the disruption of neurotransmitter synthesis and release, dysregulation of the gut–brain axis, and the induction of oxidative stress and inflammation, as summarized in Table 7.1.

The table provides an overview of the molecular and functional alterations induced by MNPs in the nervous system. These effects include the disruption of neurotransmitter systems, impairment of synaptic signaling and plasticity, induction of oxidative stress and neuroinflammatory pathways, and consequent cognitive and behavioral deficits, with potential implications for mechanisms of neurodegenerative disease.

Abbreviations: AchE (acetylcholinesterase), BDNF (brain-derived neurotrophic factor), NF-κB (nuclear factor kappa-light-chain-enhancer of activated B cells), CREB (cAMP response element-binding protein), ATF4 (activating transcription factor 4), ROS (reactive oxygen species), GSH (glutathione), IL-1β (interleukin-1 beta), TNF-α (tumor necrosis factor alpha), and IL-6 (interleukin-6).

Neurotransmitters, which are essential for synaptic communication, include acetylcholine, monoamines, amino acids, and neuropeptides. They are synthesized in presynaptic neurons, released through exocytosis, and captured by postsynaptic receptors. MNPs can interfere with these processes, altering neural signaling and inducing neurotoxicity. In particular, the regulation of neurotransmitter synthesis, release, and reuptake appears to be highly vulnerable to NP exposure, although the precise molecular mechanisms involved remain incompletely understood. Exposure to PS NPs has been shown to significantly affect the cholinergic system. Zebrafish and mice treated with MNPs displayed reduced activity of

TABLE 7.1 Molecular and functional effects of MNPs.

Factors	Main Effects	References
Neurotransmitter disruption	• ↓ AChE activity; glutamate–glutamine cycle impairment • Glutamate-driven excitotoxicity	(Kang et al., 2023)
Synaptic dysfunction and plasticity	• Impaired synaptic signaling • ↓ BDNF/CREB/ATF4 • Altered neurogenesis	(Hua et al., 2022; Parkhurst et al., 2013)
Oxidative stress	• ROS overproduction • Mitochondrial damage • ↓ GSH	(Guo et al., 2024)
Neuroinflammation	• Microglial/astrocyte activation • NF-κB signaling • ↑ TNF-α, IL-1β, IL-6	(Shan et al., 2022; Kwon et al., 2022)
Cognitive and behavioral effects	• Learning/memory deficits • Altered neuronal markers • Circadian disruption	(Qiao et al., 2021; Kang et al., 2023)
Links to neurodegeneration	• Amyloidogenic aggregation • α-Synuclein fibril propagation • Neuronal loss	(Caccamo et al., 2017; Dosunmu et al., 2007)

acetylcholinesterase (AChE), a key enzyme responsible for acetylcholine hydrolysis. This reduction leads to excessive accumulation of acetylcholine in synapses, resulting in neurobehavioral abnormalities. A possible underlying mechanism is oxidative stress, which damages organelles involved in AChE synthesis and maturation (the endoplasmic reticulum and Golgi apparatus), thereby impairing proper enzymatic function. Consistent with this hypothesis, however, hydrogen peroxide (H_2O_2) modulates both the structure and isoforms of AChE, suggesting that the interaction between ROS and MNPs may exert an antagonistic effect on the enzyme. MNPs also interfere with the glutamatergic system, the CNS's primary excitatory pathway. In zebrafish, PS exposure reduces the activity of GS, an ATP dependent enzyme essential for the glutamate–glutamine cycle and ammonia homeostasis. Decreased GS activity leads to the accumulation of extracellular glutamate, which, in turn, induces excitotoxicity by overactivating glutamatergic receptors. This process results in excessive influx of Ca^{2+} and Na^{+} into neurons, disruption of ionic homeostasis, cellular damage, and neuronal death. Moreover, glutamate is a precursor for both γ-aminobutyric acid (GABA) and GSH. Thus increased extracellular glutamate may alter GABA levels and deplete antioxidant reserves, further exacerbating oxidative stress.

7.5.2 Impairment of synaptic function and plasticity

In addition to neurotransmitter dysregulation, MNPs have been shown to impair synaptic function and neuronal plasticity directly. Excessive extracellular glutamate can hyperactivate both ionotropic and metabotropic receptors, ultimately leading to synaptic dysfunction and neuronal degeneration. Moreover, the reduced availability of GABA and GSH disrupts inhibitory signaling and weakens antioxidant defense, thereby compromising the delicate balance between excitation and inhibition within the brain (Spanaki and Plaitakis, 2012). Alterations in the cholinergic system further contribute to this imbalance, as inhibition of AChE induces neurobehavioral abnormalities and disrupts the fine regulation of neuronal circuits. In this context, recent studies on human cortical organoids exposed to PS NPs have shown significant changes in gene expression and a reduction in key neuronal markers, such as β-tubulin III, TBR1/TBR2 (T-box Brain 1/T-box Brain 2), and Nestin, highlighting the potential direct impact of MPs on neurogenesis and early neuronal development (Hua et al., 2022). Collectively, these mechanisms converge to undermine synaptic plasticity, the fundamental process underlying learning and memory. The resulting disruption of neuronal networks is consistent with the cognitive deficits and neurobehavioral alterations reported in animal models exposed to NPs.

7.6 Potential health implications

7.6.1 Cognitive and behavioral consequences

MNPs can directly alter synapse function and neuronal plasticity, compromising CNS development. Excess extracellular glutamate can hyperactivate both ionotropic and metabotropic receptors, leading to synaptic dysfunction and neuronal degeneration. Moreover, the reduced availability of GABA and GSH disrupts inhibitory transmission and weakens antioxidant defenses, compromising the balance between excitation and inhibition in the brain (Spanaki and Plaitakis, 2012). Alterations in the cholinergic system further contribute to this imbalance, as inhibition of AChE induces neurobehavioral abnormalities and disrupts the regulation of neuronal circuits. Recent studies on human cortical organoids exposed to PS MPs revealed significant changes in gene expression and reductions in key neuronal markers, including β-tubulin III, TBR1/TBR2, and Nestin, indicating a direct impact of MPs on neurogenesis and early neuronal development (Hua et al., 2022). Evidence from mouse studies further supports the neurotoxic potential of MPs and NPs. Chronic exposure to PS MPs is associated with cognitive impairments, behavioral alterations, and compromised synaptic plasticity, including reduced *Bdnf*, *TrkB* (tropomyosin receptor kinase B), and *Nr2b* (N-methyl-D-aspartate receptor subunit 2B) expression; decreased numbers of DCX- and NeuN-positive cells in the hippocampus; microglial activation; and increased proinflammatory signaling (TNF-α and IL-1β) (Qiao et al., 2021; Kang et al., 2023). At the molecular level, exposure to PS MPs reduces the expression of immediate early genes, including Arc, which are essential for synaptic plasticity and memory, while inducing neuroinflammation via microglial activation and the upregulation of TNF-α, IL-1β, and ERK signaling in the hippocampus. Ingested NPs can also alter the composition of the gut microbiota, affecting the gut–brain axis and the expression of central circadian clock genes and contributing to cognitive and behavioral disturbances (Qiao et al., 2021; Kang et al., 2023).

7.6.2 Associations with neurodevelopmental and neurodegenerative disorders

Epidemiological and toxicological data indicate that MNP exposure can trigger neurodegenerative changes, including abnormal protein accumulation, neuronal loss, and cognitive deficits. MNPs can interfere with protein folding and promote the production of aberrant amyloidogenic proteins, potentially initiating amyloidosis, a hallmark of many neurodegenerative conditions. Misfolded proteins form amyloid fibrils or large oligomeric aggregates during the dynamic process of protein fibrillation. Although the precise mechanisms underlying AD and PD remain incompletely understood, the aggregation of misfolded proteins is hypothesized to be a central pathogenic event. While these diseases often occur sporadically or due to genetic predispositions, environmental and lifestyle factors are increasingly being implicated. Epidemiological evidence across multiple populations supports a strong association between particle exposure and an increased risk of neurodegenerative disorders (Li et al., 2024).

Evidence further suggests that MNPs can disrupt native protein folding and interact with intracellular amino acids, potentially initiating amyloid protein clustering and the early stages of AD. The hypothesis that environmental factors may act as triggers of AD is supported by the very low prevalence of purely genetic cases and may help explain the sporadic onset of the disease (Dosunmu et al., 2007). Aberrant protein folding and amyloidosis are thought to increase susceptibility to neurodegeneration. In vivo studies have demonstrated that environmental MNPs can induce neuronal loss (Boldrini et al., 2018), with structural, functional, and behavioral impairments particularly affecting cholinergic and dopaminergic neurons (Bekdash, 2021).

Anionic NP contaminants also accelerate the formation and propagation of α-synuclein fibrils via high affinity interactions with the amphipathic and nonamyloid-β component domains of α-synuclein. NPs can enter neurons through clathrin-mediated endocytosis, causing mild lysosomal damage that hampers the degradation of aggregated α-synuclein. In mice, NPs interact with α-synuclein fibrils, exacerbating their spread across vulnerable, interconnected brain regions and leading to strong induction of α-synuclein inclusions in the dopaminergic neurons of the substantia nigra. These findings highlight a potential link between NP contamination and α-synuclein aggregation, with implications for Parkinson's disease and related dementias.

7.7 Conclusion

The evidence collected outlines a complex picture of the neurotoxic effects of MNPs. Findings from animal models, particularly zebrafish and rodents, have demonstrated that these particles can cross critical biological barriers, such as the intestinal epithelium and the BBB; accumulate within nervous tissues; and significantly impair neuronal function. The mechanisms underlying this toxicity are multifactorial: beyond oxidative stress and neuroinflammation, which are characterized by microglial and astrocytic activation, a substantial impact on neurotransmitter systems emerges. Studies in both zebrafish and murine models have shown that inhibition of AChE leads to cholinergic dysregulation; disruption of the glutamate–glutamine cycle, with extracellular glutamate accumulation and excitotoxicity; and perturbations of inhibitory systems, ultimately altering the excitation/inhibition balance. Moreover, alterations in synaptic plasticity and neurogenesis, including changes in the expression of key genes (e.g., *Bdnf, Creb, Atf4*) and deficits in synaptic remodeling, with downstream consequences for the maintenance of neuronal circuits and learning and memory processes, have been reported.

These findings highlight the value of zebrafish as an experimental model because of its rapid development and ability to monitor neurobehavioral alterations in vivo, as well as the value of murine models, which allows for a more detailed investigation of neurobiological and behavioral dynamics. However, both approaches present limitations and critical issues that reduce their translational applicability to humans: in zebrafish, phylogenetic distance and structural differences in the nervous system; in rodents, the heterogeneity of experimental conditions, often on the basis of acute or high-dose exposures that may not accurately reflect environmental realities. Furthermore, the intrinsic variability of the particles used (size, shape, composition, additives, and adsorbed contaminants) represents an additional source of complexity and potential bias in data interpretation. In the future, it is essential to develop integrated, multidisciplinary approaches that combine in vivo models, advanced in vitro systems, and in silico methods, along with human biomonitoring studies, to more precisely define the neurological impact of MNPs. Only through a comprehensive and translational perspective will it be possible to clarify the real consequences of chronic low-dose exposure, identify sensitive biomarkers of neurotoxicity, and establish effective strategies for prevention and risk mitigation.

References

Alimba, C.G., Faggio, C., 2019. Microplastics in the marine environment: current trends in environmental pollution and mechanisms of toxicological profile. Environ. Toxicol. Pharmacol. 18727077. 68, 61–74. Available from: https://www.elsevier.com/locate/etap. doi:10.1016/j.etap.2019.03.001.

Bekdash, R.A., 2021. The cholinergic system, the adrenergic system and the neuropathology of Alzheimer's disease. Int. J. Mol. Sci. 14220067. 22 (3), 1–18. Available from: https://www.mdpi.com/1422-0067/22/3/1273/pdf. doi:10.3390/ijms22031273.

Boldrini, M., Fulmore, C.A., Tartt, A.N., Simeon, L.R., Pavlova, I., Poposka, V., Rosoklija, G.B., Stankov, A., Arango, V., Dwork, A.J., Hen, R., Mann, J.J., 2018. Human hippocampal neurogenesis persists throughout aging. Cell Stem Cell. 18759777. 22 (4), 589. Available from: http://www.journals.elsevier.com/cell-stem-cell/. doi:10.1016/j.stem.2018.03.015.

Caccamo, A., Branca, C., Piras, I.S., Ferreira, E., Huentelman, M.J., Liang, W.S., Readhead, B., Dudley, J.T., Spangenberg, E.E., Green, K.N., Belfiore, R., Winslow, W., Oddo, S., 2017. Necroptosis activation in Alzheimer's disease. Nat. Neurosci. 15461726. 20 (9), 1236–1246. Available from: http://www.nature.com/neuro/index.html. doi:10.1038/nn.4608.

Cai, Z., Li, M., Zhu, Z., Wang, X., Huang, Y., Tianmu, L., Gong, H., Yan, M., 2023. Biological degradation of plastics and microplastics: a recent perspective on associated mechanisms and influencing factors. Microorganisms. 2076-2607. 11 (7), 1661. doi:10.3390/microorganisms11071661.

Calderón-Garcidueñas, L., Mora-Tiscareño, A., Ontiveros, E., Gómez-Garza, G., Barragán-Mejía, G., Broadway, J., Chapman, S., Valencia-Salazar, G., Jewells, V., Maronpot, R.R., Henríquez-Roldán, C., Pérez-Guillé, B., Torres-Jardón, R., Herrit, L., Brooks, D., Osnaya-Brizuela, N., Monroy, M.E., González-Maciel, A., Reynoso-Robles, R., Villarreal-Calderon, R., Solt, A.C., Engle, R.W., 2008. Air pollution, cognitive deficits and brain abnormalities: a pilot study with children and dogs. Brain Cogn. 02782626. 68 (2), 117–127. doi:10.1016/j.bandc.2008.04.008.

Campanale, C., Massarelli, C., Savino, I., Locaputo, V., Uricchio, V.F., 2020. A detailed review study on potential effects of microplastics and additives of concern on human health. Int. J. Environ. Res. Public Health. 1660-4601. 17 (4), 1212. doi:10.3390/ijerph17041212.

Chen, S., Chen, Y., Gao, Y., Han, B., Wang, T., Dong, H., Chen, L., 2023. Toxic effects and mechanisms of nanoplastics on embryonic brain development using brain organoids model. Sci. Total Environ. 00489697. 904, 166913. doi:10.1016/j.scitotenv.2023.166913.

Corona, C., Pasini, S., Liu, J., Amar, F., Greene, L.A., Shelanski, M.L., 2018. Activating transcription factor 4 (ATF4) regulates neuronal activity by controlling GABABR trafficking. J. Neurosci. 15292401. 38 (27), 6102–6113. Available from: http://www.jneurosci.org/content/38/27/6102.full.pdf. doi:10.1523/JNEUROSCI.3350-17.2018.

Das, A., 2023. The emerging role of microplastics in systemic toxicity: involvement of reactive oxygen species (ROS). Sci. Total Environ. 00489697. 895, 165076. doi:10.1016/j.scitotenv.2023.165076.

Dosunmu, R., Wu, J., Basha, M.R., Zawia, N.H., 2007. Environmental and dietary risk factors in Alzheimer's disease. Expert Rev. Neurother. 17448360. 7 (7), 887–900. Available from: http://www.future-drugs.com/doi/pdf/10.1586/14737175.7.7.887. doi:10.1586/14737175.7.7.887.

Gecegelen, E., Ucdal, M., Dogu, B.B., 2025. A novel risk factor for dementia: chronic microplastic exposure. Front. Neurol. 16642295. 16. Available from: http://www.frontiersin.org/Neurology. doi:10.3389/fneur.2025.1581109.

Guo, M., Li, Y., Niu, S., Zhang, R., Shen, X., Ma, Y., Wu, L., Wu, T., Zhang, T., Tang, M., Xue, Y., 2024. Oxidative stress-activated Nrf2 remitted polystyrene nanoplastic-induced mitochondrial damage and inflammatory response in HepG2 cells. Environ. Toxicol. Pharmacol. 13826689. 106, 104385. doi:10.1016/j.etap.2024.104385.

Guttenplan, K.A., Weigel, M.K., Prakash, P., Wijewardhane, P.R., Hasel, P., Rufen-Blanchette, U., Münch, A.E., Blum, J.A., Fine, J., Neal, M.C., Bruce, K.D., Gitler, A.D., Chopra, G., Liddelow, S.A., Barres, B.A., 2021. Neurotoxic reactive astrocytes induce cell death via saturated lipids. Nature. 14764687. 599 (7883), 102–107. Available from: http://www.nature.com/nature/index.html. doi:10.1038/s41586-021-03960-y.

Guzzetti, E., Sureda, A., Tejada, S., Faggio, C., 2018. Microplastic in marine organism: environmental and toxicological effects. Environ. Toxicol. Pharmacol. 13826689. 64, 164–171. doi:10.1016/j.etap.2018.10.009.

Hua, T., Kiran, S., Yan, L., Sang, Q.-X.A., 2022. Microplastics exposure affects neural development of human pluripotent stem cell-derived cortical spheroids. J. Hazard. Mater. 03043894. 435, 128884. doi:10.1016/j.jhazmat.2022.128884.

Hwang, K.S., Son, Y., Kim, S.S., Shin, D.S., Lim, S.H., Yang, J.Y., Jeong, H.N., Lee, B.H., Bae, M.A., 2022. Size-dependent effects of polystyrene nanoparticles (PS-NPS) on behaviors and endogenous neurochemicals in zebrafish larvae. Int. J. Mol. Sci. 14220067. 23 (18). Available from: http://www.mdpi.com/journal/ijms. doi:10.3390/ijms231810682.

Jeong, B., Baek, J.Y., Koo, J., Park, S., Ryu, Y.-K., Kim, K.-S., Zhang, S., Chung, C., Dogan, R., Choi, H.-S., Um, D., Kim, T.-K., Lee, W.S., Jeong, J., Shin, W.-H., Lee, J.-R., Kim, N.-S., Lee, D.Y., 2022. Maternal exposure to polystyrene nanoplastics causes brain abnormalities in progeny. J. Hazard. Mater. 03043894. 426, 127815. doi:10.1016/j.jhazmat.2021.127815.

Jin, H., Yang, C., Jiang, C., Li, L., Pan, M., Li, D., Han, X., Ding, J., 2022. Evaluation of neurotoxicity in BALB/c Mice following chronic exposure to polystyrene microplastics. Environ. Health Perspect. 0091-6765. 130 (10). doi:10.1289/ehp10255.

Kang, H., Zhang, W., Jing, J., Huang, D., Zhang, L., Wang, J., Han, L., Liu, Z., Wang, Z., Gao, A., 2023. The gut-brain axis involved in polystyrene nanoplastics-induced neurotoxicity via reprogramming the circadian rhythm-related pathways. J. Hazard. Mater. 03043894. 458, 131949. doi:10.1016/j.jhazmat.2023.131949.

Kihara, S., Ashenden, A., Kaur, M., Glasson, J., Ghosh, S., van der Heijden, N., Brooks, A.E.S., Mata, J.P., Holt, S., Domigan, L.J., Köper, I., McGillivray, D.J., 2021. Cellular interactions with polystyrene nanoplastics—The role of particle size and protein Corona. Biointerphases. 1934-8630. 16 (4). doi:10.1116/6.0001124.

Kwon, W., Kim, D., Kim, H.Y., Jeong, S.W., Lee, S.G., Kim, H.C., Lee, Y.J., Kwon, M.K., Hwang, J.S., Han, J.E., Park, J. K., Lee, S.J., Choi, S.K., 2022. Microglial phagocytosis of polystyrene microplastics results in immune alteration and apoptosis in vitro and in vivo. Sci. Total Environ. 18791026. 807. Available from: https://www.elsevier.com/locate/scitotenv. doi:10.1016/j.scitotenv.2021.150817.

Li, M.R., Men, S.H., Wang, Z.Y., Liu, C., Zhou, G.R., Yan, Z.G., 2024. The application of human-derived cell lines in neurotoxicity studies of environmental pollutants. Sci. Total Environ. 18791026. 912. Available from: https://www.sciencedirect.com/science/journal/00489697. doi:10.1016/j.scitotenv.2023.168839.

Liddelow, S.A., Guttenplan, K.A., Clarke, L.E., Bennett, F.C., Bohlen, C.J., Schirmer, L., Bennett, M.L., Münch, A.E., Chung, W.S., Peterson, T.C., Wilton, D.K., Frouin, A., Napier, B.A., Panicker, N., Kumar, M., Buckwalter, M.S., Rowitch, D.H., Dawson, V.L., Dawson, T.M., Stevens, B., Barres, B.A., 2017. Neurotoxic reactive astrocytes are induced by activated microglia. Nature. 14764687. 541 (7638), 481–487. Available from: http://www.nature.com/nature/index.html. doi:10.1038/nature21029.

Maher, B.A., Ahmed, I.A.M., Karloukovski, V., MacLaren, D.A., Foulds, P.G., Allsop, D., Mann, D.M.A., Torres-Jardón, R., Calderon-Garciduenas, L., 2016. Magnetite pollution nanoparticles in the human brain. Proc. Natl. Acad. Sci. USA. 0027-8424. 113 (39), 10797–10801. doi:10.1073/pnas.1605941113.

Monteggia, L.M., Barrot, M., Powell, C.M., Berton, O., Galanis, V., Gemelli, T., Meuth, S., Nagy, A., Greene, R.W., Nestler, E.J., 2004. Essential role of brain-derived neurotrophic factor in adult hippocampal function. Proc. Natl. Acad. Sci. USA. 0027-8424. 101 (29), 10827–10832. doi:10.1073/pnas.0402141101.

Oberdörster, G., Sharp, Z., Atudorei, V., Elder, A., Gelein, R., Kreyling, W., Cox, C., 2004. Translocation of inhaled ultrafine particles to the brain. Inhal. Toxicol. 16 (6-7), 437–445. doi:10.1080/08958370490439597.

Oeckinghaus, A., Ghosh, S., 2009. The NF-kappaB family of transcription factors and its regulation. Cold Spring Harb. Perspect. Biol. 19430264. 1 (4), a000034. doi:10.1101/cshperspect.a000034.

Parkhurst, C.N., Yang, G., Ninan, I., Savas, J.N., Yates, J.R., Lafaille, J.J., Hempstead, B.L., Littman, D.R., Gan, W.B., 2013. Microglia promote learning-dependent synapse formation through brain-derived neurotrophic factor. Cell. 10974172. 155 (7), 1596–1609. Available from: https://www.sciencedirect.com/journal/cell. doi:10.1016/j.cell.2013.11.030.

Pelegrini, K., Pereira, T.C.B., Maraschin, T.G., Teodoro, L.D.S., Basso, N.R.D.S., De Galland, G.L.B., Ligabue, R.A., Bogo, M.R., 2023. Micro- and nanoplastic toxicity: a review on size, type, source, and test-organism implications. Sci. Total Environ. 18791026. 878. Available from: https://www.elsevier.com/locate/scitotenv. doi:10.1016/j.scitotenv.2023.162954.

Prüst, M., Meijer, J., Westerink, R.H.S., 2020. The plastic brain: neurotoxicity of micro- and nanoplastics. Part. Fibre Toxicol. 17438977. 17 (1). Available from: http://www.particleandfibretoxicology.com/. doi:10.1186/s12989-020-00358-y.

Qiao, J., Chen, R., Wang, M., Bai, R., Cui, X., Liu, Y., Wu, C., Chen, C., 2021. Perturbation of gut microbiota plays an important role in micro/nanoplastics-induced gut barrier dysfunction. Nanoscale. 2040-3364. 13 (19), 8806–8816. doi:10.1039/d1nr00038a.

Salthammer, T., 2022. Microplastics and their additives in the indoor environment. Angew. Chem. Int. Ed. 15213773. 61 (32). Available from: http://onlinelibrary.wiley.com/journal/10.1002/(ISSN)1521-3773. doi:10.1002/anie.202205713.

Shan, S., Zhang, Y., Zhao, H., Zeng, T., Zhao, X., 2022. Polystyrene nanoplastics penetrate across the blood-brain barrier and induce activation of microglia in the brain of mice. Chemosphere. 18791298. 298. Available from: https://www.elsevier.com/locate/chemosphere. doi:10.1016/j.chemosphere.2022.134261.

Singh, A., Chauhan, A., Gaur, R., 2025. A comprehensive review on the synthesis, properties, environmental impacts, and chemiluminescence applications of polystyrene (PS). Discover Chem. 3005-1193. 2 (1). doi:10.1007/s44371-025-00125-y.

Spanaki, C., Plaitakis, A., 2012. The role of glutamate dehydrogenase in mammalian ammonia metabolism. Neurotox. Res. 1029-8428. 21 (1), 117–127. doi:10.1007/s12640-011-9285-4.

Teng, M., Zhao, X., Wang, C., Wang, C., White, J.C., Zhao, W., Zhou, L., Duan, M., Wu, F., 2022. Polystyrene nanoplastics toxicity to zebrafish: dysregulation of the brain-intestine-microbiota axis. ACS Nano. 1936086X. 16 (5), 8190–8204. Available from: http://pubs.acs.org/journal/ancac3. doi:10.1021/acsnano.2c01872.

Wang, H., Zhang, H., Su, Y., 2022. New insights into the diurnal rhythmicity of gut microbiota and its crosstalk with host circadian rhythm. Animals. 2076-2615. 12 (13), 1677. doi:10.3390/ani12131677.

Wei, W., Li, Y., Lee, M., Andrikopoulos, N., Lin, S., Chen, C., Leong, D.T., Ding, F., Song, Y., Ke, P.C., 2022. Anionic nanoplastic exposure induces endothelial leakiness. Nat. Commun. 20411723. 13 (1). Available from: http://www.nature.com/ncomms/index.html. doi:10.1038/s41467-022-32532-5.

Wright, S.L., Kelly, F.J., 2017. Plastic and human health: a micro issue? Environ. Sci. Technol. 15205851. 51 (12), 6634–6647. Available from: http://pubs.acs.org/journal/esthag. doi:10.1021/acs.est.7b00423.

Wu, T., Dejanovic, B., Gandham, V.D., Gogineni, A., Edmonds, R., Schauer, S., Srinivasan, K., Huntley, M.A., Wang, Y., Wang, T.M., Hedehus, M., Barck, K.H., Stark, M., Ngu, H., Foreman, O., Meilandt, W.J., Elstrott, J., Chang, M.C., Hansen, D.V., Carano, R.A.D., Sheng, M., Hanson, J.E., 2019. Complement C3 is activated in human AD brain and is required for neurodegeneration in mouse models of amyloidosis and tauopathy. Cell Rep. 22111247. 28 (8), 2111. Available from: http://www.sciencedirect.com/science/journal/22111247. doi:10.1016/j.celrep.2019.07.060.

Yadav, H., Sethulekshmi, S. Shriwastav, A., 2022. Estimation of microplastic exposure via the composite sampling of drinking water, respirable air, and cooked food from Mumbai, India. Environ. Res. 00139351. 214 (Pt 1), 113735. doi:10.1016/j.envres.2022.113735.

Yang, Z.S., Bai, Y.L., Jin, C.H., Na, J., Zhang, R., Gao, Y., Pan, G.W., Yan, L.J., Sun, W., 2022. Evidence on invasion of blood, adipose tissues, nervous system and reproductive system of mice after a single oral exposure: nanoplastics versus microplastics. Biomed. Environ. Sci. 08953988. 35 (11), 1025–1037. Available from: http://www.elsevier.com. doi:10.3967/bes2022.131.

Zhao, Z., Wang, X., Jiang, J., Dong, Y., Pan, Y., Guan, X., Wang, B., Gao, S., Chen, Z., Zhou, Z., 2023. Adverse effects of polystyrene nanoplastics on sea cucumber *Apostichopus japonicus* and their association with gut microbiota dysbiosis. Chemosphere. 00456535. 330, 138568. doi:10.1016/j.chemosphere.2023.138568.

CHAPTER

8

Effects of microplastics on animal reproductive physiology

Victoria Anthony Uyanga[a,b], *Nesrein M. Hashem*[c], *and Oyegunle Emmanuel Oke*[d]

[a]Department of Animal Sciences, Purdue University, West Lafayette, IN, United States
[b]Department of Agriculture and Environmental Sciences, Lincoln University of Missouri, Jefferson City, MO, United States
[c]Department of Animal and Fish Production, Alexandria University, Alexandria, Egypt
[d]Department of Animal Physiology, Federal University of Agriculture, Abeokuta, Nigeria

8.1 Introduction

Over the last two decades, global plastic production has increased by >50%, from 234 million tons in 2000 to 460 million tons in 2019, while only 9% of plastic waste has been recycled. In contrast, the remaining waste has been incinerated (19%), sent to sanitary landfills (50%), disposed of at dumpsites, burned in open pits, or leaked into the environment (22%) (OECD, 2022). With the rise in plastic pollution, the presence and abundance of smaller plastic particles, known as micro- and nanoplastics, have steadily increased in the ecosystem, directly and indirectly affecting both animal and human health.

Microplastics (MPs) are synthetic polymers with dimensions ≤5 mm that are resistant to natural degradation and, as such, may persist in the environment for several years. They are derived from the breakdown of larger plastic materials into smaller items, the attrition of textiles, the shedding of synthetic fibers, tire and road wear particles, household laundering, and microbeads used in personal care products (Bhowmik et al., 2024). Animals are exposed to MPs via various routes, including land, water, and air, where they occur in a large variety of shapes, sizes, colors, and compositions. Various anthropogenic activities release plastic materials and their particles into the environment, including plastics used as animal feed ingredients, litter materials, plastic silos, agriplastic mulching materials, and fertilizers. During grazing, farm animals readily come into contact with MPs that accumulate in forages, grasses, and surface waters (Aardema et al., 2024).

Microplastics in Animal and Human Physiology
https://doi.org/10.1016/B978-0-443-45635-0.00010-0

Upon exposure and uptake by animals, these MPs may pose physical, microbial, and chemical hazards that impact animal health and welfare. MPs can accumulate in various tissues, including the blood, digestive tract, lungs, spleen, liver, kidneys, intestines, brain, reproductive organs, and animal manure. MPs have been detected in the manure of farm animals, including sheep, swine, chickens, and cows, at concentrations ranging from 10^2 to 10^5 particles/kg (Aardema et al., 2024). The presence and accumulation of MPs in the reproductive tissues (testes and ovaries) of animals result in reproductive toxicity and dysfunction, impairing the tissue architecture, maturation, function, immune response, and redox balance within the reproductive organs and system. This poses a growing threat to the health, fertility, breeding, long-term survival, and continuity of diverse animal species.

8.2 Entry and fates of microplastics in animals

8.2.1 Overview of microplastics: Sources, types, and environmental persistence

Plastics transform into MP with a smaller particle size (<5 mm) after being released into the environment. MPs are released when synthetic fibers, primarily polyesters, are used to wash clothing, causing wear or scratches. The classification of MPs can be based on their shape, density, and chemical makeup. Thus MPs are categorized into fragments, microfibers, films, fibers, microbeads, pellets, and foams (Burns and Boxall, 2018). MPs are widely available in our environment because they can be transmitted over long distances, and common polymer types include polypropylene, polystyrene (PS), polyvinyl chloride (PVC), and polyethylene terephthalate.

8.2.2 Pathways of microplastic exposure and bioaccumulation in animals

MP can enter an animal's body through the mouth, nose, skin, or other openings. The most commonly reported route is ingestion, which is often caused by contaminated feed, water, or prey. Fish readily ingest MP because it is similar in size to plankton. MP can adhere to aquatic organisms and be retained by ubiquitous biofilms on rocks, sediments, and plant surfaces, increasing their abundance and bioavailability in marine ecosystems. While aquatic species such as fish, mollusks, and crustaceans consume MPs because they mistake them for food, terrestrial animals may be exposed through fodder, soil, and water. The major routes of MP exposure in livestock include foraging in fields contaminated with packaging debris and leftover mulching materials; plastic films (silage bags, bale wraps, and plastic bunker silos) used for animal feed storage; farm environments; contaminated water; and feed sources.

In addition, the use of plastic cutting boards and plastic bags during livestock slaughter and packaging may further expose animal tissues to plastic particles and contamination. Furthermore, birds and poultry are exposed to MP through access to plastic litter materials, contaminated feed and water sources, packaging debris from feed ingredients, and environmental plastic debris. On the basis of surface chemistry, particle size, and animal physiology, MP

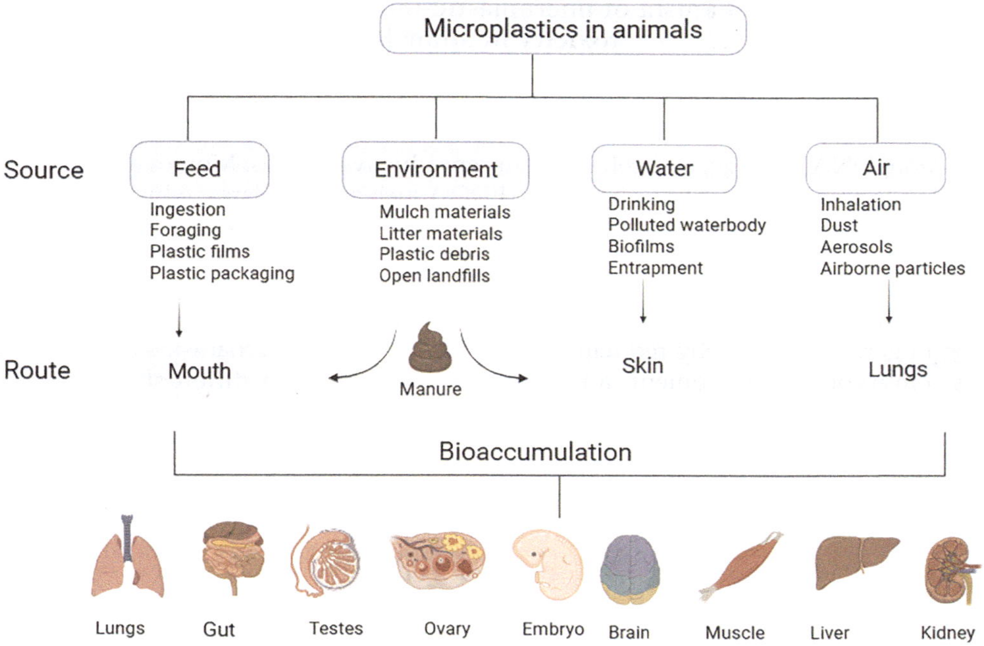

FIG. 8.1 Schematic representation of the sources, routes, and tissue accumulation of microplastics in animals. Routes through which microplastics (MPs) enter the body and accumulate in different tissues, including the digestive, respiratory, and reproductive tracts. *MP can readily enter an animal's body through contaminated feed and water, packaging materials, plastic debris, air inhalation, biofilms, litter materials, and access to open landfills. Upon entry, these MPs are translocated into the blood circulation. They are deposited in various tissues of the gut, gonads, lungs, muscle, liver, kidney, and embryonic tissues, leading to bioaccumulation and the development of MP-induced toxicity.*

may accumulate in the gastrointestinal tract after ingestion or be deposited in other tissues, including the liver, lungs, muscles, kidneys, and reproductive organs (Fig. 8.1).

8.2.3 Microplastics as vectors for other reproductive toxic substances

MPs are ubiquitous and pose a threat to the environment's equilibrium and animal health, as they can act as vectors for environmental pollutants. They can adsorb various contaminants, such as persistent heavy metals and organic pollutants, due to their high specific surface area and compact size. In addition, toxic substances and microbes may be transported by biofilms that develop on MP surfaces. Specific bacterial pathogens, such as *Escherichia coli* and *Vibrio* spp., thrive on plastic waste and are often detected in MP biofilms (Jia et al., 2024). In addition, the leaching of compounds linked to plastics, such as additives and adsorbed toxins, could increase chemical toxicity. For example, bisphenol A and phthalates have been connected to developmental abnormalities, ovarian dysfunction, altered hormone signaling, and lower sperm quality (Ullah et al., 2023). The reproductive risks associated with MP exposure are increased by this vector role, particularly in contaminated environments where several pollutants coexist.

8.3 Mechanism of microplastic-induced reproductive toxicity in animals

MP-induced reproductive toxicity in animals is a result of a complex interplay of physiological abnormalities, as summarized in Fig. 8.2. MP induces oxidative stress, leading to lipid peroxidation, DNA damage, and cell death in reproductive organs. Moreover, MP serves as a vector for endocrine-disrupting chemicals (EDCs) such as phthalates, which interfere with the hypothalamic–pituitary–gonadal (HPG) axis, causing hormonal imbalances that hinder gametogenesis, folliculogenesis, and steroidogenesis (Ullah et al., 2023). The toxicity of MP in the gonads results in structural harm, such as testicular degeneration and ovarian atresia, and impacts sperm quality and oocyte viability. In addition, MP can disrupt mitochondrial function in germ cells, thereby reducing ATP synthesis. These mechanisms adversely affect fertility, embryonic development, and reproductive success in different animal species (Fig. 8.2).

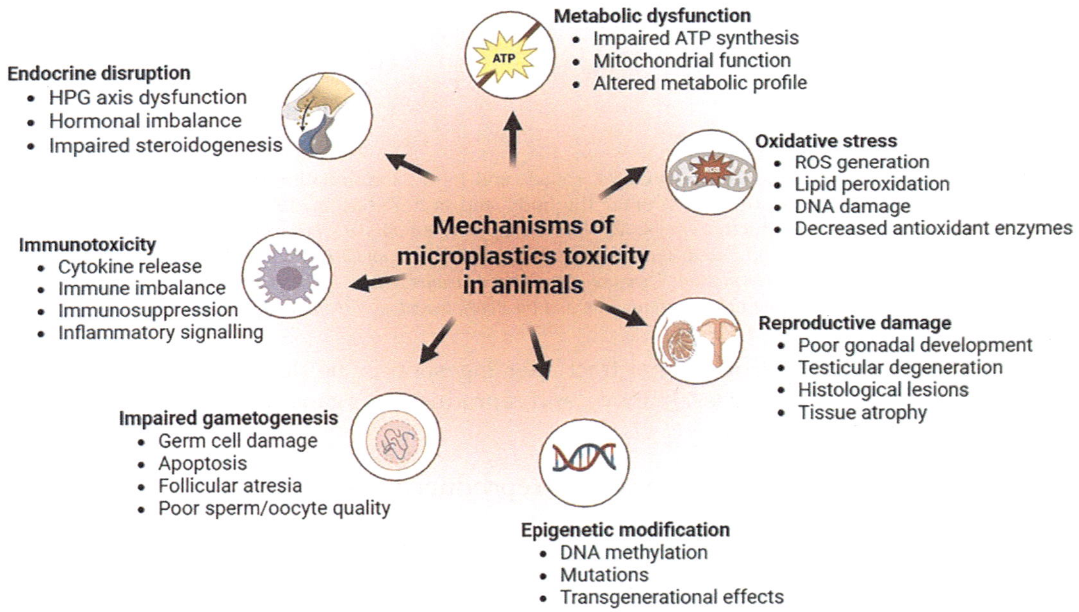

FIG. 8.2 Physiological mechanism responsible for the reproductive toxicity of microplastics in animals. Mechanisms through which microplastics (MPs) cause reproductive toxicity and dysfunction in animals. *Following exposure, MPs can translocate from the bloodstream into the reproductive system, where elevated concentrations and bioaccumulation contribute to reproductive toxicity. Mechanistically, MPs induce oxidative stress by increasing free radical production and decreasing antioxidant enzyme activity. This oxidative disturbance promotes DNA damage, metabolic dysfunction, immune suppression, inflammatory responses, and impaired ATP synthesis. These disruptions collectively contribute to dysregulation of the hypothalamic–pituitary–gonadal (HPG) axis, germ cell damage, hormonal imbalances, impaired steroidogenesis, and reproductive failure across the cellular, molecular, and tissue levels. Furthermore, MPs may trigger epigenetic modifications that could affect offspring development and reproductive health in future generations.*

8.3.1 Disruptions in gametogenesis

The toxicity of MP in the gonads disrupts germ cell mitochondrial function and causes structural damage, such as testicular degeneration, which impacts sperm quality. MP is known to alter gonadal morphology by disrupting the organization of testicular spermatids, leading to the apoptosis of spermatogenic cells. Importantly, the Nrf2/HO-1/NF-κB pathway is intimately linked to aberrant sperm quality in mice during PS exposure (Hou et al., 2021).

8.3.2 Endocrine disruption and hormonal dysregulation

EDCs are environmental substances that can interfere with normal endocrine processes, leading to adverse health effects. MPs can disrupt several hypothalamic axes because they can cross various biological membranes. Exposing testicular Leydig cells to PS downregulates the levels of thyroid-stimulating hormone, luteinizing hormone, testosterone, and free triiodothyronine (Sun et al., 2023). By disrupting the HPG axis, MPs may affect hormonal balance, gonad development, reproductive function, and overall health.

8.3.3 Oxidative stress and ROS-mediated cellular damage

Lipid peroxidation, DNA damage, and protein breakdown result from an imbalance between antioxidant and oxidant states as a result of excessive reactive oxygen species (ROS) formation (Oke et al., 2024). By generating ROS in vascular cells, the functionality and integrity of the arterial lining may be compromised, leading to endothelial cell oxidative damage. PS MP readily interacts with cell membranes, increasing intracellular ROS concentrations and reducing cell viability (Wu et al., 2019). The polyethene MP can disrupt the antioxidant status by decreasing the activities of the antioxidant enzymes superoxide dismutase, catalase, and glutathione-S-transferase, as well as glutathione and ascorbic acid levels, and increasing the malondialdehyde content (Kehinde et al., 2024). In many species, MP-induced oxidative stress is the key mechanism causing reproductive and systemic toxicity.

8.3.4 Inflammatory responses and immunotoxicity

Inflammatory responses are often elicited by the accumulation of microplastics in reproductive organs, leading to local hyperactivation or immunosuppression. Danger-associated molecular patterns and ROS produced during oxidative stress can trigger a series of inflammatory reactions in immune cells and activate toll-like receptors, which are mediators of innate immunity. According to earlier studies, MP exposure may increase the production of proinflammatory cytokines and activate immune cells, particularly macrophages (Chowdhury et al., 2024). Larger polyethylene particles (0.3–10 μm) may also stimulate the production of cytokines, including inflammatory mediators such as Tumor Necrosis Factor-alpha (TNF-α), Interleukin (IL)-1β, and IL-6. Evidently, exposure of animals to MP causes an antioxidant imbalance, altered inflammatory signaling, and disruption of immune–endocrine crosstalk, which may impair reproductive function.

8.3.5 Epigenetic modifications and transgenerational effects

Animals may exhibit transgenerational epigenetic inheritance, which may be related to the environmental effects of MP. Exposing *Daphnia magna* to pristine polymer microspheres as an aquatic model for over four generations reduced survival in subsequent generations, potentially increasing their likelihood of extinction (Martins and Guilhermino, 2018). Similarly, female and male mice exposed to polyethylene MP after birth presented alterations in the sex ratio, offspring weight, and lipid and amino acid metabolism (Park et al., 2020). This implies that MP may influence the transcriptional activity of genes involved in stress responses and reproduction, thereby impacting the well-being of subsequent generations.

8.3.6 Interactions with other environmental pollutants

In these environments, MP may coexist with other pollutants, including endocrine disruptors (e.g., dioxin) and heavy metals (e.g., mercury), which can lead to synergistic toxicity. Some chemical pollutants, such as heavy metals, phthalate esters (PAEs), perfluoroalkyl substances, persistent organic pollutants, and medicines, are carried by MP. The surfaces of MP can absorb environmental heavy metals, including chromium, lead, arsenic, and copper. When MPS interacts with other toxicants in the environment, their combined effects exceed the sum of their individual impacts, creating a synergistic effect. Once adherents have been absorbed, their toxicity can intensify when they are introduced into organisms by acting in concert with MP to induce synergistic toxicity. These chemical-rich MPs can release toxins into the host upon ingestion or absorption, thereby increasing the bioavailability of substances known to disrupt reproductive function. For example, when nickel and pyrene were added to MP, this combination increased mortality, comparable to the effects of either substance alone, indicating an additive effect (Albani et al., 2023).

8.3.7 Alterations to reproductive tissues and organs

The relatively small size of the MP allows it to transit through the cells, tissues, and organs involved in reproduction, thereby altering the architecture, morphology, and physiology of the reproductive system. Male mice exposed to varying levels of PS exhibited significant histological lesions in the testis and changes in semen biomarkers (Amereh et al., 2020). In zebrafish, PS MP decreases fecundity, alters HPG axis function, and elevates ROS levels, leading to reproductive impairments (Gupta et al., 2023). MP-induced reproductive defects are ultimately characterized by a decline in the quantity, quality, and viability of oocytes or sperm, as well as gonadal disintegration, which are frequently linked to elevated ROS levels.

8.4 Implications of microplastic exposure for the reproductive physiology of animals

MP triggers oxidative stress, activates proinflammatory molecular pathways, causes DNA mutations, and disrupts the endocrine system, which can directly injure reproductive tissues, as summarized in Table 8.1. Moreover, MP exposure may have transgenerational effects on offspring, altering their reproductive physiology.

TABLE 8.1 Types, routes, target tissues, and effects of microplastics (MPs) on the reproductive functions of different animal species.

Animals	Type of MP	Study conditions	Route	Target tissue	Effects on reproductive function	References
***Aquatic* species**						
Marine medaka *(Oryzias melastigma)*	Polystyrene microplastic	2, 20, and 200 μg/L polystyrene MP for 60 days	Waterbody	Testis and ovary	Delayed gonad maturation; decreased fecundity; altered HPG axis; oxidative stress; and histological changes	(Wang et al., 2019)
African catfish *(Clarias gariepinus)*	Virgin or phenanthrene loaded with low-density polyethylene (LDPE) fragments	Virgin LDPE (50 or 500 mg/L) with phenanthrene (10 or 100 mg/L) for 96 h	Waterbody	Brain samples	Affected expression of gonadotropin-releasing hormone and 11β-hydroxysteroid dehydrogenase type 2	(Karami et al., 2016)
Larvae carp *(Cyprinus carpio* var. *larvae)*	Polyvinyl chloride (PVC)	10%, 20%, and 30% PVC for 60 days	Diets	Ovaries, testes, and brains	Impaired gonadal development, altered gonad and brain structures, steroid hormones, and HPG axis-related genes	(Liu et al., 2023)
Zebrafish *(Danio rerio)*	Proprietary polymer	MP (2 mg/L), copper (25 μg/L, Cu25), and their mixture (Cu25 + MPs) for 30 days	Waterbody	Brain and gonad	Dysregulation of the HPG axis-related genes in the gonads	(Santos et al., 2025)
Zebrafish embryos	Polyethylene terephthalategranules and fibers	Particles of ~ 150 μm in diameter; fibers of 3–5 mm length and 20 μm diameter	Waterbody	Embryo	Accelerated blood flow speed and heart rate, and inhibited hatching percentage	(Cheng et al., 2021)
Amphibians and reptiles						
Loggerhead sea turtles *(Caretta caretta)*	Polypropylene, cotton fibers, and polyethylene were found	–	Bioaccumulation	Reproductive organs	Higher MP accumulation in male gonads than in the heart, kidneys, spleen, muscles, intestines, and fat	(Costello et al., 2025)
Snake-eyed lizard *(Ophisops elegans)*	Polyethylene terephthalate, polyacrylonitrile, polypropylene, polyethylene, polymethyl methacrylate, and polyamide	–	Bioaccumulation	MP in the gastrointestinal tract	High levels of MPcontamination were associated with extensive anthropogenic activity	(Altunışık et al., 2024)
Physalaemus cuvieri tadpoles	Polyethylene microplastics (PE MP)	MP-PE at a concentration of 60 mg/L for 7 days	Waterbody	Larvae	Cytotoxicity and mutagenic effects, increased abnormalities in nuclear erythrocytes	(da Costa Araújo et al., 2020)

Continued

TABLE 8.1 Types, routes, target tissues, and effects of microplastics (MPs) on the reproductive functions of different animal species—cont'd

Animals	Type of MP	Study conditions	Route	Target tissue	Effects on reproductive function	References
Rodents						
Kunming mice	Polyvinyl chloride (PVC)	30 $g \cdot kg^{-1}$ PVC for 28 days	Feed	Ovary, uterus, and fallopian tube; testicular tissues	Impaired testicular structure, reduction in spermatogonia, and spermatozoa in males Decreased ovarian follicles, uterus volume, fallopian tube, and muscle layer thickness in females	(Yang et al., 2024)
C57 mice	Polystyrene	Low-dose MP (25–30 μg/kg body weight/day)	Drinking water	Testes and prostate	Low-dose MPS exposure reduced sperm quality and the number of offspring born	(Gao et al., 2024)
Kunming strain mice (20 ± 1 g)	Polystyrene	Polystyrene beads at 200 mg/kg body weight	Gavage	Testis and epididymis	Polystyrene penetrated the blood–brain/testis barriers in a size-dependent manner	(Yang et al., 2022b)
C57BL mice	polyethylene terephthalate (PET MP)	PET MP solutions at concentrations of 50, 200, and 1000 nm at a dose of 8 mg/kg/day for 90 days	Intragastric administration	Testicular tissues	Low sperm quality and related hormones, inhibition of spermatogenic cell proliferation, and inflammation	(Zhang et al., 2025)
Female mice and their offspring	Polyethylene microplastics	40 mg/kg per day for 30 days	Oral administration	Oocytes	Reduced oocyte maturation and fertilization rates, oxidative stress, mitochondrial dysfunction, and DNA damage in oocytes	(Zhang et al., 2023)
Male BALB/c mice	Polystyrene microplastics	100 μg/L and 1000 μg/L polystyrene for 180 consecutive days	Drinking water	Testes, Leydig cells	Altered testicular histology, abnormal spermatogenesis, low testosterone levels, and downregulation of the LHR/cAMP/PKA/StAR pathway	(Jin et al., 2021)
Male BALB/c mice	Polystyrene microplastics	0.01, 0.1, or 1 mg/d MP	Oral gavage	Testes	Decreased sperm quantity, quality, and testosterone levels; increased oxidative stress; and activation of the p38 MAPK signaling	(Xie et al., 2020)
Specific pathogen-free (SPF) CD-1 female mice	Polystyrene microplastics	30 mg/kg body MP for 35 days	Oral gavage	Ovaries	MP decreased malondialdehyde, glutathione, mitochondrial membrane potential, and endoplasmic reticulum calcium but increased reactive oxygen species and IL-6 levels	(Liu et al., 2022)

Livestock						
Goats	Polyethylene and low-density polyethylene polymer were found	Highest MP number found in the goat farm was 250 particles/kg	Limit detection	MP detection in grass fields, feed, drinking water, feces, and milk	Fiber was the most dominant form of MP	(Wahyudi et al., 2025)
Cow	Nylon and low-density polyethylene	Average of 0.14 item g − 1	Limit detection	Liver, meat, and tripe	Potential exposure to MP through the consumption of edible cow tissues	(Bahrani et al., 2024)
Bovine oocytes/embryos	Xenoestrogen bisphenol A	1 and 10 ng/mL	In vitro test	Embryo	Decreased blastocyst rate, low quality of transferable embryos	(Choi et al., 2016)
Sheep	Nylon, polystyrene, and low-density polyethylene	Average of 0.13 item g − 1	Limit detection	Liver, meat, and tripe	Potential transfer of MP from the environment and accumulation in livestock tissues	(Bahrani et al., 2024)
Avian species						
Chicken	Bisphenol A	Bisphenol A (0.005, 0.02, and 0.1 μL/egg)	Injection	Embryo	Fetal abnormalities, congenital anomalies, chick lethargy and inflammation, paralysis, and liver malformations	(Alsharifi et al., 2023)
Chickens	Microplastics	MP were spherical and 50–100 μm	Limit detection	Eggs	MP exist in eggs, with an average content of 11.67 ± 3.98 particles/egg, and the main type was polyethylene	(Liu et al., 2022)
Chickens	Polyethylene MS	2% and 10% polyethylene MP	Feed	Ovary	Impaired ovarian function, delays in ovarian development, and egg-laying behavior	(Xu et al., 2025)
Chicken	Polystyrene	0, 1, and 100 mg/L for 28 and 42 days	Drinking water	Testicular tissue	Testicular inflammatory infiltration, interstitial hemorrhage, decreased blood–testis barrier	(Hou et al., 2022)
Quails	Polystyrene particles	200 mg/kg for 6 weeks (150 μm)	Feed	Testes	Low testicular weight, thickened spermatogenic tubule membrane	(You et al., 2025)
Muscovy female ducks	Polyvinyl chloride	Low or high dose of 1 or 10 mg/L PVC MP for 60 days (5 μm size)	Water	Ovary	PVC MPS altered ovarian morphology, follicular development, and steroidogenesis	(Ali et al., 2024)

8.4.1 Spermatogenesis and sperm quality indices

MP negatively affects the development of testicular germ and somatic cells, disrupting spermatocyte development and spermatogenesis. For example, mice exposed to PS MP exhibit increased atrophy, apoptosis, and sperm cell abscission, which are linked to the Nrf2/HO-1/NF-κB signaling pathway (Hou et al., 2021). In another study, PS MP could enter different mouse testicular compartments in vitro (Jin et al., 2021). In the same study, treatment with PS MP for 28 days decreased sperm quality and testosterone levels, which were associated with disoriented spermatogonia, multinucleated gonadotrophic cells, and disruption of the blood–testis barrier, as indicated by histopathological examination. PS MP decreases sperm count and motility, increases sperm abnormalities, and reduces serum testosterone concentration (Xie et al., 2020). In bulls, MP was detected in epididymal sperm at a mean concentration of 0.37 μg/mL. This concentration of PS MP could reduce sperm motility and impair sperm fertilization but did not affect sperm acrosome integrity or induce oxidative stress (Grechi et al., 2024).

8.4.2 Ovarian function and follicular development

MP can adversely affect the reproductive health of female animals, particularly by disrupting ovarian function and follicular development. The tiny MP particles can accumulate in ovarian tissues, leading to oxidative stress, inflammation, and hormonal imbalances that impair folliculogenesis and the ovarian reserve. In female mice, PS MP caused microstructural abnormalities, including the development of ovarian cysts, decreased granulosa layer thickness in secondary follicles, fewer growing follicles, increased ovarian collagen content, granulosa apoptosis, and oviduct dilatation (Wei et al., 2022). In rats, MP decreases the levels of anti-Müllerian hormone, a marker of follicle reserve, thereby reducing the ovarian follicle reserve and shortening the reproductive lifespan of females (Gao et al., 2021). Furthermore, MP increases fibrotic processes in the ovaries by inducing oxidative stress and lipid peroxidation while decreasing the levels of antioxidant enzymes (superoxide dismutase, catalase, and glutathione peroxidase). Studies in rats have indicated that ROS accumulation leads to granulosa apoptosis, followed by follicular atresia, which may result in anovulation and infertility (Gao et al., 2021). MP and associated chemical pollutants may further affect oocyte maturation, reducing the likelihood of successful fertilization. In this context, the levels of PS MP generated in the follicular fluid are sufficient to inhibit the growth and maturation of bovine oocytes (Grechi et al., 2022). The negative impacts of MP and/or cochemicals have also been reported in aquatic animals, where phthalates, plasticizers, etc., were detected in sea turtle egg yolk, leading to hatching failure and retarded embryonic development (Savoca et al., 2021).

8.4.3 Fertilization and embryonic development

MP can cause physical damage or induce oxidative stress in sperm cells, impairing the success and outcomes of fertilization. For example, embryos derived from bovine sperm exposed to PS MPs exhibited reduced blastocyst formation, increased apoptosis, and elevated free radical production (Grechi et al., 2024). MP often interacts with toxic additives and pollutants (e.g., bisphenol A, phthalates, and heavy metals) that can interfere with embryonic gene expression and cellular differentiation. The detection of bisphenol A has been linked to reduced cleavage rates, impaired blastocyst development, and changes in gene expression in bovine embryos (Choi et al., 2016). Alsharifi et al. (2023) reported that the injection of fertilized eggs with bisphenol A induced malformation of developing fetuses on days 10 and 21 of incubation. In poultry, exposure to bisphenol A in chickens led to malformed reproductive organs and increased embryo mortality. In aquatic animals, MP can adhere to the chorion and reduce oxygen exchange, leading to embryonic mortality. For example, polyethylene terephthalate granular particles and fibers exhibit an affinity for the zebrafish embryonic chorion, inhibiting embryonic development (Cheng et al., 2021).

8.4.4 Gestation and parturition

The effects of MP on gestation and parturition in animals are an emerging area of research, and while comprehensive data are still lacking, preliminary studies suggest potential adverse impacts. MP or associated chemicals can cross placental barriers in some animals, potentially causing developmental toxicity. Exposures have been linked to reduced fetal viability, developmental abnormalities, and growth retardation. In late-gestation-exposed female rats, Fournier et al. (2020) reported that PS particles can be transferred to fetal tissues, including the heart, lungs, liver, kidneys, and brain. Moreover, Aghaei et al. (2022)) reported that exposure to PS MP altered the placental metabolic profile, potentially affecting fetal development. In mice, PS MP has the potential to increase embryo resorption, induce spontaneous abortion, reduce the number and diameter of uterine arterioles, and reduce pregnancy outcomes via immune-mediated disturbances (Hu et al., 2021). Despite the placenta-mediated effects on gestation and fetal development, the negative impacts of MP can occur via other biological pathways. Importantly, MP disrupts the HPG axis, deters reproductive health, decreases fertility, and induces transgenerational defects (Wang et al., 2019).

8.4.5 Postnatal development

In rodent models, MP exposure during gestation causes developmental abnormalities in newborns, such as lowered birth weights, altered steroid hormones, and delayed maturation of gonads. Exposing pregnant mice to prenatal PS reduces testicular weight, damages the reproductive epithelium, and decreases the offspring's sperm count (Huang et al., 2022). In addition, in pregnant mice exposed to PS MP, newborns exhibited metabolic disorders, including changes in serum triglyceride and cholesterol concentrations (Luo et al., 2019). Maternal exposure to polyethylene MP for 30 days impaired the growth and fertility of offspring in mice (Zhang et al., 2023). Interestingly, MP may also affect the mental functions and social communication skills of offspring exposed to MP during their pre- and postnatal life. PS micro- and nanoparticles in pregnant mice induce oxidative stress in the brains of fetuses and apoptosis in the fetal thalamus, leading to anxiety-like behaviors and decreased gamma-aminobutyric acid levels in offspring (Yang et al., 2022a).

8.5 Comparative studies of microplastic exposure in various animal species

8.5.1 Aquatic species

MP induces reproductive dysfunction in aquatic animals by altering the regulation of the HPG axis, which is a critical control mechanism for reproduction. Following MP exposure, the transcription levels of hormones related to the HPG axis are downregulated in the brain and gonads of African catfish (*Clarias gariepinus*), larval carp (*Cyprinus carpio*) (Liu et al., 2023), and zebrafish (*Danio rerio*) (Santos et al., 2025). In addition, exposing marine medakas (*Oryzias melastigma*) to PS MP results in tissue bioaccumulation, oxidative stress, delayed gonad maturation, decreased fecundity, and negative regulation of the HPG axis in female fishes. Importantly, parental exposure can lead to transgenerational effects, including delayed incubation time, reduced hatching rate, decreased heart rate, and decreased body length in offspring (Wang et al., 2019). As such, exposing aquatic animals to environmentally relevant concentrations of MP adversely affects their reproductive efficiency, thereby posing a significant risk to aquatic populations, as documented in Table 8.1.

8.5.2 Mammals

Extensive studies on mice have shown that MP causes reproductive dysfunction with notable sex-dependent effects on the male and female reproductive systems. In male animals, both low and high levels of MP exposure are associated with spermatogenic dysfunction, reduced sperm quality, testicular developmental disorders, disruption of the blood–testis barrier, oxidative stress, and inflammatory responses. The exposure of testicular tissues to polyethylene terephthalate MP inhibits spermatogenic cell proliferation, induces apoptosis, and activates the MAPK signaling pathway, which is responsible for the inflammatory response (Zhang et al., 2025). PVC, a common MP, also damages the reproductive and digestive systems of male and female mice, with notable sex-dependent effects. In male animals, PVC impairs testicular structure and inhibits the development of sperm cells in the seminiferous tubules. In female mice, ovarian follicles, uterine volume, fallopian tube volume, and ovarian tissues are affected by PVC toxicity, leading to reproductive damage (Yang et al., 2024). Mechanistically, MP affects the male reproductive system by disrupting the blood–testis barrier, inducing testicular malformations and spermatogenic dysfunctions, and activating ROS-associated inflammatory pathways, including the Nrf2/HO-1/NF-κB, p38 MAPK, and MAPK/Nrf2 signaling pathways. In females, MP affects ovarian structures by inducing granulosa apoptosis, decreasing ovarian volume, reducing oocyte production, and acting through the ROS-related Wnt/β-catenin and NLRP3/Caspase-1 pathways to cause reproductive dysfunction (Yuan et al., 2022).

8.5.3 Birds and poultry

Chronic MP exposure is associated with redox imbalance, immune dysfunction, and reproductive abnormalities in avian species. Feeding quail a PS MP reduces testicular weight and compromises the testicular structure by thickening the basal membrane of the spermatogenic tubule and increasing the abundance of deformed cells (You et al., 2025). Similarly, providing PS MP through drinking water to chickens causes testicular tissue damage and impairs the integrity of the blood–testis barrier (Hou et al., 2021). Mechanistically, PS MP disrupts redox balance by inhibiting the Nrf2–Keap1 pathway and activating NF-κB signaling, thereby inducing oxidative stress, inflammation, and apoptosis in reproductive tissues. Recent evidence indicates that high accumulation of PVC MP (10 mg/L) in duck ovaries alters ovarian morphology, follicular development, and the immunosignaling of steroidogenic proteins, contributing to reproductive disorders (Ali et al., 2024). In addition, coexposure to MP and other organic contaminants, such as PAEs, can impair ovarian function and egg-laying behavior in chickens, with the potential for bioaccumulation in eggs (Xu et al., 2025). Overall, the ingestion and bioaccumulation of MP in birds and poultry are increasing concerns, with significant implications for animal production and food safety. MP has detrimental effects on avian health and welfare by inducing oxidative stress, activating inflammatory responses, and altering the gut microbiota balance (Table 8.1).

8.5.4 Amphibians and reptiles

Plastic ingestion is a significant challenge for sea turtles, and snakes are commonly found to have swallowed plastic containers and bottle caps or to be entangled in plastic bags (Altunışık et al., 2024). MP are found in all major organs (kidney, heart, skeletal muscle, subcutaneous fat, liver, spleen, gastrointestinal, and reproductive organs) of the Mediterranean loggerhead sea turtle (*Caretta caretta*), with the reproductive organs containing the highest concentration of MP (Costello et al., 2025). Within the gut of the snake-eyed lizard (*Ophisops elegans*), six distinct MPs, namely, polyethylene terephthalate, polyacrylonitrile, polypropylene, polyethylene, poly(methyl methacrylate), and polyamide, have been detected (Altunışık et al., 2024). These MPs may pose extensive health risks since their abundance in reproductive organs can cause toxicity and further disrupt the endocrine system, impacting reproductive efficiency. MPs are a potential threat to amphibians, including frogs, toads, salamanders, and caecilians, given their important role in energy supply and their vulnerability in the food chain. Common MP types reported in wild amphibians include polyethylene, polyester, polyethylene terephthalate, and PS. Importantly, MP exposure may affect the survival and reproduction of amphibians through cellular, genetic, and mutagenic effects (da Costa Araújo et al., 2020). MPs accumulate at various life stages, including tadpoles and adults, and may influence tadpole genetic heritage. However, their impacts on amphibian metamorphosis and reproduction remain unclear.

8.6 Methodologies for assessing the reproductive toxicity of microplastics in animals

A multidisciplinary approach that combines different technologies is necessary for a proper assessment of MP reproductive toxicity. Owing to their complexities (such as cocontaminants, chemical additives, polymer types, and particle sizes), standardized and reliable methods are necessary for MP examination.

8.6.1 In vivo and in vitro study models

In vivo and in vitro methods have been used to evaluate the toxicity of MP in the reproductive system. The animals commonly used in in vivo studies to assess the reproductive toxicity of MP include livestock species, invertebrates (e.g., oysters), fish (e.g., medakas), and rodents. In vitro studies often use cell models to investigate the mechanisms underlying reproductive system abnormalities that occur in vivo. The marker for oocyte nuclear maturation, the polar body extrusion rate, is lower in mice exposed to PS during in vitro oocyte maturation (Yang et al., 2023). Furthermore, ovarian and granulosa cells exhibit increased oxidative stress and an inflammatory response following exposure to MP, whereas granulosa cells also exhibit increased apoptosis (Liu et al., 2022). In vitro studies provide controlled conditions for investigating cellular and molecular pathways and minimize the use of animals in research. However, they lack the systemic interactions present in a whole animal.

8.6.2 Biomarkers as physiological endpoints of microplastic toxicity

The identification of metabolic biomarkers associated with MP exposure is essential for demonstrating how these particles interfere with metabolic processes and alter physiological functions. The ingestion and bioaccumulation of MP have been linked to the activation of antioxidant defense mechanisms. The accumulation of ROS can cause oxidative damage to essential biomolecules if they are not effectively neutralized, as they interact with biological components. For this reason, oxidative stress biomarkers, such as malondialdehyde, and detoxification enzymes, such as glutathione S-transferase, are frequently used as biochemical indicators of oxidative stress induced by pollutants. Hormonal assays can also be used to assess reproductive endocrine disruptions, in addition to gamete quality, which involves computer-assisted sperm analysis in males and fertilization potential, oocyte maturity stage, and diameter in females. Thus reproductive toxicity can be assessed via several endpoints and biomarkers to evaluate functional abnormalities in animal reproduction.

8.6.3 Morphological and histopathological assessments

Morphological analysis can be used to assess the impact of MP exposure by examining microscopic and macroscopic changes and damage to reproductive tissues. Functional alterations in reproductive organs result from morphological changes in the reproductive system. For example, in male mice, PS MP causes germ cell death, which is characterized by morphological changes such as nuclear rupture and pyknosis (Fang et al., 2024).

8.6.4 Gene and protein expression

Exposure to specific environmental contaminants and stressors, such as MP, can change an organism's gene expression. In a recent study, rats fed oral PS MP for 180 days presented changes in testicular histology, aberrant spermatogenesis, and disrupted testosterone levels, which were mediated by the downregulation of the LHR/cAMP/PKA/StAR pathway and by luteinizing hormone. In addition, exposing granulosa cells to 500-nm PS for 24 h increased apoptosis, NLRP3, and caspase-1 cleavage (Hou et al., 2021). These findings point to oxidative stress and inflammatory responses in the gonads following MP exposure, confirming that analyzing gene and protein expression can provide a mechanistic understanding of how MP interferes with reproduction.

8.6.5 Omics approaches

By using metabolomics, proteomics, and transcriptomics to determine changes in metabolites, proteins, and gene expression at the molecular level, omics techniques enable the creation of large-scale datasets. These markers can vary depending on the pathological condition, growth stage, or physiological state of the animals, as studies have confirmed that exposing animals to MP changes the expression and relative abundance of genes and proteins related to reproduction, inflammation, oxidative stress, and apoptosis (Jin et al., 2022). Following exposure to PS MP, the proteome of adult *Crassostrea gigas* oocytes revealed alterations in arginine kinase (downregulated) and severin (upregulated), which are linked to cytoskeletal dynamics and ATP metabolism (Sussarellu et al., 2016). These findings provide further evidence regarding the detrimental impacts of MP on oocyte maturation, fertilization, and embryo development. Extensive studies adopting omic approaches for MP research are limited; however, a comprehensive assessment of MP effects on reproductive disruption can be obtained through multiomic studies correlating metabolite, protein, and gene changes.

8.7 Policy implications and mitigation strategies

The growing concern surrounding MP pollution necessitates a comprehensive approach to mitigate its impacts on human, animal, and environmental health. The literature emphasizes a multipronged strategy that integrates policy changes, waste management improvements, and public awareness initiatives to address this pressing issue effectively. International efforts and initiatives, such as the upcoming Global Plastics Treaty, ICRI Call on Microbeads Worldwide 2016, UNEA Resolutions 1/6 and 2/11 Worldwide 2016, SDG14 Worldwide 2017, and the UN Plan, aimed to monitor and regulate MP production, use, and contamination; encourage advocacy, awareness, and intersectoral collaboration to minimize MP pollution. Policies focusing on the responsible use of MP by consumers (e.g., the polluter pays principle, the user/consumer/beneficiary pays approach, and deposit–refund programs) are also crucial for reducing plastic use in daily life and preventing irresponsible usage.

In addition to policies, mitigation strategies offer rapid, effective interventions to minimize MP-related health hazards alongside regulatory policies. Dietary, genetic, and pharmacological interventions offer promising strategies to mitigate MP-related health risks and improve reproductive performance in animals. In animal models, antioxidants and anti-inflammatory compounds (resveratrol, beta-carotene, quercetin, and vitamin C), probiotics, and prebiotics can mitigate the negative impacts of MP on animal health. Both lycopene and chlorella can ameliorate MP-induced reproductive dysfunction by improving sperm quality parameters, reproductive hormonal levels, and histological configuration. In addition to nutritional mitigation strategies, genetic studies and gene editing tools (e.g., CRISPR–Cas9) focused on understanding and manipulating the biological pathways involved in MPS uptake, bioaccumulation, and toxicity may present effective and sustainable solutions through identifying genetic variants related to resistance or susceptibility to MP-induced damage or modifying genes related to endocytosis, detoxification enzymes, or cellular repair mechanisms, thereby reducing MPS internalization or enhancing clearance (Liu et al., 2023). Finally, pharmacological strategies involve the use of drugs or compounds that can counteract the oxidative stress and inflammation induced by MP exposure (e.g., N-acetylcysteine, vitamin C, or vitamin E). They can bind and facilitate the removal of hydrophobic pollutants or heavy metals adsorbed onto MPs (chelating agents), inhibit MP uptake via cellular pathways, or promote autophagy and cellular repair, thereby serving as therapeutic options. For example, N-acetylcysteine, a potential therapeutic agent against MP-induced reproductive toxicity, could counter the effects of PS MP on spermatogenesis by disrupting the blood–testis barrier via activating the overlapping with the m-AAA protease 1 homolog (OMA1)–DAP3-binding cell death enhancer 1 (DELE1)–heme-regulated inhibitor kinase (HRI) signaling pathway.

8.8 Future directions and emerging concerns

Green-dependent strategies are among the future directions for reducing MP release into the environment. These strategies focus primarily on the bioremediation of MP residuals in the environment, the replacement of MP products with biodegradable polymers, and the use of MP alternatives (Agarwal, 2020). Biological degradation of MP is one of the most innovative strategies. The use of microorganisms (e.g., *Zalerion maritimum, Ideonella sakaiensis* 201-F6, *Rhodococcus* sp., *Aspergillus* sp., *Pseudomonas and Alcanivorax, Vibrio alginolyticus, Muricauda* sp., *Lactobacillus plantarum* (MTCC 4461), *Actinomycetes, Rhodobacteraceae, Mucor rouxii* NRRL 1835, *Streptomyces bacteria,* and *Aspergillus flavus*) and photodegradation are among the effective mitigation strategies (Munhoz et al., 2023). Other strategies include using biodegradable and compostable materials sourced from natural substances such as starch, cellulose, or biopolymers (e.g., polylactic acid) that break down naturally in the environment, reducing long-term pollution and advancing the circular economy concept, since such materials can originate from biomass and waste.

While there are numerous efforts to reduce and cure the negative impacts of MP on whole ecosystem elements, there are significant concerns and gaps in knowledge, including the following: (1) the persistent, long-term, sublethal effects of MP on physiological changes, cumulative toxicity, or alterations in reproductive and immune functions over time, which require more advanced research; (2) advancements in analytical techniques and molecular markers for MP toxicity tolerance; (3) studies on the diversity of organisms affected by MP—from microorganisms to humans—while understanding species-specific responses; (4) reference information on validating toxicity across multiple taxa and developing standardized testing protocols tailored to different species; and (5) a lack of practical tools to measure the impacts of the policies (Usman et al., 2022).

Addressing these emerging concerns requires a multidisciplinary effort that combines advanced analytical methods, ecological and toxicological studies, and integrative health frameworks. Prioritizing long-term, species-specific, and mechanistic research will enhance our understanding of MP impacts, ultimately guiding practical regulatory actions and public health interventions to mitigate their risks in the environment and across the biosphere. The interconnectedness of human, animal, and environmental health underscores the importance of adopting a One Health framework in MP research. This holistic perspective considers how the MP in environmental matrices impacts reproductive health across species, including humans. Integrating epidemiological studies, ecological monitoring, and laboratory investigations can reveal transmission pathways and cumulative effects. This approach promotes interdisciplinary collaboration, informing policies that simultaneously protect reproductive health and ecosystem integrity.

8.9 Conclusion

Reproductive efficiency is an important trait for the survival of both human and animal species. The key to this is gonadal function and germ cell quality, which are negatively affected by MP exposure. MP causes reproductive dysfunction with notable sex-dependent effects on the male and female reproductive systems. Moreover, MP exposure is associated with structural damage, functional abnormalities, oxidative stress, inflammation, and apoptosis in reproductive tissues. However, investigations into the reproductive impacts of MP in food-producing animals are limited, which could provide timely information to support breeding, fertility, and production outcomes. Owing to its complexity, a multidisciplinary approach is essential for the proper evaluation of MP-induced reproductive toxicity. Importantly, adopting relevant policies, fostering intersectoral collaboration, and implementing dietary, genetic, and pharmacological interventions would offer promising strategies to mitigate MP-related health risks affecting reproductive outcomes in animals.

References

Aardema, H., Vethaak, A.D., Kamstra, J.H., Legler, J., 2024. Farm animals as a critical link between environmental and human health impacts of micro-and nanoplastics. Microplast. Nanoplast. 26624966. 4 (1). Available from: https://www.springer.com/journal/43591. doi:10.1186/s43591-024-00082-w.

Agarwal, S., 2020. Biodegradable polymers: present opportunities and challenges in providing a microplastic-free environment. Macromol. Chem. Phys. 1022-1352. 221 (6). doi:10.1002/macp.202000017.

Aghaei, Z., Mercer, G.V., Schneider, C.M., Sled, J.G., Macgowan, C.K., Baschat, A.A., Simpson, A.J., Simpson, M.J., 2022. Maternal exposure to polystyrene microplastics alters placental metabolism in mice. Metabolomics 19 (1), 1.

Albani, G., Asiedu, D., Abrokwah, S., Jónasdóttir, S.H., Nielsen, T.G., Acheampong, E., Ruiz, L.H., Ekumah, B., Koski, M., 2023. Synergistic and additive effects of microplastic, nickel and pyrene on survival, reproduction, and egestion of a tropical copepod. Aquat. Toxicol. 18791514. 265. Available from: http://www.elsevier.com/wps/find/journaldescription.cws_home/505509/description#description. doi:10.1016/j.aquatox.2023.106737.

Ali, W., Chen, Y., Shah, M.G., Buriro, R.S., Sun, J., Liu, Z., Zou, H., 2024. Ferroptosis: first evidence in premature duck ovary induced by polyvinyl chloride microplastics. Sci. Total Environ. 18791026. 933. Available from: https://www.sciencedirect.com/science/journal/00489697. doi:10.1016/j.scitotenv.2024.173032.

Alsharifi, A., Alsultan, R., Mohammed, R., 2023. Morphological and histological impacts of bisphenol a on chicken embryos. J. Life Bio Sci. Res. 2708-1818. 4 (02), 71–76. doi:10.38094/jlbsr402108.

Altunışık, A., Yıldız, M.Z., Tatlı, H.H., 2024. Microplastic accumulation in a lizard species: observations from the terrestrial environments. Environ. Pollut. 18736424. 0269-7491. 359, 124754. Available from: https://www.

sciencedirect.com/science/article/pii/S0269749124014684; https://www.sciencedirect.com/science/journal/02697491. doi:10.1016/j.envpol.2024.124754.

Amereh, F., Babaei, M., Eslami, A., Fazelipour, S., Rafiee, M., 2020. The emerging risk of exposure to nano(micro) plastics on endocrine disturbance and reproductive toxicity: from a hypothetical scenario to a global public health challenge. Environ. Pollut. 02697491. 261, 114158. doi:10.1016/j.envpol.2020.114158.

Bahrani, F., Mohammadi, A., Dobaradaran, S., De-la-Torre, G.E., Arfaeinia, H., Ramavandi, B., Saeedi, R., Tekle-R öttering, A., 2024. Occurrence of microplastics in edible tissues of livestock (cow and sheep). Environ. Sci. Pollut. Res. 16147499. 31 (14), 22145–22157. Available from: https://www.springer.com/journal/11356. doi:10.1007/s11356-024-32424-9.

Bhowmik, A., Saha, G., Saha, S.C., 2024. Microplastics in animals: the silent invasion. Pollutants. 26734672. 4 (4), 490–497. Available from: https://www.mdpi.com/journal/pollutants. doi:10.3390/pollutants4040033.

Burns, E.E., Boxall, A.B.A., 2018. Microplastics in the aquatic environment: evidence for or against adverse impacts and major knowledge gaps. Environ. Toxicol. Chem. 15528618. 37 (11), 2776–2796. Available from: http://onlinelibrary.wiley.com/journal/10.1002/(ISSN)1552-8618. doi:10.1002/etc.4268.

Cheng, H., Feng, Y., Duan, Z., Duan, X., Zhao, S., Wang, Y., Gong, Z., Wang, L., 2021. Toxicities of microplastic fibers and granules on the development of zebrafish embryos and their combined effects with cadmium. Chemosphere. 00456535. 269, 128677. doi:10.1016/j.chemosphere.2020.128677.

Choi, B.I., Harvey, A.J., Green, M.P., 2016. Bisphenol a affects early bovine embryo development and metabolism that is negated by an oestrogen receptor inhibitor. Sci. Rep. 2045-2322. 6 (1), 29318. Available from https://www.nature.com/srep/index.html. doi:10.1038/srep29318.

Chowdhury, S.R., Dey, A., Mondal, S., Gautam, M.K., 2024. Environmental microplastics and nanoplastics: effects on cardiovascular system. Toxicol. Anal. Clin. 23520078. 36 (2), 145–157. Available from: https://www.sciencedirect.com/science/journal/23520078. doi:10.1016/j.toxac.2023.11.006.

Costello, L., Zetterström, A., Gardner, P., Crespo-Picazo, J.L., Bussy, C., Kane, I., Shiels, H.A., 2025. Microplastics accumulate in all major organs of the Mediterranean loggerhead sea turtle (*Caretta caretta*). Mar. Environ. Res. 18790291. 208. Available from: https://www.sciencedirect.com/science/journal/01411136. doi:10.1016/j.marenvres.2025.107100.

da Costa Araújo, A.P., de Melo, N.F.S., de Oliveira Junior, A.G., Rodrigues, F.P., Fernandes, T., de Andrade Vieira, J. E., Rocha, T.L., Malafaia, G., 2020. How much are microplastics harmful to the health of amphibians? A study with pristine polyethylene microplastics and *Physalaemus cuvieri*. J. Hazard. Mater. 18733336. 382. Available from: https://www.elsevier.com/locate/jhazmat. doi:10.1016/j.jhazmat.2019.121066.

Fang, Q., Wang, C., Xiong, Y., 2024. Polystyrene microplastics induce male reproductive toxicity in mice by activating spermatogonium mitochondrial oxidative stress and apoptosis. Chem. Biol. Interact. 00092797. 396, 111043. doi:10.1016/j.cbi.2024.111043.

Fournier, S.B., D'Errico, J.N., Adler, D.S., Kollontzi, S., Goedken, M.J., Fabris, L., Yurkow, E.J., Stapleton, P.A., 2020. Nanopolystyrene translocation and fetal deposition after acute lung exposure during late-stage pregnancy. Part. Fibre Toxicol. 17438977. 17 (1). Available from: http://www.particleandfibretoxicology.com/. doi:10.1186/s12989-020-00385-9.

Gao, D., Zhang, C., Guo, H., Huan, X., Liu, H., Wang, Z., Bin, X., Gang, W., 2024. Low-dose polystyrene microplastics exposure impairs fertility in male mice with high-fat diet-induced obesity by affecting prostate function. Environ. Pollut. 0269-7491. 346, 123567. Available from https://www.sciencedirect.com/science/article/pii/S0269749124002811. doi:10.1016/j.envpol.2024.123567.

Gao, X., Hassan, I., Peng, Y., Huo, S., Ling, L., 2021. Behaviors and influencing factors of the heavy metals adsorption onto microplastics: a review. J. Clean. Prod. 09596526. 319, 128777. doi:10.1016/j.jclepro.2021.128777.

Grechi, N., Ferronato, G.A., Devkota, S., Ferraz, M.A.M.M., 2024. Microplastics are detected in bull epididymal sperm and polystyrene microparticles impair sperm fertilization. Biol. Reprod. 0006-3363. 111 (6), 1341–1351. doi:10.1093/biolre/ioae154.

Grechi, N., Franko, R., Rajaraman, R., Stöckl, J.B., Trapphoff, T., Dieterle, S., Fröhlich, T., Noonan, M.J., Ferraz, M.D. A.M.M., 2022. Microplastics are present in women's and cows' follicular fluid and polystyrene microplastics compromise bovine oocyte function in vitro. bioRxiv. 26928205. Available from: https://www.biorxiv.org. doi:10.1101/2022.11.04.514813.

Gupta, P., Mahapatra, A., Suman, A., Ray, S.S., Malafaia, G., Singh, R.K., 2023. Polystyrene microplastics disrupt female reproductive health and fertility via sirt1 modulation in zebrafish (*Danio rerio*). J. Hazard. Mater. 18733336. 460. Available from: https://www.elsevier.com/locate/jhazmat. doi:10.1016/j.jhazmat.2023.132359.

Hou, B., Wang, F., Liu, T., Wang, Z., 2021. Reproductive toxicity of polystyrene microplastics: in vivo experimental study on testicular toxicity in mice. J. Hazard. Mater. 03043894. 18733336. 405 (October), 124028. doi:10.1016/j.jhazmat.2020.124028.

Hou, L., Wang, D., Yin, K., Zhang, Y., Hongmin, L., Guo, T., Junbo, L., Zhao, H., Xing, M., 2022. Polystyrene microplastics induce apoptosis in chicken testis via crosstalk between NF-κB and Nrf2 pathways. Comp. Biochem. Physiol. C Toxicol. Pharmacol. 1532-0456. 262, 109444. Available from https://www.sciencedirect.com/science/article/pii/S153204562200179X. doi:10.1016/j.cbpc.2022.109444.

Huang, T., Zhang, W., Lin, T., Liu, S., Sun, Z., Liu, F., Yuan, Y., Xiang, X., Kuang, H., Yang, B. 2022 Maternal exposure to polystyrene nanoplastics during gestation and lactation induces hepatic and testicular toxicity in male mouse offspring. Food Chem. Toxicol. 0278-6915. 160, 112803.

Jia, J., Liu, Q., Zhao, E., Li, X., Xiong, X., Wu, C., 2024. Biofilm formation on microplastics and interactions with antibiotics, antibiotic resistance genes and pathogens in aquatic environment. Eco-Environ. Health. 27729850. 3 (4), 516–528. Available from: https://www.sciencedirect.com/science/journal/27729850. doi:10.1016/j.eehl.2024.05.003.

Jianing, H., Qin, X., Zhang, J., Zhu, Y., Zeng, W., Lin, Y., Liu, X., 2021. Polystyrene microplastics disturb maternal-fetal immune balance and cause reproductive toxicity in pregnant mice. Reprod. Toxicol. 08906238. 106, 42–50. doi:10.1016/j.reprotox.2021.10.002.

Jiaping, X., Cheng, Z., Wang, Y., Zhang, Q., Zhu, H., Zhao, H., Yao, Y., Wang, L., Sun, H., 2025. Microplastics containing plasticizers affect chicken reproduction via disturbing ovary DNA homologous recombination activity. J. Environ. Sci. 10010742. doi:10.1016/j.jes.2025.05.009

Jin, H., Ma, T., Sha, X., Liu, Z., Zhou, Y., Meng, X., Chen, Y., Han, X., Ding, J., 2021. Polystyrene microplastics induced male reproductive toxicity in mice. J. Hazard. Mater. 18733336. 401 (May 2020), 123430. Available from. doi:10.1016/j.jhazmat.2020.123430.

Jin, H., Yan, M., Pan, C., Liu, Z., Sha, X., Jiang, C., Li, L., Pan, M., Li, D., Han, X., Ding, J., 2022. Chronic exposure to polystyrene microplastics induced male reproductive toxicity and decreased testosterone levels via the LH-mediated LHR/cAMP/PKA/StAR pathway. Part. Fibre Toxicol. 1743-8977. 19 (1). doi:10.1186/s12989-022-00453-2.

Karami, A., Romano, N., Galloway, T., Hamzah, H., 2016. Virgin microplastics cause toxicity and modulate the impacts of phenanthrene on biomarker responses in African catfish (*Clarias gariepinus*). Environ. Res. 00139351. 151, 58–70. doi:10.1016/j.envres.2016.07.024.

Kehinde, S.A., Fatokun, T.P., Olajide, A.T., Praveena, S.M., Sokan-Adeaga, A.A., Adekunle, A.P., Fouad, D., Papadakis, M., 2024. Impact of polyethylene microplastics exposure on kallikrein-3 levels, steroidal-thyroidal hormones, and antioxidant status in murine model: protective potentials of naringin. Sci. Rep. 20452322. 14 (1). Available from: https://www.nature.com/srep/. doi:10.1038/s41598-024-74637-5.

Liu, Q., Chen, Z., Chen, Y., Yang, F., Yao, W., Xie, Y. 2022 Microplastics contamination in eggs: detection, occurrence and status. Food Chem. 0308-8146. 397, 133771.

Liu, X., Liang, C., Fan, J., Zhou, M., Chang, Z., Li, L., 2023. Polyvinyl chloride microplastics induce changes in gene expression and organ histology along the HPG axis in *Cyprinus carpio* var. larvae. Aquat. Toxicol. 18791514. 258. Available from: http://www.elsevier.com/wps/find/journaldescription.cws_home/505509/description#description. doi:10.1016/j.aquatox.2023.106483.

Luo, T., Zhang, Y., Wang, C., Wang, X., Zhou, J., Shen, M., Zhao, Y., Zhengwei, F., Jin, Y., 2019. Maternal exposure to different sizes of polystyrene microplastics during gestation causes metabolic disorders in their offspring. Environ. Pollut. 02697491. 255 (Pt 1), 113122. doi:10.1016/j.envpol.2019.113122.

Martins, A., Guilhermino, L., 2018. Transgenerational effects and recovery of microplastics exposure in model populations of the freshwater cladoceran *Daphnia magna* Straus. Sci. Total Environ. 00489697. 631-632, 421–428. doi:10.1016/j.scitotenv.2018.03.054.

Munhoz, D.R., Harkes, P., Beriot, N., Larreta, J., Basurko, O.C., 2023. Microplastics: a review of policies and responses. Microplastics. 26738929. 2 (1), 1–26. Available from: https://www.mdpi.com/journal/microplastics. doi:10.3390/microplastics2010001.

OECD, 2022. Global Plastics Outlook: Economic Drivers, Environmental Impacts and Policy Options. OECD Publishing, Paris. doi:10.1787/de747aef-en.

Oke, O.E., Akosile, O.A., Oni, A.I., Opowoye, I.O., Ishola, C.A., Adebiyi, J.O., Odeyemi, A.J., Adjei-Mensah, B., Uyanga, V.A., Abioja, M.O., 2024. Oxidative stress in poultry production. Poult. Sci. 00325791. 103 (9), 104003. doi:10.1016/j.psj.2024.104003.

Park, E.J., Han, J.S., Park, E.J., Seong, E., Lee, G.H., Kim, D.W., Son, H.Y., Han, H.Y., Lee, B.S., 2020. Repeated-oral dose toxicity of polyethylene microplastics and the possible implications on reproduction and development of the next generation. Toxicol. Lett. 18793169. 324, 75–85. Available from: https://www.elsevier.com/locate/toxlet. doi:10.1016/j.toxlet.2020.01.008.

Santos, D., Cabecinha, E., Luzio, A., Bellas, J., Monteiro, S.M., 2025. Long-term effects of individual and combined exposure to microplastics and copper in zebrafish hypothalamic-pituitary-gonadal axis – a multi-biomarker evaluation. J. Environ. Manage. 03014797. 378, 124770. Available from https://www.sciencedirect.com/science/article/pii/S0301479725007467. doi:10.1016/j.jenvman.2025.124770.

Savoca, D., Arculeo, M., Vecchioni, L., Cambera, I., Visconti, G., Melfi, R., Arizza, V., Piccionello, A.P., Buscemi, S., Pace, A., 2021. Can phthalates move into the eggs of the loggerhead sea turtle Caretta caretta? The case of the nests on the Linosa Island in the Mediterranean Sea. Mar. Pollut. Bull. 0025326X. 168, 112395. doi:10.1016/j.marpolbul.2021.112395.

Sun, Z., Wen, Y., Zhang, F., Fu, Z., Yuan, Y., Kuang, H., Kuang, X., Huang, J., Zheng, L., Zhang, D., 2023. Exposure to nanoplastics induces mitochondrial impairment and cytomembrane destruction in Leydig cells. Ecotoxicol. Environ. Saf. 01476513. 255, 114796. doi:10.1016/j.ecoenv.2023.114796.

Sussarellu, R., Suquet, M., Thomas, Y., Lambert, C., Fabioux, C., Pernet, M.E.J., Goïc, N.L., Quillien, V., Mingant, C., Epelboin, Y., Corporeau, C., Guyomarch, J., Robbens, J., Paul-Pont, I., Soudant, P., Huvet, A., 2016. Oyster reproduction is affected by exposure to polystyrene microplastics. Proc. Natl. Acad. Sci. U. S. A. 10916490. 113 (9), 2430–2435. Available from: http://www.pnas.org/content/113/9/2430.full.pdf. doi:10.1073/pnas.1519019113.

Ullah, S., Ahmad, S., Guo, X., Ullah, S., Ullah, S., Nabi, G., Wanghe, K., 2023. A review of the endocrine disrupting effects of micro and nano plastic and their associated chemicals in mammals. Front. Endocrinol. 1664-2392. 13. doi:10.3389/fendo.2022.1084236.

Usman, S., Razis, A.F.A., Shaari, K., Azmai, M.N.A., Saad, M.Z., Isa, N.M., Nazarudin, M.F., 2022. The burden of microplastics pollution and contending policies and regulations. Int. J. Environ. Res. Public Health. 16604601. 19 (11). Available from: https://www.mdpi.com/1660-4601/19/11/6773/pdf?version=1654081760. doi:10.3390/ijerph19116773.

Wahyudi, A., Setyobudi, R.H., Iswahyudi, I., Garfansa, M.P., Savitri, E.S., Prahardika, B.A., Patmawati, P., Putri, R. D., Dodi, D.S., Putra, N.S., Yunita, E., Wahono, S.K., 2025. Determination and quantification of microplastics in goat farms. Bioresour. Technol. Rep. 2589014X. 31. Available from: https://www.sciencedirect.com/science/journal/2589014X. doi:10.1016/j.biteb.2025.102199.

Wang, J., Yuejiao, L., Lin, L., Zheng, M., Zhang, X., Tian, H., Wang, W., Shaoguo, R., 2019. Polystyrene microplastics cause tissue damages, sex-specific reproductive disruption and transgenerational effects in marine medaka (*Oryzias melastigma*). Environ. Pollut. 02697491. 254 (Pt B), 113024. Available from: https://www.sciencedirect.com/science/article/pii/S0269749119311261. doi:10.1016/j.envpol.2019.113024.

Wei, Z., Wang, Y., Wang, S., Xie, J., Han, Q., Chen, M., 2022. Comparing the effects of polystyrene microplastics exposure on reproduction and fertility in male and female mice. Toxicology. 0300483X. 465, 153059. doi:10.1016/j.tox.2021.153059.

Wu, B., Wu, X., Liu, S., Wang, Z., Chen, L., 2019. Size-dependent effects of polystyrene microplastics on cytotoxicity and efflux pump inhibition in human Caco-2 cells. Chemosphere. 00456535. 221, 333–341. doi:10.1016/j.chemosphere.2019.01.056.

Xie, X., Deng, T., Duan, J., Xie, J., Yuan, J., Chen, M., 2020. Exposure to polystyrene microplastics causes reproductive toxicity through oxidative stress and activation of the p38 MAPK signaling pathway. Ecotoxicol. Environ. Saf. 01476513. 190, 110133. doi:10.1016/j.ecoenv.2019.110133.

Yang, D., Zhu, J., Zhou, X., Pan, D., Nan, S., Yin, R., Chen, J., 2022a. Polystyrene micro-and nano-particle coexposure injures fetal thalamus by inducing ROS-mediated cell apoptosis. Environ. Int. 166, 107362.

Yang, J., Kamstra, J., Legler, J., Aardema, H., 2023. The impact of microplastics on female reproduction and early life. Anim. Reprod. 1984-3143. 20 (2). doi:10.1590/1984-3143-ar2023-0037.

Yang, Y.K.X., Ge, S.J., Su, Q.L., Chen, J.J., Wu, J., Kang, K., 2024. Effects of polyvinyl chloride microplastics on the reproductive system, intestinal structure, and microflora in male and female mice. Vet. Sci. 23067381. 11 (10). Available from http://www.mdpi.com/journal/vetsci/about. doi:10.3390/vetsci11100488.

Yang, Z.S., Bai, Y.L., Jin, C.H., Na, J., Zhang, R., Gao, Y., Pan, G.W., Yan, L.J., Sun, W., 2022b. Evidence on invasion of blood, adipose tissues, nervous system and reproductive system of mice after a single oral exposure: nanoplastics versus microplastics. Biomed. Environ. Sci. 08953988. 35 (11), 1025–1037. Available from: http://www.elsevier.com. doi:10.3967/bes2022.131.

You, X., Zhang, Z., Tian, G., Zhang, Y., Pei, Y., Wu, Y., Li, G., Wang, Q., Yang, Y., 2025. The impact of polyethylene microplastics exposure on the, growth performance, reproductive performance, antioxidant capacity, and intestinal microbiota of quails. Ecotoxicol. Environ. Saf. 01476513. 296, 118218. doi:10.1016/j.ecoenv.2025.118218.

Yuan, Y., Qin, Y., Wang, M., Xu, W., Chen, Y., Zheng, L., Chen, W., Luo, T., 2022. Microplastics from agricultural plastic mulch films: a mini-review of their impacts on the animal reproductive system. Ecotoxicol. Environ. Saf. 10902414. 244. Available from: http://www.elsevier.com/inca/publications/store/6/2/2/8/1/9/index.htt. doi:10.1016/j.ecoenv.2022.114030.

Zhang, X., Ma, T., Sun, J., Fang, Y., Wang, Y., Li, X., Zhao, Z., Su, N., Liu, M., Bai, J., Liu, J., Wu, J., 2025. Polyethylene terephthalate (PET) microplastics induced male reproductive toxicity in mice by activating the p38-MAPK pathway. Ecotoxicol. Environ. Saf. 01476513. 302, 118756. doi:10.1016/j.ecoenv.2025.118756.

Zhang, Y., Wang, X., Zhao, Y., Zhao, J., Yu, T., Yao, Y., Zhao, R., Yu, R., Liu, J., Su, J., 2023. Reproductive toxicity of microplastics in female mice and their offspring from induction of oxidative stress. Environ. Pollut. 02697491. 327, 121482. doi:10.1016/j.envpol.2023.121482.

CHAPTER

9

Epidemiological trends related to microplastics and environmental exposure

Khudrotul Nisa Indriyasari[a], Agoes Soegianto[b], and Eko Prasetyo Kuncoro[b]

[a]Mathematics and Natural Sciences Doctoral Program, Universitas Airlangga, Surabaya, Indonesia
[b]Department of Biology, Universitas Airlangga, Surabaya, Indonesia

9.1 Introduction

Microplastic (MP) pollution has been identified as an element of the triple planetary crisis, documented across diverse environments, from the depths of the Mariana Trench to the peak of Mount Everest and from marine biota to the human placenta. In the hadal zone of the trench, the MP levels are much higher than those in the open-ocean subsurface waters. In bottom waters, they range from 2.06 to 13.51 items/L, and in sediments, they range from 200 to 2200 items/L, which is higher than that reported in most other deep-sea records (Peng et al., 2018). MPs are found in stream water (approximately 1 item/L) and snow (up to 95 items/L), even at 8440 m on Everest. This is one of the highest levels ever recorded (Napper et al., 2020; Wang et al., 2025). The widespread presence of MPs in the environment can be attributed to multiple anthropogenic sources, including agricultural, industrial, landfill, and urban activities. The plastic waste released into the environment undergoes physical, chemical, and biological degradation, resulting in the formation of smaller fragments known as secondary MPs. These particles are transported through surface runoff and rivers, ultimately reaching marine environments. Once in the ocean, their distribution is influenced by wind patterns and sea currents, facilitating their dispersion across various regions. Some aquatic organisms ingest MPs, whereas others settle on the seabed. Over time, sediment disturbances can resuspend these particles, allowing them to reenter the water column and potentially be ingested by marine biota (Fig. 9.1). This widespread presence has also coincided with

https://doi.org/10.1016/B978-0-443-45635-0.00003-3

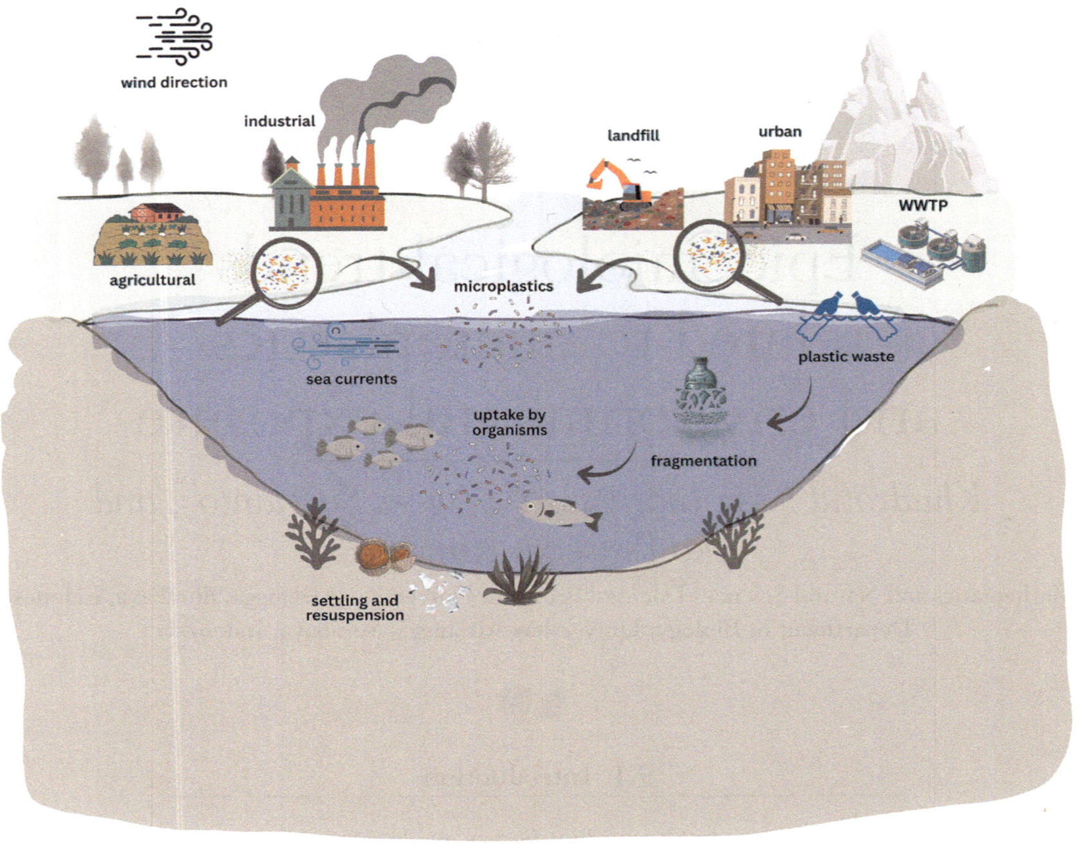

FIG. 9.1 Overview of microplastic sources and their transport through aquatic systems to the ocean. Particles are transported through surface runoff and rivers, ultimately reaching marine environments. Once in the ocean, their distribution is influenced by wind patterns and sea currents, facilitating their dispersion across various regions. Some aquatic organisms ingest microplastics, whereas others settle on the seabed. Over time, sediment disturbances can resuspend these particles, allowing them to reenter the water column and potentially be ingested by marine biota.

a significant rise in environmental release over the past few decades, with current measurements showing increasing MP loads in oceans, rivers, soils, and the atmosphere.

Biologically, zooplankton, such as copepods in Tampa Bay, consume MPs at a rate of 15.38 items per 1000 individuals (Fibbe et al., 2023), and fish larvae in the northern South China Sea have encounter rates as high as 143% (Sun et al., 2017). Human exposure assessments indicate that dietary intake contributes approximately 39,000–52,000 ingested MP particles annually, increasing to approximately 74,000–121,000 particles when inhalation is also considered (Cox et al., 2019). Recent findings suggest that MPs may penetrate human tissues; notably, MPs have been identified in the placenta (Kim et al., 2025), prompting concerns about possible developmental effects, although the underlying mechanisms related to outcomes such as intrauterine growth restriction are still under investigation (Ali-Hassanzadeh et al., 2025). Micro/nanoplastics in carotid atheromas are linked to an ~4.5-fold increased risk of myocardial infarction, stroke, or death over ~3 years, highlighting a potential cardiovascular risk (Marfella et al., 2024). The increasing number of environmental reservoirs, measurable human exposure, presence in vital tissues, and development of clinical correlations render MPs a growing public health issue that requires immediate monitoring and risk evaluation.

It is essential to examine MPs from an epidemiological standpoint, given their presence in the air we breathe (Wright et al., 2020), the water we drink (Yang et al., 2024), the food we eat (De-la-Torre, 2020), and even within human tissues (Yang et al., 2024). Because this is so common, exposure is unavoidable, raising concerns about potential health effects. The Global Burden of Disease program estimated that exposure to airborne particulate matter caused approximately 4.2 million premature deaths in 2015 (Forouzanfar et al., 2016), highlighting the significant public health risks associated with inhaled particles. The precise role of MPs remains unknown, but they may be an emerging factor in this issue (Wang et al., 2020). Experimental evidence has demonstrated that MPs can provoke inflammation, oxidative stress, and immune dysregulation, mechanisms associated with chronic conditions, including inflammatory bowel disease (IBD), cardiovascular disorders, and specific cancers (Cheng et al., 2024; Han et al., 2024; Mahmud et al., 2024). However, connecting these biological processes to health outcomes at the population level is still difficult. Finding MPs in people is difficult because of the risk of contamination and the lack of standardized protocols (Primpke et al., 2020). Most studies use environmental proxies rather than direct biomarkers of exposure, making it hard to establish that specific factors cause specific diseases. These limitations underscore the critical need for comprehensive epidemiological studies, bolstered by reliable detection methods, extended cohort surveillance, and authenticated biomarkers, to elucidate risks and protect public health.

This urgency becomes clearer when framed against historical shifts in disease patterns. Before the middle of the 20th century, mortality was dominated by infectious diseases such as tuberculosis, pneumonia, and gastrointestinal infections, reflecting inadequate sanitation and limited medical care. As industrialization has advanced, chronic noncommunicable diseases—including cancer, diabetes, and cardiovascular disease—have risen to prominence, driven by lifestyle changes, urbanization, and exposure to industrial pollutants (Formichi et al., 2025; Ozdemir et al., 2024). The prevalence of cancer, once relatively rare, has increased with exposure to occupational hazards, tobacco, and chemical by-products of industrial activity (Zhuikova et al., 2020). Within this continuum, large-scale plastic production since the 1950s has introduced a novel exposure pathway—MPs—with outputs projected to double by 2040 and potentially quadruple to 650 million tons annually by 2050 (Fayshal, 2024); this burden is set to intensify. While their role in disease etiology is still being defined, MPs may represent an emerging class of inhaled and ingested pollutants with carcinogenic or co-carcinogenic potential, adding to the complex interplay of environmental drivers shaping today's chronic disease landscape.

The epidemiological relevance of MPs becomes most evident when considered alongside the growing global burden of chronic diseases. In recent decades, the global incidence of IBD and colorectal cancer (CRC) has steadily increased (Dharni et al., 2024), with early-onset CRC among those born after 1960 emerging as a particularly concerning trend

(Aldhaleei et al., 2024; Chen and Yang, 2024; Read and Sylla, 2020). Such shifts cannot be attributed solely to genetic factors but instead point to environmental and lifestyle influences (Hayes, 2021; Takada et al., 2023). MPs, which are now detected in human stool, blood, placenta, and even tumor tissues (Horvatits et al., 2022), represent one of these emerging exposures. Their presence in the gastrointestinal tract coincides with alterations in the gut microbiota, chronic inflammation, and mucosal damage, mechanisms directly implicated in IBD and CRC pathogenesis (Cheng et al., 2024; Dennis et al., 2025; Fournier et al., 2023). Although causal relationships remain to be established, the detection of MPs in human tissues alongside biological pathways consistent with IBD and CRC pathogenesis suggests a potential association, even though current epidemiological evidence remains too limited for firm conclusions. Situating MPs within these wider epidemiological patterns strengthens the rationale for incorporating MP monitoring into population health research.

Building on the preceding discussion, this chapter synthesizes current evidence on the epidemiological dimensions of MP exposure and its effects on human health. It examines pathways of contamination across environments and entry routes into the human body; reviews emerging associations with gastrointestinal, metabolic, reproductive, immune, and neurological outcomes; and considers early epidemiological signals from tissue and clinical studies. The aim is not only to summarize existing knowledge but also to identify the critical gaps that remain, particularly in population-level research, where long-term risks are still poorly defined. In doing so, this chapter seeks to provide a foundation for future investigations and encourage new approaches to understanding the role of MPs in contemporary disease burdens.

9.2 Pathways exposing humans to MPs

Human exposure to MPs occurs through several pathways that differ in their contributions to the internal dose and potential health relevance. Ingestion and inhalation are currently regarded as the dominant routes, supported by environmental occurrence, human activity patterns, and biomonitoring evidence. In contrast, dermal contact is generally considered a minor but biologically plausible pathway. The following subsections critically review the available evidence for each pathway, highlighting dominant sources and uptake mechanisms.

9.2.1 Ingestion: Major routes for human exposure

Ingestion is usually considered the primary mechanism of human exposure to MPs, with annual intake estimated at 39,000–52,000 particles and over 120,000 when inhalation is considered (Cox et al., 2019). MPs have been found in many different foods, such as seafood (Daniel et al., 2020; Werorilangi et al., 2025), sea salt (Amqam et al., 2024), bottled and tap water (Gambino et al., 2022), beer, honey, rice, sugar, and even fresh fruits and vegetables (Jasso–Salcedo et al., 2024; Mondal et al., 2025). Mollusks, crabs, and small fish are especially important nutritional sources because people often eat them whole, including their gastrointestinal tracts, where MPs accumulate (Hermabessiere et al., 2023). According to Van Cauwenberghe and Janssen (2014), people in Europe who eat mollusks may consume between 1800 and 11,000 MPs each year. In Asia, where seafood is more commonly eaten, this number may reach 55,000 MPs per year (Danopoulos et al., 2020). In Indonesia, where shellfish consumption is high, the predicted intake from green mussels alone may reach approximately 775,000 MP items per person per year (Irnidayanti et al., 2023), whereas the consumption of cockles (*Anadara granosa*) may contribute up to 156,000 MP items per adult per year (Rahmatin et al., 2024). Upon entering the digestive system, MPs can be absorbed by transcytosis in enterocytes, internalized by phagocytes, or absorbed through passive diffusion or active endocytic routes, including clathrin-mediated endocytosis and fast endophilin-mediated endocytosis (DeLoid et al., 2024), eventually entering the circulatory and lymphatic systems (Prata, 2023).

9.2.2 Inhalation: MPs in airborne particles

Inhalation is increasingly recognized as a key route of exposure to MPs, along with ingestion (Fig. 9.2). Airborne MPs come from a variety of sources, such as shedding textile fibers, wearing out tires, road dust, breaking down urban waste, and resuspending domestic

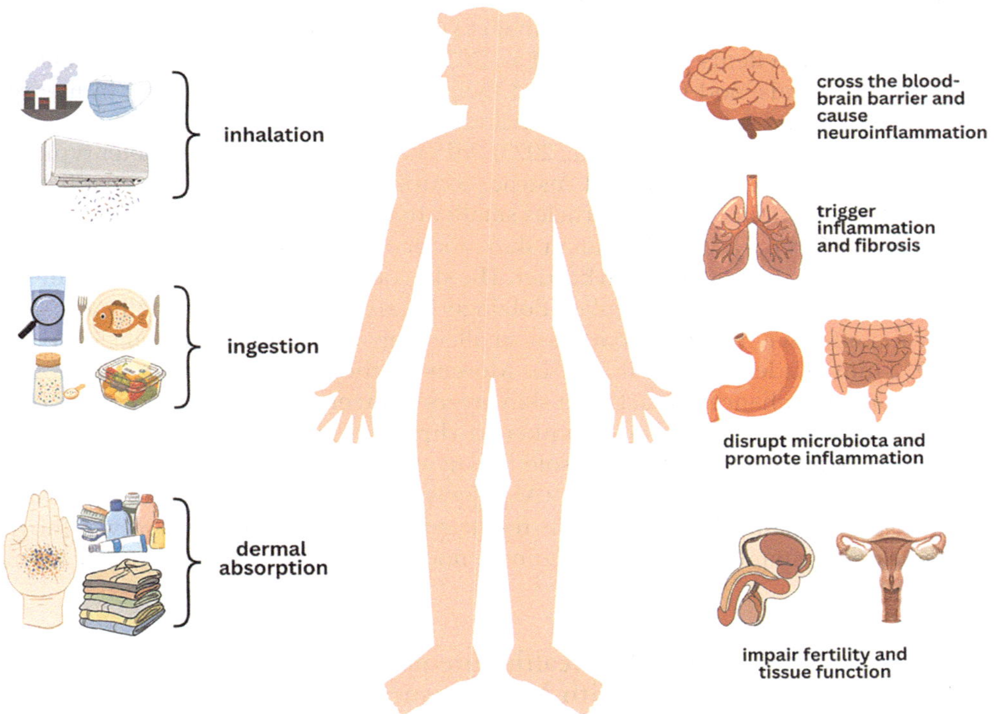

FIG. 9.2 Key routes of human microplastic exposure and their potential impacts on the brain, lungs, gastrointestinal tract, and reproductive system. Illustration showing how microplastics enter the human body and affect major organs. Exposure occurs through inhalation, ingestion, and dermal contact, leading to impacts on the brain, lungs, gut, and reproductive system.

dust (Rednikin et al., 2024; Zhao et al., 2023). While studies on airborne MPs have shown inconsistent patterns, with several reporting significantly higher indoor concentrations than outdoor concentrations (Jahanzaib et al., 2025; Lin et al., 2025), it is estimated that adults can breathe up to 3000 MPs a day (Liao et al., 2021). Once inhaled, their small size allows them to pass through epithelial barriers into the circulation, where they can then reach organs such as the liver, kidneys, and brain (Rajendran and Chandrasekaran, 2023; Ziani et al., 2023). Particles in the 5–10-μm range are worrisome, as alveolar macrophages can engulf them, facilitating their easier translocation across alveolar and gastrointestinal barriers into the bloodstream and lymphatic system (Dzierżyński et al., 2024). Moreover, particles measuring 1–2.5 μm typically reach the terminal bronchiole, where significant build-up and tissue damage occur, as frequently observed in centrilobular emphysema (Schraufnagel, 2020). MPs of larger size that accumulate in the upper respiratory tract are often removed by mucociliary clearance, in which mucus captures the particles and facilitates their movement toward the nose or gastrointestinal tract for removal (Chen et al., 2025).

9.2.3 Dermal contact: Minor exposure from household and consumer products

Compared with ingestion and inhalation, dermal contact is typically considered a minor pathway for MP exposure; however, it remains significant because of its use in cosmetics, textiles, and household items (Chaudhari et al., 2024). MPs are highly likely to be detected on the human scalp, hands, and facial skin, and their penetration is significantly influenced by particle size, shape, and surface chemistry, including hydrophobicity, surface charge, and the presence of functional groups (Hou et al., 2023). MPs exceeding 1 μm in diameter typically do not permeate the skin efficiently and instead accumulate on the surface, creating a film (Lademann et al., 2004). In contrast, particles smaller than 2 μm have been shown to penetrate the dermal layer within 1 h in a 3D human skin model (Song et al., 2024), indicating that MPs below this threshold can cross the skin barrier. The skin offers various penetration pathways—intercellular, intracellular, and follicular (Bolzinger et al., 2012)—with the follicular channel facilitating the effective ingress of particles smaller than 1.5 μm into hair follicles to depths surpassing 2 mm (Lademann et al., 2004). Moreover, the development of a nanoparticle protein corona may enhance contact with skin layers (Hou et al., 2023), and upon absorption, MPs and their additives (e.g., polybrominated diphenyl ethers or possibly carcinogenic additives such as butylated hydroxyanisole (Indriyasari et al., 2023) may infiltrate systemic circulation via the cutaneous vasculature (Akpojevwe Abafe et al., 2024), thus increasing the hazardous load. Current evidence suggests restricted dermal absorption; however, the mechanisms and extent remain inadequately defined, underscoring the need for further research.

9.3 Possible health outcomes associated with MP exposure

This section synthesizes current epidemiological, biomonitoring, and experimental evidence on health outcomes potentially associated with MP exposure. The discussion spans respiratory, gastrointestinal, neurological, reproductive, and metabolic end points, with particular attention given to the strength, consistency, and limitations of human data relative to supporting mechanistic findings. Emphasis is placed on outcomes for which converging lines of evidence suggest biological plausibility while clearly distinguishing established occupational observations from emerging signals in the general population.

9.3.1 Pulmonary inflammation and fibrotic lung disease

Links between inhaled synthetic polymer particles and respiratory disease were well-documented before the MP era. Sentinel outbreaks of interstitial lung disease among nylon flock and textile workers in North America revealed a distinct clinicopathologic entity later termed "flock worker's lung," characterized by lymphocytic bronchiolitis, interstitial pneumonitis, and fibrosis (Kern, 1998; Kern et al., 1998). These case series and retrospective cohort studies established that inhaled synthetic microfibers are causal exposures, demonstrating that chronic occupational exposure to polymer dust can lead to severe pulmonary disease. In the general population, biomonitoring has since confirmed the deposition of MPs in lung tissue, with μFTIR and Raman analyses identifying fibers and fragments—including sub10 μm particles—capable of persisting in the lower airways (Jenner et al., 2022; Riaz et al., 2025). Mechanistic studies add biological plausibility: human alveolar cells exposed to polystyrene micro/nanoplastics are prone to oxidative stress, proinflammatory signaling, and organelle dysfunction (Shahzadi et al., 2023), whereas in mice, polystyrene MPs are associated with increased respiratory resistance, reduced lung compliance, and the influx of inflammatory cells, including tumor necrosis factor alpha (TNF-α), interleukin 1 beta (IL-1β), interleukin 5 (IL-5), interleukin 6 (IL-6), and macrophage inflammatory protein 2-alpha (MIP-2α) (Balkrishna et al., 2025). Sunlight-degraded polyethylene terephthalate (PET) has similar effects, increasing airway resistance and lung inflammation (Ishihara et al., 2025). Other studies reported that PSMPs drive oxidative stress, apoptosis, and collagen deposition in lung tissue, culminating in fibrotic changes (Cao et al., 2023). The convergence of occupational case studies, direct detection of MPs in human lung tissue, and consistent inflammatory responses in experimental models suggest that MPs may be risk factors for chronic respiratory morbidity. However, it should be noted that the strongest epidemiological evidence comes from small occupational cohorts with exposures far exceeding those of the general population, and direct data on inhalation-related lung effects remain limited and cannot easily be extended to environmental exposures.

9.3.2 Gastrointestinal disorders and CRC

Definitive epidemiological evidence on the effects of MPs on the gastrointestinal tract is lacking. However, signals from human biomonitoring studies, along with experimental findings in animal models and in vitro systems, suggest that long-term exposure to MPs may lead to gastrointestinal disorders ranging from gastritis and IBD to malignancies such as gastric cancer and CRC (Zhao et al., 2024). Biomonitoring first established that MPs are routinely ingested and excreted, with stool samples from participants in multiple countries containing up to nine polymer types, including polypropylene (PP) and PET (Schwabl et al., 2019). Yan et al. (2022) reported significantly higher fecal MP loads in IBD patients than in controls, with burdens correlating with disease severity and dominated by PET and polyamide, averaging ~42 items/g dry stool compared with ~28 items/g in controls. Early data also suggest that the fecal MP burden is associated with microbiome disruption (Gao et al., 2025), which is consistent with experimental evidence that MPs modulate the gut microbial composition, deplete beneficial taxa, enrich potential pathogens, and thereby contribute to dysbiosis and downstream intestinal or systemic inflammation (Gałęcka et al., 2025).

The evidence linking MPs to CRC remains preliminary but increasingly consistent, coinciding with the rising global incidence of early-onset CRC—particularly among cohorts born after 1960—where environmental and lifestyle factors are considered major drivers (Siegel et al., 2020). In a study of 11 adults from northeastern Peninsular Malaysia, MPs were detected in colectomy samples at mean concentrations of ~28 particles/g, which were primarily composed of polycarbonate, polyamide, and PP (Ibrahim et al., 2021). Recent analyses suggest that tumoral tissues carry higher MP loads—such as polyethylene and polymethyl methacrylate—than adjacent nontumoral tissues do, although larger studies with appropriate controls are still needed to clarify any potential link with CRC (Cetin et al., 2023). A systematic review (Mashayekhi-Sardoo et al., 2025) reported 25.9–32.2 MPs/g of tumor tissue, most frequently polyamide, polycarbonate, and PP, but highlighted major limitations, including the small number of studies, limited sample sizes, reliance mainly on cross-sectional designs, and the absence of robust case–control, cohort, or longitudinal evidence. These findings indicate that MPs may serve as potential environmental co-exposure factors in gastrointestinal disease; however, the existing evidence is insufficient and methodologically constrained to establish causal relationships, highlighting the pressing necessity for extensive, contamination-controlled, prospective epidemiological studies.

9.3.3 Neurotoxicity and brain impacts

Recent evidence shows that MP exposure is possibly associated with neurotoxic outcomes, although current findings remain preliminary and are largely derived from experimental models with limited direct human data. Nihart et al. (2025) provided direct evidence of MP contamination in the human frontal cortex, with median concentrations of 3345 $\mu g\ g^{-1}$ in 2016, 4917 $\mu g\ g^{-1}$ in 2024, and markedly higher levels in dementia patients (26,076 $\mu g\ g^{-1}$). Direct evidence of cerebral deposition was also reported by Amato-Lourenço et al. (2024), who reported 43.8% PP particles in the olfactory bulbs of 8 out of 15 patients at autopsy, suggesting that inhaled particles may bypass the blood–brain barrier (BBB) via the olfactory nerve. This mechanistic plausibility is reinforced by Cho et al. (2024), who demonstrated that engineered BBB models showed size-dependent translocation of polystyrene MPs, with smaller particles (0.2 μm) causing pronounced barrier disruption, particularly under inflammatory conditions. In experimental systems, Ehsanifar and Yavari (2025) synthesized in vivo and in vitro studies showing that MPs and nanoplastics trigger oxidative stress, microglial activation, acetylcholinesterase inhibition, and neuronal apoptosis, resulting in behavioral disturbances across species. Functional consequences in genetically susceptible hosts were reported by Gaspar et al. (2025), who found that short-term MP exposure in humanized APOE4 mice—the major Alzheimer's risk allele—produced sex-dependent impairments in recognition memory and altered astrocytic and microglial markers, suggesting a synergistic effect of environmental plastic exposure and genetic predisposition. Collectively, these studies converge on a model in which MPs can access the brain via olfactory or systemic routes, breach the BBB, and elicit inflammatory and neurodegenerative processes. However, limitations—including small autopsy series, reliance on polystyrene bead models, high-dose animal exposures, and the absence of longitudinal human cohorts—underscore that these findings remain associative. The accumulating evidence nonetheless indicates that chronic MP exposure could contribute to neurotoxicity and neurodegenerative risk in human populations.

9.3.4 Reproductive and developmental health

The direct detection of MPs has established strong evidence of internal exposure in almost all major human reproductive system compartments, including the placenta, testis, semen, endometrium, follicular fluid, amniotic fluid, and breast milk (Halfar et al., 2023; Hu et al., 2024; Hunt et al., 2024; Montano et al., 2025; Qin et al., 2024; Ragusa et al., 2022; Han et al., 2024). The presence of these compounds in the placenta has been confirmed by Raman and Pyrolysis-Gas Chromatography/Mass Spectrometry (Py-GC/MS) analyses, initially in small series (Ragusa et al., 2022) and more recently in all 62 samples studied, with concentrations ranging from 6.5 to 685 μg/g (Garcia et al., 2024). In contrast, longitudinal analyses have shown a temporal increase in placental burden over the past 15 years (Weingrill et al., 2023). MPs have also been identified in the female reproductive tract, including uterine fibroids (Xu et al., 2024), the endometrium (Qin et al., 2024), and endometrial polyps, where burdens are significantly greater than those in normal tissue, with in vitro work showing that polystyrene MPs stimulate stromal cell proliferation via PI3K/AKT signaling (He and Zhang, 2025). During assisted reproduction, MPs < 10 μm in size were detected in ovarian follicular fluid from 14 of 18 in vitro fertilization (IVF) patients, at an average of 2191 particles/mL, although no correlation was detected with AMH levels, fertilization, miscarriage, or live birth outcomes (Montano et al., 2025).

In males, MPs have consistently been detected in semen and urine across various cohorts (Han et al., 2024) and quantified in all analyzed testicular samples, yielding a mean burden of 328 μg/g (Hu et al., 2024). After controlling for confounding variables, this study provides the first epidemiological evidence that increased exposure to mixed MPs, especially PTFE, is significantly associated with lower sperm motility, count, and concentration (Han et al., 2024). The integration of tissue detection and experimental data bolsters biological plausibility, although such outcome associations remain scarce in women, where existing studies are predominantly descriptive or yield null results in small IVF cohorts. Moreover, canine models corroborate the accumulation of MPs in the testis, suggesting exploratory associations with sperm parameters (Hu et al., 2024). In vivo rodent studies further substantiate these concerns by showing that chronic MP exposure reduces sperm count and motility, disrupts the estrous cycle, and impairs ovarian folliculogenesis and embryo implantation (Fang et al., 2024; Ilechukwu et al., 2022; Wei et al., 2022). While larger prospective studies are necessary to collect data on human female outcomes, this expanding body of evidence indicates that MPs are not only detectable in essential reproductive tissues but also may compromise male reproductive health.

9.3.5 Metabolic disruptions and obesity

Epidemiologically, there are signals—such as diet-associated MP exposure and obesity risk—but no extensive or causal evidence from humans is currently available. The most convincing evidence comes from animal and mechanistic studies demonstrating that MPs disrupt glucose metabolism, alter lipid balance, affect how gut bacteria function, modify insulin signaling, and impair fat tissue function (Shruti et al., 2025). Rodents exposed to polystyrene MPs exhibit hepatic steatosis, insulin resistance, adipocyte hypertrophy, and impaired thermogenesis (Roh et al., 2024; Tao et al., 2024; Han et al., 2024; Zhao et al., 2024). Furthermore, in vivo animal studies revealed that MPs modify gut microbiota composition, cause intestinal barrier dysfunction, and facilitate endotoxin transfer, thereby initiating TLR4/NF-κB-mediated inflammatory signaling and reducing insulin sensitivity (Okamura et al., 2023; Hou et al., 2023). In addition, MPs interfere with adipose tissue regulation by altering beige fat function, lowering uncoupling protein 1 (UCP1) levels, and increasing lipid accumulation (Han et al., 2024; Zhao et al., 2024). While direct evidence from humans is lacking, the mechanistic plausibility suggests that MPs may contribute to metabolic disruption as potential environmental obesogens.

9.4 Scarcity and insufficiency of population-level data on epidemiological studies linked to MPs

Although MPs are becoming acknowledged as an environmental health issue, the epidemiological evidence remains highly constrained. Thus far, the strongest evidence for human health consequences (Table 9.1) comes from occupational inhalation investigations, with respiratory effects (e.g., in the lungs of flock workers) as the only consistently identified pathology. In contrast, evidence for systemic effects in an epidemiological context—such as gastrointestinal, reproductive, neurotoxic, or metabolic impacts—remains limited and of insufficient quality, primarily owing to the use of small biomonitoring cohorts, cross-sectional tissue studies, and extrapolation from in vitro experiments and high-dose animal models, without addressing co-pollutant exposures associated with MPs, rather than relying on robust prospective epidemiological designs. These gaps indicate multiple challenges:

- The lack of dependable exposure assessment tools for oral, inhaled, and dermally absorbed MPs in population settings;
- Confounding exposures within occupational cohorts that hinder exclusive attribution to MPs; and
- The absence of extensive long-term cohort data, which obstructs a thorough evaluation of chronic outcomes.

Collectively, these constraints indicate that the present results cannot be extrapolated to the broader population. Consequently, assertions regarding the systemic health effects of MPs in humans should be approached with caution: although experimental and animal

TABLE 9.1 Evidence base for possible human health outcomes associated with microplastic exposure, organized by study type.

Health outcome	Study type	Notes/results	Reference(s)
Pulmonary	Human case series and retrospective cohort	Occupational nylon/textile workers; interstitial lung disease ("flock worker's lung")	(Kern, 1998; Kern et al., 1998)
	Human biomonitoring	MPs were detected in 11 out of 13 lung tissues, with an average of 0.69 ± 0.84 MP/g after contamination adjustment	(Jenner et al., 2022)
	In vitro (A549 human alveolar cells)	Exposure to MP and NP typically 7.5–30 μg/cm^2 for 2 h–96 h, resulting in oxidative stress, proinflammatory signaling, apoptosis, mitochondrial dysfunction, and morphological changes	(Shahzadi et al., 2023)
	In vivo (mice)	4 mg/kg, 3x/week × 21 days, PS MPs resulting in airway hyperresponsiveness, pulmonary inflammation, elevated proinflammatory cytokines, upregulated inflammatory gene expression, and histopathological signs of leukocyte infiltration	(Balkrishna et al., 2025)
	In vivo (mice)	40 mg/kg, daily × 21 days, PSMPs (1–5 μm vs. 10–20 μm) resulting in TLR2/NF-κB-mediated lung injury, especially with smaller MPs	(Cao et al., 2023)
Gastrointestinal	Human biomonitoring/cross-sectional	Higher fecal MP loads in IBD patients (42.8 items/g) vs. the healthy group (28 items/g); fecal MP concentration positively correlated with IBD disease activity	(Yan et al., 2022)
	Human biomonitoring (tissue)	MPs were detected in colorectal cancer tissues (an average of 28.1 items/g in 11 patients), but no statistical correlation was performed	(Ibrahim et al., 2021)
	Human biomonitoring (tissue)	MPs detected in tumoral colon tissue (702.7 items/g), nontumoral colon tissue (207.8 items/g), and control tissue (218.3 items/g)	(Cetin et al., 2023)
	Human biomonitoring	All 8/8 stool samples contained MPs, with a median of 20 items per 10 g of stool	(Schwabl et al., 2019)
	Human biomonitoring, animal experiment	MPs are detectable in human blood and are associated with gut microbiome dysbiosis, enrichment of virulence and quorum-sensing genes, and microbial enzymes linked to MP biodegradation. Chronic PS MP exposure in mice (80 mg/kg/day, 14 weeks) produced similar disruptions	(Gao et al., 2025)
	In vitro/in vivo	Gut microbiota modulation, dysbiosis, inflammation	(Gałęcka et al., 2025)

Continued

TABLE 9.1 Evidence base for possible human health outcomes associated with microplastic exposure, organized by study type—cont'd

Health outcome	Study type	Notes/results	Reference(s)
Reproductive	Human biomonitoring (placenta)	MPs detected in the placenta (6.5–685 μg/g), showing a temporal increase	(Ragusa et al., 2022; Garcia et al., 2024)
	Human biomonitoring (testis, semen)	MPs quantified in semen/testis; associations with reduced sperm motility/count	(Hu et al., 2024; Han et al., 2024)
	Human biomonitoring (follicular fluid)	MPs < 10 μm detected in IVF patients	(Montano et al., 2025)
	Human biomonitoring (endometrium, fibroids, polyps)	Higher burdens in diseased vs. normal tissue	(Qin et al., 2024; Xu et al., 2024)
	In vitro	MPs stimulate stromal cell proliferation (PI3K/AKT pathway)	(He and Zhang, 2025)
	In vivo (rodents, canines)	Reduced sperm motility, impaired folliculogenesis, disrupted cycles	(Fang et al., 2024; Wei et al., 2022)
Neurotoxicity	Human biomonitoring (autopsy, cortex)	MPs were detected in the frontal cortex; higher levels were found in dementia cases	(Nihart et al., 2025)
	Human biomonitoring (autopsy, olfactory bulb)	MPs in 8/15 subjects; possible olfactory nerve pathway	(Amato-Lourenço et al., 2024)
	In vitro (BBB model)	Size-dependent translocation, barrier disruption	(Cho et al., 2024)
	In vivo (rodents)	Oxidative stress, microglial activation, and memory deficits	(Gaspar et al., 2025; Ehsanifar and Yavari, 2025)
Metabolic	–	No direct human evidence available	–
	In vivo (rodents)	Hepatic steatosis, insulin resistance, adipocyte hypertrophy	(Roh et al., 2024; Tao et al., 2024; Zhao et al., 2024)
	In vivo (rodents)	Gut microbiota disruption, intestinal barrier dysfunction, and inflammation	(Okamura et al., 2023; Hou et al., 2023)
	Mechanistic review	MPs as potential environmental obesogens	(Shruti et al., 2025)

Note: *A549*, human alveolar adenocarcinoma cell line #549; *BBB*, blood–brain barrier; *IBD*, inflammatory bowel disease; *IVF*, in vitro fertilization; *MPs*, microplastics; *NF-κB*, nuclear factor kappa-light-chain-enhancer of activated B cells; *NPs*, nanoplastics; *PI3K/AKT* pathway, phosphoinositide 3-kinase/protein kinase B signaling pathway; *PS MPs*, polystyrene microplastics; *TLR2*, Toll-like receptor 2.

studies suggest credible biological mechanisms, human epidemiological data are inadequate to confirm causal relationships or validate risks beyond the respiratory system, highlighting the critical necessity for standardized biomonitoring and longitudinal human research.

9.5 Future directions and research priorities

MP epidemiology needs to move from descriptive discovery to causal inference in the future, which calls for methodological harmonization, thorough population-level research, and translational integration with toxicology. The lack of defined exposure measurements that enable meaningful comparisons across studies and populations is the greatest obstacle to risk assessment, not the lack of detection. Therefore

- The creation and global implementation of standardized protocols for human MP evaluation, including the required quality assurance measures (field blanks, procedural blanks, and spike recoveries), should be the top priority. Building global confidence in reported values requires interlaboratory validation efforts.
- Nested case–control designs with recurrent biospecimen sampling and long-term cohort studies are urgently needed. In other words, we require long-term population studies that follow individuals over years; collect blood, stool, urine, and tissue samples regularly; and use those samples to assess who becomes ill and who remains well. These studies are far more robust than the small cross-sectional studies conducted today. To increase biological plausibility and pathway connection, exposure assessment with mechanistic biomarkers, such as oxidative stress, systemic inflammation, gut microbiome diversity, and endocrine disruption, should be incorporated.
- Since learning from the lungs of flock workers and the synthetic textile industries already shows that inhaling polymers can cause chronic respiratory disease, epidemiological designs should take occupational exposures into account.
- Studies need to go beyond compartments of individual exposure to integrated risk frameworks that connect estimates of disease burden, human exposure, and environmental burden. Policymakers can be guided toward mitigation by forecasting population-level impacts via modeled exposure–response connections in conjunction with Global Burden of Disease metrics.
- Cooperation across disciplines is crucial. To ensure that MP research not only enhances mechanistic understanding but also produces useful data for risk assessment and regulation, chemists, toxicologists, epidemiologists, physicians, and policy specialists must collaborate. Instead of moving forward with prevention and intervention, the field risks remaining stuck in descriptive case reporting in the absence of such integration.

9.6 Conclusion

MPs have evolved from being perceived as ecological pollutants to emerging as public health issues, with increasing evidence of their presence in human tissues and their potential involvement in chronic disease mechanisms. The existing data elucidate potential mechanisms linking MPs to pulmonary inflammation, gastrointestinal disorders, neurotoxicity, reproductive dysfunction, and metabolic disruption. Nonetheless, despite increasing biological plausibility and experimental validation, epidemiological evidence in human populations is still limited, fragmented, and methodologically constrained. The majority of findings originate from occupational cohorts, case reports, or limited biomonitoring studies that have not yet determined causality. Major obstacles include the lack of standardized detection methodologies, insufficient long-term cohort studies, and limited global monitoring of exposure. To move the field forward, we need standardized analytical protocols within epidemiological study designs, strict contamination control, the use of mechanistic biomarkers, and studies that look ahead across different groups of people. It is important to fill these gaps to clarify causal links and provide people with information they can use to assess risks, establish regulations, and protect public health as the amount of plastic in the world grows rapidly.

References

Abafe, O.A., Harrad, S., Abdallah, M.A.-E., 2024. Assessment of human dermal absorption of flame retardant additives in polyethylene and polypropylene microplastics using 3D human skin equivalent models. Environ. Int. 01604120. 186, 108635. doi:10.1016/j.envint.2024.108635.

Aldhaleei, W.A., Wallace, M.B., Bhagavathula, A.S., 2024. Trends and age–period–cohort effect on the incidence of early-onset colorectal cancer (20–44 Years) from 1990 to 2021 in the United States. Cancers. 20726694. 16 (16). Available from: http://www.mdpi.com/journal/cancers/. doi:10.3390/cancers16162883.

Ali-Hassanzadeh, M., Arefinia, N., Ghoreshi, Z.A.S., Askarpour, H., Mashayekhi-Sardoo, H., 2025. The effects of exposure to microplastics on female reproductive health and pregnancy outcomes: a systematic review and meta-analysis. Reprod. Toxicol. 18731708. 135. Available from: https://www.sciencedirect.com/science/journal/08906238. doi:10.1016/j.reprotox.2025.108932.

Amato-Lourenço, L.F., Dantas, K.C., Júnior, G.R., Paes, V.R., Ando, R.A., De Oliveira Freitas, R., Da Costa, O.M.M. M., Rabelo, R.S., Soares Bispo, K.C., Carvalho-Oliveira, R., Mauad, T., 2024. Microplastics in the olfactory bulb of the human brain. JAMA Netw. Open. 25743805. 7 (9), e2440018. Available from: https://jamanetwork.com/journals/jamanetworkopen. doi:10.1001/jamanetworkopen.2024.40018.

Amqam, H., Natsir, M.F., Yusriani, Z.F., 2024. Microplastic contamination in Indonesian consumable salts. J. Sea Res. 13851101. 198. Available from: https://www.sciencedirect.com/science/journal/13851101. doi:10.1016/j.seares.2024.102475.

Balkrishna, A., Tiwari, A., Sinha, S., Kumari, A., Gohel, V., Dev, R., Bhattacharya, K., Varshney, A., 2025. Polystyrene microplastic induced airway hyper-responsiveness, and pulmonary inflammation are mitigated by bronchom treatment in murine model of lung disease. Biomed. Pharmacother. 07533322. 187, 118122. doi:10.1016/j.biopha.2025.118122.

Bolzinger, M.A., Briançon, S., Pelletier, J., Chevalier, Y., 2012. Penetration of drugs through skin, a complex rate-controlling membrane. Curr. Opin. Colloid Interface Sci. 18790399. 17 (3), 156–165. doi:10.1016/j.cocis.2012.02.001.

Cao, J., Xu, R., Geng, Y., Xu, S., Guo, M., 2023. Exposure to polystyrene microplastics triggers lung injury via targeting toll-like receptor 2 and activation of the NF-κB signal in mice. Environ. Pollut. 02697491. 320, 121068. doi:10.1016/j.envpol.2023.121068.

Cetin, M., Demirkaya Miloglu, F., Kilic Baygutalp, N., Ceylan, O., Yildirim, S., Eser, G., Gul, H.I., 2023. Higher number of microplastics in tumoral colon tissues from patients with colorectal adenocarcinoma. Environ. Chem. Lett. 16103661. 21 (2), 639–646. Available from: https://www.springer.com/journal/10311. doi:10.1007/s10311-022-01560-4.

Chaudhari, S., Parmar, H., Samnani, P., 2024. Microplastics in personal care products and cosmetics. In: Microplastics Pollution and Its Remediation. Springer Nature, India, pp. 171–188. Available from: https://link.springer.com/10.1007/978-981-97-4068-0. doi:10.1007/978-981-97-4068-0_9.

Chen, C.Y., Chen, S.Y., Liao, C.M., 2025. Regional and population-scale trends in human inhalation exposure to airborne microplastics: implications for health risk assessment. Environ. Pollut. 18736424. 371. Available from: https://www.sciencedirect.com/science/journal/02697491. doi:10.1016/j.envpol.2025.125950.

Chen, J., Yang, W., 2024. Comparing cohort and period trends of early-onset colorectal cancer: a global analysis. JNCI Cancer Spectr. 2515-5091. 8 (4). doi:10.1093/jncics/pkae052.

Cheng, Y., Yang, Y., Bai, L., Cui, J., 2024. Microplastics: an often-overlooked issue in the transition from chronic inflammation to cancer. J. Transl. Med. 14795876. 22 (1). Available from: https://translational-medicine.biomedcentral.com. doi:10.1186/s12967-024-05731-5.

Cho, Y., Seo, E.U., Hwang, K.S., Kim, H., Choi, J., Kim, H.N., 2024. Evaluation of size-dependent uptake, transport and cytotoxicity of polystyrene microplastic in a blood-brain barrier (BBB) model. Nano Converg. 21965404. 11 (1). Available from: https://nanoconvergencejournal.springeropen.com. doi:10.1186/s40580-024-00448-z.

Cox, K.D., Covernton, G.A., Davies, H.L., Dower, J.F., Juanes, F., Dudas, S.E., 2019. Human consumption of microplastics. Environ. Sci. Technol. 15205851. 53 (12), 7068–7074. Available from: http://pubs.acs.org/journal/esthag. doi:10.1021/acs.est.9b01517.

Daniel, D.B., Ashraf, P.M., Thomas, S.N., 2020. Microplastics in the edible and inedible tissues of pelagic fishes sold for human consumption in Kerala, India. Environ. Pollut. 18736424. 266. Available from: https://www.journals.elsevier.com/environmental-pollution. doi:10.1016/j.envpol.2020.115365.

Danopoulos, E., Jenner, L.C., Twiddy, M., Rotchell, J.M., 2020. Microplastic contamination of seafood intended for human consumption: a systematic review and meta-analysis. Environ. Health Perspect. 15529924. 128 (12). Available from: https://ehp.niehs.nih.gov/doi/pdf/10.1289/EHP7171. doi:10.1289/EHP7171.

De-la-Torre, G.E., 2020. Microplastics: an emerging threat to food security and human health. J. Food Sci. Technol. 09758402. 57 (5), 1601–1608. Available from: http://www.springerlink.com/content/121580/. doi:10.1007/s13197-019-04138-1.

DeLoid, G.M., Yang, Z., Bazina, L., Kharaghani, D., Sadrieh, F., Demokritou, P., 2024. Mechanisms of ingested polystyrene micro-nanoplastics (MNPs) uptake and translocation in an in vitro tri-culture small intestinal epithelium. J. Hazard. Mater. 03043894. 473, 134706. doi:10.1016/j.jhazmat.2024.134706.

Dennis, J., Arulraj, D., Mistri, T.K., 2025. Unseen toxins: exploring the human health consequences of micro and nanoplastics. Toxicol. Rep. 22147500. 14, 101955. doi:10.1016/j.toxrep.2025.101955.

Dharni, K., Singh, A., Sharma, S., Midha, V., Kaur, K., Mahajan, R., Dulai, P.S., Sood, A., 2024. Trends of inflammatory bowel disease from the Global Burden of Disease Study (1990–2019). Indian J. Gastroenterol. 09750711. 43 (1), 188–198. Available from: https://www.springer.com/journal/12664. doi:10.1007/s12664-023-01430-z.

Dzierżyński, E., Gawlik, P.J., Puźniak, D., Flieger, W., Jóźwik, K., Teresiński, G., Forma, A., Wdowiak, P., Baj, J., Flieger, J., 2024. Microplastics in the human body: exposure, detection, and risk of carcinogenesis: a state-of-the-art review. Cancers. 2072-6694. 16 (21), 3703. doi:10.3390/cancers16213703.

Ehsanifar, M., Yavari, Z., 2025. Neurotoxicity following exposure to micro and nanoplastics. OBM Neurobiol. 25734407. 09 (01), 1–20. doi:10.21926/obm.neurobiol.2501277.

Fang, Q., Wang, C., Xiong, Y., 2024. Polystyrene microplastics induce male reproductive toxicity in mice by activating spermatogonium mitochondrial oxidative stress and apoptosis. Chem. Biol. Interact. 00092797. 396, 111043. doi:10.1016/j.cbi.2024.111043.

Fayshal, M.A., 2024. Current practices of plastic waste management, environmental impacts, and potential alternatives for reducing pollution and improving management. Heliyon. 24058440. 10 (23), e40838. doi:10.1016/j.heliyon.2024.e40838.

Fibbe, M.C., Carroll, D., Gowans, S., Siuda, A.N.S., 2023. Ingestion of microplastics by copepods in Tampa Bay Estuary, FL. Front. Ecol. Evol. 2296701X. 11. Available from: https://www.journal.frontiersin.org/journal/ecology-and-evolution. doi:10.3389/fevo.2023.1143377.

Formichi, C., Caprio, S., Nigi, L., Dotta, F., 2025. The impact of environmental pollution on metabolic health and the risk of non-communicable chronic metabolic diseases in humans. Nutr. Metab. Cardiovasc. Dis. 09394753. 35 (6), 103975. doi:10.1016/j.numecd.2025.103975.

Forouzanfar, M.H., Afshin, A., Alexander, L.T., Biryukov, S., Brauer, M., Cercy, K., Charlson, F.J., Cohen, A.J., Dandona, L., Estep, K., Ferrari, A.J., Frostad, J.J., Fullman, N., Godwin, W.W., Griswold, M., Hay, S.I., Kyu, H.H., Larson, H.J., Lim, S.S., Liu, P.Y., Lopez, A.D., Lozano, R., Marczak, L., Mokdad, A.H., Moradi-Lakeh, M., Naghavi, M., Reitsma, M.B., Roth, G.A., Sur, P.J., Vos, T., Wagner, J.A., Wang, H., Zhao, Y., Zhou, M., Barber, R.M., Bell, B., Blore, J.D., Casey, D.C., Coates, M.M., Cooperrider, K., Cornaby, L., Dicker, D., Erskine, H.E., Fleming, T., Foreman, K., Gakidou, E., Haagsma, J.A., Johnson, C.O., Kemmer, L., Ku, T., Leung, J., Masiye, F., Millear, A., Mirarefin, M., Misganaw, A., Mullany, E., Mumford, J.E., Ng, M., Olsen, H., Rao, P., Reinig, N., Roman, Y., Sandar, L., Santomauro, D.F., Slepak, E.L., Sorensen, R.J.D., Thomas, B.A., Vollset, S.E., Whiteford, H.A., Zipkin, B., Murray, C.J.L., Mock, C.N., Anderson, B.O., Futran, N.D.,

Anderson, H.R., Bhutta, Z.A., Nisar, M.I., Akseer, N., Krueger, H., Gotay, C.C., Kissoon, N., Kopec, J.A., Pourmalek, F., Burnett, R., Abajobir, A.A., Knibbs, L.D., Veerman, J.L., Lalloo, R., Scott, J.G., Alam, N.K.M., Gouda, H.N., Guo, Y., McGrath, J.J., Jeemon, P., Dandona, R., Goenka, S., Kumar, G.A., Gething, P.W., Bisanzio, D., Deribew, A., Darby, S.C., Ali, R., Bennett, D.A., Jha, V., Kinfu, Y., McKee, M., Murthy, G.V.S., Pearce, N., Stöckl, H., Duan, L., Jin, Y., Li, Y., Liu, S., Wang, L., Ye, P., Liang, X., Azzopardi, P., Patton, G.C., Meretoja, A., Alam, K., Borschmann, R., Colquhoun, S.M., Weintraub, R.G., Szoeke, C.E.I., Ademi, Z., Taylor, H. R., Wijeratne, T., Batis, C., Barquera, S., Campos-Nonato, I.R., Contreras, A.G., Cuevas-Nasu, L., De, V., Gomez-Dantes, H., Heredia-Pi, I.B., Medina, C., Mejia-Rodriguez, F., Montañez Hernandez, J.C., Razo-García, C.A., Rivera, J.A., Rodríguez-Ramírez, S., Sánchez-Pimienta, T.G., Servan-Mori, E.E., Shamah, T., Mensah, G.A., Hoff, H.J., Neal, B., Driscoll, T.R., Kemp, A.H., Leigh, J., Mekonnen, A.B., Bhatt, S., Fürst, T., Piel, F.B., Rodriguez, A., Hutchings, S.J., Majeed, A., Soljak, M., Salomon, J.A., Thorne-Lyman, A.L., Ajala, O.N., Bärnighausen, T., Cahill, L.E., Ding, E.L., Farvid, M.S., Khatibzadeh, S., Wagner, G.R., Shrime, M.G., Fitchett, J. R.A., Aasvang, G.M., Savic, M., Abate, K.H., Gebrehiwot, T.T., Gebremedhin, A.T., Abbafati, C., Abbas, K.M., Abd-Allah, F., Abdulle, A.M., Abera, S.F., Melaku, Y.A., Abyu, G.Y., Betsu, B.D., Hailu, G.B., Tekle, D.Y., Yalew, A.Z., Abraham, B., Abu-Raddad, L.J., Adebiyi, A.O., Adedeji, I.A., Adou, A.K., Adsuar, J.C., Agardh, E. E., Rehm, J., Badawi, A., Popova, S., Agarwal, A., Ahmad, A., Akinyemiju, T.F., Schwebel, D.C., Singh, J.A., Al-Aly, Z., Aldhahri, S.F., Altirkawi, K.A., Terkawi, A.S., Aldridge, R.W., Tillmann, T., Alemu, Z.A., Tegegne, T.K., Alkerwi, A., Alla, F., Guillemin, F., Allebeck, P., Rabiee, R.H.S., Fereshtehnejad, S.M., Kivipelto, M., Carrero, J.J., Weiderpass, E., Havmoeller, R., Sindi, S., Alsharif, U., Alvarez, E., Alvis-Guzman, N., Amare, A.T., Ciobanu, L.G., Taye, B.W., Amberbir, A., Amegah, A.K., Amini, H., Karema, C.K., Ammar, W., Harb, H.L., Amrock, S.M., Andersen, H.H., Antonio, C.A.T., Faraon, E.J.A., Anwari, P., Ärnlöv, J., Larsson, A., Artaman, A., Asayesh, H., Asghar, R.J., Assadi, R., Atique, S., Avokpaho, E.F.G.A., Awasthi, A., Ayala, B.P., Bacha, U., Bahit, M.C., Balakrishnan, K., Barac, A., Barker-Collo, S.L., Del Pozo-cruz, B., Mohammed, S., Barregard, L., Petzold, M., Barrero, L.H., Basu, S., Del, L.C., Bazargan-Hejazi, S., Beardsley, J., Bedi, N., Beghi, E., Sheth, K.N., Bell, M.L., Huang, J.J., Bello, A.K., Santos, I.S., Bensenor, I.M., Lotufo, P.A., Berhane, A., Wolfe, C.D., Bernabé, E., Roba, H.S., Beyene, A.S., Hassen, T.A., Mesfin, Y.M., Bhala, N., Bhansali, A., Biadgilign, S., Bikbov, B., Bjertness, E., Htet, A.S., Boufous, S., Degenhardt, L., Resnikoff, S., Calabria, B., Bourne, R.R.A., Brainin, M., Brazinova, A., Majdan, M., Shen, J., Breitborde, N.J.K., Brenner, H., Schöttker, B., Broday, D.M., Brugha, T.S., Brunekreef, B., Kromhout, H., Butt, Z.A., van Donkelaar, A., Martin, R.V., Cárdenas, R., Carpenter, D.O., Castañeda-Orjuela, C.A., Castillo, J., Castro, R.E., Catalá-López, F., Chang, J., Chiang, P.P., Chibalabala, M., Chimed-Ochir, O., Jiang, Y., Takahashi, K., Chisumpa, V.H., Mapoma, C.C., Chitheer, A.A., Choi, J.J., Christensen, H., Christopher, D.J., Cooper, L.T., Crump, J.A., Poulton, R.G., Damasceno, A., Dargan, P.I., Das Neves, J., Davis, A.C., Newton, J.N., Steel, N., Davletov, K., de Castro, E.F., De, D., Dellavalle, R.P., Des, D.C., Dharmaratne, S.D., Dhillon, P.K., Lal, D.K., Zodpey, S., Diaz-Torné, C., Dorsey, E.R., Doyle, K.E., Dubey, M., Rahman, M.H.U., Ram, U., Singh, A., Yadav, A. K., Duncan, B.B., Kieling, C., Schmidt, M.I., Elyazar, I., Endries, A.Y., Ermakov, S.P., Eshrati, B., Farzadfar, F., Kasaeian, A., Parsaeian, M., Esteghamati, A., Hafezi-Nejad, N., Sheikhbahaei, S., Fahimi, S., Malekzadeh, R., Roshandel, G., Sepanlou, S.G., Hassanvand, M.S., Heydarpour, P., Rahimi-Movaghar, V., Yaseri, M., Farid, T.A., Khan, A.R., Farinha, C.S.E.S., Faro, A., Feigin, V.L., Fernandes, J.G., Fischer, F., Foigt, N., Shiue, I., Fowkes, F.G.R., Franklin, R.C., Garcia-Basteiro, A.L., Geleijnse, J.M., Jibat, T., Gessner, B.D., Tefera, W., Giref, A.Z., Haile, D., Manamo, W.A.A., Giroud, M., Gishu, M.D., Martinez-Raga, J., Gomez-Cabrera, M.C., Gona, P., Goodridge, A., Gopalani, S.V., Goto, A., Inoue, M., Gugnani, H.C., Gupta, R., Gutiérrez, R.A., Orozco, R., Halasa, Y.A., Undurraga, E.A., Hamadeh, R.R., Hamidi, S., Handal, A.J., Hankey, G.J., Hao, Y., Harikrishnan, S., Haro, J.M., Hernández-Llanes, N.F., Hoek, H.W., Tura, A.K., Horino, M., Horita, N., Hosgood, H.D., Hoy, D.G., Hsairi, M., Hu, G., Husseini, A., Huybrechts, I., Iburg, K.M., Idrisov, B.T., Kwan, G.F., Ileanu, B.V., Pana, A., Kawakami, N., Shibuya, K., Jacobs, T.A., Jacobsen, K.H., Jahanmehr, N., Jakovljevic, M.B., Jansen, H.A.F., Jassal, S.K., Stein, M.B., Javanbakht, M., Jayaraman, S.P., Jayatilleke, A.U., Jee, S.H., Jonas, J.B., Kabir, Z., Kalkonde, Y., Kamal, R., She, J., Kan, H., Karch, A., Karimkhani, C., Kaul, A., Kazi, D.S., Keiyoro, P.N., Parry, C.D., Kengne, A.P., Matzopoulos, R., Wiysonge, C.S., Stein, D.J., Mayosi, B.M., Keren, A., Khader, Y.S., Khan, E.A., Khan, G., Khang, Y.H., Won, S., Khera, S., Tavakkoli, M., Khoja, T.A.M., Khubchandani, J., Kim, C., Kim, D., Kimokoti, R.W., Kokubo, Y., Koul, P.A., Koyanagi, A., Kravchenko, M., Varakin, Y.Y., Kuate, B., Kuchenbecker, R.S., Kucuk, B., Kuipers, E.J., Lallukka, T., Shiri, R., Meretoja, T.J., Lan, Q., Latif, A.A., Lawrynowicz, A.E.B., Leasher, J.L., Levi, M., Li, X., Liang, J., Lloyd, B.K., Logroscino, G., Lunevicius, R., Pope, D., Mahdavi, M., Malta, D.C., Marcenes, W., Matsushita, K., Nachega, J.B., Tran, B.X., Meaney, P.A.,

Mehari, A., Tedla, B.A., Memish, Z.A., Mendoza, W., Mensink, G.B.M., Mhimbira, F.A., Miller, T.R., Mills, E.J., Mohammadi, A., Mola, G.L.D., Monasta, L., Morawska, L., Norman, R.E., Mori, R., Mozaff, D., Shi, P., Werdecker, A., Mueller, U.O., Paternina, A.J., Westerman, R., Seedat, S., Naheed, A., Nangia, V., Nassiri, N., Nguyen, Q.L., Nkamedjie, P.M., Norheim, O.F., Norrving, B., Nyakarahuka, L., Obermeyer, C.M., Ogbo, F.A., Oh, I., Oladimeji, O., Sartorius, B., Olusanya, B.O., Olivares, P.R., Olusanya, J.O., Opio, J.N., Oren, E., Ortiz, A., Ota, E., Mahesh, P.A., Park, E., Patel, T., Patil, S.T., Patten, S.B., Wang, J., Pereira, D.M., Cortinovis, M., Giussani, G., Perico, N., Remuzzi, G., Pesudovs, K., Phillips, M.R., Pillay, J.D., Plass, D., Tobollik, M., Polinder, S., Pond, C.D., Pope, C.A., Prasad, N.M., Qorbani, M., Radfar, A., Rafay, A., Rana, S.M., Rahman, M., Rahman, S.U., Rajsic, S., Rai, R.K., Raju, M., Ranganathan, K., Refaat, A.H., Rehm, C.D., Ribeiro, A.L., Rojas-Rueda, D., Roy, A., Satpathy, M., Tandon, N., Rothenbacher, D., Saleh, M.M., Sanabria, J.R., Sanchez-Riera, L., Sanchez-Niño, M.D., Sarmiento-Suarez, R., Sawhney, M., Schmidhuber, J., Schneider, I.J.C., Schutte, A.E., Silva, D.A.S., Shahraz, S., Shin, M., Shaheen, A., Shaikh, M.A., Sharma, R., Shigematsu, M., Yoon, S., Shishani, K., Sigfusdottir, I.D., Singh, P.K., Silveira, D.G.A., Silverberg, J.I., Yano, Y., Soneji, S., Stranges, S., Steckling, N., Sreeramareddy, C.T., Stathopoulou, V., Stroumpoulis, K., Sunguya, B.F., Swaminathan, S., Sykes, B. L., Tabarés-Seisdedos, R., Talongwa, R.T., Tanne, D., Tuzcu, E.M., Thakur, J., Shaddick, G., Thomas, M.L., Thrift, A.G., Thurston, G.D., Thomson, A.J., Topor-Madry, R., Topouzis, F., Towbin, J.A., Uthman, O.A., Tobe-Gai, R., Tsilimparis, N., Tsala, Z., Tyrovolas, S., Ukwaja, K.N., van Os, J., Vasankari, T., Venketasubramanian, N., Violante, F.S., Waller, S.G., Uneke, C.J., Wang, Y., Weichenthal, S., Woolf, A.D., Xavier, D., Xu, G., Yakob, B., Yip, P., Kesavachandran, C.N., Montico, M., Ronfani, L., Yu, C., Zaidi, Z., Yonemoto, N., Younis, M.Z., Wubshet, M., Zuhlke, L.J., Zaki, M.E., Zhu, J., 2016. Global, regional, and national comparative risk assessment of 79 behavioural, environmental and occupational, and metabolic risks or clusters of risks, 1990–2015: a systematic analysis for the Global Burden of Disease Study 2015. Lancet. 1474547X. 388 (10053), 1659–1724. Available from: http://www.journals.elsevier.com/the-lancet/. doi:10.1016/S0140-6736(16)31679-8.

Fournier, E., Leveque, M., Ruiz, P., Ratel, J., Durif, C., Chalancon, S., Amiard, F., Edely, M., Bezirard, V., Gaultier, E., Lamas, B., Houdeau, E., Lagarde, F., Engel, E., Etienne-Mesmin, L., Blanquet-Diot, S., Mercier-Bonin, M., 2023. Microplastics: what happens in the human digestive tract? First evidences in adults using in vitro gut models. J. Hazard. Mater. 03043894. 442, 130010. doi:10.1016/j.jhazmat.2022.130010.

Gałęcka, I., Rychlik, A., Całka, J., 2025. Influence of selected dosages of plastic microparticles on the porcine fecal microbiome. Sci. Rep. 2045-2322. 15 (1). doi:10.1038/s41598-024-80337-x.

Gambino, I., Bagordo, F., Grassi, T., Panico, A., De Donno, A., 2022. Occurrence of microplastics in tap and bottled water: current knowledge. Int. J. Environ. Res. Public Health. 1660-4601. 19 (9), 5283. doi:10.3390/ijerph19095283.

Gao, B., Chen, L., Wu, L., Zhang, S., Zhao, S., Mo, Z., Chen, Z., Tu, P., 2025. Association between microplastics and the functionalities of human gut microbiome. Ecotoxicol. Environ. Saf. 01476513. 290, 117497. doi:10.1016/j.ecoenv.2024.117497.

Garcia, M.A., Liu, R., Nihart, A., Hayek, E.E., Castillo, E., Barrozo, E.R., Suter, M.A., Bleske, B., Scott, J., Forsythe, K., Gonzalez-Estrella, J., Aagaard, K.M., Campen, M.J., 2024. Quantitation and identification of microplastics accumulation in human placental specimens using pyrolysis gas chromatography mass spectrometry. Toxicol. Sci. 10960929. 199 (1), 81–88. Available from: http://toxsci.oxfordjournals.org/. doi:10.1093/toxsci/kfae021.

Gaspar, L., Bartman, S., Tobias-Wallingford, H., Coppotelli, G., Ross, J.M., 2025. Short-term exposure to polystyrene microplastics alters cognition, immune, and metabolic markers in an APOE genotype and sex-dependent manner. bioRxiv. 26928205. Available from: https://www.biorxiv.org. doi:10.1101/2025.05.02.651942.

Halfar, J., Čabanová, K., Vávra, K., Delongová, P., Motyka, O., Špaček, R., Kukutschová, J., Šimetka, O., Heviánková, S., 2023. Microplastics and additives in patients with preterm birth: the first evidence of their presence in both human amniotic fluid and placenta. Chemosphere. 00456535. 343, 140301. doi:10.1016/j.chemosphere.2023.140301.

Han, M., Liu, H., Zhu, T., Tang, S., Li, Y., Zhu, C., Zhou, Z., Jiang, Q., 2024. Toxic effects of micro(nano)-plastics on terrestrial ecosystems and human health. TrAC Trends Anal. Chem. 01659936. 172, 117517. doi:10.1016/j.trac.2023.117517.

Hayes, R.B., 2021. Advances in understanding early-onset colorectal cancer. Cancer Epidemiol. Biomarkers Prev. 15387755. 30 (10), 1775–1777. Available from: https://cebp.aacrjournals.org/content/30/10/1775. doi:10.1158/1055-9965.EPI-21-0844.

Hermabessiere, L., Thaysen, C., Sherlock, C., Choy, C.A., Rochman, C.M., 2023. Little evidence for bioaccumulation or biomagnification of microplastics in a deep-sea food web. Mar. Ecol. Prog. Ser. 0171-8630. 724, 17–32. doi:10.3354/meps14458.

Horvatits, T., Tamminga, M., Liu, B., Sebode, M., Carambia, A., Fischer, L., Püschel, K., Huber, S., Fischer, E.K., 2022. Microplastics detected in cirrhotic liver tissue. eBioMedicine. 23523964. 82, 104147. doi:10.1016/j.ebiom.2022.104147.

Hou, Y., Tu, S., Zhao, X., Li, G., Li, N., Zou, A., 2023. An integrative method for evaluating the biological effects of nanoparticle-protein Corona. Biochim. Biophys. Acta Gen. Subj. 03044165. 1867 (3), 130300. doi:10.1016/j.bbagen.2022.130300.

Hu, C.J., Garcia, M.A., Nihart, A., Liu, R., Yin, L., Adolphi, N., Gallego, D.F., Kang, H., Campen, M.J., Yu, X., 2024. Microplastic presence in dog and human testis and its potential association with sperm count and weights of testis and epididymis. Toxicol. Sci. 10960929. 200 (2), 235–240. Available from: http://toxsci.oxfordjournals.org/. doi:10.1093/toxsci/kfae060.

Hunt, K., Davies, A., Fraser, A., Burden, C., Howell, A., Buckley, K., Harding, S., Bakhbakhi, D., 2024. Exposure to microplastics and human reproductive outcomes: a systematic review. BJOG. 1470-0328. 131 (5), 675–683. doi:10.1111/1471-0528.17756.

Ibrahim, Y.S., Tuan Anuar, S., Azmi, A.A., Wan Mohd Khalik, W.M.A., Lehata, S., Hamzah, S.R., Ismail, D., Ma, Z.F., Dzulkarnaen, A., Zakaria, Z., Mustaffa, N., Tuan Sharif, S.E., Lee, Y.Y., 2021. Detection of microplastics in human colectomy specimens. JGH Open. 23979070. 5 (1), 116–121. Available from: http://onlinelibrary.wiley.com/journal/10.1002/(ISSN)2397-9070. doi:10.1002/jgh3.12457.

Ilechukwu, I., Ehigiator, B.E., Ben, I.O., Okonkwo, C.J., Olorunfemi, O.S., Modo, U.E., Ilechukwu, C.E., Ohagwa, N.J., 2022. Chronic toxic effects of polystyrene microplastics on reproductive parameters of male rats. Environ. Anal. Health Toxicol. 26719525. 37 (2). Available from: https://www.eaht.org/upload/pdf/eaht-37-2-e2022015pdf. doi:10.5620/eaht.2022015Japan.

Indriyasari, K.N., Soegianto, A., Irawan, B., Rahmatin, N.M., Marchellina, A., Mukholladun, W., Payus, C.M., 2023. The presence of microplastics and plasticizers in different tissues of mullet (*Mugil cephalus*) along the east java coast in Indonesia. Water, Air, Soil Pollut. 15732932. 234 (9). Available from: https://www.springer.com/journal/11270. doi:10.1007/s11270-023-06623-y.

Irnidayanti, Y., Soegianto, A., Brabo, A.H., Abdilla, F.M., Indriyasari, K.N., Rahmatin, N.M., Putranto, T.W.C., Payus, C.M., 2023. Microplastics in green mussels (*Perna viridis*) from Jakarta Bay, Indonesia, and the associated hazards to human health posed by their consumption. Environ. Monit. Assess. 15732959. 195 (7). Available from: https://www.springer.com/journal/10661. doi:10.1007/s10661-023-11535-9.

Ishihara, Y., Kajino, M., Iwamoto, Y., Nakane, T., Nabetani, Y., Okuda, T., Kono, M., Okochi, H., 2025. Impact of artificial sunlight aging on the respiratory effects of polyethylene terephthalate microplastics through degradation-mediated terephthalic acid release in male mice. Toxicol. Sci. 1096-6080. 203 (2), 242–252. doi:10.1093/toxsci/kfae135.

Jahanzaib, M., Sharma, S., Park, D., 2025. Microplastics comparison of indoor and outdoor air and ventilation rate effect in outskirts of the Seoul metropolitan city. Emerg. Contam. 24056650. 11 (1), 100408. doi:10.1016/j.emcon.2024.100408.

Jasso-Salcedo, A.B., Díaz-Cruz, C.A., Rivera-Vallejo, C.C., Jiménez-Regalado, E.J., Aguirre-Loredo, R.Y., 2024. Human consumption of microplastics via food type and habits: recent review. Water, Air, Soil Pollut. 15732932. 235 (2). Available from: https://www.springer.com/journal/11270. doi:10.1007/s11270-024-06920-0.

Jenner, L.C., Rotchell, J.M., Bennett, R.T., Cowen, M., Tentzeris, V., Sadofsky, L.R., 2022. Detection of microplastics in human lung tissue using μFTIR spectroscopy. Sci. Total Environ. 18791026. 831. Available from: https://www.elsevier.com/locate/scitotenv. doi:10.1016/j.scitotenv.2022.154907.

Kern, D.G., Crausman, R.S., Durand, K.T.H., Nayer, A., Kuhn, C., 1998. Flock worker's lung: chronic interstitial lung disease in the nylon flocking industry. Ann. Intern. Med. 00034819. 129 (4), 261–272. Available from: http://annals.org/issues.aspx. doi:10.7326/0003-4819-129-4-199808150-00001.

Kern, D.G., 1998. The unexpected result of an investigation of an outbreak of occupational lung disease. Int. J. Occup. Environ. Health. 10773525. 4 (1), 19–32. Available from: http://www.tandfonline.com/loi/yjoh20#.Vxm2Kk1f3cs. doi:10.1179/oeh.1998.4.1.19.

Kim, J., Chen, M., White, R.S., 2025. Microplastics and the placenta: a call to action for perinatal research. Am. J. Perinatol. 10988785. Available from: https://www.thieme.de/de/american-journal-perinatology/journal-information-9405.htm. doi:10.1055/a-2657-6249.

Lademann, J., Schaefer, H., Otberg, N., Teichmann, A., Blume-Peytavi, U., Sterry, W., 2004. Penetration von Mikropartikeln in die menschliche Haut. Der Hautarzt. 0017-8470. 55 (12), 1117–1119. doi:10.1007/s00105-004-0841-1.

Liao, Z., Ji, X., Ma, Y., Lv, B., Huang, W., Zhu, X., Fang, M., Wang, Q., Wang, X., Dahlgren, R., Shang, X., 2021. Airborne microplastics in indoor and outdoor environments of a coastal city in Eastern China. J. Hazard. Mater. 03043894. 417, 126007. doi:10.1016/j.jhazmat.2021.126007.

Lin, K.T., Chen, I.A., Chen, K.Y., Jung, C.C., 2025. Temporal variations and characteristics of microplastics in indoor and outdoor air. Aerosol Air Qual. Res. 20711409. 25 (1). Available from: https://link.springer.com/journal/44408. doi:10.1007/s44408-025-00001-5.

Mahmud, F., Sarker, D.B., Jocelyn, J.A., Sang, Q.X.A., 2024. Molecular and cellular effects of microplastics and nanoplastics: focus on inflammation and senescence. Cells. 20734409. 13 (21). Available from: https://www.mdpi.com/journal/cells. doi:10.3390/cells13211788.

Marfella, R., Prattichizzo, F., Sardu, C., Fulgenzi, G., Graciotti, L., Spadoni, T., D'Onofrio, N., Scisciola, L., Grotta, R. L., Frigé, C., Pellegrini, V., Municinò, M., Siniscalchi, M., Spinetti, F., Vigliotti, G., Vecchione, C., Carrizzo, A., Accarino, G., Squillante, A., Spaziano, G., Mirra, D., Esposito, R., Altieri, S., Falco, G., Fenti, A., Galoppo, S., Canzano, S., Sasso, F.C., Matacchione, G., Olivieri, F., Ferraraccio, F., Panarese, I., Paolisso, P., Barbato, E., Lubritto, C., Balestrieri, M.L., Mauro, C., Caballero, A.E., Rajagopalan, S., Ceriello, A., D'Agostino, B., Iovino, P., Paolisso, G., 2024. Microplastics and nanoplastics in atheromas and cardiovascular events. N. Engl. J. Med. 0028-4793. 390 (10), 900–910. doi:10.1056/nejmoa2309822.

Mashayekhi-Sardoo, H., Ghoreshi, Z.A.S., Askarpour, H., Arefinia, N., Ali-Hassanzadeh, M., 2025. The clinical relevance of microplastic exposure on colorectal cancer: a systematic review. Cancer Epidemiol. 1877783X. 97. Available from: https://www.sciencedirect.com/science/journal/18777821. doi:10.1016/j.canep.2025.102840.

Mondal, R., Sarkar, D.J., Bhattacharyya, S., Raja, R., Chaudhuri, P., Biswas, J.K., Das, B.K., 2025. Health risk assessment of microplastics contamination in the daily diet of South Asian countries. J. Hazard. Mater. 18733336. 494. Available from: https://www.sciencedirect.com/science/journal/03043894. doi:10.1016/j.jhazmat.2025.138527.

Montano, L., Raimondo, S., Piscopo, M., Ricciardi, M., Guglielmino, A., Chamayou, S., Gentile, R., Gentile, M., Rapisarda, P., Conti, G.O., Ferrante, M., Motta, O., 2025. First evidence of microplastics in human ovarian follicular fluid: an emerging threat to female fertility. Ecotoxicol. Environ. Saf. 01476513. 291, 117868. doi:10.1016/j.ecoenv.2025.117868.

Napper, I.E., Davies, B.F.R., Clifford, H., Elvin, S., Koldewey, H.J., Mayewski, P.A., Miner, K.R., Potocki, M., Elmore, A.C., Gajurel, A.P., Thompson, R.C., 2020. Reaching new heights in plastic pollution—preliminary findings of microplastics on Mount Everest. One Earth. 25903322. 3 (5), 621–630. Available from: https://www.cell.com/one-earth. doi:10.1016/j.oneear.2020.10.020.

Nihart, A.J., Garcia, M.A., El Hayek, E., Liu, R., Olewine, M., Kingston, J.D., Castillo, E.F., Gullapalli, R.R., Howard, T., Bleske, B., Scott, J., Gonzalez-Estrella, J., Gross, J.M., Spilde, M., Adolphi, N.L., Gallego, D.F., Jarrell, H.S., Dvorscak, G., Zuluaga-Ruiz, M.E., West, A.B., Campen, M.J., 2025. Bioaccumulation of microplastics in decedent human brains. Nat. Med. 1546170X. 31 (4), 1114–1119. Available from: https://www.nature.com/nm/. doi:10.1038/s41591-024-03453-1.

Okamura, T., Hamaguchi, M., Hasegawa, Y., Hashimoto, Y., Majima, S., Senmaru, T., Ushigome, E., Nakanishi, N., Asano, M., Yamazaki, M., Sasano, R., Nakanishi, Y., Seno, H., Takano, H., Fukui, M., 2023. Oral exposure to polystyrene microplastics of mice on a normal or high-fat diet and intestinal and metabolic outcomes. Environ. Health Perspect. 0091-6765. 131 (2). doi:10.1289/ehp11072.

Ozdemir, C., Kucuksezer, U.C., Ogulur, I., Pat, Y., Yazici, D., Ardicli, S., Akdis, M., Nadeau, K., Akdis, C.A., 2024. Lifestyle changes and industrialization in the development of allergic diseases. Curr. Allergy Asthma Rep. 15346315. 24 (7), 331–345. Available from: https://www.springer.com/journal/11882. doi:10.1007/s11882-024-01149-7.

Peng, X., Chen, M., Chen, S., Dasgupta, S., Xu, H., Ta, K., Du, M., Li, J., Guo, Z., Bai, S., 2018. Microplastics contaminate the deepest part of the world's ocean. Geochem. Perspect. Lett. 24103403. 9, 1–5. doi:10.7185/geochemlet.1829.

Prata, J.C., 2023. Microplastics and human health: integrating pharmacokinetics. Crit. Rev. Environ. Sci. Technol. 15476537. 53 (16), 1489–1511. Available from: https://www.tandf.co.uk/journals/titles/10643389.asp. doi:10.1080/10643389.2023.2195798.

Primpke, S., Christiansen, S.H., Cowger, W., De Frond, H., Deshpande, A., Fischer, M., Holland, E.B., Meyns, M., O'Donnell, B.A., Ossmann, B.E., Pittroff, M., Sarau, G., Scholz-Böttcher, B.M., Wiggin, K.J., 2020. Critical assessment of analytical methods for the harmonized and cost-efficient analysis of microplastics. Appl. Spectrosc. 19433530. 74 (9), 1012–1047. Available from: https://journals.sagepub.com/loi/asp. doi:10.1177/0003702820921465.

Qin, X., Cao, M., Peng, T., Shan, H., Lian, W., Yu, Y., Shui, G., Li, R., 2024. Features, potential invasion pathways, and reproductive health risks of microplastics detected in human uterus. Environ. Sci. Technol. 0013-936X. 58 (24), 10482–10493. doi:10.1021/acs.est.4c01541.

Ragusa, A., Matta, M., Cristiano, L., Matassa, R., Battaglione, E., Svelato, A., De Luca, C., D'Avino, S., Gulotta, A., Rongioletti, M.C.A., Catalano, P., Santacroce, C., Notarstefano, V., Carnevali, O., Giorgini, E., Vizza, E., Familiari, G., Nottola, S.A., 2022. Deeply in plasticenta: presence of microplastics in the intracellular compartment of human placentas. Int. J. Environ. Res. Public Health. 16604601. 19 (18). Available from: http://www.mdpi.com/journal/ijerph. doi:10.3390/ijerph191811593.

Rahmatin, N.M., Soegianto, A., Irawan, B., Payus, C.M., Indriyasari, K.N., Marchellina, A., Mukholladun, W., Irnidayanti, Y., 2024. The spatial distribution and physico-chemical characteristic of microplastics in the sediment and cockle (*Anadara granosa*) from the coastal waters of East Java, Indonesia, and the health hazards associated with cockle consumption. Mar. Pollut. Bull. 18793363. 198. Available from: https://www.sciencedirect.com/science/journal/0025326X. doi:10.1016/j.marpolbul.2023.115906.

Rajendran, D., Chandrasekaran, N., 2023. Journey of micronanoplastics with blood components. RSC Adv. 2046-2069. 13 (45), 31435–31459. doi:10.1039/d3ra05620a.

Read, B., Sylla, P., 2020. Aggressive colorectal cancer in the young. Clin. Colon Rectal Surg. 1531-0043. 33 (05), 298–304. doi:10.1055/s-0040-1713747.

Rednikin, A.R., Frank, Y.A., Rozhin, A.O., Vorobiev, D.S., Fakhrullin, R.F., 2024. Airborne microplastics: challenges, prospects, and experimental approaches. Federation Atmosphere. 20734433. 15 (11). Available from: https://www.mdpi.com/journal/atmosphere. doi:10.3390/atmos15111380.

Riaz, H.H., Lodhi, A.H., Munir, A., Zhao, M., Ali, M.H., Sauret, E., Gu, Y.T., Islam, M.S., 2025. Breath of pollutants: how breathing patterns influence microplastic accumulation in the human lung. Int. J. Multiph. Flow. 03019322. 185. Available from: https://www.sciencedirect.com/science/journal/03019322. doi:10.1016/j.ijmultiphaseflow.2025.105156.

Roh, Y., Kim, J., Song, H., Seol, A., Kim, T., Park, E., Park, K., Lim, S., Wang, S., Jung, Y., Kim, H., Lim, Y., Hwang, D., 2024. Impact of the oral administration of polystyrene microplastics on hepatic lipid, glucose, and amino acid metabolism in C57BL/6Korl and C57BL/6-Lepem1hwl/Korl Mice. Int. J. Mol. Sci. 1422-0067. 25 (9), 4964. doi:10.3390/ijms25094964.

Schraufnagel, D.E., 2020. The health effects of ultrafine particles. Exp. Mol. Med. 20926413. 52 (3), 311–317. Available from: https://www.nature.com/emm/. doi:10.1038/s12276-020-0403-3.

Schwabl, P., Köppel, S., Königshofer, P., Bucsics, T., Trauner, M., Reiberger, T., Liebmann, B., 2019. Detection of various microplastics in human stool. Ann. Intern. Med. 0003-4819. 171 (7), 453–457. doi:10.7326/m19-0618.

Shahzadi, C., Di Serafino, A., Aruffo, E., Mascitelli, A., Di Carlo, P., 2023. A549 as an in vitro model to evaluate the impact of microplastics in the air. Biology. 2079-7737. 12 (9), 1243. doi:10.3390/biology12091243.

He, S., Zhang, Y., 2025. Detection and quantification of microplastics in endometrial polyps and their role in polyp formation. Reprod. Toxicol. 08906238. 132, 108757. doi:10.1016/j.reprotox.2024.108757.

Shruti, V.C., Kutralam-Muniasamy, G., Pérez-Guevara, F., Alvarez, R.C., 2025. Exploring the micro- and nanoplastics–diabetes nexus: shattered barriers, toxic links, and methodological horizons. Environ. Pollut. 02697491. 375, 126319. doi:10.1016/j.envpol.2025.126319.

Siegel, R.L., Miller, K.D., Goding Sauer, A., Fedewa, S.A., Butterly, L.F., Anderson, J.C., Cercek, A., Smith, R.A., Jemal, A., 2020. Colorectal cancer statistics, 2020. CA Cancer J. Clin. 15424863. 70 (3), 145–164. Available from: http://onlinelibrary.wiley.com/journal/10.3322/(ISSN)1542-4863. doi:10.3322/caac.21601.

Song, G.B., Nam, J., Ji, S., Woo, G., Park, S., Kim, B., Hong, J., Choi, M.G., Kim, S., Lee, C., Lim, W., Yoon, S., Kim, J.M., Choi, W.J., Choi, M.J., Koh, H.R., Lim, T.G., Hong, S., 2024. Deciphering the links: fragmented polystyrene as a driver of skin inflammation. J. Hazard. Mater. 18733336. 480. Available from: https://www.sciencedirect.com/science/journal/03043894. doi:10.1016/j.jhazmat.2024.135815.

Sun, X., Qingjie, L., Zhu, M., Liang, J., Zheng, S., Zhao, Y., 2017. Ingestion of microplastics by natural zooplankton groups in the northern South China Sea. Mar. Pollut. Bull. 0025326X. 115 (1-2), 217–224. doi:10.1016/j.marpolbul.2016.12.004.

Takada, K., Hotta, K., Kishida, Y., Ito, S., Imai, K., Ono, H., 2023. Comprehensive analysis of early-onset colorectal cancer: a review. J. Anus. Rectum Colon. 2432-3853. 7 (4), 241–249. doi:10.23922/jarc.2023-032.

Tao, J., Deng, P., Lin, M., Chen, C., Ma, Q., Yang, L., Zhang, W., Luo, Y., Chen, S., Pi, H., Zhou, Z., Yu, Z., 2024. Long-term exposure to polystyrene microplastics induces hepatotoxicity by altering lipid signatures in C57BL/6J mice. Chemosphere. 18791298. 347. Available from: https://www.elsevier.com/locate/chemosphere. doi:10.1016/j.chemosphere.2023.140716.

Van Cauwenberghe, L., Janssen, C.R., 2014. Microplastics in bivalves cultured for human consumption. Environ. Pollut. 18736424. 193, 65–70. Available from: https://www.journals.elsevier.com/environmental-pollution. doi:10.1016/j.envpol.2014.06.010.

Wang, Y., Xu, X., Wu, J., Dong, Z., Su, Y., Huang, G., Guo, Q., Feng, Z., Jia, H., Zhao, J., Xing, B., 2025. Microplastics and nanoplastics on Mt. Everest. Cell Rep. Sustain. 29497906. 2 (8), 100467. doi:10.1016/j.crsus.2025.100467.

Wang, Z., Lin, T., Chen, W., 2020. Occurrence and removal of microplastics in an advanced drinking water treatment plant (ADWTP). Sci. Total Environ. 00489697. 700, 134520. doi:10.1016/j.scitotenv.2019.134520.

Wei, Z., Wang, Y., Wang, S., Xie, J., Han, Q., Chen, M., 2022. Comparing the effects of polystyrene microplastics exposure on reproduction and fertility in male and female mice. Toxicology. 0300483X. 465, 153059. doi:10.1016/j.tox.2021.153059.

Weingrill, R.B., Lee, M.J., Benny, P., Riel, J., Saiki, K., Garcia, J., Oliveira, L.F.A.D.M., Fonseca, E.J.D.S., Souza, S.T.D., D'Amato, F.D.O.S., Silva, U.R., Dutra, M.L., Marques, A.L.X., Borbely, A.U., Urschitz, J., 2023. Temporal trends in microplastic accumulation in placentas from pregnancies in Hawai'i. Environ. Int. 18736750. 180. Available from: https://www.elsevier.com/locate/envint. doi:10.1016/j.envint.2023.108220.

Werorilangi, S., Wicaksono, E.A., Afdal, M., Sari, K., Nimzet, R., Samad, W., Mawadda, R., Gosalam, S., Fachruddin, L., Massinai, A., Faizal, A., 2025. Ingested microplastics: a comparative analysis of contaminated shellfish from two sites in the Makassar Strait. Environ. Monit. Assess. 15732959. 197 (4). Available from: https://www.springer.com/journal/10661. doi:10.1007/s10661-025-13804-1.

Wright, S.L., Ulke, J., Font, A., Chan, K.L.A., Kelly, F.J., 2020. Atmospheric microplastic deposition in an urban environment and an evaluation of transport. Environ. Int. 01604120. 136, 105411. doi:10.1016/j.envint.2019.105411.

Xu, H., Dong, C., Yu, Z., Hu, Z., Yu, J., Ma, D., Yao, W., Qi, X., Ozaki, Y., Xie, Y., 2024. First identification of microplastics in human uterine fibroids and myometrium. Environ. Pollut. 02697491. 360, 124632. doi:10.1016/j.envpol.2024.124632.

Yan, Z., Liu, Y., Zhang, T., Zhang, F., Ren, H., Zhang, Y., 2022. Analysis of microplastics in human feces reveals a correlation between fecal microplastics and inflammatory bowel disease status. Environ. Sci. Technol. 0013-936X. 56 (1), 414–421. doi:10.1021/acs.est.1c03924.

Yang, L., Kang, S., Luo, X., Wang, Z., 2024. Microplastics in drinking water: a review on methods, occurrence, sources, and potential risks assessment. Environ. Pollut. 02697491. 348, 123857. doi:10.1016/j.envpol.2024.123857.

Zhao, J., Adiele, N., Gomes, D., Malovichko, M., Conklin, D.J., Ekuban, A., Luo, J., Gripshover, T., Watson, W.H., Banerjee, M., Smith, M.L., Rouchka, E.C., Xu, R., Zhang, X., Gondim, D.D., Cave, M.C., O'Toole, T.E., 2024. Obesogenic polystyrene microplastic exposures disrupt the gut-liver-adipose axis. Toxicol. Sci. 10960929. 198 (2), 210–220. Available from: http://toxsci.oxfordjournals.org/. doi:10.1093/toxsci/kfae013.

Zhao, X., Zhou, Y., Liang, C., Song, J., Yu, S., Liao, G., Zou, P., Tang, K.H.D., Wu, C., 2023. Airborne microplastics: occurrence, sources, fate, risks and mitigation. Sci. Total Environ. 18791026. 858. Available from: https://www.elsevier.com/locate/scitotenv. doi:10.1016/j.scitotenv.2022.159943.

Zhuikova, L.D., Choynzonov, E.L., Ananina, O.A., Lyakhova, N.P., Pikalova, L.V., 2020. Lung cancer incidence in various cities of the world (review). Voprosy Onkologii. 05073758. 66 (3), 239–246. Available from: https://www.niioncologii.ru/science/pio. doi:10.37469/0507-3758-2020-66-3-239-246.

Ziani, K., Ioniță-Mîndrican, C.B., Mititelu, M., Neacşu, S.M., Negrei, C., Moroşan, E., Drăgănescu, D., Preda, O.T., 2023. Microplastics: a real global threat for environment and food safety: a state of the art review. Nutrients. 20726643. 15 (3). Available from: http://www.mdpi.com/journal/nutrients/. doi:10.3390/nu15030617.

CHAPTER

10

Interaction mechanisms for physical and chemical damage by microplastics

Ruby Ruby[a], *Dimple Dimple*[a], *Poonam Mundlia*[a], *Rajesh Kumar*[b], *and Arup Giri*[a]

[a]Department of Zoology, Baba Mastnath University, Asthal Bohar, Rohtak, Haryana, India
[b]Department of Biosciences, Himachal Pradesh University, Shimla, India

10.1 Introduction

The pervasive distribution of microplastics (MPs) across marine, freshwater, terrestrial, and atmospheric systems has transformed them from emerging contaminants into persistent global stressors (Jolaosho et al., 2025). Defined as plastic particles typically smaller than 5 mm, MPs originate from both primary sources (manufactured small particles) and secondary fragmentation of larger plastic debris. Their physicochemical diversity, encompassing variations in polymer composition, particle size, morphology, crystallinity, and surface chemistry, governs environmental behavior and biological reactivity (Wei et al., 2026). Unlike conventional dissolved pollutants, MPs function simultaneously as particulate stressors and chemical carriers, enabling complex interaction mechanisms that extend beyond simple exposure–response paradigms (Mulky et al., 2026).

Environmental conditioning substantially alters the properties of MPs after release. Weathering processes, including ultraviolet (UV) radiation, thermooxidative degradation, hydrolysis, and mechanical abrasion, alter polymer chains, generating surface cracks, oxygen-containing functional groups, and increased surface area (Duan et al., 2021). Photoaging and fragmentation not only reduce the particle size but also increase the reactivity and sorption capacity. Concurrently, interactions with natural organic matter, microorganisms, and biomolecules promote the formation of eco-coronas and biocoronas (Wang et al., 2023; Xu et al., 2024). These surface coatings redefine particle identity, influencing aggregation behavior, transport dynamics, and biological recognition. Consequently, the toxicity profile of environmentally aged particles may differ markedly from that of pristine polymers, underscoring the importance of transformation processes in shaping biological outcomes (Zhao et al., 2025). Exposure pathways reflect the ubiquity of MPs across environmental compartments. Ingestion,

Microplastics in Animal and Human Physiology
https://doi.org/10.1016/B978-0-443-45635-0.00019-7

inhalation, and dermal contact represent primary routes of entry into organisms. Once internalized, particle bioavailability depends strongly on size, shape, and surface characteristics (Bhardwaj et al., 2025). Smaller particles exhibit greater translocation potential across epithelial barriers, whereas fibrous or irregular morphologies may increase mechanical retention and localized tissue stress. Cellular uptake mechanisms include endocytosis, phagocytosis, and passive diffusion for nanoscale fractions, leading to intracellular localization within endosomal–lysosomal systems or the cytosol. The physicochemical traits acquired during environmental aging further modulate internal distribution and persistence (Cahn et al., 2024).

Upon contact with biological interfaces, MPs initiate physical interaction mechanisms. Mechanical abrasion and epithelial irritation may compromise barrier integrity, altering tight junction stability and increasing permeability (Verma et al., 2024). Disruption of the plasma membrane structure perturbs ionic gradients and membrane-associated signaling processes. At the subcellular level, particle accumulation can disrupt cytoskeletal organization and organelle architecture, thereby promoting structural stress independent of chemical toxicity (Wang et al., 2023). These mechanical effects are particularly relevant for fibrous or irregular particles that can induce frictional damage or localized pressure within confined biological spaces.

Beyond physical perturbation, MPs can cause chemical damage through leaching and vector-mediated transport. Many polymers contain additives such as plasticizers, stabilizers, flame retardants, and colorants that may be released under physiological conditions (Kochanek et al., 2025). Degradation processes further generate low-molecular-weight by-products with potential biological activity. In addition, MPs adsorb hydrophobic organic contaminants, metals, and other environmental pollutants from surrounding matrices. Once internalized, desorption may occur under changing pH, ionic strength, or enzymatic conditions, enhancing localized chemical exposure (Vranic et al., 2025). The combined presence of intrinsic additives and sorbed contaminants can produce synergistic toxic effects, complicating risk assessment frameworks (Li et al., 2025).

At the cellular level, these physical and chemical stimuli converge to activate integrated stress responses. Excessive production of reactive oxygen species (ROS) disrupts redox homeostasis, whereas inflammatory signaling pathways amplify immune activation (Singh and Nandave, 2024). Mitochondrial dysfunction and endoplasmic reticulum (ER) stress compromise bioenergetic balance and protein homeostasis, potentially leading to progression toward apoptosis under sustained exposure. These interconnected processes illustrate that MP toxicity arises not from a single mechanism but from a dynamic network of interacting pathways influenced by particle properties and environmental history.

Despite substantial progress, significant uncertainties remain regarding dose–response relationships, chronic low-level exposure effects, and the comparative contributions of particle-mediated versus chemical-mediated toxicity. Variability in experimental design, particle characterization, and exposure conditions further complicates cross-study comparisons (Bretz and Bornkamp, 2025). A comprehensive understanding of the interaction mechanisms, spanning environmental transformation, biological entry, mechanical interference, chemical release, and integrated cellular stress responses, is therefore essential for advancing mechanistic toxicology and informing environmental risk evaluation (Olanrewaju and Ford Versypt, 2025).

10.2 Environmental conditioning and particle transformation

When MPs are released into natural environments, they do not remain inert or chemically pristine. Instead, they undergo severe, continuous physical, chemical, and biological degradation. Researchers have reported that these degradations alter particle characteristics, such as size, morphology, surface chemistry, and reactivity, thereby modulating their interactions with biota and biological systems (Luo et al., 2025).

10.2.1 Polymer composition, size, shape, and surface properties

The intrinsic properties of MPs are largely determined by their polymer composition, which is related to environmental persistence, density, and surface chemistry. Common polymers such as polyethene, polypropylene, polystyrene, polyvinyl chloride, and polyethene terephthalate differ in terms of crystallinity, hydrophobicity, and chemical stability, influencing their environmental distribution and interaction potential. Low-density polymers tend to remain suspended or buoyant, whereas denser polymers preferentially accumulate in sediments and soils (Jolaosho et al., 2025). Particle size critically influences surface area, mobility, and biological relevance. Smaller MPs exhibit higher surface area-to-volume ratios and greater reactivity, contaminant adsorption, and bioavailability (Jeong et al., 2024).

Particle shape and surface morphology also modulate environmental behavior and biological interactions. MPs occur as fragments, fibers, films, and pellets, with irregular or fibrous forms exhibiting higher mechanical interaction potential than smooth, spherical particles do. Surface properties, including roughness, charge, and functional groups, evolve through environmental exposure, altering aggregation behavior, binding to contaminants, and interactions with biological membranes.

10.2.2 Weathering, photoaging, and fragmentation processes

Once released, MPs undergo complex weathering processes, including UV radiation, thermal cycling, mechanical abrasion, oxidative reactions, and microbial activity. Photoaging induced by UV exposure initiates polymer chain scission, cleavage, oxidation, and the formation of oxygen-containing functional groups, such as carbonyls and hydroxyls. These processes increase surface polarity, roughness, and brittleness, thereby accelerating fragmentation into smaller particles.

Mechanical forces generated by wave action, sediment transport, and bioturbation further contribute to physical fragmentation, particularly for aged and embrittled polymers. Moreover, weathering enhances the release of additives and degradation by-products and affects the ability of MPs to adsorb environmental contaminants (Harraq et al., 2022). Collectively, these transformation processes shift MPs from relatively inert materials to chemically active particles with enhanced environmental reactivity, altered aggregation behavior, and increased potential for biological interactions.

10.2.3 Formation of eco-coronas and biocoronas

In natural waters, MPs rapidly adsorb ecological macromolecules such as humic acids, fulvic acids, extracellular polymeric substances, proteins, and polysaccharides, forming an "eco-corona" (Yang et al., 2025). This dynamic coating modifies the surface charge, hydrophobicity, colloidal stability, and aggregation behavior. Eco-corona formation is primarily driven by electrostatic interactions, hydrophobic forces, hydrogen bonding, van der Waals interactions, and π–π stacking, depending on the polymer type and environmental conditions (Yang et al., 2025). Humic substances, which are rich in aromatic and oxygen-containing functional groups, strongly bind to MP surfaces and can either stabilize particles through steric and electrostatic repulsion or promote heteroaggregation *via* molecular bridging mechanisms (Yang et al., 2025).

Upon entry into biological systems, MPs acquire a biocorona composed of proteins, lipids, carbohydrates, and metabolites derived from biological fluids (Rex M et al., 2023). This biocorona redefines the biological identity of MPs by regulating cellular recognition, uptake, immune interactions, and toxicity. Experimental evidence has demonstrated that biocorona formation can attenuate oxidative stress, reduce membrane damage, and modulate enzymatic inhibition. However, in some contexts, it may prolong particle retention or facilitate Trojan-horse transport of cocontaminants (Camil Rex et al., 2025). Thus environmental conditioning and corona formation are not passive processes but active determinants of MP behavior, facilitating transport, bioavailability, and biological effects across environmental and physiological interfaces.

10.3 Biological entry and internal exposure pathways

The adverse effects of MPs depend not only on environmental transformation but also on their ability to enter biological systems and interact with internal tissues. Exposure pathways are influenced by particle characteristics, organism physiology, and environmental conditions, ultimately affecting the internal dose and toxicity outcomes.

10.3.1 Routes of exposure and internal availability

Ingestion remains one of the primary routes of MP exposure for most organisms, including aquatic invertebrates, fish, birds, and mammals. MPs may be intentionally ingested because of their size and shape resemblance to food particles or unintentionally through contaminated food and water (Jeong et al., 2024). In the terrestrial environment, the inhalation of airborne MP fibers and fragments is increasingly recognized as a major exposure route. It may constitute a relatively greater risk to public health than exposure to contaminated water, food, or soil does (Zhao et al., 2023).

Dermal exposure, although considered less significant, may occur through direct contact with contaminated water, soil, or consumer products, particularly when skin integrity is compromised. Following exposure, internal availability depends on particle size, surface properties, and the efficiency of biological barriers. Studies have reported that larger MPs typically remain within the gastrointestinal tract and are excreted, whereas smaller particles may cross epithelial barriers and enter systemic circulation (Bhattacharyya et al., 2025).

10.3.2 Size- and shape-dependent barrier interactions

Biological barriers, such as the intestinal epithelium, respiratory mucosa, and placental interface, act as critical filters that regulate MP translocation. Particle size plays a decisive role in barrier penetration, with smaller particles exhibiting higher translocation efficiency (Schwarzfischer and Rogler, 2022). Micro- and nanoplastics can cross epithelial layers *via* paracellular transport or transcytosis, while larger MPs are more likely to remain confined to luminal spaces (Prata, 2023).

Particle shape also influences barrier interactions. Irregularly shaped MPs with sharp edges and high surface curvature can directly damage cell membranes and compromise barrier integrity (Cheng et al., 2024). Experimental evidence has indicated that smooth, high-density polyethene particles (>25 μm) elicit immune activation in the absence of cytotoxicity. In contrast, roughened particles induce increased cytotoxicity, hemolysis, and ROS production, highlighting the role of surface irregularities in modulating cytotoxic and immunological effects (Choi et al., 2021). Surface charge and hydrophobicity further modulate interactions with mucosal surfaces and epithelial cells. Studies have shown that positively charged or highly hydrophobic particles exhibit stronger adhesion to cell membranes, increasing the likelihood of cellular uptake and barrier disruption (Sadique et al., 2025).

10.3.3 Cellular uptake mechanisms and intracellular localization

At the cellular level, MPs can be internalized through multiple mechanisms, including phagocytosis, macropinocytosis, clathrin-mediated endocytosis, and caveolae-dependent pathways (Liu et al., 2021). The predominant uptake route depends on particle size, shape, surface chemistry, and cell type. Once internalized, MPs may localize to endosomes and lysosomes, where they can induce membrane destabilization, oxidative stress, and lysosomal dysfunction. Smaller particles have been reported to escape endolysosomal compartments and interact directly with intracytoplasmic organelles such as mitochondria and the ER (Ragusa et al., 2022). Such interactions can disrupt cellular homeostasis, impair energy metabolism, and trigger inflammatory and apoptotic signaling pathways (Hernández et al., 2025). In some cases, persistent intracellular MPs may evade degradation and accumulate over time, leading to chronic cellular stress (Domenech et al., 2021).

10.4 Physical interaction mechanisms

MPs can damage biological tissues, cells, and organelles through direct mechanical and structural interactions.

10.4.1 Mechanical abrasion and epithelial injury

Owing to their small size and irregular surfaces, MPs can mechanically damage epithelial tissues, compromising barrier integrity and triggering inflammation. In aquatic organisms, ingestion causes gastrointestinal lesions, whereas in humans, inhalation or dermal exposure may provoke epithelial irritation. Mechanical abrasion from tires, brake pads, and footwear generates airborne MPs (<5 mm) that accumulate in the environment and can be inhaled, posing respiratory risks (Ketian et al., 2024). UV radiation and thermal oxidation, combined with soil abrasion, further accelerate MP formation, alter polymer chemistry, and increase toxicity. An *in vitro* study by Choi et al. (2020) revealed that polystyrene microfragments induce cell membrane damage and hemolysis in fibroblasts and red blood cells, with effects amplified by higher concentrations and surface roughness, highlighting the strong correlation between MP roughness and cellular injury.

10.4.2 Disruption of membrane integrity and barrier function

MPs and nanoplastics destabilize cell membranes, impair tight junctions, and increase epithelial permeability, enabling systemic circulation and organ deposition (Fig. 10.1). They cross barriers such as the gut, placenta, and potentially the blood–brain barrier, inducing oxidative stress, inflammation, and immune activation, with positively charged particles being particularly disruptive. Barrier dysfunction in the gut, skin, respiratory, and ocular tissues allows harmful metabolites and bacteria to enter circulation, amplifying tissue injury and systemic inflammation, and potentially affecting several organs (Thapliyal et al., 2025; Zhang et al., 2026)

10.4.3 Subcellular interference and organelle dysfunction

MPs can be internalized by cells *via* endocytosis and passive diffusion and subsequently accumulate in key organelles, including lysosomes, mitochondria, the ER, and the nucleus. Their presence disrupts normal organelle function, promotes oxidative stress, impairs autophagy, and destabilizes the cytoskeleton, collectively compromising cellular homeostasis (Hua and Wang, 2022). Specifically, MPs impair mitochondrial activity, induce ER stress, and compromise lysosomal integrity. These disturbances can initiate apoptosis, provoke inflammatory responses, and lead to metabolic dysregulation (Li et al., 2025) (Fig. 10.1).

10.5 Chemical damage and vector-mediated toxicity

MPs are not inert; they carry hazardous additives and can adsorb environmental pollutants, acting as "chemical shuttles" that transfer toxic substances such as POPs and heavy metals into living tissues.

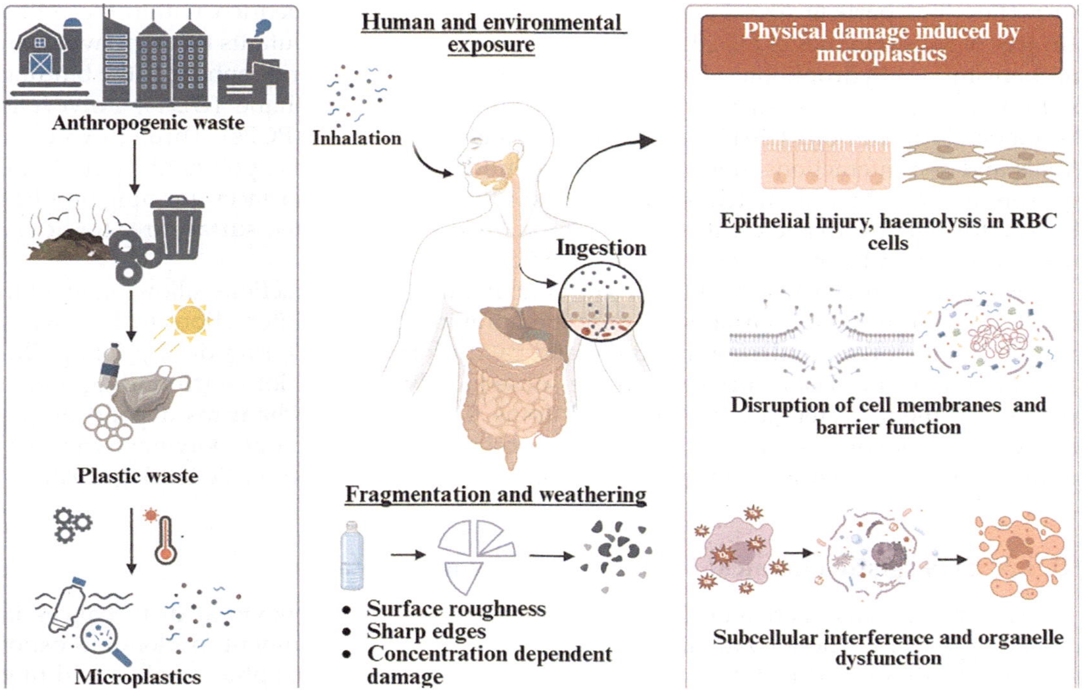

FIG. 10.1 Schematic representation of microplastic generation through environmental weathering and physical and chemical degradation of plastic materials. Following environmental release, microplastics may enter the human body *via* ingestion or inhalation. Owing to their small size, irregular morphology, and surface physicochemical properties, microplastics can interact with biological systems, inducing mechanical damage to epithelial cells, disrupting cellular membranes and barrier integrity, and interfering with subcellular structures.

10.5.1 Leaching of additives and polymer degradation products

Plastic additives (plasticizers, flame retardants, stabilizers, and antioxidants) leach from MPs because they are not chemically bound to the polymer matrix. Compounds such as phthalates, bisphenol A, and perfluorinated chemicals persist in the environment, disrupt endocrine function, and bioaccumulate (Iftikhar et al., 2024).

Physicochemical factors, including solubility, molecular size, temperature, pH, and polymer permeability, influence leaching. It is enhanced by physical (abrasion, fragmentation, and weathering) and chemical (photodegradation, UV radiation, and hydrolysis) degradation. Once leached, these chemicals migrate into water, soil, food, or onto plastic surfaces, posing ecological and health risks.

10.5.2 Adsorption and release of environmental contaminants

The attachment of pollutants to MPs depends on surface structure, polymer type [*e.g.*, Polyethylene (PE), Polystyrene (PS), or Polypropyline (PP)], surface area, chemical properties, and particle size. Toxic chemicals are often found embedded within the pitted surfaces of MPs (Holmes et al., 2014). MPs mainly interact with pollutants through two main mechanisms: (1) hydrophobic adsorption, as their low surface polarity enhances the binding of lipophilic pollutants such as Dichloro-Diphenyl-Trichloroethane (DDT), Polycyclic Aromatic Hydrocarbons (PAHs), and Polychlorinated Biphenyls (PCBs). Through van der Waals and π–π interactions, noncovalent compounds adhere to toxic pollutants and (2) are incorporated into biofilm matrices. Microbes rapidly colonize MP surfaces, forming biofilms that can further absorb and create surface pits on MPs, increasing their surface area for heavy metal sorption and enhancing their hydrophobicity.

These pollutants can adsorb onto MPs through noncovalent interactions, allowing them to detach easily from plastic surfaces. As described previously, these interactions include weak van der Waals forces, hydrogen bonding, and cavity formation. During desorption, pollutants migrate from plastic surfaces and are subsequently released into surrounding media such as sediment, water, soil, or biological tissue. Desorption mechanisms depend on the nature of the adsorption sites and are influenced by factors such as polymer type, pH, salinity, and the presence of inorganic or organic ligands (Iftikhar et al., 2024) (Fig. 10.2).

10.5.3 Synergistic chemical stress induced by MP surfaces

Heavy metals adsorbed onto MPs cause physiological and behavioral disturbances in aquatic organisms. These contaminated MPs persist in the environment, prolonging exposure and facilitating transfer through the food web to higher trophic levels, including humans. Coexposure often results in synergistic effects, such as increased oxidative stress, developmental abnormalities, and metabolic disruption, beyond the effects of individual contaminants (Li et al., 2025).

Heavy metals can act as cationic bridges (*e.g.*, Ca^{2+} and Mg^{2+}), enhancing the adsorption of anionic contaminants such as Per- and Polyfluoroalkyl Substances (PFAS) onto MP surfaces, while other organic chemicals may compete for sorption sites. Sorption mechanisms include electrostatic attraction, ion exchange, hydrogen bonding, hydrophobic interactions, and π–π bonding (Wang et al., 2025).

At the cellular level, combined exposure to MPs and PFAS overwhelms antioxidant defenses, inducing oxidative stress, genotoxicity, neurotoxicity, and metabolic dysfunction. MPs also function as vectors for PFAS, altering their environmental fate and transport. Overall, coexposure amplifies toxicity across trophic levels, affecting processes from algal photosynthesis to lipid metabolism and neurogenesis in animals, with significant ecological and human health implications (Adeleye et al., 2024; Wang et al., 2025) (Fig. 10.2).

10.6 Integrated cellular stress responses

MP exposure disrupts cellular homeostasis through tightly interconnected oxidative, inflammatory, and organelle-mediated mechanisms. Particle internalization and surface reactivity promote excessive ROS generation, overwhelming antioxidant defenses and inducing lipid peroxidation, protein oxidation, and DNA damage (Mahmud et al., 2024). Redox

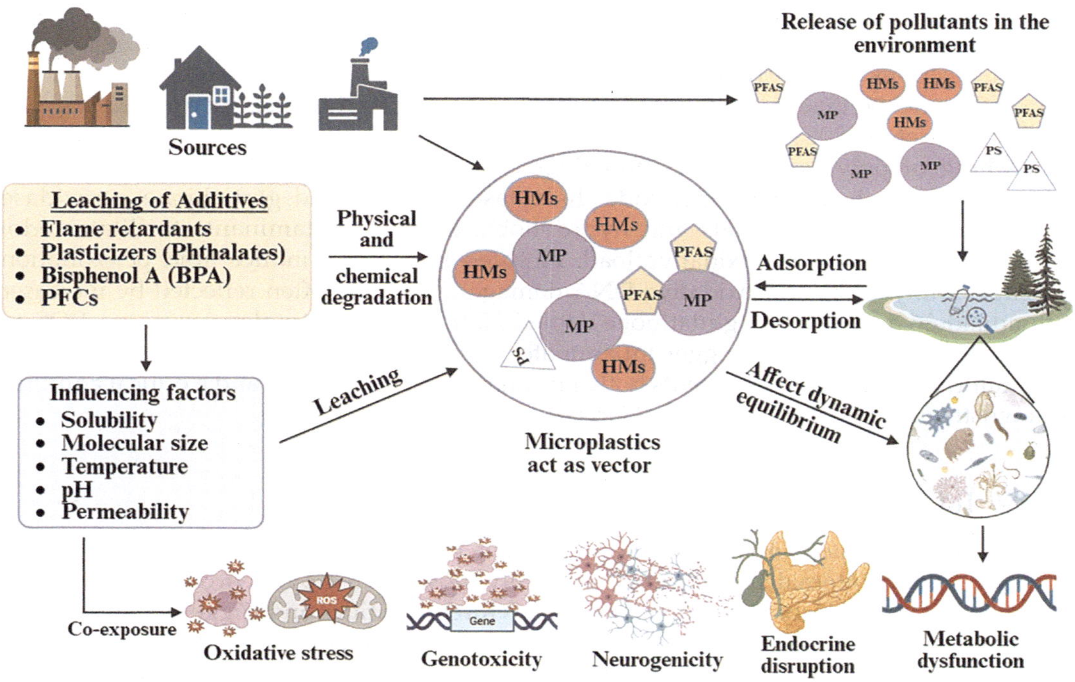

FIG. 10.2 **Schematic illustration of vector-mediated toxicity associated with MPs.** In environmental matrices, plastic additives, including plasticizers, flame retardants, stabilizers, and antioxidants, may leach from microplastics because they are not covalently bound to the polymer matrix. Microplastics can also adsorb and concentrate environmental contaminants, including heavy metals and organic pollutants. The extent of contaminant adsorption is influenced by the type of polymer, surface morphology, surface area, particle size, and physicochemical properties, as well as by environmental factors such as pH and salinity. Upon uptake by aquatic organisms, contaminant-laden microplastics may induce physiological alterations, including oxidative stress, genotoxicity, neurotoxicity, and metabolic dysfunction.

imbalance activates NF-κB- and MAPK-dependent inflammatory pathways, increasing the expression of proinflammatory cytokines and sustaining immune activation. Concurrently, mitochondrial dysfunction, characterized by a loss of membrane potential, impaired ATP synthesis, and ER stress, amplifies apoptotic signaling *via* calcium dysregulation and unfolded protein responses (Li et al., 2025; Selvaraj et al., 2025). These integrated stress networks determine the transition from adaptive responses to irreversible cellular injury under MP exposure.

10.6.1 Oxidative stress and redox imbalance

Oxidative stress is a central and reproducible outcome of MP exposure across experimental models. Following internalization, particles that accumulate in endolysosomal compartments or the cytosol stimulate excessive ROS production through surface reactivity and associated chemical constituents. Environmentally aged particles exhibit enhanced redox-active functional groups and increased surface area, thereby intensifying electron transfer reactions. Concurrent membrane perturbation and ionic imbalance promote calcium influx, destabilize mitochondrial electron transport, and further increase ROS generation (Zhang et al., 2025).

Leaching of polymer additives and degradation products can suppress antioxidant defense systems, including superoxide dismutase, catalase, and glutathione peroxidase. Additionally, adsorbed metals and hydrophobic organic contaminants facilitate redox cycling, compounding the oxidative load. Elevated ROS levels induce lipid peroxidation, protein carbonylation, and oxidative DNA damage, which are often reflected by increased malondialdehyde levels and glutathione depletion. In addition to molecular injury, ROS act as signaling mediators that trigger inflammatory cascades and apoptosis. Persistent redox disequilibrium, therefore, represents both a primary toxic mechanism and a critical amplifier of downstream cellular dysfunction (Garcia-Llorens et al., 2025).

10.6.2 Inflammatory and immune signaling pathways

MP exposure induces inflammatory and immune responses, particularly under chronic or repeated contact. Internalized particles, especially those altered by environmental aging or coated with eco- or biocoronas, are recognized as foreign entities and trigger innate immune signaling. Cellular recognition mechanisms initiate downstream pathways that coordinate the production of inflammatory mediators and the activation of immune cells (Vanetti et al., 2025).

The accumulation of ROS serves as a central regulatory signal, activating redox-sensitive transcription factors such as the NF-κB and MAPK pathways. These signaling cascades increase the expression of proinflammatory cytokines, including interleukins and tumor necrosis factor, thereby amplifying local inflammatory responses (Chen et al., 2025; Hong et al., 2024). Lysosomal destabilization caused by particle overload may further activate inflammasome complexes, promoting cytokine maturation and sustained immune signaling (Feng et al., 2024).

Chemical components associated with MPs intensify these processes. Released additives, polymer degradation products, and adsorbed environmental contaminants can engage nuclear receptors and xenobiotic response elements, prolonging inflammatory activation (John et al., 2023). The combined influence of mechanical irritation, oxidative imbalance, and chemical stress generates a persistent proinflammatory microenvironment. The magnitude and duration of immune responses are strongly influenced by particle size, morphology, surface chemistry, and aging state, underscoring the role of physicochemical properties in shaping immunotoxic outcomes (Dash et al., 2025).

10.6.3 Mitochondrial and ER dysfunction

Mitochondria are highly susceptible to MP-induced oxidative and chemical perturbations. Elevated ROS levels and sustained calcium influx disrupt mitochondrial membrane potential by impairing electron transport chain complexes, particularly complexes I and III, leading to enhanced electron leakage and superoxide generation (Das, 2023). This dysfunction reduces ATP production and promotes the opening of the mitochondrial permeability transition pore, facilitating cytochrome *c* release and activation of caspase-dependent apoptotic pathways. Alterations in mitochondrial dynamics, including excessive fission and reduced fusion, further compromise structural integrity and metabolic efficiency (Lee, 2025; Videla et al., 2022).

ER stress develops concurrently due to the accumulation of misfolded or oxidatively modified proteins. Activation of the unfolded protein response through PERK, ATF6, and IRE1 signaling initially attempts to restore proteostasis; however, prolonged stress induces C/EBP homologous protein (CHOP)-mediated apoptosis. Chemical additives and degradation products may disturb ER membrane lipid composition and calcium storage capacity, exacerbating protein misfolding (Moreews and Karlsson, 2025; Ranzato and Martinotti, 2026). Functional coupling between mitochondria and the ER intensifies cellular injury, as dysregulated calcium exchange amplifies ROS production and disrupts metabolic coordination, promoting progressive cytotoxic outcomes (Zhao and Sheng, 2025).

10.7 Conclusion

The interaction of MPs with biological systems is governed by a complex interplay between particle physicochemical properties and environmentally induced transformations. As detailed in this chapter, processes such as weathering, photooxidation, fragmentation, and corona formation substantially modify MP surfaces, thereby influencing their reactivity, biological recognition, and interaction potential. These modifications determine how MPs interact with biological interfaces and shape their behavior after internalization. The biological impacts of MPs arise through closely linked physical and chemical interaction mechanisms. Physical effects include mechanical abrasion of epithelial surfaces, destabilization of cellular membranes, impairment of barrier integrity, and disruption of subcellular structures. In parallel, chemical toxicity is mediated through the leaching of polymer additives and the transport of adsorbed environmental contaminants, intensifying localized chemical stress following biological uptake. These interacting stressors converge at the cellular level, triggering oxidative stress, inflammatory signaling pathways, and dysfunction of key organelles such as mitochondria and the ER, ultimately compromising cellular homeostasis. By integrating environmental transformation processes with biological interaction pathways, this chapter highlights the multifactorial nature of MP-induced damage. The synthesis presented emphasizes that observed biological effects result from the combined influence of particle-mediated physical stress and chemically driven toxicity. This mechanistic perspective provides a coherent framework for interpreting the impacts of MPs across biological systems and contributes to a deeper understanding of their ecological and health-related significance.

AI disclosure

During the preparation of this work, the author(s) used ChatGPT to assist in the compilation of rough content. However, the content was cross-checked, verified, rephrased, and aligned with scholarly literature. After using this tool/service, the author(s) reviewed and edited the content as needed and take(s) full responsibility for the content of the publication.

References

Adeleye, A.T., Bahar, M.M., Megharaj, M., Fang, C., Rahman, M.M., 2024. The Unseen Threat of the Synergistic Effects of Microplastics and Heavy Metals in Aquatic Environments: a Critical Review. Curr. Pollut. Rep. 21986592. 10 (3), 478–497. Available from: https://www.springer.com/journal/40726. doi:10.1007/s40726-024-00298-7.

Bhardwaj, G., Abdulkadhim, M., Joshi, K., Wankhede, L., Das, R.K., Brar, S.K., 2025. Exposure Pathways, Systemic Distribution, and Health Implications of Micro- and Nanoplastics in Humans. Appl. Sci. 2076-3417. 15 (16), 8813. doi:10.3390/app15168813.

Bhattacharyya, S., Greer, M.L., Salehi, M., 2025. Impact of micro- and nanoplastics exposure on human health: focus on neurological effects from ingestion. Front. Public Health. 22962565. 13. Available from: http://journal.frontiersin.org/journal/public-health/section/public-health-education-and-promotion#archive. doi:10.3389/fpubh.2025.1681776.

Bretz, F., Bornkamp, B., 2025. Thomas Dumortier, Dose-response characterization: a key to success in drug development. Clin. Trials. 1740-7745. 22 (4), 384–392. doi:10.1177/17407745251350289.

Cahn, D., Stern, A., Buckenmeyer, M., Wolf, M., Duncan, G.A., 2024. Extracellular Matrix Limits Nanoparticle Diffusion and Cellular Uptake in a Tissue-Specific Manner. ACS Nano. 1936086X. 18 (46), 32045–32055. Available from: http://pubs.acs.org/journal/ancac3. doi:10.1021/acsnano.4c10381.

Chen, Y.S., Tian, H.X., Rong, D.C., Wang, L., Chen, S., Zeng, J., Xu, H., Mei, J., Wang, L.Y., Liou, Y.L., Zhou, H.H., 2025. ROS homeostasis in cell fate, pathophysiology, and therapeutic interventions. Mol. Biomed. 26628651. 6 (1). Available from: https://www.springer.com/journal/43556. doi:10.1186/s43556-025-00338-8.

Cheng, Y., Yang, Y., Bai, L., Cui, J., 2024. Microplastics: an often-overlooked issue in the transition from chronic inflammation to cancer. J. Transl. Med. 14795876. 22 (1). Available from: https://translational-medicine.biomedcentral.com. doi:10.1186/s12967-024-05731-5.

Choi, D., Bang, J., Kim, T., Yoogyeong, O., Hwang, Y., Hong, J., 2020. In vitro chemical and physical toxicities of polystyrene microfragments in human-derived cells. J. Hazard. Mater. 03043894. 400, 123308. doi:10.1016/j.jhazmat.2020.123308.

Choi, D., Hwang, J., Bang, J., Han, S., Kim, T., Yoogyeong, O., Hwang, Y., Choi, J., Hong, J., 2021. In vitro toxicity from a physical perspective of polyethylene microplastics based on statistical curvature change analysis. Sci. Total Environ. 00489697. 752, 142242. doi:10.1016/j.scitotenv.2020.142242.

Das, A., 2023. The emerging role of microplastics in systemic toxicity: involvement of reactive oxygen species (ROS). Sci. Total Environ. 00489697. 895, 165076. doi:10.1016/j.scitotenv.2023.165076.

Dash, U.C., Bhol, N.K., Swain, S.K., Samal, R.R., Nayak, P.K., Raina, V., Panda, S.K., Kerry, R.G., Duttaroy, A.K., Jena, A.B., 2025. Oxidative stress and inflammation in the pathogenesis of neurological disorders: mechanisms and implications. Acta Pharm. Sin. B. 22113843. 15 (1), 15–34. Available from: https://www.sciencedirect.com/science/journal/22113835. doi:10.1016/j.apsb.2024.10.004.

Domenech, J., de Britto, M., Velázquez, A., Pastor, S., Hernández, A., Marcos, R., Cortés, C., 2021. Long-Term Effects of Polystyrene Nanoplastics in Human Intestinal Caco-2 Cells. Biomolecules. 2218-273X. 11 (10), 1442. doi:10.3390/biom11101442.

Duan, J., Bolan, N., Li, Y., Ding, S., Atugoda, T., Vithanage, M., Sarkar, B., Tsang, D.C.W., Kirkham, M.B., 2021. Weathering of microplastics and interaction with other coexisting constituents in terrestrial and aquatic environments. Water Res. 18792448. 196. Available from: www.elsevier.com/locate/watres. doi:10.1016/j.watres.2021.117011.

Feng, Y., Hongying, F., Zhang, X., Liu, S., Wei, X., 2024. Lysosome toxicities induced by nanoparticle exposure and related mechanisms. Ecotoxicol. Environ. Saf. 01476513. 286, 117215. doi:10.1016/j.ecoenv.2024.117215.

Garcia-Llorens, G., Ouardi, M.E., Valls-Belles, V., 2025. Oxidative Stress Fundamentals: unraveling the Pathophysiological Role of Redox Imbalance in Non-Communicable Diseases. Appl. Sci. 2076-3417. 15 (18), 10191. doi:10.3390/app151810191.

Harraq, A.A., Brahana, P.J., Arcemont, O., Zhang, D., Valsaraj, K.T., Bharti, B., 2022. Effects of Weathering on Microplastic Dispersibility and Pollutant Uptake Capacity. ACS Environ. Au. 26942518. 2 (6), 549–555. Available from: https://pubs.acs.org/page/aeacc4/about.html. doi:10.1021/acsenvironau.2c00036.

Hernández, A.F., Lacasaña, M., Tsatsakis, A.M., Docea, A.O., 2025. Cellular and Molecular Mechanisms of Micro- and Nanoplastics Driving Adverse Human Health Effects. Toxics. 23056304. 13 (11). Available from http://www.mdpi.com/journal/toxics. doi:10.3390/toxics13110921.

Holmes, L.A., Turner, A., Thompson, R.C., 2014. Interactions between trace metals and plastic production pellets under estuarine conditions. Mar. Chem. 03044203. 167, 25–32. Available from: www.elsevier.com/inca/publications/store/5/0/3/3/4/9. doi:10.1016/j.marchem.2014.06.001.

Hong, Y., Boiti, A., Vallone, D., Foulkes, N.S., 2024. Reactive Oxygen Species Signaling and Oxidative Stress: transcriptional Regulation and Evolution. Antioxidants. 20763921. 13 (3). Available from: www.mdpi.com/journal/antioxidants. doi:10.3390/antiox13030312.

Hua, X., Wang, D., 2022. Cellular Uptake, Transport, and Organelle Response After Exposure to Microplastics and Nanoplastics: current Knowledge and Perspectives for Environmental and Health Risks. Rev. Environ. Contam. Toxicol. 0179-5953. 260 (1). doi:10.1007/s44169-022-00013-x.

Iftikhar, A., Qaiser, Z., Sarfraz, W., Ejaz, U., Aqeel, M., Rizvi, Z.F., Khalid, N., 2024. Understanding the leaching of plastic additives and subsequent risks to ecosystems. Water Emerg. Contam. Nanoplast. 28312597. 3 (1). Available from: https://f.oaes.cc/xmlpdf/694f760b-d4b1-41ae-b9d5-a4bdf381ec19/wecn2058_down.pdf?v=70. doi:10.20517/wecn.2023.58.

Jeong, E., Lee, J.-Y., Redwan, M., 2024. Animal exposure to microplastics and health effects: a review. Emerg. Contam. 24056650. 10 (4), 100369. doi:10.1016/j.emcon.2024.100369.

John, K.I., Omorogie, M.O., Bayode, A.A., Adeleye, A.T., Helmreich, B., 2023. Environmental microplastics and their additives—a critical review on advanced oxidative techniques for their removal. Chem. Pap. 13369075. 77 (2), 657–676. Available from: https://www.springer.com/journal/11696. doi:10.1007/s11696-022-02505-5.

Jolaosho, T.L., Rasaq, M.F., Omotoye, E.V., Araomo, O.V., Adekoya, O.S., Abolaji, O.Y., Hungbo, J.J., 2025. Microplastics in freshwater and marine ecosystems: occurrence, characterization, sources, distribution dynamics, fate, transport processes, potential mitigation strategies, and policy interventions. Ecotoxicol. Environ. Saf. 10902414. 294. Available from: https://www.sciencedirect.com/science/journal/01476513. doi:10.1016/j.ecoenv.2025.118036.

Ketian, L., Kunhao, Y., Zhang, Y., Haixu, D., Sioutas, C., Wang, Q., 2024. Unveiling the mechanism secret of abrasion emissions of particulate matter and microplastics. Sci. Rep. 2045-2322. 14 (1). doi:10.1038/s41598-024-74137-6.

Kochanek, A., Grąz, K., Potok, H., Gronba-Chyła, A., Kwaśny, J., Wiewiórska, I., Ciuła, J., Basta, E., Łapiński, J., 2025. Micro- and Nanoplastics in the Environment: current State of Research, Sources of Origin, Health Risks, and Regulations—A Comprehensive Review. Toxics. 2305-6304. 13 (7), 564. doi:10.3390/toxics13070564.

Lee, J., 2025. Mitochondrial Dysfunction and Cell Death: Mechanisms and Implications in Pathology. IntechOpen. 10.5772/intechopen.1010111.

Li, B., Ming, H., Qin, S., Nice, E.C., Dong, J., Du, Z., Huang, C., 2025. Redox regulation: mechanisms, biology and therapeutic targets in diseases. Signal Transduct. Target. Ther. 20593635. 10 (1). Available from: https://www.nature.com/sigtrans. doi:10.1038/s41392-024-02095-6.

Liu, L., Kexin, X., Zhang, B., Yiyuan, Y., Zhang, Q., Jiang, W., 2021. Cellular internalization and release of polystyrene microplastics and nanoplastics. Sci. Total Environ. 00489697. 779, 146523. doi:10.1016/j.scitotenv.2021.146523.

Luo, H., Zhen, L., Zeng, Y., Dongqin, H., Sun, J., Zhang, A., Pan, X., 2025. Degradation of microplastics in the natural environment: a comprehensive review on process, mechanism, influencing factor and leaching behavior. J. Environ. Manag. 03014797. 391, 126429. doi:10.1016/j.jenvman.2025.126429.

Wei, M., Liang, X., Ding, C., Yingying, Y., Jiji, L., 2026. From Enrichment to Fate: transport, Transformation, and Fate of Micro- and Nanoplastics in Marine Environments. Toxics. 2305-6304. 14 (2), 120. doi:10.3390/toxics14020120.

Mahmud, F., Sarker, D.B., Jocelyn, J.A., Sang, Q.X.A., 2024. Molecular and Cellular Effects of Microplastics and Nanoplastics: focus on Inflammation and Senescence. Cells. 20734409. 13 (21). Available from https://www.mdpi.com/journal/cells. doi:10.3390/cells13211788.

Moreews, M., Karlsson, M.C.I., 2025. Endoplasmic reticulum stress: a key player in immune cell regulation and autoimmune disorders. Semin. Immunol. 10963618. 78. Available from: https://www.sciencedirect.com/science/journal/10445323. doi:10.1016/j.smim.2025.101954.

Mulky, L., Ronak, R., Somashekara, D., 2026. Pollution to Solution: understanding and Addressing Microplastic Contamination in the Environment. Environ. Qual. Manag. 15206483. 35 (3). Available from: http://onlinelibrary.wiley.com/journal/10.1002/(ISSN)1520-6483. doi:10.1002/tqem.70307.

Olanrewaju, K.B., Ford Versypt, A.N., 2025. Roles of chemical species transport and transformation in the biophysics of human pathophysiology. Npj Biol. Phys. Mech. 3004-863X. 2 (1). doi:10.1038/s44341-025-00025-3.

Prata, J.C., 2023. Microplastics and human health: integrating pharmacokinetics. Crit. Rev. Environ. Sci. Technol. 15476537. 53 (16), 1489–1511. Available from: www.tandf.co.uk/journals/titles/10643389.asp. doi:10.1080/10643389.2023.2195798.

Ragusa, A., Matta, M., Cristiano, L., Matassa, R., Battaglione, E., Svelato, A., De Luca, C., D'Avino, S., Gulotta, A., Rongioletti, M.C.A., Catalano, P., Santacroce, C., Notarstefano, V., Carnevali, O., Giorgini, E., Vizza, E., Familiari, G., Nottola, S.A., 2022. Deeply in Plasticenta: presence of Microplastics in the Intracellular Compartment of Human Placentas. Int. J. Environ. Res. Public Health. 16604601. 19 (18). Available from: http://www.mdpi.com/journal/ijerph. doi:10.3390/ijerph191811593.

Ranzato, E., Martinotti, S., 2026. The Interplay Between Ca2+ Homeostasis, Endoplasmic Reticulum Stress, and the Unfolded Protein Response in Human Diseases. Cells. 2073-4409. 15 (4), 352. doi:10.3390/cells15040352.

C. Rex M A. Debroy, M.J. Nirmala, A. Mukherjee, 2023. Ecotoxicological significance of bio-Corona formation on micro/nanoplastics in aquatic organisms. RSC Adv. 20462069. 13 (33), 22905–22917. Available from: http://pubs.rsc.org/en/journals/journal/ra. doi:10.1039/d3ra04054b.

Camil Rex, M., Bairoliya, A., Gairola, S., Sureshkumar, P., Niranjan, V., Nag, A., Mukherjee, A., 2025. Engineered biocorona on microplastics as a toxicity mitigation strategy in marine environment: experiments with a marine crustacean Artemia salina. J. Hazard. Mater. Adv. 27724166. 17, 100558. doi:10.1016/j.hazadv.2024.100558.

Sadique, S.A., Konarova, M., Niu, X., Szilagyi, I., Nirmal, N., Li, L., 2025. Impact of microplastics and nanoplastics on human Health: emerging evidence and future directions. Emerg. Contam. 24056642. 11 (3). Available from: https://www.sciencedirect.com/science/journal/24056650. doi:10.1016/j.emcon.2025.100545.

Schwarzfischer, M., Rogler, G., 2022. The Intestinal Barrier—Shielding the Body from Nano- and Microparticles in Our Diet. Metabolites. 2218-1989. 12 (3), 223. doi:10.3390/metabo12030223.

Selvaraj, N.R., Nandan, D., Nair, B.G., Nair, V.A., Venugopal, P., Aradhya, R., 2025. Oxidative Stress and Redox Imbalance: common Mechanisms in Cancer Stem Cells and Neurodegenerative Diseases. Cells. 20734409. 14 (7). Available from https://www.mdpi.com/journal/cells. doi:10.3390/cells14070511.

Singh, S., Nandave, M., 2024. Cellular Stress, Survival Mechanism and Adaptation Adaptation under Stressful Environments through Biological Adjustments and Interventions. Springer Nature, India, pp. 243–262. Available from https://link.springer.com/10.1007/978-981-99-7652-2. doi:10.1007/978-981-99-7652-2_16.

Thapliyal, C., Negi, S., Nagarkoti, S., Daverey, A., 2025. Mechanistic insight into potential toxic effects of microplastics and nanoplastics on human health. Discov. Appl. Sci. 3004-9261. 7 (6). doi:10.1007/s42452-025-07214-8.

Vanetti, C., Broggiato, M., Pezzana, S., Clerici, M., Fenizia, C., 2025. Effects of microplastics on the immune system: how much should we worry? Immunol. Lett. 01652478. 272, 106976. doi:10.1016/j.imlet.2025.106976.

Verma, J., Devi, S., Yadav, M., 2024. Microplastics and Microbe Interaction on Human and Animal Health Microplastics Pollution and Its Remediation. Springer Nature, India, pp. 215–234. Available from https://link.springer.com/10.1007/978-981-97-4068-0. doi:10.1007/978-981-97-4068-0_11.

Videla, L.A., Marimán, A., Ramos, B., José Silva, M., Del Campo, A., 2022. Standpoints in mitochondrial dysfunction: underlying mechanisms in search of therapeutic strategies. Mitochondrion. 18728278. 63, 9–22. Available from: https://www.journals.elsevier.com/mitochondrion. doi:10.1016/j.mito.2021.12.006.

Vranic, S., Kurapati, R., Kostarelos, K., Bianco, A., 2025. Biological and environmental degradation of two-dimensional materials. Nat. Rev. Chem. 2397-3358. 9 (3), 173–184. doi:10.1038/s41570-024-00680-5.

Wang, H., Zhu, J., Yuan, H., Wang, J., Zeng, N., Zhan, X., 2023. Photoaging process and mechanism of four commonly commercial microplastics. J. Hazard. Mater. 03043894. 451, 131151. doi:10.1016/j.jhazmat.2023.131151.

Wang, P., Shi, Y.Z., Guan, Q., 2025. The Microplastic–PFAS Nexus: from Co-Occurrence to Combined Toxicity in Aquatic Environments. Toxics. 23056304. 13 (12). Available from: http://www.mdpi.com/journal/toxics. doi:10.3390/toxics13121041.

Xu, Y., Ou, Q., van der Hoek, J.P., Liu, G., Lompe, K.M., 2024. Photo-oxidation of Micro- and Nanoplastics: physical, Chemical, and Biological Effects in Environments. Environ. Sci. Technol. 15205851. 58 (2), 991–1009. Available from: http://pubs.acs.org/journal/esthag. doi:10.1021/acs.est.3c07035.

Yang, H., Chen, Z., Kong, L., Xing, H., Yang, Q., Jun, W., 2025. A Review of Eco-Corona Formation on Micro/Nanoplastics and Its Effects on Stability, Bioavailability, and Toxicity. Water. 2073-4441. 17 (8), 1124. doi:10.3390/w17081124.

Zhang, X., Wei, F., Haiyan, F., Guo, H., 2025. Characterisation of environmentally persistent free radicals and their contributions to oxidative potential and reactive oxygen species in sea spray and size-resolved ambient particles. Npj Clim. Atmos. Sci. 2397-3722. 8 (1). doi:10.1038/s41612-025-00911-6.

Zhang, X., Zheng, P., Yang, M., Yin Huang, E.L., Liu, A., Zhang, H., Guo, J., 2026. Emerging mechanisms of microplastic-induced skin diseases: a perspective from the gut–skin axis. J. Transl. Med. 1479-5876. 24 (1). doi:10.1186/s12967-025-07300-w.

Zhao, J., Na, D.H., Garcia, S.J., 2025. Surface modification of natural porous silica microparticles to control the loading and release of organic corrosion inhibitors in coatings. NPG Asia Mater. 18844057. 17 (1). Available from: https://www.nature.com/am/. doi:10.1038/s41427-025-00608-5.

Zhao, W.-B., Sheng, R., 2025. The correlation between mitochondria-associated endoplasmic reticulum membranes (MAMs) and Ca2+ transport in the pathogenesis of diseases. Acta Pharmacol. Sin. 1671-4083. 46 (2), 271–291. doi:10.1038/s41401-024-01359-9.

Zhao, X., Zhou, Y., Liang, C., Song, J., Yu, S., Liao, G., Zou, P., Tang, K.H.D., Wu, C., 2023. Airborne microplastics: occurrence, sources, fate, risks and mitigation. Sci. Total Environ. 18791026. 858. Available from: www.elsevier.com/locate/scitotenv. doi:10.1016/j.scitotenv.2022.159943.

CHAPTER

11

Microplastic interference at the cellular level: Mechanisms of action

Cristiana Roberta Multisanti[a], *Oksana Stoliar*[b], *and Caterina Faggio*[c,d]

[a]Department of Veterinary Sciences, University of Messina, Messina, Italy
[b]Ternopil Volodymyr Hnatiuk National Pedagogical University, Ternopil, Ukraine
[c]Department of Chemical, Biological, Pharmaceutical, and Environmental Sciences, University of Messina, Messina, Italy
[d]Department of Eco-sustainable Marine Biotechnology, Stazione Zoologica Anton Dohrn, Naples, Italy

11.1 Introduction

The word "plastic" encompasses a range of properties that have made these materials essential in daily life and ubiquitous across environments. These characteristics include malleability and flexibility, as well as various advantages of plastic objects, such as durability, low cost, lightweight, and corrosion resistance. These characteristics make plastic commercially advantageous, but they also contribute to its persistence and accumulation in natural environments.

Since the 1950s, when global plastic production was estimated at 1.5 million tons, production has increased. This increase has led to the release of a massive volume of waste into environmental compartments and raised concerns about its long-term impact (Burgos-Aceves et al., 2024). Indeed, the massive, often uncontrolled use of plastic has led to its accumulation across all ecosystems. Moreover, research has demonstrated that the COVID-19 pandemic has worsened plastic pollution by increasing waste from disposable masks, gloves, and protective gear (Parashar and Hait, 2021).

In the environment, plastic items degrade, producing smaller particles. These tiny particles are small enough to enter living organisms and disrupt biological systems. These invisible pollutants are recognized as potential disruptors at several biological levels, with the potential to cross biological barriers and interfere with essential physiological cellular processes (Wang et al., 2022). As a result, aquatic organisms are continuously exposed to plastic particles through various routes, including filtration, ingestion, and

Microplastics in Animal and Human Physiology
https://doi.org/10.1016/B978-0-443-45635-0.00016-1

direct contact (Pothiraj et al., 2023), whereas terrestrial and aerial species are exposed through contaminated food, soil, and airborne dust (Sucharitakul et al., 2024). Human exposure occurs mainly through contaminated food and water (Kannan and Vimalkumar, 2021). In fact, food webs are the main drivers of these exposure pathways, highlighting how easily plastic particles first enter natural ecosystems and then organisms (Mossotto et al., 2025). In this context, bioaccumulation and biomagnification factors across trophic levels amplify the concentration and therefore the effects of plastic particles, as predators consuming contaminated prey are exposed to relatively high particle loads and associated chemical contaminants (Krause et al., 2021). For these reasons, the "One Health" perspective is fundamental, as the consequences of plastic pollution extend to all environmental compartments, including ecosystems, animal welfare, and human health.

11.2 Overview of plastic debris classification

The adverse effects of plastic particles on organisms are influenced by several factors, including their origin, shape, size, density, and chemical composition. Moreover, these characteristics determine the bioavailability and mobility of particles within natural ecosystems, especially aquatic ones (Wright et al., 2013).

11.2.1 Size-based classification

The classification of plastic particles on the basis of their size provides a standardized framework for assessing their environmental distribution and biological effects (Gigault et al., 2021):

- Macroplastics: > 25 mm;
- Mesoplastics: 5–25 mm;
- Microplastics (MPs): < 5 mm;
- Nanoplastics (NPs): < 100 nm.

11.2.2 Source-based classification

On the basis of their origin, plastic particles can be classified into primary and secondary particles (LI et al., 2016). The first class comprises particles intended for inclusion in products, including pharmaceutical and personal care products, as well as agricultural additives (Zicarelli et al., 2022). Primary plastic particles are continuously released into the environment through the regular use and disposal of these products. Examples include microbeads in toothpaste, shampoo, and shower gel, as well as polymeric beads or seeds used in agriculture (Burgos-Aceves et al., 2024). The second class includes particles resulting from the degradation of larger plastic debris in the ecosystem.

11.2.3 Shape-based classification

Plastic debris can also be classified by shape. The main categories include the following:

- Beads, granules, microspheres, pellets, and spheres: These type of particles are solid, rounded, or cylindrical in shape and are intentionally added to cosmetics and cleaning products (Zicarelli et al., 2022).
- **Fibers, microfibers, and filaments:** The shape of these particles is elongated with a length that exceeds the width, and they are mainly released from synthetic fabrics during washing. These particles are characterized by high density, leading to sinking and sediment accumulation (Hartmann et al., 2019).
- **Films:** These are thin particles whose flexibility distinguishes them from rigid fragments. They are smaller than other types and typically originate from packaging and agricultural mulching films (Song et al., 2024).
- **Foam and fragments:** Plastic particles are irregularly shaped and are derived mainly from the degradation of plastic products, such as bags and polystyrene packaging (Ngo et al., 2019).

11.2.4 Polymer-type classification

The increase in global plastic production has introduced a wide variety of synthetic polymers, including polyethylene (PE), low-density polyethylene, high-density polyethylene, polypropylene (PP), polyvinyl chloride (PVC), polystyrene (PS), polyurethane, and polyethylene terephthalate (PET), into the environment (Liu et al., 2019).

11.3 Physicochemical properties of plastic particles relevant to cellular interactions

This paragraph explores the different effects of MPs/NPs on the basis of the characteristics mentioned above, with a focus on additives used in the industrial production of plastic objects and on the interactions of plastic particles with other contaminants in the environment.

11.3.1 Size, shape, polymer type, and surface chemistry

The size of plastic particles is a relevant factor in evaluating their interactions with biological systems, as it influences both their internalization capacity and the pathways involved in this process. Furthermore, particle size modulates tissue biodistribution and the type and intensity of toxic responses (Chen et al., 2024). Differences in size translate into distinct toxicological profiles. NPs are more likely to interfere with cellular processes. For example, they induce significantly greater mortality in fish embryos than MPs do (Kim et al., 2022). However, size-dependent effects are not limited to aquatic species. In mammalian cell models, NPs show a greater capacity for internalization and more substantial cytotoxic and proinflammatory effects (Wang et al., 2019), as well as a greater ability to stimulate the production of immune cytokines and chemokines than their larger MP counterparts do (Hwang et al., 2020).

Similarly, different shapes have been demonstrated to modulate the behavior of plastic particles. This difference was confirmed in an in vitro study using Caco-2 intestinal epithelial cells, which revealed that spherical particles and fibrous fragments had similar stimulatory effects on the antioxidant response. However, subcytotoxic changes were observed mainly with spherical particles (Saenen et al., 2023).

The type of polymer also affects the toxicity of plastic particles. Polymers such as PE, PP, and PS are among the most abundant in the environment (Banaee et al., 2025). PE is highly reactive because exposure of adult fish to this polymer causes molecular disturbances, including the activation of cytochrome P450 enzymes and alterations in vitellogenin expression, which are endpoints that reflect interference with metabolic and reproductive pathways. Owing to these molecular changes, behavioral alterations may be influenced. PE is linked to increased levels of antioxidant enzymes (Banaee et al., 2023), damage to mechanosensory receptors in the lateral line system of fish, and morphological changes in their gastrointestinal tract (Kim et al., 2022). PS has been demonstrated to induce morphological and metabolic alterations and activate prooxidative and inflammatory pathways (Mahmud et al., 2024). In contrast, PVC is less reactive in biological systems, as its effects on aquatic organisms are minimal in terms of survival, movement, and stress biomarkers (de Mello Pereira et al., 2023).

The surface chemistry of plastic particles is another relevant factor in their interactions with cells. The physicochemical changes induced by exposure to environmental agents, such as oxidation, biofilm formation, and the adsorption of persistent organic pollutants and metals, can alter the reactivity and affinity of particles for biological macromolecules (Banaee et al., 2025).

11.3.2 Plastic product additives

During industrial production, a wide range of additives are employed to improve the performance of plastic items (Burgos-Aceves et al., 2021). These additives are generally mixed with the plastic matrix without forming covalent bonds, and the identity of the polymers determines the potential additives they may release during degradation, which, in turn, may increase their toxicity (Banaee et al., 2025).

Brominated flame retardants are the main additives used in plastics to reduce their flammability (Bridson et al., 2021). Among these, the three most common are polybrominated diphenyl ethers, hexabromocyclododecanes, and tetrabromobisphenol-A (TBBPA), which are widely recognized for their hepatic toxicity in fish (Rochman et al., 2013; Herrera et al., 2022). In particular, TBBPA, unlike other flame retardants, is chemically bound to the polymer. Despite this, it has been detected in multiple environmental compartments and tissues (Yang et al., 2012).

In addition, additives include plasticizers, which are added to make plastics more flexible and workable and to reduce fragility. These compounds are known for their lipophilic properties (LI et al., 2016) and easy release from the polymer matrix (Porretti et al., 2022). These compounds can mimic or alter the functions of hormones in fish, leading to endocrine disruption. Among these are phthalates, which include phthalic acid esters. These effects have been associated with several adverse effects on fish development, with transcriptional alterations in genes involved in development. These changes are also reflected at a relatively high functional level as abnormal behavior in exposed organisms (Qian et al., 2020). The toxicity of dibutyl phthalate and diethyl phthalate has been associated with significant alterations in ionic homeostasis in marine organisms, as evidenced by reduced levels of Na^+, K^+, and Cl^-; inhibition of the activity of Na^+/K^+-ATPase in the gills and brain; and neurophysiological impairment. Other additives, such as polycyclic aromatic hydrocarbons and bisphenol A, have been linked to the onset of oxidative stress and alterations in antioxidant systems. This imbalance between oxidative stress and detoxification capacity can alter blood biochemical parameters in aquatic species (Vázquez-Rowe et al., 2021).

Thermal stabilizers and UV stabilizers are widely used to protect plastic materials from high temperatures and UV radiation, respectively. Their role is to reduce polymer degradation and limit discoloration and mechanical integrity loss (Kim et al., 2022). Among these, specific UV stabilizers, such as UV-234, UV-327, and BP-12, have been studied for their persistence and bioaccumulation rates in fish tissues (Hasegawa et al., 2022).

11.3.3 Adsorption and transport of co-contaminants

Plastic particles are also recognized for their ability to act as vectors for other contaminants. Indeed, they can efficiently adsorb heavy metals, pesticides, and other chemicals usually present in the surrounding environment because of their own properties and surface chemistry modifications (Impellitteri et al., 2023; Gholamhosseini et al., 2024). This ability to carry harmful compounds highlights the dual role of MPs/NPs in aquatic ecosystems, both as physical particles and as vectors of chemical stressors (Banihashemi et al., 2022).

These other pollutants can then be released into tissues, leading to bioaccumulation and potentially synergistic (Li et al., 2021), neutral, or inhibitory effects.

11.4 Cellular uptake of plastic particles

Depending on their size, plastic particles can cross biological barriers, enter cells through various pathways, and interfere with biological functions. In this context, there is clear evidence that MPs and NPs are ubiquitous in human biological fluids and tissues and can cross biological barriers such as the placenta, blood–brain barrier, and blood–testicular barrier (Bai et al., 2024). The intracellular transfer of MPs/NPs occurs in two stages. First, the particle surface interacts with the cell. This event can be both specific, through binding to specific membrane receptors (Rohrbach et al., 2020), or nonspecific, through interactions, such as electrostatic forces (Forest et al., 2015). In the second stage, particles can be passively internalized or endocytosed.

One factor that strongly influences the likelihood that a particle will be internalized by cells is the formation of a surface coating. In the aquatic environment, this process occurs readily, with particles quickly acquiring a coating of biomolecules and environmental compounds on their surfaces, a process known as "eco-corona" formation, which changes the physicochemical properties of the cell (Łazarski et al., 2025).

11.4.1 Passive membrane penetration

The passive penetration of plastic particles into cells depends on the particle size, surface modifications, and physicochemical properties of the polymer (Ruenraroengsak and Tetley, 2015).

In this context, PS and PE can spontaneously enter the bilayer. The adsorbed lipids form a monolayer crown that reduces their hydrophobicity and confers a negative charge. In contrast, PET penetrates membranes more efficiently with minimal lipid absorption due to its lower hydrophobicity. Other polymers, such as PP and PVC, dissolve within the bilayer, with slower translocation but greater membrane disruption. Their morphology also influences their cellular uptake. Indeed, rounder NPs cross membranes more efficiently, whereas elongated fibers are often oriented along their longitudinal axis during penetration, making translocation more difficult.

11.4.2 Endocytosis

Plastic particles can cross cellular membranes mainly through active endocytosis, during which they interact with the membrane to be internalized into endocytic vesicles. In this context, particle size determines the amount and rate of endocytosis. Moreover, these processes start with physicochemical interactions between the particle and the membrane (Li et al., 2021).

Endocytic pathways are mainly protein-mediated, such as clathrin-dependent endocytosis and caveolin-dependent endocytosis. In the former, particles are bound to receptors that can be found in clathrin-coated membrane invaginations. Then, they fuse with early endosomes, and the cargo can be destined for lysosomal degradation (Li et al., 2021). Caveolin-mediated endocytosis is characterized by caveolae, which are flask-shaped structures enriched in caveolins and cholesterol-rich lipid domains. Caveolae vesicles are directed toward caveosomes and then to the Golgi apparatus or the endoplasmic reticulum (ER) (Ying et al., 2025).

Macropinocytosis is a less selective endocytic pathway that involves the nonspecific internalization of extracellular fluids and the materials suspended in them. This process is driven by the reorganization of the actin cytoskeleton, which generates membrane protrusions that fold in on themselves, engulfing large volumes in vesicles called macropinosomes (Griffiths et al., 2022).

11.4.3 Phagocytosis

Phagocytosis occurs in specialized phagocytic cells that remove pathogens and debris. Moreover, it is mediated by specific receptors (Freeman and Grinstein, 2014). This interaction activates signaling cascades that lead to actin polymerization. This promotes the extension of membrane protrusions that engulf the particle. With respect to plastic particles, phagocytosis suggests that immune cells may perceive them as foreign bodies, thereby stimulating immune responses.

11.5 Cellular effects of plastic particles and physiological consequences

The effects of the cellular uptake of plastic particles are reflected at all biological levels, right down to the physiological consequences for the entire organism.

11.5.1 Oxidative stress and redox imbalance

When plastic particles come into contact with cellular structures, they can disrupt redox homeostasis, leading to the formation of reactive oxygen species (ROS) (Martyniuk et al., 2022) and reactive nitrogen species (RNS) (Kadac-Czapska et al., 2024). This imbalance between ROS production and altered antioxidant defenses is one of the first and most relevant cellular responses to plastic exposure (Liang et al., 2021).

Oxidative stress induced by plastic particles can damage lipids, proteins, and DNA. In addition, it can activate inflammatory responses that promote adaptive adjustments or apoptosis, depending on the intensity and persistence of the stress (Rizzo et al., 2023).

Generally, the leading cause of these alterations is excessive ROS production. This overproduction necessitates the availability of key antioxidant defense enzymes, such as superoxide dismutase, catalase, glutathione peroxidase, glutathione S-transferase, and glutathione reductase. Antioxidant system failure in response to ROS can lead to increased lipid peroxidation, which can be quantified by measuring malondialdehyde (MDA). In this context, recent studies have demonstrated that increased Na^+/K^+-ATPase activity is associated with higher MDA levels than those observed under physiological conditions. This finding provides evidence that cells attempt to compensate for ionic imbalances by increasing Na^+/K^+-ATPase activity (Rangasamy et al., 2022).

Excessively increased and sustained exposure to ROS causes oxidative damage and can eventually result in cell death by destroying lipids, proteins, nucleic acids, and other cellular constituents. On the contrary, moderately increased ROS levels can stimulate adaptive responses (Rizzo et al., 2025).

11.5.2 Disruption of cytoplasmic membranes and organelles

Exposure to plastic can also interfere with cytoplasmic membranes and organelles (Hua and Wang, 2022). Among these structures, mitochondria are among the most important, as they are responsible for producing cellular energy and maintaining physiological cell metabolism. To ensure proper mitochondrial function, they must have intact membranes, and the complexes of the electron transport chain (ETC) must work in coordination. Exposure to plastic particles can disrupt this balance by altering lipid membranes, their organization, and membrane fluidity. These impairments increase the potential for electron leakage from the mitochondrial ETC, resulting in excessive ROS production (Félix et al., 2023). Notably, plastic particles can inhibit mitochondrial complexes I and IV, thereby reducing electron flow. This can lead to a decrease in the proton gradient and trigger ATP synthesis (Seward et al., 2025). Moreover, owing to excessive ROS and low ATP levels, the Ca^{2+} potential can be destabilized (Kawamata and Manfredi, 2010). In addition, mitochondrial integrity is essential for key processes such as cell signaling, apoptosis, and cellular adaptation to stress. These impairments can affect multiple levels of biological organization. For example, a cell may switch from aerobic respiration to glycolysis (Vazquez et al., 2010).

In addition to mitochondria, peroxisomes and the ER can be affected by plastic particles. Peroxisomal dysfunction impairs fatty acid β-oxidation and hydrogen peroxide detoxification (Pan et al., 2017). In addition, the interaction of nitric oxide and superoxide radicals can induce the formation of RNS. Consequently, RNS participate in the nitration of tyrosine residues in proteins, which affects cell signaling and metabolism (Radi, 2013). In addition, stress to the ER caused by plastic particles can activate the unfolded protein response and stimulate additional ROS production (Wang et al., 2023).

11.5.3 Inflammation and immune dysregulation

Exposure to plastic particles induces inflammatory responses in organisms. In this context, increasing ROS production activates redox-sensitive transcription factors, including NF-κB, Nrf2, and AP-1. NF-κB stimulates proinflammatory cytokine production; Nrf2 coordinates the expression of antioxidant defenses; and AP-1 contributes to both stress responses and the regulation of apoptosis-related genes. In parallel, plastic particles can interact with toll-like receptors (TLRs), thereby inducing additional signaling pathways. Moreover, the inflammasome NLRP3 plays a key role in inflammatory processes by integrating signals from oxidative stress and TLRs. In addition, apoptotic mechanisms may be activated, as caspase activation and ER stress-related pathways amplify inflammatory signaling via thioredoxin-interacting protein (TXNIP), which acts as a sensor linking redox damage to NLRP3 activation (Cui et al., 2024). Chronic activation of these pathways can induce long-term immune and inflammatory dysfunctions (Rizzo et al., 2025).

These findings emerge from studies using aquatic model organisms and human cells, highlighting that these mechanisms are shared across species.

11.5.4 Genotoxicity and DNA damage

Genotoxicity refers to the compromised integrity of genetic material in cells. These effects, due to exposure to plastic particles (Menezes et al., 2024), are consequences of both ROS overproduction, which promotes DNA damage (da Costa Araújo et al., 2020), and the activation of intracellular signaling cascades, which influence genomic stability and cell fate.

Plastic particles can interfere with the mitotic spindle and compromise chromosome segregation (Xuan et al., 2023). If they reach the nucleus, they can amplify their mutagenic potential by directly interfering with DNA structure and function (Li et al., 2021). In this context, direct DNA damage, alterations in gene expression, and chromosomal abnormalities can result from structural chromosome damage or errors in chromosome segregation during cell division (Tagorti and Kaya, 2022). Moreover, the gene expression of stress proteins, detoxification enzymes, and immune factors is influenced following exposure to plastic particles (Reddy et al., 2025). These changes are often associated with tissue damage, suggesting that genotoxicity is a systemic event (Tang, 2025).

11.5.5 Cell death pathways

Exposure to MPs has been shown to induce programmed cell death pathways in cellular models. While small NPs mainly induce necrosis, larger particles mainly induce apoptosis (Wang et al., 2022).

Exposure to plastic particles leads to excessive ROS production, which, in turn, activates intracellular signaling cascades, such as the MAPK pathway. In addition, they alter the expression of genes involved in apoptotic cascades, including those of the Bcl-2 family and caspases (Wang et al., 2022). One of the primary targets of this signaling axis is hypoxia-inducible factor 1α, which integrates oxidative stress with apoptotic signaling. Studies on fish indicate that plastic exposure can modulate the expression of cell death regulators and alter Ca^{2+} homeostasis, thereby further activating proteolytic enzymes, such as calpains. Moreover, miRNA expression can be influenced. Dysregulation of these noncoding RNAs can cause an imbalance between proapoptotic and antiapoptotic factors, thereby increasing programmed cell death (Li et al., 2021).

11.6 Experimental strategies for assessing the risk of plastic particles

Risk assessment following exposure to MPs/NPs should integrate experimental strategies to elucidate their interactions at different biological levels. For example, studies on model organisms can provide key insights into the impact of plastic particles on human health, whereas knowledge from human toxicology can be applied to ecotoxicological scenarios. This two-way exchange promotes multidisciplinary collaboration among specialists, including ecologists, veterinarians, toxicologists, and medical researchers.

A method for obtaining comprehensive cytotoxicity results from plastic particle exposure requires a complementary approach that integrates in vivo and in vitro data. Each of these in silico studies offers unique perspectives. Moreover, owing to the heterogeneity of plastic particles, the integration of multiple methodologies is needed (Arslan Yüce et al., 2025). In this context, the adverse outcome pathway (AOP) framework provides a clear overview by linking molecular causes to key events leading to adverse outcomes, which describe a predisease condition. In the context of MPs/NPs, AOPs indicate that ROS generation often represents the molecular initiating step, leading to inflammation, mitochondrial dysfunction, and cytotoxicity. These cellular effects can then induce higher-level consequences, such as reduced growth, reproductive failure, impaired immunity, and increased mortality (Jeong and Choi, 2019).

11.6.1 In vivo models

An advantage of traditional in vivo analyses is the ability to link molecular and cellular mechanisms to consequences at higher levels of biological organization (González-Acedo et al., 2021).

In addition, in vivo models allow the investigation of complex parameters, such as growth, reproduction, behavior, immunocompetence, and survival, which are relevant endpoints from an ecological perspective (Tang and Zhou, 2025). In addition, they ensure greater environmental realism, as exposure can occur through pathways (ingestion, respiration, or contact) that simulate realistic conditions. Indeed, in vivo studies can analyze bioaccumulation, biomagnification, and indirect effects within trophic networks (Wang et al., 2019), as well as synergistic interactions among plastic particles, chemical contaminants, and environmental factors (Ali et al., 2025). Studies have also evaluated changes in cellular functions following in vivo exposure. Specifically, in model organisms exposed to MPs/NPs, cell isolation protocols are implemented starting from target organs, and subsequent analyses are performed on isolated cells to evaluate changes in their physiological processes. Naturally, in vivo methods must employ ethical and responsible practices in experimental design. In this context, the application of the 3Rs principles (reduction, refinement, and replacement), initially proposed by Russell and Burch, is essential to minimize animal use and suffering while ensuring scientific validity. To avoid this issue, various alternative strategies and nontraditional model systems have been developed.

11.6.2 In vitro studies

In vitro methods are advantageous models for investigations because they can reproduce the complexity of a living organism, linking toxicokinetic concepts with toxicodynamic and molecular interactions (Testai et al., 2021).

This approach represents integration and an alternative to traditional in vivo studies, as it dramatically reduces the ethical concerns associated with animal testing. This approach aligns with the 3Rs framework, as it limits the use of whole organisms (Tagkalidou et al., 2024), enabling researchers to study key mechanisms of toxicity without causing pain, suffering, or mortality in laboratory animals. Employing cell-based systems enables us to effectively study the mechanisms of action by which plastic particles interfere with cellular and molecular processes, allowing comprehensive analyses of specific pathways involved. Furthermore, in vitro analyses often offer greater sensitivity and reproducibility, allowing the isolation of specific pathways that the complexity of whole-organism responses may mask. Therefore they minimize ethical concerns and offer mechanistic insights that can complement in vivo testing.

In addition, three-dimensional (3D) cell models are increasingly used to study the effects of plastic particles, as they better replicate the architecture, polarity, and microenvironment of tissues in vivo than conventional two-dimensional cultures do (Rafazi et al., 2024). Spheroids and organoids help assess particle penetration, intracellular localization, and toxicity within an organized system. In addition, 3D models facilitate long-term exposure studies and the evaluation of intercellular communication (Rizzo et al., 2023, 2025).

11.6.3 In silico approaches

In addition to in vivo and in vitro studies, in silico approaches are increasingly used to assess the risks associated with MNs/NPs. Methodologies such as computational modeling and quantitative structure–activity relationships enable predictions of outcomes from exposure to realistic environmental scenarios. These tools are fast, efficient, and low-cost, enabling the analysis of the properties of complex substances by computer processing of their molecular structures, thereby providing predictive information. Quantitative Structure–Property Relationship (QSPR) models have already been applied to study the sorption capacity of plastic particles and have provided valuable insights into their interactions with persistent and toxic chemical contaminants. These approaches are enhanced by advanced imaging techniques, high-throughput screening, and omics technologies (transcriptomics, proteomics, and metabolomics), which enable the detection of significant alterations induced by MPs/NPs at the cellular and molecular levels (Baspakova et al., 2025). Therefore these tools can be valuable additions, enabling the reconstruction of causal chains between initial molecular events and their consequences at the organism and population levels (Wang et al., 2023).

11.7 Future perspectives and research gaps

While research on the cytotoxic effects of MPs is ubiquitous, the inconsistencies, lack of uniformity, and reproducibility crisis in this field are worsening. The use of implausible assumptions in study design has led to a disconnect in the literature, limiting the comparability of study results. For example, while laboratory studies report MP/NP concentrations used in experiments, field studies have reported concentrations that are an order of magnitude lower (Burns and Boxall, 2018). In addition, the MPs/NPs used in experimental research are uniform, smooth, and small, whereas real-world scenarios involve diverse microfibers and unsorted MPs/NPs. A metaanalysis of the literature revealed that only 17% of the measured concentrations and 20% of the measured particle sizes in the laboratory are similar to those reported in field studies (Weis and Palmquist, 2021).

Another widely highlighted limitation is the lack of standardization in experimental procedures. Exposure protocols and analytical methodologies are often insufficiently described. Cowger et al. (2020) presented detailed guidelines to increase reproducibility and comparability in MP studies. Without shared criteria for particle size, shape, chemical composition, and dosimetry, each investigation provides answers that are difficult to relate to a standard set of conditions. For example, Tang (2025) noted the wide variability in dose units and explicitly recommended the use of uniform units.

This methodological disorder also translates into heterogeneous toxicological endpoints. There are no standardized biomarkers for the effects of MPs, so each study chooses different indicators. Tang and Zhou (2025) noted that the lack of common biomarkers complicates the identification of "critical" doses. Indeed, various effects have been identified at very different exposure levels. Some endpoints (such as acute mortality) require unrealistically high experimental doses, precisely because plastic particles have low intrinsic toxicity. In the absence of a shared reference, the results remain fragmentary and difficult to generalize.

11.8 Conclusion

MPs and NPs pose a significant threat to the environment and all forms of life, from the cellular level to the entire ecosystem. Studies conducted on humans and animals have confirmed that internalized plastic particles can hinder fundamental physiological functions, cause cellular damage by disrupting organelles that contain plastic, trigger and alter inflammatory and immune responses, and affect genomic stability. These effects are influenced by the chemical and physical characteristics of the particles, the presence of additives that can be further released into the environment, and the adsorption of cocontaminants, which increases their toxicological potential.

Despite growing efforts by the scientific community, some critical gaps remain. Current studies often use unrealistic concentrations and particle types that do not accurately reflect environmental conditions, and the lack of standardized methodologies continues to hinder comparisons across studies. Furthermore, the absence of shared biomarkers and end points limits our ability to define thresholds of biological relevance. These limitations underscore the need for harmonized experimental strategies that integrate in vivo, in vitro, and in silico approaches, together with innovative models such as 3D cell systems, to provide more realistic and predictive information.

References

Ali, W., Jeong, H., Kim, D.H., Lee, J.S., Zinck, P., Souissi, S., Lee, J.S., 2025. Adverse effects of environmentally relevant microplastics on in vivo endpoints, oxidative stress, and mitogen-activated protein kinase signaling pathway and multixenobiotic resistance system in the marine rotifer Brachionus plicatilis. Sci. Total Environ. 18791026. 958. Available from: https://www.sciencedirect.com/science/journal/00489697. doi:10.1016/j.scitotenv.2024.178027.

Bai, C.-L., Wang, D., Luan, Y.-L., Huang, S.-N., Liu, L.-Y., Guo, Y., 2024. A review on micro- and nanoplastics in humans: implication for their translocation of barriers and potential health effects. Chemosphere 361. 10.1016/j.chemosphere.2024.142424

Banaee, M., Multisanti, C.R., Impellitteri, F., Piccione, G., Faggio, C., 2025. Environmental toxicology of microplastic particles on fish: a review. Comp. Biochem. Physiol. C Toxicol. Pharmacol. 15320456. 287, 110042. doi:10.1016/j.cbpc.2024.110042.

Banaee, M., Zeidi, A., Sinha, R., Faggio, C., 2023. Individual and Combined Toxic Effects of Nano-ZnO and Polyethylene Microplastics on Mosquito Fish (Gambusia holbrooki). Water. 2073-4441. 15 (9), 1660. doi:10.3390/w15091660.

Banihashemi, E.A., Soltanian, S., Gholamhosseini, A., Banaee, M., 2022. Effect of microplastics on Yersinia ruckeri infection in rainbow trout (Oncorhynchus mykiss). Environ. Sci. Pollut. Res. 16147499. 29 (8), 11939–11950. Available from: https://link.springer.com/journal/11356. doi:10.1007/s11356-021-16517-3.

Baspakova, A., Zare, A., Suleimenova, R., Berdygaliev, A.B., Karimsakova, B., Tussupkaliyeva, K., Mussin, N.M., Zhilisbayeva, K.R., Tanideh, N., Tamadon, A., 2025. An updated systematic review about various effects of microplastics on cancer: a pharmacological and in-silico based analysis. Mol. Aspects Med. 00982997. 101, 101336. doi:10.1016/j.mam.2024.101336.

Bridson, J.H., Gaugler, E.C., Smith, D.A., Northcott, G.L., Gaw, S., 2021. Leaching and extraction of additives from plastic pollution to inform environmental risk: a multidisciplinary review of analytical approaches. J. Hazard. Mater. 18733336. 414. Available from: www.elsevier.com/locate/jhazmat. doi:10.1016/j.jhazmat.2021.125571.

Burgos-Aceves, M.A., Abo-Al-Ela, H.G., Faggio, C., 2021. Physiological and metabolic approach of plastic additive effects: immune cells responses. J. Hazard. Mater. 18733336. 404. Available from: www.elsevier.com/locate/jhazmat. doi:10.1016/j.jhazmat.2020.124114.

Burgos-Aceves, M.A., Banaee, M., Vazzana, I., Betancourt-Lozano, M., González-Mille, D.J., Aliko, V., Faggio, C., Ilizaliturri-Hernández, C.A., 2024. Effect of emerging pollutants on the gut microbiota of freshwater animals: focusing on microplastics and pesticides. Sci. Total Environ. 18791026. 948. Available from: https://www.sciencedirect.com/science/journal/00489697. doi:10.1016/j.scitotenv.2024.174809.

Burns, E.E., Boxall, A.B.A., 2018. Microplastics in the aquatic environment: evidence for or against adverse impacts and major knowledge gaps. Environ. Toxicol. Chem. 15528618. 37 (11), 2776–2796. Available from: http://onlinelibrary.wiley.com/journal/10.1002/(ISSN)1552-8618. doi:10.1002/etc.4268.

Chen, J., Liang, Q., Zheng, Y., Lei, Y., Gan, X., Mei, H., Bai, C., Wang, H., Jingjuan, J., Dong, Q., Song, Y., 2024. Polystyrene nanoplastics induced size-dependent developmental and neurobehavioral toxicities in embryonic and juvenile zebrafish. Aquat. Toxicol. 0166445X. 267, 106842. doi:10.1016/j.aquatox.2024.106842.

Cowger, W., Booth, A.M., Hamilton, B.M., Thaysen, C., Primpke, S., Munno, K., Lusher, A.L., Dehaut, A., Vaz, V.P., Liboiron, M., Devriese, L.I., Hermabessiere, L., Rochman, C., Athey, S.N., Lynch, J.M., De Frond, H., Gray, A., Jones, O.A.H., Brander, S., Steele, C., Moore, S., Sanchez, A., Nel, H., 2020. Reporting Guidelines to Increase the Reproducibility and Comparability of Research on Microplastics. Appl. Spectrosc. 19433530. 74 (9), 1066–1077. Available from: https://journals.sagepub.com/loi/asp. doi:10.1177/0003702820930292.

Cui, J., Zhu, M., Sun, X., Yang, J., Guo, M., 2024. Microplastics induced endoplasmic reticulum stress to format an inflammation and cell death in hepatocytes of carp (Cyprinus carpio). Aquat. Toxicol. 0166445X. 269, 106870. doi:10.1016/j.aquatox.2024.106870.

da Costa Araújo, A.P., de Andrade Vieira, J.E., Malafaia, G., 2020. Toxicity and trophic transfer of polyethylene microplastics from Poecilia reticulata to Danio rerio. Sci. Total Environ. 18791026. 742. Available from: www.elsevier.com/locate/scitotenv. doi:10.1016/j.scitotenv.2020.140217.

de Mello Pereira, D., Mazon, S.C., Mendes, E.J., Brunetto, R., Ozelame, B., Zembruski, F.S., Dalcin, A.L.F., Marsaro, I. B., Aguiar, G.P., Antônio Lutinski, J., Tavella, R.A., da Silva Júnior, F.M.R., Oliveira, J.V., Müller, L.G., Antônio Fiori, M., Sachett, A., Siebel, A.M., 2023. Recycled polyvinyl chloride microplastics: investigation of environmentally relevant concentrations on toxicity in adult zebrafish. Journal of Toxicology and Environmental Health, Part A. 1528-7394. 86 (11), 347–360. doi:10.1080/15287394.2023.2203154.

Félix, L., Carreira, P., Peixoto, F., 2023. Effects of chronic exposure of naturally weathered microplastics on oxidative stress level, behaviour, and mitochondrial function of adult zebrafish (Danio rerio). Chemosphere. 00456535. 310, 136895. doi:10.1016/j.chemosphere.2022.136895.

Forest, V., Cottier, M., Pourchez, J., 2015. Electrostatic interactions favor the binding of positive nanoparticles on cells: a reductive theory. Nano Today. 17480132. 10 (6), 677–680. doi:10.1016/j.nantod.2015.07.002.

Freeman, S.A., Grinstein, S., 2014. Phagocytosis: receptors, signal integration, and the cytoskeleton. Immunol. Rev. 1600065X. 262 (1), 193–215. Available from: http://onlinelibrary.wiley.com/journal/10.1111/(ISSN)1600-065X. doi:10.1111/imr.12212Canada.

Gholamhosseini, A., Banaee, M., Zeidi, A., Multisanti, C.R., Faggio, C., 2024. Individual and combined impact of microplastics and lead acetate on the freshwater shrimp (Caridina fossarum): biochemical effects and physiological responses. J. Contam. Hydrol. 18736009. 262. Available from: https://www.sciencedirect.com/science/journal/01697722. doi:10.1016/j.jconhyd.2024.104325.

Gigault, J., Hadri, H.E., Nguyen, B., Grassl, B., Rowenczyk, L., Tufenkji, N., Feng, S., Wiesner, M., 2021. Nanoplastics are neither microplastics nor engineered nanoparticles. Nat. Nanotechnol. 1748-3387. 16 (5), 501–507. doi:10.1038/s41565-021-00886-4.

González-Acedo, A., García-Recio, E., Illescas-Montes, R., Ramos-Torrecillas, J., Melguizo-Rodríguez, L., 2021. Víctor Javier Costela-Ruiz, Evidence from in vitro and in vivo studies on the potential health repercussions of micro- and nanoplastics. Chemosphere. 00456535. 280, 130826. doi:10.1016/j.chemosphere.2021.130826.

Griffiths, G., Gruenberg, J., Marsh, M., Wohlmann, J., Jones, A.T., Parton, R.G., 2022. Nanoparticle entry into cells; the cell biology weak link. Adv. Drug Deliv. Rev. 18728294. 188. Available from: www.elsevier.com/locate/drugdeliv. doi:10.1016/j.addr.2022.114403.

Hartmann, N.B., Hüffer, T., Thompson, R.C., Hassellöv, M., Verschoor, A., Daugaard, A.E., Rist, S., Karlsson, T., Brennholt, N., Cole, M., Herrling, M.P., Hess, M.C., Ivleva, N.P., Lusher, A.L., Wagner, M., 2019. Are We Speaking the Same Language? Recommendations for a Definition and Categorization Framework for Plastic Debris. Environ. Sci. Technol. 15205851. 53 (3), 1039–1047. Available from: http://pubs.acs.org/journal/esthag. doi:10.1021/acs.est.8b05297.

Hasegawa, T., Mizukawa, K., Yeo, B.G., Sekioka, T., Takada, H., Nakaoka, M., 2022. The significance of trophic transfer of microplastics in the accumulation of plastic additives in fish: an experimental study using brominated flame retardants and UV stabilizers. Mar. Pollut. Bull. 0025326X. 185 (Pt B), 114343. doi:10.1016/j.marpolbul.2022.114343.

Herrera, A., Acosta-Dacal, A., Pérez Luzardo, O., Martínez, I., Rapp, J., Reinold, S., Montesdeoca-Esponda, S., Montero, D., Gómez, M., 2022. Bioaccumulation of additives and chemical contaminants from environmental microplastics in European seabass (Dicentrarchus labrax). Sci. Total Environ. 00489697. 822, 153396. doi:10.1016/j.scitotenv.2022.153396.

Hua, X., Wang, D., 2022. Cellular Uptake, Transport, and Organelle Response After Exposure to Microplastics and Nanoplastics: current Knowledge and Perspectives for Environmental and Health Risks. Rev. Environ. Contam. Toxicol. 0179-5953. 260 (1). doi:10.1007/s44169-022-00013-x.

Hwang, J., Choi, D., Han, S., Jung, S.Y., Choi, J., Hong, J., 2020. Potential toxicity of polystyrene microplastic particles. Sci. Rep. 20452322. 10 (1). Available from: www.nature.com/srep/index.html. doi:10.1038/s41598-020-64464-9.

Impellitteri, F., Yunko, K., Martyniuk, V., Matskiv, T., Lechachenko, S., Khoma, V., Mudra, A., Piccione, G., Stoliar, O., Faggio, C., 2023. Physiological and biochemical responses to caffeine and microplastics in Mytilus galloprovincialis. Sci. Total Environ. 00489697. 890, 164075. doi:10.1016/j.scitotenv.2023.164075.

Jeong, J., Choi, J., 2019. Adverse outcome pathways potentially related to hazard identification of microplastics based on toxicity mechanisms. Chemosphere. 00456535. 231, 249–255. doi:10.1016/j.chemosphere.2019.05.003.

Kadac-Czapska, K., Ośko, J., Knez, E., Grembecka, M., 2024. Microplastics and Oxidative Stress—Current Problems and Prospects. Antioxidants. 2076-3921. 13 (5), 579. doi:10.3390/antiox13050579.

Kannan, K., Vimalkumar, K., 2021. A Review of Human Exposure to Microplastics and Insights Into Microplastics as Obesogens. Front. Endocrinol. 1664-2392. 12. doi:10.3389/fendo.2021.724989.

Kawamata, H., Manfredi, G., 2010. Mitochondrial dysfunction and intracellular calcium dysregulation in ALS. Mech. Ageing Dev. 00476374. 131 (7-8), 517–526. doi:10.1016/j.mad.2010.05.003.

Kim, S.A., Kim, L., Kim, T.H., An, Y.J., 2022. Assessing the size-dependent effects of microplastics on zebrafish larvae through fish lateral line system and gut damage. Mar. Pollut. Bull. 18793363. 185. Available from: www.elsevier.com/locate/marpolbul. doi:10.1016/j.marpolbul.2022.114279.

Krause, S., Baranov, V., Nel, H.A., Drummond, J.D., Kukkola, A., Hoellein, T., Smith, G.H.S., Lewandowski, J., Bonnet, B., Packman, A.I., Sadler, J., Inshyna, V., Allen, S., Allen, D., Simon, L., Mermillod-Blondin, F., Lynch, I., 2021. Gathering at the top? Environmental controls of microplastic uptake and biomagnification in freshwater food webs. Environ. Pollut. 18736424. 268. Available from: https://www.journals.elsevier.com/environmental-pollution. doi:10.1016/j.envpol.2020.115750.

Łazarski, G., Rajtar, N., Romek, M., Jamróz, D., Rawski, M., Kepczynski, M., 2025. Interaction of Polystyrene Nanoplastic with Lipid Membranes. J. Phys. Chem. B. 1520-6106. 129 (16), 4110–4122. doi:10.1021/acs.jpcb.5c00738.

Li, M., Wu, D., Wu, D., Guo, H., Han, S., 2021. Influence of polyethylene-microplastic on environmental behaviors of metals in soil. Environ. Sci. Pollut. Res. 16147499. 28 (22), 28329–28336. Available from: https://link.springer.com/journal/11356. doi:10.1007/s11356-021-12718-y.

LI, W.C., TSE, H.F., FOK, L., 2016. Plastic waste in the marine environment: a review of sources, occurrence and effects. Sci. Total Environ. 00489697. 566-567, 333–349. doi:10.1016/j.scitotenv.2016.05.084.

Liang, Y., Yang, X., Wang, Y., Liu, R., Haifeng, G., Mao, L., 2021. Influence of polystyrene microplastics on rotifer (Brachionus calyciflorus) growth, reproduction, and antioxidant responses. Aquat. Ecol. 1386-2588. 55 (3), 1097–1111. doi:10.1007/s10452-021-09885-y.

Liu, K., Tianning, W., Wang, X., Song, Z., Zong, C., Wei, N., Daoji, L., 2019. Consistent Transport of Terrestrial Microplastics to the Ocean through Atmosphere. Environ. Sci. Technol. 0013-936X. 53 (18), 10612–10619. doi:10.1021/acs.est.9b03427.

Mahmud, F., Sarker, D.B., Jocelyn, J.A., Sang, Q.X.A., 2024. Molecular and Cellular Effects of Microplastics and Nanoplastics: focus on Inflammation and Senescence. Cells. 20734409. 13 (21). Available from https://www.mdpi.com/journal/cells. doi:10.3390/cells13211788.

Martyniuk, V., Khoma, V., Matskiv, T., Baranovsky, V., Orlova-Hudim, K., Gylytė, B., Symchak, R., Matciuk, O., Gnatyshyna, L., Manusadžianas, L., Stoliar, O., 2022. Indication of the impact of environmental stress on the responses of the bivalve mollusk Unio tumidus to ibuprofen and microplastics based on biomarkers of reductive stress and apoptosis. Comp. Biochem. Physiol. C Toxicol. Pharmacol. 15320456. 261, 109425. doi:10.1016/j.cbpc.2022.109425.

Menezes, M., de Mello, F.T., Ziegler, L., Wanderley, B., Gutiérrez, J.M., Dias, J.D., 2024. Revealing the hidden threats: genotoxic effects of microplastics on freshwater fish. Aquat. Toxicol. 18791514. 276. Available from: https://www.sciencedirect.com/science/journal/0166445X. doi:10.1016/j.aquatox.2024.107089.

Mossotto, C., Maganza, A., Gabetti, A., Esposito, G., Faggio, C., Renzi, M., Kazmi, S.S.U.H., Bozzetta, E., Prearo, M., Pastorino, P., 2025. Tackling micro(nano)plastic pollution in aquaculture systems. Curr. Opin. Chem. Eng. 48. 10.1016/j.coche.2025.101104

Ngo, P.L., Pramanik, B.K., Shah, K., Roychand, R., 2019. Pathway, classification and removal efficiency of microplastics in wastewater treatment plants. Environ. Pollut. 18736424. 255. Available from: https://www.journals.elsevier.com/environmental-pollution. doi:10.1016/j.envpol.2019.113326.

Pan, H., Li, L.Y., Li, J.M., Wang, W.L., Limbu, S.M., Degrace, P., Li, D.L., Du, Z.Y., 2017. Inhibited fatty acid β-oxidation impairs stress resistance ability in Nile tilapia (Oreochromis niloticus). Fish Shellfish Immunol. 10959947. 68, 500–508. Available from. http://www.elsevier.com/inca/publications/store/6/2/2/8/3/2/index.htt. doi:10.1016/j.fsi.2017.07.058.

Parashar, N., Hait, S., 2021. Plastics in the time of COVID-19 pandemic: protector or polluter? Sci. Total Environ. 00489697. 759, 144274. doi:10.1016/j.scitotenv.2020.144274.

Porretti, M., Arrigo, F., Di Bella, G., Faggio, C., 2022. Impact of pharmaceutical products on zebrafish: an effective tool to assess aquatic pollution. Comp. Biochem. Physiol. C Toxicol. Pharmacol. 15320456. 261, 109439. doi:10.1016/j.cbpc.2022.109439.

Pothiraj, C., Gokul, T.A., Kumar, K.R., Ramasubramanian, A., Palanichamy, A., Venkatachalam, K., Pastorino, P., Barcelò, D., Balaji, P., Faggio, C., 2023. Vulnerability of microplastics on marine environment: a review. Ecol. Indic. 1470160X. 155. Available from: http://www.elsevier.com/locate/ecolind. doi:10.1016/j.ecolind.2023.111058.

Qian, L., Liu, J., Lin, Z., Chen, X., Yuan, L., Shen, G., Yang, W., Wang, D., Huang, Y., Pang, S., Xiyan, M., Wang, C., Yingren, L., 2020. Evaluation of the spinal effects of phthalates in a zebrafish embryo assay. Chemosphere. 00456535. 249, 126144. doi:10.1016/j.chemosphere.2020.126144.

Radi, R., 2013. Protein Tyrosine Nitration: biochemical Mechanisms and Structural Basis of Functional Effects. Acc. Chem. Res. 0001-4842. 46 (2), 550–559. doi:10.1021/ar300234c.

Rafazi, P., Haghi-Aminjan, H., Bagheri, Z., Rahimifard, M., 2024. Microplastics and cancer progression: a comparative study of 2D and 3D gastric cancer models using ISO Compliant protocols. Results Eng. 25901230. 24, 103329. doi:10.1016/j.rineng.2024.103329.

Rangasamy, B., Malafaia, G., Maheswaran, R., 2022. Evaluation of antioxidant response and Na+-K+-ATPase activity in zebrafish exposed to polyethylene microplastics: shedding light on a physiological adaptation. J. Hazard. Mater. 03043894. 426, 127789. doi:10.1016/j.jhazmat.2021.127789.

Reddy, S.R., Bangeppagari, M., Lee, S.J., 2025. Immune–Epigenetic Effects of Environmental Pollutants: mechanisms, Biomarkers, and Transgenerational Impact. Curr. Issues Mol. Biol. 14673045. 47 (9). Available from: https://www.mdpi.com/journal/cimb. doi:10.3390/cimb47090703.

Rizzo, M.G., Briglia, M., Zammuto, V., Morganti, D., Faggio, C., Impellitteri, F., Multisanti, C.R., Graziano, A.C.E., 2025. Innovation in Osteogenesis Activation: role of Marine-Derived Materials in Bone Regeneration. Curr. Issues Mol. Biol. 14673045. 47 (3). Available from: https://www.mdpi.com/journal/cimb. doi:10.3390/cimb47030175.

Rizzo, M.G., Palermo, N., Alibrandi, P., Sciuto, E.L., Gaudio, C.D., Filardi, V., Fazio, B., Caccamo, A., Oddo, S., Calabrese, G., Conoci, S., 2023. Physiologic Response Evaluation of Human Foetal Osteoblast Cells within Engineered 3D-Printed Polylactic Acid Scaffolds. Biology. 2079-7737. 12 (3), 424. doi:10.3390/biology12030424.

Rochman, C.M., Hoh, E., Kurobe, T., Teh, S.J., 2013. Ingested plastic transfers hazardous chemicals to fish and induces hepatic stress. Sci. Rep. 20452322. 3. doi:10.1038/srep03263.

Rohrbach, A., Meyer, T., Stelzer, E.H.K., Kress, H., 2020. Measuring Stepwise Binding of Thermally Fluctuating Particles to Cell Membranes without Fluorescence. Biophysical Society, Germany Biophysical Journal. 15420086. 118 (8), 1850–1860. Available from: http://www.sciencedirect.com/science/journal/00063495. doi:10.1016/j.bpj.2020.03.005.

Ruenraroengsak, P., Tetley, T.D., 2015. Differential bioreactivity of neutral, cationic and anionic polystyrene nanoparticles with cells from the human alveolar compartment: robust response of alveolar type 1 epithelial cells. Part. Fibre Toxicol. 17438977. 12 (1). Available from: http://www.particleandfibretoxicology.com/. doi:10.1186/s12989-015-0091-7.

Saenen, N.D., Witters, M.S., Hantoro, I., Tejeda, I., Ethirajan, A., Van Belleghem, F., Smeets, K., 2023. Polystyrene Microplastics of Varying Sizes and Shapes Induce Distinct Redox and Mitochondrial Stress Responses in a Caco-2 Monolayer. Antioxidants. 20763921. 12 (3). Available from: www.mdpi.com/journal/antioxidants. doi:10.3390/antiox12030739.

Seward, D.M., Keegan, C.D., Davey, G.P., 2025. Polystyrene nanoplastics target electron transport chain complexes in brain mitochondria. J. Hazard. Mater. 30510600. 1, 100003. doi:10.1016/j.hazmp.2025.100003.

Song, Z., Zhao, L., Junguo, B., Tang, Q., Wang, G., Yuxiang, L., 2024. Classification of Degradable Mulch Films and Their Promotional Effects and Limitations on Agricultural Production. Agriculture. 2077-0472. 14 (8), 1235. doi:10.3390/agriculture14081235.

Sucharitakul, P., Wu, W.M., Zhang, Y., Peng, B.Y., Gao, J., Wang, L., Hou, D., 2024. Exposure Pathways and Toxicity of Microplastics in Terrestrial Insects. Environ. Sci. Technol. 15205851. 58 (27), 11887–11900. Available from: http://pubs.acs.org/journal/esthag. doi:10.1021/acs.est.4c02842.

Tagkalidou, N., Multisanti, C.R., Bleda, M.J., Bedrossiantz, J., Prats, E., Faggio, C., Barata, C., Raldúa, D., 2024. Analyzing the Effects of Age, Time of Day, and Experiment on the Basal Locomotor Activity and Light-Off Visual Motor Response Assays in Zebrafish Larvae. Toxics. 23056304. 12 (5). Available from http://www.mdpi.com/journal/toxics. doi:10.3390/toxics12050349.

Tagorti, G., Kaya, B., 2022. Genotoxic effect of microplastics and COVID-19: the hidden threat. Chemosphere. 00456535. 286 (Pt 3), 131898. doi:10.1016/j.chemosphere.2021.131898.

Tang, K.H.D., 2025. Genotoxicity of Microplastics on Living Organisms: effects on Chromosomes, DNA and Gene Expression. Environments. 2076-3298. 12 (1), 10. doi:10.3390/environments12010010.

Tang, K.H.D., Zhou, J., 2025. Ecotoxicity of Biodegradable Microplastics and Bio-based Microplastics: a Review of in vitro and in vivo Studies. Environ. Manage. 75, 663–679. doi:10.1007/s00267-024-02106-w.

Testai, E., Bechaux, C., Buratti, F.M., Darney, K., Di Consiglio, E., Kasteel, E.E.J., Kramer, N.I., Lautz, L.S., Santori, N., Skaperda, Z., Kouretas, D., Turco, L., Vichi, S., 2021. Modelling human variability in toxicokinetic and toxicodynamic processes using Bayesian meta-analysis, physiologically-based modelling and in vitro systems. EFSA Support. Publ. 23978325. 18 (4). doi:10.2903/sp.efsa.2021.en-6504.

Vázquez-Rowe, I., Ita-Nagy, D., Kahhat, R., 2021. Microplastics in fisheries and aquaculture: implications to food sustainability and safety. Curr. Opin. Green Sustain. Chem. 24522236. 29, 100464. doi:10.1016/j.cogsc.2021.100464.

Vazquez, A., Liu, J., Zhou, Y., Oltvai, Z.N., 2010. Catabolic efficiency of aerobic glycolysis: the Warburg effect revisited. BMC Syst. Biol. 17520509. 4. Available from: http://www.biomedcentral.com/1752-0509/4/58. doi:10.1186/1752-0509-4-58.

Wang, W., Gao, H., Jin, S., Ruijing, L., Guangshui, N., 2019. The ecotoxicological effects of microplastics on aquatic food web, from primary producer to human: a review. Ecotoxicol. Environ. Saf. 01476513. 173, 110–117. doi:10.1016/j.ecoenv.2019.01.113.

Wang, W., Zhang, J., Qiu, Z., Cui, Z., Ningqi, L., Xin, L., Wang, Y., Zhang, H., Zhao, C., 2022. Effects of polyethylene microplastics on cell membranes: a combined study of experiments and molecular dynamics simulations. J. Hazard. Mater. 03043894. 429, 128323. doi:10.1016/j.jhazmat.2022.128323.

Wang, Y., Yang, W., Qikun, P., Sun, P., Zhao, W., Liu, M., Li, Y., 2023. Aquatic toxicity of tire microplastics on marine and freshwater organisms: an in silico approach. Chemosphere. 00456535. 313, 137523. doi:10.1016/j.chemosphere.2022.137523.

Weis, J.S., Palmquist, K.H., 2021. Reality Check: experimental Studies on Microplastics Lack Realism. Appl. Sci. 2076-3417. 11 (18), 8529. doi:10.3390/app11188529.

Wright, S.L., Thompson, R.C., Galloway, T.S., 2013. The physical impacts of microplastics on marine organisms: a review. Environ. Pollut. 02697491. 178, 483–492. doi:10.1016/j.envpol.2013.02.031.

Xuan, L., Xiao, L., Huang, R., 2023. The geno-toxicological impacts of microplastic (MP) exposure on health: mechanistic pathways and research trends from a Chinese perspective. Environ. Sci. Process. Impacts. 2050-7887. 25 (1), 26–36. doi:10.1039/d2em00301e.

Yang, S., Wang, S., Liu, H., Yan, Z., 2012. Tetrabromobisphenol A: tissue distribution in fish, and seasonal variation in water and sediment of Lake Chaohu, China. Environ. Sci. Pollut. Res. 0944-1344. 19 (9), 4090–4096. doi:10.1007/s11356-012-1023-9.

Ying, H., Jiang, S., Ying, X., Zhang, Y., Zhang, Q., Zhou, W., Liang, J., Wenhui, S., 2025. Comparative analysis of reproductive toxicity of polystyrene-nanoplastics and polystyrene-microplastics in rat Sertoli cells. Andrology 2047–2919. doi:10.1111/andr.70115.

Yüce, P.A., Günal, A.Ç., Erkmen, B., Yurdakok-Dikmen, B., Çağan, A.S., Çırak, T., Başaran Kankılıç, G., Seyfe, M., Filazi, A., Tavşanoğlu, Ü.N., 2025. In vitro and in vivo effects of commercial and environmental microplastics on Unio delicatus. Ecotoxicology. 15733017. 34 (1), 22–37. Available from: https://www.springer.com/journal/10646. doi:10.1007/s10646-024-02807-2.

Zicarelli, G., Multisanti, C.R., Falco, F., Faggio, C., 2022. Evaluation of toxicity of Personal Care Products (PCPs) in freshwaters: zebrafish as a model. Environ. Toxicol. Pharmacol. 13826689. 94, 103923. doi:10.1016/j.etap.2022.103923.

CHAPTER

12

Molecular impacts of microplastics

Auroshree Sadhu and Shubhajit Saha

Ecotoxicology Research Laboratory, Department of Zoology, The University of Burdwan, Burdwan, West Bengal, India

12.1 Introduction

Plastics are now omnipresent in our daily lives, with applications in practically every industry and economic area (Nayak et al., 2025). It is an obligatory component for commodities and packaging, including surgical equipment, fabrics, cosmetics, pharmaceuticals, and face masks (Napper et al., 2015; Duis and Coors, 2016; Dris et al., 2017; Lebreton and Andrady, 2019; Li et al., 2021). Owing to their many uses, such as their great durability, impermeable nature, affordability, and ease of manufacturing with low energy requirements, plastics are used excessively (Napper et al., 2015; Duis and Coors, 2016; Dris et al., 2017; Lebreton and Andrady, 2019; Li et al., 2021). These characteristics make them especially well-suited for producing food containers, various plastic products, and medical equipment (such as intravenous bags, syringes, prosthetic joints, medical apparatus packaging, artificial appendages, and tissue engineering) (North and Halden, 2013). Plastics release long-chain polymer segments into the surrounding environment from industrialized human sources. Plastic output has increased from 1.7 million tons in 1950 to 280 million metric tons (MMt) in 2012 and 2013, respectively, and has since quadrupled from 37 MMt in 2018 to 86 MMt in 2050, highlighting the long-lasting character of the material (Duis and Coors, 2016; Kan et al., 2023).

The spread of microplastics (MPs) and nanoplastics (NPs) in the environment, whether directly or indirectly, can expose organisms to toxicants (Blettler et al., 2018). Some studies have shown that MPs and NPs have negative impacts on a variety of taxa, including zooplankton, fish, mice, and nematodes (Qu et al., 2019; Boyle and Örmeci, 2020; Ding et al., 2021; Li et al., 2021; Wang et al., 2021; Yang et al., 2021; Zantis et al., 2022). In many animal models, exposure to polystyrene (PS) MPs (PS MPs) causes developmental, neurological, reproductive, metabolic, and immunological problems (Zou et al., 2020; Chen et al., 2021; Yang et al., 2021; Dolar et al., 2022; Li et al., 2022). These pollutants, which originate from chief sources (e.g., microbeads in personal care products) or secondary degradation of larger plastics, act as trajectories for contaminants, including pesticides (e.g., carbamates, organophosphates, neonicotinoids, biopesticides), surfactants, and polycyclic aromatic hydrocarbons (Chakraborty et al., 2024, 2025;

Microplastics in Animal and Human Physiology
https://doi.org/10.1016/B978-0-443-45635-0.00018-5

Mukherjee et al., 2025; Sharma et al., 2025). Chronic exposure to UV-aged PS MPs may disrupt neurotransmission, leading to neurotoxicity, as evidenced by reduced locomotor behaviors (Chen et al., 2021). NP toxicity has been observed not only in the parental cohort but also in the progeny of organisms such as nematodes (Li et al., 2022; Zhao et al., 2023). Specific chemical surface changes, including amino acid residue modifications, resulted in greater transgenerational toxicity to germline development and reproductive capacity than did the use of pure PS NPs (Sun et al., 2021). Research has shown that MPs and NPs can be found in various organs of environmental animals (including gills, hearts, kidneys, stomachs, livers, and muscles) (Collard et al., 2017; Kolandhasamy et al., 2018; Prata et al., 2022). MP abundance ranges from 8.8 items/g (guts) to 5.2 items/g (gills) in fish captured on the coast (Su et al., 2019). NPs may permeate biological membranes and accumulate in the organs of organisms at relatively high trophic levels, such as humans (Chae et al., 2018). After entering the body, certain MPs and NPs may be transported from the bloodstream to several organs, including the brain, heart, kidneys, and reproductive system (Deng et al., 2017; Duan et al., 2020). These impacts threaten food security, economic livelihoods, and the ecological integrity of aquatic ecosystems (Mandal et al., 2024; Sadhu et al., 2024; Mukherjee et al., 2025). Effective policy frameworks are essential for mitigating plastic pollution, a global challenge that threatens ecosystems, human health, and the blue economy (Sadhu et al., 2025), with MPs (1 μm–5 mm) and NPs (<0.1 μm) posing significant risks owing to their persistence and ability to act as contaminant vectors.

12.2 Entry and distribution of MPs within biological systems

12.2.1 Routes of exposure

MPs are pervasive in the environment, as they are found in all ecosystem components (Enyoh et al., 2019; Nayak et al., 2025). As a consequence, biota interact with each other through soil, water, or air. The bioavailability and, consequently, the toxicological impact of MPs are determined by their size, density, abundance, shape, and color (Verla et al., 2019). For the past 5 years, scientists have concentrated on the transport and impacts of MPs on the food chain. Various species are now being examined to determine which species are more sensitive to MP exposure (Rochman et al., 2015). Cunningham and Sigwart published a metaanalysis that included experimental MP exposure studies ($n = 128$) on marine and freshwater species (Cunningham and Sigwart, 2019). MP exposure was categorized as high (organ damage, decreased reproductive output, and mortality) once the concentration was greater than 100 MPs/L for water and greater than 100 MPs/kg (for sediment) or low (transient behavioral changes that include elevated respiration, diminished feeding, and decreased energy) when the concentration was ≤ 100 MPs/L for water and ≤ 100 MPs/kg for sediment. The threshold between the two groups was determined by the maximum reported ambient surface water concentrations (Burns and Boxall, 2018). Although MP concentrations may change over time, the 100 MPs/L criterion is intended to reflect the maximum levels of MPs documented at any given time (Cunningham and Sigwart, 2019). As a result, because the numerical concentration of MPs varies with size, different cut-off sizes provide different results. Many animals, however, are directly exposed to MPs, either by selective targeting of plastic objects or through unintentional absorption by filtering or predatory activity (Lusher et al., 2015). According to the current literature, all types of exposure to MPs fall into four broad categories: entanglement, contact, ingestion, and inhalation. Entanglement and contact are similar; however, this study distinguished them according to the method of exposure. Contact includes attachment or absorption through skin pores, whereas entanglement is characterized by MPs remaining attached to the organism's surface, causing the organism to sink if submerged in water.

12.2.2 Absorption and translocation across biological barriers

12.2.2.1 The gut barrier

A healthy gut barrier should prevent germs and other materials from entering the bloodstream. Nevertheless, investigations of fluorescent MPs in both mice and zebrafish have shown that MPs pass through the gut lumen and accumulate in gastrointestinal cells and tissues in a dimension- and shape-dependent way (Deng et al., 2017; Qiao et al., 2019; Lee et al., 2022). How MPs travel from the stomach lumen into tissues and throughout the body is unclear. MPs interact with epithelial villi and disrupt the monolayer despite being internalized, indicating that they traverse the intestinal barrier via paracellular transport (Stock et al., 2021; Paul et al., 2023). Absorption is the fundamental function of the gut; thus MP internalization by intestinal epithelial cells has been examined as an alternative absorption mechanism (Stock et al., 2019; Wu et al., 2019; Stock et al., 2021). Confocal imaging has revealed that MPs voluntarily adhere to cell surfaces in vitro and that MPs up to 10 μm in size are internalized after 12 h of exposure via intestinal cell absorption (Stock et al., 2019; Ramsperger et al., 2020). Some evidence suggests that smaller MPs can be transferred across the cell via lysosomes; however, this process has yet to be replicated (Wu et al., 2019). In Caco-2 cells, inhibiting the ATP-binding cassette transporter increased the accumulation of 0.1-μm MPs within the cells, suggesting that this efflux pump may help remove small MPs (Wu et al., 2019).

12.2.2.2 Lung barrier

According to reports, an adult inhales approximately 21 MP particles per day (Prata, 2018). Various particle features, including size and density, affect the extent of their accumulation in the respiratory system. Particles with lower densities penetrate deeply into the lungs. Upon settling, MPs are cleared by macrophages or transported to the circulatory or lymphatic system, leading to their translocation. Finally, the large number of these small MP fragments in the respiratory system triggers the release of chemical substances, increases particle permeability, and induces chronic lung inflammation (Geiser et al., 2005).

12.2.2.3 Blood–brain barrier

MPs that flow in the blood eventually contact the blood–brain barrier (BBB), which is prone to damage from environmental contaminants and chronic inflammation. Shan et al. (2022) reported that exposure to MPs in a BBB model (in vitro) decreased the expression of tight junction proteins (e.g., zonulin and occludin). This effect was demonstrated in vivo, with MP exposure increasing BBB permeability in the hypothalamus, hippocampus, and mouse brain cortex (Jin et al., 2022; Lee et al., 2022; Shan et al., 2022). MPs up to 10 μm in size infiltrate the cerebellum, cortex, and hippocampus of mice in a dose-dependent manner after only 24 h of exposure (Jin et al., 2022; Lee et al., 2022; Gaspar et al., 2023). Although this research is mostly based on animal models, MP penetration into brain tissue has been detected in situ.

12.2.2.4 *Placental barrier*

MPs and NPs penetrate the maternal–fetal barrier, influencing placental function and fetal development. The buildup of these particles in the placenta has been demonstrated to cause inflammatory responses, oxidative stress, and epigenetic changes. They also bioaccumulate in the testes, affecting spermatogenesis, spermatozoa shape, testosterone production, body weight, and inflammation, as observed in mice (Anifowoshe et al., 2025).

There are four modes of transportation through the placental barrier: active transport, passive diffusion, phagocytosis or pinocytosis, and biotransformation (Syme et al., 2004). The specific mechanism by which MPs and NPs breach the placental barrier remains to be determined. Various nanosized materials, including silica, gold, and titanium dioxide, have been shown in animal experiments to cross the placental barrier and cause toxicity; thus the placental translocation of most nanosized plastic fragments in humans is a cause for concern (Yamashita et al., 2011). Little is known about the presence, composition, distribution, and types of nanoparticles in the human placenta owing to the inability of current analytical methods to detect nanoparticles in complex environmental and biological materials (Brachner et al., 2020; Mandal et al., 2024).

12.2.3 Intracellular uptake mechanisms

Endosomes and autophagosomes can transport MPs and NPs, which then enter lysosomes and are dispersed into various organelles, including mitochondria, the nucleus, and the endoplasmic reticulum (ER). The mode of cellular absorption is the most important determinant of the intracellular transport pathways of NPs/MPs (Chou et al., 2011; Foroozandeh and Aziz, 2018). After the entrance of NPs/MPs into cells via endocytosis, endocytic vesicles are generated and merge with the early endosome. Certain NPs/MPs remain in early endosomes, moving slowly before maturing and differentiating into late endosomes (Behzadi et al., 2017; Foroozandeh and Aziz, 2018). Finally, late endosomes combine with lysosomes.

In addition, a few NPs/MPs were internalized via passive diffusion or discharged into the cytoplasm. By triggering autophagy, MPs are enclosed in autophagosomes, which are then merged with lysosomes. Others may be located in the cytoplasm and interact with other organelles, leading to malfunction (Behzadi et al., 2017).

12.3 Molecular mechanisms of MP-induced toxicity

12.3.1 Oxidative stress induction

The properties of MPs and NPs influence oxidative stress within the cell. For example, size is a significant physicochemical feature. Following the administration of PS MPs, reactive oxygen species (ROS) generation in Caco-2 cells increased in a size-dependent manner, leading to oxidative stress (Wang et al., 2020). Compared with 1-μm PS NPs, 100-nm PS NPs led to significantly greater oxidative stress in nematode worms (Yang and Wang, 2022). The surface characteristics of MPs and NPs play an important role in generating oxidative stress by increasing free radical production. They can be matured by heat and photoinitiated oxidative breakdown, which can produce environmentally persistent hydroxyl, singlet oxygen, and superoxide anion radicals (Alimi et al., 2022; Saha et al., 2023, 2024). These findings suggest that increased free radical production during the aging of MPs and NPs is directly linked to the generation of oxidative stress within cells. Dichlorofluorescein diacetate staining revealed that 100 mg/L pristine or UV-aged PS NPs dramatically elevated ROS levels, indicating that oxidative stress was induced in A549 cells (human alveolar type II epithelium) (Shi et al., 2021, 2022). Compared with pristine PS NPs, UV-accelerated aged NPs reduced cell viability and increased ROS formation in a dose- and aging-dependent manner (Shi et al., 2021, 2022).

12.3.2 Chronic inflammation and its pathological implications

Chronic inflammation caused by MP exposure can exacerbate oxidative stress. Inflammatory mediators promote the generation of ROS, which play a role in the immune reaction. As a result, a persistent cycle in which inflammation promotes oxidative stress and oxidative stress exacerbates inflammation may emerge (Zhang et al., 2022). Inflammation increases oxidative stress. Inflammatory cells, such as macrophages and neutrophils, can produce large amounts of ROS in response to infections and tissue injury. Inflammatory proteins, including interleukin-1 beta (IL-1β), tumor necrosis factor-alpha (TNF-α), and interleukin 6 (IL-6), may increase ROS production in cells, leading to oxidative stress in chronic inflammation. Moreover, ROS can directly trigger signaling pathways by oxidizing substances, including lipids and DNA, resulting in an inflammatory response. NF-κB controls the production of proinflammatory genes, including chemokines and cytokines, as well as adhesion molecules, which can lead to tissue damage, fibrosis, and cancer (Reuter et al., 2010; Cheng et al., 2022). Chronic oxidative stress, along with inflammation, harms tissues and hinders healing systems. ROS, as well as inflammatory mediators, destroy endothelial cells, increasing vascular permeability and hastening fibrosis. Cancer may be viewed as a form of persistent inflammation (Fig. 12.1). In cancer, both inflammation and oxidative

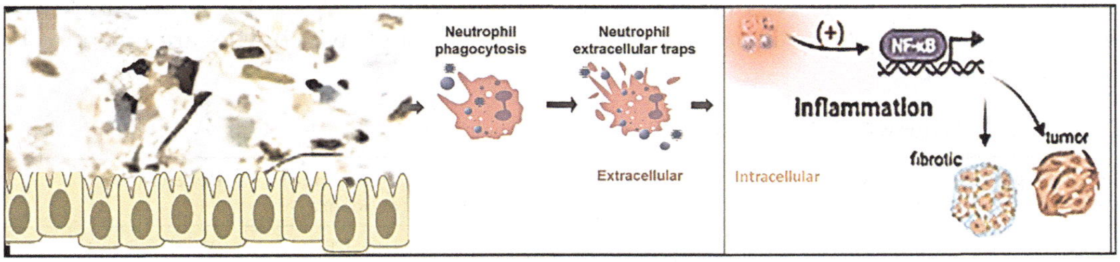

FIG. 12.1 Inflammatory response to microplastic accumulation. Microplastics penetrate the epithelial barrier, leading to neutrophil activation, phagocytosis, and the formation of neutrophil extracellular traps (NETs). This immunological activity triggers the NF-κB pathway. This signaling pathway causes persistent inflammation that might contribute to fibrosis and cancer formation.

stress induce tumor initiation, development, and progression via the same mechanisms as previously described. Furthermore, ROS cause DNA damage and transgene expression, trigger protooncogenes, and inactivate tumor suppressor genes (Gorrini et al., 2013).

12.3.3 Cellular damage and dysfunction

12.3.3.1 Alterations in cell membrane disruption and permeability

MPs and NPs may engage with the plasma membrane after feeding and inhalation, as well as during systemic circulation, determining their toxicity directly. Hydrophobic effects are thought to be a common route by which NPs are absorbed by cell membranes before entering cells (Jing and Zhu, 2011). Cellular absorption occurs through two mechanisms: endocytosis and passive diffusion, both of which contribute to the induction of MP and NP cytotoxicity.

Clathrin-mediated endocytosis is the primary mechanism by which cells absorb chemicals. NPs might surpass clathrin-coated plasma membranes before being released into the cytoplasm via vesicles (Praefcke and McMahon, 2004). PS particles ranging in size from 20 to 120 nm might be endocytosed in kidney cells via a clathrin-mediated mechanism (Fazlollahi et al., 2011). Other investigations revealed that both 40- and 200-nm carboxylated PSs were internalized via a clathrin-dependent mechanism in brain astrocytoma cells and a caveolae-dependent endocytosis-associated pathway in lung cancer cells (Dos Santos et al., 2011). In caveolae-mediated endocytosis, NPs can be internalized by binding to caveolin, which induces cell membrane invagination and maintains structural stability (Pelkmans and Helenius, 2002; Behzadi et al., 2017). NPs might also be encapsulated in vesicles formed during micropinocytosis, which are large plasma membrane invaginations containing the extracellular matrix (Lim and Gleeson, 2011). MP endocytosis is typically mediated by micropinocytosis and phagocytosis (Rejman et al., 2004; Kuhn et al., 2014). In several situations, the uptake of MPs and NPs involves multiple endocytic routes. The 190-nm PSs may enter the lung epithelium via phagocytosis along with clathrin-mediated endocytosis (Ji et al., 2021).

Certain MPs and NPs may enter cells via passive diffusion. The movement of 20-nm PS NPs through the rat alveolar epithelium was quicker than that through artificial bilayers (Yacobi et al., 2010). The passive diffusion of NPs is determined by the surface characteristics of the cell membrane, including the tension modulus, saturation kinetics, and degree of bending. In addition, the interface of the NPs with the plasma membrane was associated with the diffusion of the NPs (Thake et al., 2013; Contini et al., 2018). However, micropinocytosis and phagocytosis are the only pathways through which MPs are taken up by the cell (Table 12.1).

TABLE 12.1 Pathways for the cellular uptake of only microplastics.

Cellular uptake pathway	Type of plastic	Size	Alteration	Exposure concentration	Duration of exposure	References
Micropinocytosis	PS MPs	1 μm	Pristine	N.A.	6 and 18 h	(Firdessa et al., 2014)
Phagocytosis	PE MPs	1–5 μm	Pristine	0.05, 5, and 50 mg L^{-1}	48 h	(Abihssira-García et al., 2020)

Note: *N.A.*, not available; *PS MPs*, polystyrene microplastics; *PE MPs*, polyethylene microplastics.

12.3.3.2 *Dysfunction of mitochondria*

Mitochondria are essential for energy synthesis in cells and are a major source of ROS throughout normal metabolism. MPs can affect mitochondrial function via obstruction of ETC complex activity, resulting in electron seepage from complex I (NADH dehydrogenase) and complex III (cytochrome bc1) and the generation of superoxide anions (Li et al., 2022; Das, 2023; Gu et al., 2023; Zhang et al., 2024). This mitochondrial malfunction not only increases ROS levels but also impairs ATP production by reducing oxidative phosphorylation, leading to cellular energy deficiencies. MP exposure has been shown to reduce the mitochondrial membrane potential and increase mitochondrial permeability, both of which lead to apoptosis and cell death (Woo et al., 2023; Rong et al., 2024). Furthermore, NPs may stimulate p66Shc, which increases mitochondrial oxidative stress (Zhao et al., 2025). In neurons, NPs activate mitochondrial autophagy via the AMPK/ULK1 pathway, thereby increasing ROS levels (Halimu et al., 2022). Exposure to MPs can also cause disproportionate expansion of the permeability transition pore in mitochondria, leading to the release of cytochrome C and triggering a caspase cascade that ultimately leads to programmed cell death. MPs can cause intestinal inflammation, leading to elevated levels of IL-6 and TNF-α. This can worsen mitochondrial oxidative damage and trigger systemic inflammatory responses.

Nevertheless, it is unclear whether this mechanism involves monoamine oxidases (MAO) or mitochondrial NOX4. It was hypothesized that the surface charge of MPs influences their association with the mitochondrial membrane, thereby affecting the ROS-generating pathway (Huang et al., 2023). However, this theory must be experimentally verified.

12.3.3.3 *Stress in the ER*

Exposure to MPs may induce ER stress, triggering the unfolded protein response (UPR). UPR stimulation can increase ROS production by disrupting redox equilibrium and restoring normal cellular function (Chen et al., 2024). Sustained ER stress may activate apoptotic pathways, worsening oxidative damage. This phenomenon is especially important, as this stress can trigger many cell survival or bereavement pathways linked to inflammation and oxidative stress, resulting in tissue impairment (Wang et al., 2021; Wu et al., 2024). Food pollutants have been shown in studies to be potentially hazardous in neurodegenerative illnesses, as they stimulate proapoptotic gene transcription by increasing C/EBP homologous protein (CHOP) expression. The c-Jun N-terminal kinase (JNK) pathway was activated, leading to apoptosis. The breakdown of calcium homeostasis causes ER malfunction.

Furthermore, oxidative stress induced by pollutants stimulates the generation of ROS, which affect ER membranes and protein-folding machineries. The inflammatory response triggers the NF-κB pathway, leading to increased ER stress and neuronal death. This highlights the need to limit exposure to pollutants to safeguard neuronal health (Peng et al., 2022).

12.3.3.4 *Cytoskeletal disruption*

The cytoskeleton is made up of microtubules, microfilaments, and intermediate filaments, all of which contribute to the maintenance of normal cell shape and structure, as well as cell activity. MPs and NPs can penetrate cells and disrupt the cytoskeleton. With Actin Red staining, PS NP application resulted in actin fiber breakdown, reduced stress fiber production, and decreased actin network density (Li et al., 2021). Furthermore, exposure to 20-nm PS

NPs altered the gene expression of Rho-A and synaptopodin, which are associated with cytoskeleton stability (Li et al., 2021). Exposure to PS MPs in rat ovaries decreased the expression of the α-tubulin and dishevelled-associated activator of morphogenesis 1 (DAAM-1) proteins, which are involved in cytoskeleton development (Haddadi et al., 2022). PS MP exposure in zebra mussels elevated tubulin β chain (A0A194AQ74), actin-associated protein 2/3 complex subunit 2 (K1PPK1), myosin-Ie (K1QZI3), and dynein light chain roadblock (A0A210QUP6) and decreased tropomyosin 1 (H6BD84) and α-actinin (K1RH58) levels (Magni et al., 2019). As a result, MPs and NPs can influence cytoskeletal organization and function by regulating the expression of associated proteins (Table 12.2).

12.3.3.5 *Altered cellular proliferation and apoptosis*

Apoptosis caused by mitochondrial dysfunction is a common route. On the contrary, undue ROS generation induces cell death. It has been demonstrated that ROS accumulation can induce cell death via the p53 pathway (Chandra et al., 2000; Baumgartner et al., 2007). The mitochondria might activate caspases, promoting cell death (Chandra et al., 2000; Baumgartner et al., 2007). In combination with propidium iodide (PI)–annexin V staining, flow cytometry revealed that treatment with PS NPs led to a time-dependent increase in cell death (Li et al., 2022). Following ROS generation in PS NP-treated splenic cells, caspase-9, an apoptosis-associated enzyme, was also upregulated (Li et al., 2022). Compared with exposure to 20-nm virgin PS NPs, exposure to 20-nm amino-modified PS NPs increased the production of apoptosis-related proteins (Li et al., 2022). These findings reveal a link between apoptosis, NP features (such as surface charge), and exposure time.

Mitochondrial autophagy is a key process of apoptosis (Doherty and Baehrecke, 2018; Cai et al., 2019). Cell viability did not change significantly following prior administration of the caspase inhibitor z-VAD; however, it increased after administration of the autophagy inhibitor 3 mA, demonstrating that the NPs triggered autophagy activation in the context of mitochondrial injury (Yang and Wang, 2022). Exposing the mouse spermatocyte line GC-2

TABLE 12.2 Effect of the MP-induced disruption of the cytoskeleton on the nucleus.

Materials	Size	Exposure concentration	Duration of exposure	Antagonistic effect on the nucleus	References
PS	5 μm	0.1 mg day^{-1}	24–26 days	Diminished expression of α-tubulin, as well as DAAM-1 protein	(Haddadi et al., 2022)
PS	1, 10 μm	4×10^6 particles L^{-1}	6 days	Upregulation of cytoskeleton-related proteins expression	(Magni et al., 2019)
PS	500 nm	0.26 mg L^{-1}	4 weeks	Reduced dispersal of the F-actin cytoskeleton, downregulation of *Rac1*, *Arp2/3*, and *Ras*	(Han et al., 2021)
Mixture of PP, PE, PA, and PEVA	<100 μm	N.A.	7 and 14 days	Enhanced microtubule-associated protein light chain 3 (LC3)	(Missawi et al., 2022)

Note: *N.A.*, not available, *PA*, polyamide, *PE*, polyethylene, *PS*, polystyrene, *PEVA*, polyethylene vinyl acetate.

cells to 5-μm PS MPs activated the mitochondrial PINK1/Parkin autophagy pathway, as evidenced by the upregulation of PINK1 and Parkin expression (Li et al., 2022). Furthermore, mitochondrial damage causes the discharge of cytochrome C, which induces programmed cell death. Exposure to PS MPs intensifies ROS accumulation, influences Mfn2 expression along with Drp-1 associated with the fission of mitochondria, and increases Bax (the apoptotic protein) levels, as well as cytochrome C and autophagy-associated gene expression (Pink1, Parkin, LC3II, and p62), suggesting that PS MPs can induce cytotoxicity via mitochondrial damage followed by apoptosis (Pan et al., 2021).

MP exposure can trigger a chain of events. This process begins with oxidative stress and organelle damage (mitochondria, the ER, and lysosomes), which progresses to DNA damage. It eventually compromises critical biological barriers (gut, lung, brain, and placenta) that protect major organ systems. MPs can cause cellular and systemic harm, with exposure linked to a range of negative health effects.

12.3.4 Genotoxicity and mutagenicity

Genotoxic effects refer to the capacity of a chemical or environmental factor to induce genetic harm in cells. This damage frequently manifests as direct DNA damage, alterations in genetic expression, and chromosomal abnormalities, which can result from chromosomal structural disruption or a breakdown in chromosomal segregation throughout cell division.

12.3.4.1 *Direct DNA damage (single-/double-strand breaks)*

In particular, studies have demonstrated that contact with MPs can lead to the formation of DNA adducts and various forms of DNA damage, contributing to genetic instability (Tagorti and Kaya, 2022). Base oxidation, double-strand breaks, single-strand breaks, and crosslinking are all forms of DNA damage. Although cells have DNA repair mechanisms such as nucleotide excision repair and base excision repair, cellular dysfunction persists. This damage is frequently caused by reactive species produced through oxidative stress. The buildup of DNA damage can cause mutations, cellular senescence, oncogenesis, and genomic instability, increasing the risk of cancer and chronic illnesses. Furthermore, genotoxicity may affect future generations (Demir, 2021; Baj et al., 2022; Li et al., 2024; Wang et al., 2024). PS MPs damage plant DNA via oxidative stress (antioxidant system imbalances and ROS bursts), epigenetic interference, and direct physical damage. Nanoscale PS MPs infiltrate the cell, inducing DNA strand breakage, overproduction of ROS that attack bases, and blockade of repair enzyme activity (Fig. 12.2). PS MPs can serve as pollutant transporters, causing greater harm. The transgenerational genetic impacts require additional investigation (Kaur et al., 2022).

12.3.4.2 *Chromosomal aberrations (clastogenesis, aneugenesis)*

The chromosomal impact of MPs refers to the direct or indirect destructive effects of MPs on chromosomes, which can cause physical or chemical damage or impair chromosomal segregation throughout mitosis or meiosis (Tagorti and Kaya, 2022). The former, known as clastogenesis, includes structural alterations or breaks in chromosomes that result in deletions, duplications, translocations, inversions, or the creation of acentric (no centromere) or dicentric (two centromeres) chromosomes. The latter, also known as aneugenesis, is characterized by

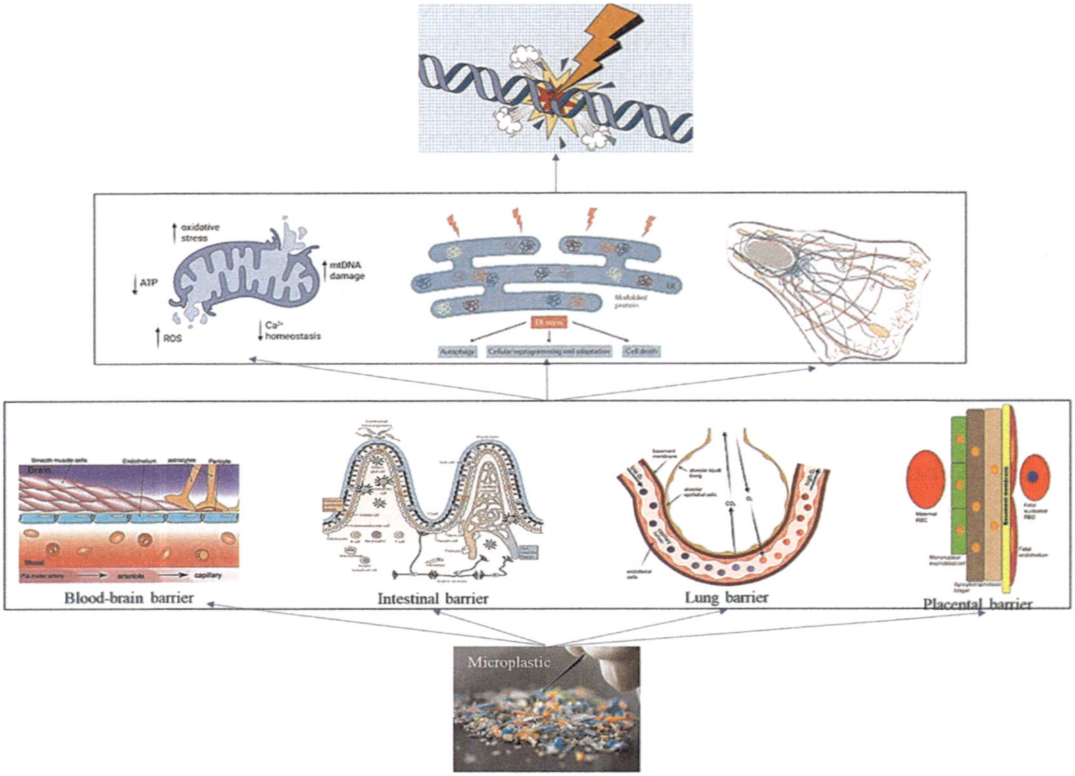

FIG. 12.2 Fate and pathway of microplastic transport through human barriers. Microplastic exposure can trigger a chain of events. This process begins with oxidative stress and organelle damage (mitochondria, the ER, and lysosomes), which progresses to DNA damage. It eventually compromises critical biological barriers (gut, lung, brain, and placenta) that protect major organ systems. Microplastics can cause cellular and systemic harm, with exposure linked to a range of negative health effects.

chromosomal abnormalities resulting from aneuploidy and the loss or gain of chromosomes. Aneugenesis-induced aneuploidy can disrupt normal cellular function, leading to defective division, senescence, or cell death (Çobanoğlu et al., 2021). Chromosomal influences can lead to the formation of a micronucleus, a tiny extranuclear structure that arises during cell division. In contrast, a chromosome or chromosomal fragment does not correctly integrate directly into one of the daughter nuclei. It can be produced by clastogenesis or aneugenesis, which represent chromosomal damage or missegregation during cell division (Sun et al., 2021).

12.4 Conclusion

MPs are both primary (e.g., cosmetic microparticles) and secondary (degraded plastic particles) particles that aggregate systemically in tissues such as the liver, lungs, and intestines through breathing, food consumption, and skin contact. The ability of MPs to impair human well-being can be investigated through both in vitro and in vivo research approaches (e.g., enzymatic degradation, advanced biodegradable materials, and nano-enhanced wastewater treatment) and through in vivo research practices (e.g., the use of antioxidants to alleviate MP impairment in the human body). Oxidative stress pathways are mostly associated with health concerns. Mechanistic investigations have revealed that MPs disrupt mitochondrial ETS, induce ER stress, trigger phagocytic inflammatory responses, and increase ROS, thereby damaging the cell's antioxidant defense. Redox imbalance causes protein carbonylation, lipid peroxidation, and DNA damage (changes in biomarkers: Malondialdehyde (MDA↑), Glutathione (GSH↓), and 8-Hydroxy-2′-deoxyguanosine (8-OHdG↑), which stimulate pathways such as NF-κB to release proinflammatory proteins, producing the "oxidative–inflammatory vicious cycle axis." Future studies should focus on analyzing the ROS signaling network via multiomics techniques, designing precise antioxidant delivery systems, developing comprehensive risk evaluation criteria for environmental toxicology and nutritional epidemiology, and introducing legislative measures that support MP control through dietary plans. Only via interdisciplinary collaboration can we successfully address the ongoing health hazard posed by MP pollution.

References

Abihssira-García, I.S., Park, Y., Kiron, V., Olsvik, P.A., 2020. Fluorescent microplastic uptake by immune cells of atlantic salmon (*Salmo salar* L.). Front. Environ. Sci. 2296665X. 8. Available from: journal.frontiersin.org/journal/environmental-science. doi:10.3389/fenvs.2020.560206.

Alimi, O.S., Claveau-Mallet, D., Kurusu, R.S., Lapointe, M., Bayen, S., Tufenkji, N., 2022. Weathering pathways and protocols for environmentally relevant microplastics and nanoplastics: what are we missing? J. Hazard. Mater. 18733336. 423. Available from: https://www.elsevier.com/locate/jhazmat. doi:10.1016/j.jhazmat.2021.126955.

Anifowoshe, A.T., Akhtar, M.N., Majeed, A., Singh, A.S., Ismail, T.F., Nongthomba, U., 2025. Microplastics: a threat to fetoplacental unit and reproductive systems. Toxicol. Rep. 22147500. 14. Available from: https://www.sciencedirect.com/science/journal/22147500. doi:10.1016/j.toxrep.2025.101938.

Baj, J., Dring, J.C., Czeczelewski, M., Kozyra, P., Forma, A., Flieger, J., Kowalska, B., Buszewicz, G., Teresiński, G., 2022. Derivatives of plastics as potential carcinogenic factors: the current state of knowledge. Cancers. 20726694. 14 (19). Available from: http://www.mdpi.com/journal/cancers/. doi:10.3390/cancers14194637.

Baumgartner, H.K., Gerasimenko, J.V., Thorne, C., Ashurst, L.H., Barrow, S.L., Chvanov, M.A., Gillies, S., Criddle, D. N., Tepikin, A.V., Petersen, O.H., Sutton, R., Watson, A.J.M., Gerasimenko, O.V., 2007. Caspase-8-mediated apoptosis induced by oxidative stress is independent of the intrinsic pathway and dependent on cathepsins. Am. J. Physiol. Gastrointest. Liver Physiol. 15221547. 293(1), G296. Available from: http://ajpgi.physiology.org/cgi/reprint/293/1/G296. doi:10.1152/ajpgi.00103.2007.

Behzadi, S., Serpooshan, V., Tao, W., Hamaly, M.A., Alkawareek, M.Y., Dreaden, E.C., Brown, D., Alkilany, A.M., Farokhzad, O.C., Mahmoudi, M., 2017. Cellular uptake of nanoparticles: journey inside the cell. Chem. Soc. Rev. 14604744. 46 (14), 4218–4244. Available from: http://pubs.rsc.org/en/journals/journal/cs. doi:10.1039/c6cs00636a.

Blettler, M.C.M., Abrial, E., Khan, F.R., Sivri, N., Espinola, L.A., 2018. Freshwater plastic pollution: recognizing research biases and identifying knowledge gaps. Water Res. 18792448. 143, 416–424. Available from: https://www.elsevier.com/locate/watres. doi:10.1016/j.watres.2018.06.015.

Boyle, K., Örmeci, B., 2020. Microplastics and nanoplastics in the freshwater and terrestrial environment: a review. Water. 2073-4441. 12 (9), 2633. doi:10.3390/w12092633.

Brachner, A., Fragouli, D., Duarte, I.F., Farias, P.M.A., Dembski, S., Ghosh, M., Barisic, I., Zdzieblo, D., Vanoirbeek, J., Schwabl, P., Neuhaus, W., 2020. Assessment of human health risks posed by nano-and microplastics is currently not feasible. Int. J. Environ. Res. Public Health. 16604601. 17 (23), 1–10. Available from: https://www.mdpi.com/1660-4601/17/23/8832/pdf. doi:10.3390/ijerph17238832.

Burns, E.E., Boxall, A.B.A., 2018. Microplastics in the aquatic environment: evidence for or against adverse impacts and major knowledge gaps. Environ. Toxicol. Chem. 15528618. 37 (11), 2776–2796. Available from: http://onlinelibrary.wiley.com/journal/10.1002/(ISSN)1552-8618. doi:10.1002/etc.4268.

Cai, J., Shi, G., Zhang, Y., Zheng, Y., Yang, J., Liu, Q., Gong, Y., Yu, D., Zhang, Z., 2019. Taxifolin ameliorates DEHP-induced cardiomyocyte hypertrophy via attenuating mitochondrial dysfunction and glycometabolism disorder in chicken. Environ. Pollut. 18736424. 255. Available from: https://www.journals.elsevier.com/environmental-pollution. doi:10.1016/j.envpol.2019.113155.

Chae, Y., Kim, D., Kim, S.W., An, Y.J., 2018. Trophic transfer and individual impact of nano-sized polystyrene in a four-species freshwater food chain. Sci. Rep. 20452322. 8 (1). Available from: https://www.nature.com/srep/index.html. doi:10.1038/s41598-017-18849-y.

Chakraborty, D., Mandal, A.H., Ghosh, S., Sadhu, A., Das, D., Saha, N.C., Saha, S., 2024. The physiological effects of acute and sub-lethal exposure to phenol on antioxidant enzyme activity in the freshwater sludge worm *Tubifex tubifex*. Toxicol. Rep. 22147500. 13. Available from: https://www.sciencedirect.com/science/journal/22147500. doi:10.1016/j.toxrep.2024.101717.

Chakraborty, D., Ghosh, S., Mandal, A.H., Roy, T., Sadhu, A., Paria, D.S., Saha, N.C., Sopjani, M., Saha, S., Faggio, C., 2025. Sodium Dodecylbenzenesulfonate-induced oxidative stress and tissue damage in *Tubifex tubifex*: a comprehensive toxicological study. Environ. Toxicol. Pharmacol. 18727077. 118. Available from: https://www.sciencedirect.com/science/journal/13826689. doi:10.1016/j.etap.2025.104778.

Chandra, J., Samali, A., Orrenius, S., 2000. Triggering and modulation of apoptosis by oxidative stress. Free Radic. Biol. Med. 08915849. 29 (3-4), 323–333. Available from: https://www.elsevier.com/locate/freeradbiomed. doi:10.1016/S0891-5849(00)00302-6.

Chen, H., Hua, X., Yang, Y., Wang, C., Jin, L., Dong, C., Chang, Z., Ding, P., Xiang, M., Li, H., Yu, Y., 2021. Chronic exposure to UV-aged microplastics induces neurotoxicity by affecting dopamine, glutamate, and serotonin neurotransmission in *Caenorhabditis elegans*. J. Hazard. Mater. 18733336. 419. Available from: https://www.elsevier.com/locate/jhazmat. doi:10.1016/j.jhazmat.2021.126482.

Chen, Y., Liu, Y., Li, Y., Yao, C., Qu, J., Tang, J., Chen, G., Han, Y., 2024. Acute exposure to polystyrene nanoplastics induces unfolded protein response and global protein ubiquitination in lungs of mice. Ecotoxicol. Environ. Saf. 10902414. 280. Available from: https://www.sciencedirect.com/science/journal/01476513. doi:10.1016/j.ecoenv.2024.116580.

Cheng, W., Xiaolan, L., Zhou, Y., Hengyi, Y., Xie, Y., Guo, H., Wang, H., Yan, L., Feng, Y., Wang, Y., 2022. Polystyrene microplastics induce hepatotoxicity and disrupt lipid metabolism in the liver organoids. Sci. Total Environ. 00489697. 806 (Pt 1), 150328. doi:10.1016/j.scitotenv.2021.150328.

Chou, L.Y.T., Ming, K., Chan, W.C.W., 2011. Strategies for the intracellular delivery of nanoparticles. Chem. Soc. Rev. 14604744. 40 (1), 233–245. doi:10.1039/c0cs00003e.

Çobanoğlu, H., Belivermiş, M., Sıkdokur, E., Kılıç, Ö., Çayır, A., 2021. Genotoxic and cytotoxic effects of polyethylene microplastics on human peripheral blood lymphocytes. Chemosphere. 18791298. 272. Available from: https://www.elsevier.com/locate/chemosphere. doi:10.1016/j.chemosphere.2021.129805.

Collard, F., Gilbert, B., Compère, P., Eppe, G., Das, K., Jauniaux, T., Parmentier, E., 2017. Microplastics in livers of European anchovies (*Engraulis encrasicolus*, L.). Environ. Pollut. 18736424. 229, 1000–1005. Available from: https://www.elsevier.com/inca/publications/store/4/0/5/8/5/6. doi:10.1016/j.envpol.2017.07.089.

Contini, C., Schneemilch, M., Gaisford, S., Quirke, N., 2018. Nanoparticle–membrane interactions. J. Exp. Nanosci. 17458099. 13 (1), 62–81. Available from: http://www.tandf.co.uk/journals/titles/17458080.asp. doi:10.1080/17458080.2017.1413253.

Cunningham, E.M., Sigwart, J.D., 2019. Environmentally accurate microplastic levels and their absence from exposure studies. Integr. Comp. Biol. 15577023. 59 (6), 1485–1496. Available from: http://icb.oxfordjournals.org/. doi:10.1093/icb/icz068.

Das, A., 2023. The emerging role of microplastics in systemic toxicity: involvement of reactive oxygen species (ROS). Sci. Total Environ. 00489697. 895, 165076. doi:10.1016/j.scitotenv.2023.165076.

Demir, E., 2021. Adverse biological effects of ingested polystyrene microplastics using *Drosophila melanogaster* as a model in vivo organism. J. Toxicol. Environ. Health A. 1528-7394. 84 (16), 649–660. doi:10.1080/15287394.2021.1913684.

Deng, Y., Zhang, Y., Lemos, B., Ren, H., 2017. Tissue accumulation of microplastics in mice and biomarker responses suggest widespread health risks of exposure. Sci. Rep. 20452322. 7. Available from: https://www.nature.com/srep/index.html. doi:10.1038/srep46687.

Ding, Y., Zhang, R., Li, B., Du, Y., Li, J., Tong, X., Wu, Y., Ji, X., Zhang, Y., 2021. Tissue distribution of polystyrene nanoplastics in mice and their entry, transport, and cytotoxicity to GES-1 cells. Environ. Pollut. 18736424. 280. Available from: https://www.journals.elsevier.com/environmental-pollution. doi:10.1016/j.envpol.2021.116974.

Doherty, J., Baehrecke, E.H., 2018. Life, death and autophagy. Nat. Cell Biol. 14764679. 20 (10), 1110–1117. Available from: http://www.nature.com/ncb/index.html. doi:10.1038/s41556-018-0201-5.

Dolar, A., Drobne, D., Dolenec, M., Marinšek, M., Kokalj, A.J., 2022. Time-dependent immune response in *Porcellio scaber* following exposure to microplastics and natural particles. Sci. Total Environ. 18791026. 818. Available from: https://www.elsevier.com/locate/scitotenv. doi:10.1016/j.scitotenv.2021.151816.

Dos Santos, T., Varela, J., Lynch, I., Salvati, A., Dawson, K.A., 2011. Effects of transport inhibitors on the cellular uptake of carboxylated polystyrene nanoparticles in different cell lines. PLoS ONE. 19326203. 6 (9). Available from: http://www.plosone.org/article/fetchObjectAttachment.action?uri=info%3Adoi%2F10.1371%2Fjournal.pone.0024438&representation=PDF. doi:10.1371/journal.pone.0024438Ireland.

Dris, R., Gasperi, J., Mirande, C., Mandin, C., Guerrouache, M., Langlois, V., Tassin, B., 2017. A first overview of textile fibers, including microplastics, in indoor and outdoor environments. Environ. Pollut. 18736424. 221, 453–458. Available from: https://www.journals.elsevier.com/environmental-pollution. doi:10.1016/j.envpol.2016.12.013.

Duan, Z., Duan, X., Zhao, S., Wang, X., Wang, J., Liu, Y., Peng, Y., Gong, Z., Wang, L., 2020. Barrier function of zebrafish embryonic chorions against microplastics and nanoplastics and its impact on embryo development. J. Hazard. Mater. 18733336. 395. Available from: https://www.elsevier.com/locate/jhazmat. doi:10.1016/j.jhazmat.2020.122621.

Duis, K., Coors, A., 2016. Microplastics in the aquatic and terrestrial environment: sources (with a specific focus on personal care products), fate and effects. Environ. Sci. Eur. 21904715. 28 (1), 1–25. Available from: http://www.springer.com/environment/journal/12302. doi:10.1186/s12302-015-0069-y.

Enyoh, C.E., Verla, A.W., Verla, E.N., Ibe, F.C., Amaobi, C.E., 2019. Airborne microplastics: a review study on method for analysis, occurrence, movement and risks. Environ. Monit. Assess. 15732959. 191 (11). Available from: https://link.springer.com/journal/10661. doi:10.1007/s10661-019-7842-0.

Fazlollahi, F., Angelow, S., Yacobi, N.R., Marchelletta, R., Yu, A.S.L., Hamm-Alvarez, S.F., Borok, Z., Kim, K.J., Crandall, E.D., 2011. Polystyrene nanoparticle trafficking across MDCK-II. Nanomed. Nanotechnol. Biol. Med. 15499642. 7 (5), 588–594. doi:10.1016/j.nano.2011.01.008.

Firdessa, R., Oelschlaeger, T.A., Moll, H., 2014. Identification of multiple cellular uptake pathways of polystyrene nanoparticles and factors affecting the uptake: relevance for drug delivery systems. Eur. J. Cell Biol. 16181298. 93 (8-9), 323–337. Available from: http://www.journals.elsevier.com/european-journal-of-cell-biology/. doi:10.1016/j.ejcb.2014.08.001.

Foroozandeh, P., Aziz, A.A., 2018. Insight into cellular uptake and intracellular trafficking of nanoparticles. Nanoscale Res. Lett. 1556276X. 13. Available from: http://www.springer.com/materials/nanotechnology/journal/11671. doi:10.1186/s11671-018-2728-6.

Gaspar, L., Bartman, S., Coppotelli, G., Ross, J.M., 2023. Acute exposure to microplastics induced changes in behavior and inflammation in young and old mice. Int. J. Mol. Sci. 14220067. 24 (15). Available from: http://www.mdpi.com/journal/ijms. doi:10.3390/ijms241512308.

Geiser, M., Rothen-Rutishauser, B., Kapp, N., Schürch, S., Kreyling, W., Schulz, H., Semmler, M., Hof, V.I., Heyder, J., Gehr, P., 2005. Ultrafine particles cross cellular membranes by nonphagocytic mechanisms in lungs and in cultured cells. Environ. Health Perspect. 15529924. 113 (11), 1555–1560. doi:10.1289/ehp.8006.

Gorrini, C., Harris, I.S., Mak, T.W., 2013. Modulation of oxidative stress as an anticancer strategy. Nat. Rev. Drug Discov. 14741784. 12 (12), 931–947. doi:10.1038/nrd4002.

Gu, Y., Xu, D., Liu, J., Chen, Y., Wang, J., Song, Y., Sun, B., Xia, B., 2023. Bioaccumulation of functionalized polystyrene nanoplastics in sea cucumber *Apostichopus japonicus* (Selenka, 1867) and their toxic effects on oxidative stress, energy metabolism and mitochondrial pathway. Environ. Pollut. 18736424. 319. Available from: https://www.journals.elsevier.com/environmental-pollution. doi:10.1016/j.envpol.2023.121015.

Haddadi, A., Kessabi, K., Boughammoura, S., Rhouma, M.B., Mlouka, R., Banni, M., Messaoudi, I., 2022. Exposure to microplastics leads to a defective ovarian function and change in cytoskeleton protein expression in rat. Environ.

Sci. Pollut. Res. 16147499. 29 (23), 34594–34606. Available from: https://link.springer.com/journal/11356. doi:10.1007/s11356-021-18218-3.

Halimu, G., Zhang, Q., Liu, L., Zhang, Z., Wang, X., Gu, W., Zhang, B., Dai, Y., Zhang, H., Zhang, C., Xu, M., 2022. Toxic effects of nanoplastics with different sizes and surface charges on epithelial-to-mesenchymal transition in A549 cells and the potential toxicological mechanism. J. Hazard. Mater. 18733336. 430. Available from: https://www.elsevier.com/locate/jhazmat. doi:10.1016/j.jhazmat.2022.128485.

Han, Y., Zhou, W., Tang, Y., Shi, W., Shao, Y., Ren, P., Zhang, J., Xiao, G., Sun, H., Liu, G., 2021. Microplastics aggravate the bioaccumulation of three veterinary antibiotics in the thick shell mussel *Mytilus coruscus* and induce synergistic immunotoxic effects. Sci. Total Environ. 18791026. 770. Available from: https://www.elsevier.com/locate/scitotenv. doi:10.1016/j.scitotenv.2021.145273.

Huang, Y., Liang, B., Li, Z., Zhong, Y., Wang, B., Zhang, B., Du, J., Ye, R., Xian, H., Min, W., Yan, X., Deng, Y., Feng, Y., Bai, R., Fan, B., Yang, X., Huang, Z., 2023. Polystyrene nanoplastic exposure induces excessive mitophagy by activating AMPK/ULK1 pathway in differentiated SH-SY5Y cells and dopaminergic neurons in vivo. Part. Fibre Toxicol. 17438977. 20 (1). Available from: https://particleandfibretoxicology.biomedcentral.com. doi:10.1186/s12989-023-00556-4.

Ji, Y., Wang, Y., Shen, D., Kang, Q., Chen, L., 2021. Mucin Corona delays intracellular trafficking and alleviates cytotoxicity of nanoplastic-benzopyrene combined contaminant. J. Hazard. Mater. 18733336. 406. Available from: https://www.elsevier.com/locate/jhazmat. doi:10.1016/j.jhazmat.2020.124306.

Jin, H., Yang, C., Jiang, C., Li, L., Pan, M., Li, D., Han, X., Ding, J., 2022. Evaluation of neurotoxicity in BALB/c mice following chronic exposure to polystyrene microplastics. Environ. Health Perspect. 15529924. 130 (10). Available from: http://ehp.niehs.nih.gov/. doi:10.1289/EHP10255.

Jing, B., Zhu, Y., 2011. Disruption of supported lipid bilayers by semihydrophobic nanoparticles. J. Am. Chem. Soc. 15205126. 133 (28), 10983–10989. doi:10.1021/ja2040305.

Kan, M., Wang, C., Zhu, B., Chen, W.Q., Liu, Y., Ren, Y., Xu, M., 2023. Seven decades of plastic flows and stocks in the United States and pathways toward zero plastic pollution by 2050. J. Ind. Ecol. 15309290. 27 (6), 1538–1552. Available from: http://onlinelibrary.wiley.com/journal/10.1111/(ISSN)1530-9290. doi:10.1111/jiec.13427.

Kaur, M., Ming, X., Wang, L., 2022. Cyto–genotoxic effect causing potential of polystyrene micro-plastics in terrestrial plants. Nanomaterials. 2079-4991. 12 (12), 2024. doi:10.3390/nano12122024.

Kolandhasamy, P., Su, L., Li, J., Qu, X., Jabeen, K., Shi, H., 2018. Adherence of microplastics to soft tissue of mussels: a novel way to uptake microplastics beyond ingestion. Sci. Total Environ. 18791026. 610-611, 635–640. Available from: https://www.elsevier.com/locate/scitotenv. doi:10.1016/j.scitotenv.2017.08.053.

Kuhn, D.A., Vanhecke, D., Michen, B., Blank, F., Gehr, P., Petri-Fink, A., Rothen-Rutishauser, B., 2014. Different endocytotic uptake mechanisms for nanoparticles in epithelial cells and macrophages. Beilstein J. Nanotechnol. 21904286. 5 (1), 1625–1636. Available from: http://www.beilstein-journals.org/bjnano/content/pdf/2190-4286-5-174.pdf. doi:10.3762/bjnano.5.174.

Lebreton, L., Andrady, A., 2019. Future scenarios of global plastic waste generation and disposal. Palgrave Commun. 20551045. 5 (1). Available from: https://www.palgrave-journals.com/palcomms/. doi:10.1057/s41599-018-0212-7.

Lee, C.W., Hsu, L.F., Wu, I.L., Wang, Y.L., Chen, W.C., Liu, Y.J., Yang, L.T., Tan, C.L., Luo, Y.H., Wang, C.C., Chiu, H. W., Yang, T.C.K., Lin, Y.Y., Chang, H.A., Chiang, Y.C., Chen, C.H., Lee, M.H., Peng, K.T., Huang, C.C.Y., 2022. Exposure to polystyrene microplastics impairs hippocampus-dependent learning and memory in mice. J. Hazard. Mater. 18733336. 430. Available from: https://www.elsevier.com/locate/jhazmat. doi:10.1016/j.jhazmat.2022.128431.

Li, B., Liang, W., Liu, Q.X., Fu, S., Ma, C., Chen, Q., Su, L., Craig, N.J., Shi, H., 2021. Fish ingest microplastics unintentionally. Environ. Sci. Technol. 15205851. 55 (15), 10471–10479. Available from: http://pubs.acs.org/journal/esthag. doi:10.1021/acs.est.1c01753.

Li, X., Huang, Y., Zu, D., Liu, H., He, H., Bao, Q., He, Y., Liang, C., Luo, G., Teng, Y., Shi, Y., Ye, Z., Cheng, X., 2024. PMMA nanoplastics induce gastric epithelial cellular senescence and cGAS-STING-mediated inflammation via ROS overproduction and NHEJ suppression. Ecotoxicol. Environ. Saf. 10902414. 287. Available from: https://www.sciencedirect.com/science/journal/01476513. doi:10.1016/j.ecoenv.2024.117284.

Li, Y., Xu, M., Zhang, Z., Halimu, G., Li, Y., Li, Y., Gu, W., Zhang, B., Wang, X., 2022. In vitro study on the toxicity of nanoplastics with different charges to murine splenic lymphocytes. J. Hazard. Mater. 18733336. 424. Available from: https://www.elsevier.com/locate/jhazmat. doi:10.1016/j.jhazmat.2021.127508.

Lim, J.P., Gleeson, P.A., 2011. Macropinocytosis: an endocytic pathway for internalising large gulps. Immunol. Cell Biol. 14401711. 89 (8), 836–843. doi:10.1038/icb.2011.20.

Lusher, A.L., Hernandez-Milian, G., O'Brien, J., Berrow, S., O'Connor, I., Officer, R., 2015. Microplastic and macroplastic ingestion by a deep diving, oceanic cetacean: the true's beaked whale *Mesoplodon mirus*. Environ. Pollut. 18736424. 199, 185–191. Available from: https://www.elsevier.com/inca/publications/store/4/0/5/8/5/6. doi:10.1016/j.envpol.2015.01.023.

Magni, S., Torre, C.D., Garrone, G., D'Amato, A., Parenti, C.C., Binelli, A., 2019. First evidence of protein modulation by polystyrene microplastics in a freshwater biological model. Environ. Pollut. 02697491. 250, 407–415. doi:10.1016/j.envpol.2019.04.088.

Mandal, A.H., Ghosh, S., Adhurjya, D., Chatterjee, P., Samajdar, I., Mukherjee, D., Dhara, K., Saha, N.C., Piccione, G., Multisanti, C.R., Saha, S., Faggio, C., 2024. Exploring the impact of zinc oxide nanoparticles on fish and fish-food organisms: a review. Aquac. Rep. 23525134. 36. Available from: https://www.sciencedirect.com/science/journal/23525134. doi:10.1016/j.aqrep.2024.102038.

Missawi, O., Venditti, M., Cappello, T., Zitouni, N., Marco, G.D.E., Boughattas, I., Bousserrhine, N., Belbekhouche, S., Minucci, S., Maisano, M., Banni, M., 2022. Autophagic event and metabolomic disorders unveil cellular toxicity of environmental microplastics on marine polychaete *Hediste diversicolor*. Environ. Pollut. 02697491. 302, 119106. doi:10.1016/j.envpol.2022.119106.

Mukherjee, D., Ghosh, S., Mandal, A.H., Saha, N.C., Faggio, C., Pastorino, P., Saha, S., 2025. Silent threats beneath the surface: unraveling the impact of organophosphate toxicity on fish. Sci. Total Environ. 18791026. 985. Available from: https://www.sciencedirect.com/science/journal/00489697. doi:10.1016/j.scitotenv.2025.179725.

Napper, I.E., Bakir, A., Rowland, S.J., Thompson, R.C., 2015. Characterisation, quantity and sorptive properties of microplastics extracted from cosmetics. Mar. Pollut. Bull. 18793363. 99 (1-2), 178–185. Available from: https://www.elsevier.com/locate/marpolbul. doi:10.1016/j.marpolbul.2015.07.029.

Nayak, S., Nayak, P., Soren, S., Nayak, S., Pradhan, S.P., Sharma, S.N., Muduli, N., Saha, S., Dash, S.N., Patnaik, L., 2025. First evidence of microplastics burden in surface waters of Budhabalanga Estuary, Chandipur, Eastern India: potential threat to aquatic ecosystem. Appl. Biochem. Biotechnol. 15590291. 197 (5), 3447–3465. Available from: https://www.springer.com/journal/12010. doi:10.1007/s12010-025-05189-1.

North, E.J., Halden, R.U., 2013. Plastics and environmental health: the road ahead. Rev. Environ. Health. 00487554. 28 (1), 1–8. doi:10.1515/reveh-2012-0030.

Pan, L., Yu, D., Zhang, Y., Zhu, C., Yin, Q., Hu, Y., Zhang, X., Yue, R., Xiong, X., 2021. Polystyrene microplastics-triggered mitophagy and oxidative burst via activation of PERK pathway. Sci. Total Environ. 18791026. 781. Available from: https://www.elsevier.com/locate/scitotenv. doi:10.1016/j.scitotenv.2021.146753.

Paul, M.B., Böhmert, L., Hsiao, I.L., Braeuning, A., Sieg, H., 2023. Complex intestinal and hepatic in vitro barrier models reveal information on uptake and impact of micro-, submicro- and nanoplastics. Environ. Int. 18736750. 179. Available from: https://www.elsevier.com/locate/envint. doi:10.1016/j.envint.2023.108172.

Pelkmans, L., Helenius, A., 2002. Endocytosis via caveolae. Traffic 3 (5), 311–320. doi:10.1034/j.1600-0854.2002.30501.x.

Peng, Y., Gu, T., Zhong, T., Xiao, Y., Sun, Q., 2022. Endoplasmic reticulum stress in metabolic disorders: opposite roles of phytochemicals and food contaminants. Curr. Opin. Food Sci. 22147993. 48. Available from: http://www.journals.elsevier.com/current-opinion-in-food-science/. doi:10.1016/j.cofs.2022.100913.

Praefcke, G.J.K., McMahon, H.T., 2004. The dynamin superfamily: universal membrane tubulation and fission molecules? Nat. Rev. Mol. Cell Biol. 5 (2), 133–147. doi:10.1038/nrm1313.

Prata, J.C., 2018. Airborne microplastics: consequences to human health? Environ. Pollut. 18736424. 234, 115–126. Available from: https://www.elsevier.com/inca/publications/store/4/0/5/8/5/6. doi:10.1016/j.envpol.2017.11.043.

Prata, J.C., da Costa, J.P., Duarte, A.C., Rocha-Santos, T., 2022. Suspected microplastics in Atlantic horse mackerel fish (*Trachurus trachurus*) captured in Portugal. Mar. Pollut. Bull. 18793363. 174. Available from: https://www.elsevier.com/locate/marpolbul. doi:10.1016/j.marpolbul.2021.113249.

Qiao, R., Deng, Y., Zhang, S., Wolosker, M.B., Zhu, Q., Ren, H., Zhang, Y., 2019. Accumulation of different shapes of microplastics initiates intestinal injury and gut microbiota dysbiosis in the gut of zebrafish. Chemosphere. 18791298. 236. Available from: http://www.elsevier.com/locate/chemosphere. doi:10.1016/j.chemosphere.2019.07.065.

Qu, M., Qiu, Y., Kong, Y., Wang, D., 2019. Amino modification enhances reproductive toxicity of nanopolystyrene on gonad development and reproductive capacity in nematode *Caenorhabditis elegans*. Environ. Pollut. 18736424. 254.

Available from: https://www.journals.elsevier.com/environmental-pollution. doi:10.1016/j.envpol.2019.112978.

Ramsperger, A.F.R.M., Narayana, V.K.B., Gross, W., Mohanraj, J., Thelakkat, M., Greiner, A., Schmalz, H., Kress, H., Laforsch, C., 2020. Environmental exposure enhances the internalization of microplastic particles into cells. Sci. Adv. 2375-2548. 6 (50). doi:10.1126/sciadv.abd1211.

Rejman, J., Oberle, V., Zuhorn, I.S., Hoekstra, D., 2004. Size-dependent internalization of particles via the pathways of clathrin-and caveolae-mediated endocytosis. Biochem. J. 377 (1), 159–169. doi:10.1042/BJ20031253.

Reuter, S., Gupta, S.C., Chaturvedi, M.M., Aggarwal, B.B., 2010. Oxidative stress, inflammation, and cancer: how are they linked? Free Radic. Biol. Med. 08915849. 49 (11), 1603–1616. doi:10.1016/j.freeradbiomed.2010.09.006.

Rochman, C.M., Tahir, A., Williams, S.L., Baxa, D.V., Lam, R., Miller, J.T., Teh, F.C., Werorilangi, S., Teh, S.J., 2015. Anthropogenic debris in seafood: plastic debris and fibers from textiles in fish and bivalves sold for human consumption. Sci. Rep. 20452322. 5. Available from: https://www.nature.com/srep/index.html. doi:10.1038/srep14340.

Rong, J., Yuan, C., Yin, X., Wu, X., He, F., Wang, Y., Leung, K.S.Y., Lin, S., 2024. Co-exposure of polystyrene nanoplastics and copper induces development toxicity and intestinal mitochondrial dysfunction in vivo and in vitro. Sci. Total Environ. 18791026. 930. Available from: https://www.sciencedirect.com/science/journal/00489697. doi:10.1016/j.scitotenv.2024.172681.

Sadhu, A., Mandal, A.H., Ghosh, S., Lakdawala, P., Saha, S., 2024. Dimethoate-induced toxicity: a comprehensive review of its aquatic environmental hazards. Toxicol. Environ. Health Sci. 22337784. 16 (3), 243–258. Available from: https://www.springer.com/journal/13530. doi:10.1007/s13530-024-00228-6.

Sadhu, A., Ghosh, S., Mandal, A.H., Nayak, S., Saha, N.C., Mitra, A., Saha, S., 2025. Effects of Covid-19 pandemic on blue economy in India: impacts, opportunities and challenges. Proc. Zool. Soc. 09746919. 78 (3), 251–266. Available from: https://www.springer.com/journal/12595. doi:10.1007/s12595-025-00580-6.

Saha, N.C., Banerjee, P., Chatterjee, A., Bhattacharya, R., Saha, S., Pastorino, P., 2024. Haematological, biochemical, enzymological changes and mitochondrial dysfunction of liver in freshwater climbing perch *Anabas testudineus* during their acute and chronic exposure to sodium fluoride. Environ. Toxicol. Pharmacol. 18727077. 106. Available from: https://www.sciencedirect.com/science/journal/13826689. doi:10.1016/j.etap.2023.104360.

Saha, S., Saha, N.C., Chatterjee, A., Banerjee, P., Garai, P., Sharma, P., Patnaik, L., Nayak, S., Dhara, K., Chukwuka, A. V., Faggio, C., 2023. Integrated multi-biomarker responses in *Mozambique tilapia, Oreochromis mossambicus* under acute and chronic Diazinon® exposures. Chem. Ecol. 10290370. 39 (3), 235–255. Available from: https://www.tandf.co.uk/journals/titles/02757540.asp. doi:10.1080/02757540.2023.2178649.

Shan, S., Zhang, Y., Zhao, H., Zeng, T., Zhao, X., 2022. Polystyrene nanoplastics penetrate across the blood-brain barrier and induce activation of microglia in the brain of mice. Chemosphere. 18791298. 298. Available from: https://www.elsevier.com/locate/chemosphere. doi:10.1016/j.chemosphere.2022.134261.

Sharma, P., Saha, S., Chatterjee, S., Mandal, A.H., Ghosh, S., Saha, N.C., Piccione, G., Impellitteri, F., Faggio, C., 2025. Toxicological and physiological impact and bioremediation strategies for polycyclic aromatic hydrocarbons (PAHs). Chem. Ecol. 10290370. 41 (6), 843–866. Available from: https://www.tandf.co.uk/journals/titles/02757540.asp. doi:10.1080/02757540.2025.2490035.

Shi, Q., Tang, J., Liu, X., Liu, R., 2021. Ultraviolet-induced photodegradation elevated the toxicity of polystyrene nanoplastics on human lung epithelial A549 cells. Environ. Sci. Nano. 20518161. 8 (9), 2660–2675. Available from: http://pubs.rsc.org/en/journals/journal/en. doi:10.1039/d1en00465d.

Shi, Q., Tang, J., Liu, R., Wang, L., 2022. Toxicity in vitro reveals potential impacts of microplastics and nanoplastics on human health: a review. Crit. Rev. Environ. Sci. Technol. 15476537. 52 (21), 3863–3895. Available from: https://www.tandf.co.uk/journals/titles/10643389.asp. doi:10.1080/10643389.2021.1951528.

Stock, V., Böhmert, L., Lisicki, E., Block, R., Cara-Carmona, J., Pack, L.K., Selb, R., Lichtenstein, D., Voss, L., Henderson, C.J., Zabinsky, E., Sieg, H., Braeuning, A., Lampen, A., 2019. Uptake and effects of orally ingested polystyrene microplastic particles in vitro and in vivo. Arch. Toxicol. 14320738. 93 (7), 1817–1833. Available from: https://www.link.springer.de/link/service/journals/00204/index.htm. doi:10.1007/s00204-019-02478-7.

Stock, V., Laurisch, C., Franke, J., Dönmez, M.H., Voss, L., Böhmert, L., Braeuning, A., Sieg, H., 2021. Uptake and cellular effects of PE, PP, PET and PVC microplastic particles. Toxicol. Vitro. 18793177. 70. Available from: https://www.elsevier.com/locate/toxinvit. doi:10.1016/j.tiv.2020.105021.

Su, L., Deng, H., Li, B., Chen, Q., Pettigrove, V., Wu, C., Shi, H., 2019. The occurrence of microplastic in specific organs in commercially caught fishes from coast and estuary area of east China. J. Hazard. Mater. 18733336. 365, 716–724. Available from: https://www.elsevier.com/locate/jhazmat. doi:10.1016/j.jhazmat.2018.11.024.

Sun, L., Liao, K., Wang, D., 2021. Comparison of transgenerational reproductive toxicity induced by pristine and amino modified nanoplastics in *Caenorhabditis elegans*. Sci. Total Environ. 18791026. 768. Available from: https://www.elsevier.com/locate/scitotenv. doi:10.1016/j.scitotenv.2020.144362.

Syme, M.R., Paxton, J.W., Keelan, J.A., 2004. Drug transfer and metabolism by the human placenta. Clin. Pharmacokinet. 43 (8), 487–514. doi:10.2165/00003088-200443080-00001.

Tagorti, G., Kaya, B., 2022. Genotoxic effect of microplastics and COVID-19: the hidden threat. Chemosphere. 00456535. 286 (Pt 3), 131898. doi:10.1016/j.chemosphere.2021.131898.

Thake, T.H.F., Webb, J.R., Nash, A., Rappoport, J.Z., Notman, R., 2013. Permeation of polystyrene nanoparticles across model lipid bilayer membranes. Soft Matter. 17446848. 9 (43), 10265–10274. doi:10.1039/c3sm51225h.

Verla, A.W., Enyoh, C.E., Verla, E.N., Nwarnorh, K.O., 2019. Microplastic–toxic chemical interaction: a review study on quantified levels, mechanism and implication. SN Appl. Sci. 2523-3963. 1 (11). doi:10.1007/s42452-019-1352-0.

Wang, K., Du, Y., Li, P., Guan, C., Zhou, M., Wu, L., Liu, Z., Huang, Z., 2024. Nanoplastics causes heart aging/myocardial cell senescence through the Ca2+/mtDNA/cGAS-STING signaling cascade. J. Nanobiotechnol. 14773155. 22 (1). Available from: https://jnanobiotechnology.biomedcentral.com. doi:10.1186/s12951-024-02375-x.

Wang, Q., Bai, J., Ning, B., Fan, L., Sun, T., Fang, Y., Wu, J., Li, S., Duan, C., Zhang, Y., Liang, J., Gao, Z., 2020. Effects of bisphenol A and nanoscale and microscale polystyrene plastic exposure on particle uptake and toxicity in human Caco-2 cells. Chemosphere. 18791298. 254. Available from: https://www.elsevier.com/locate/chemosphere. doi:10.1016/j.chemosphere.2020.126788.

Wang, S., Liu, H., Qu, M., Wang, D., 2021. Response of tyramine and glutamate related signals to nanoplastic exposure in *Caenorhabditis elegans*. Ecotoxicol. Environ. Saf. 10902414. 217. Available from: http://www.elsevier.com/inca/publications/store/6/2/2/8/1/9/index.htt. doi:10.1016/j.ecoenv.2021.112239.

Woo, J.H., Seo, H.J., Lee, J.Y., Lee, I., Jeon, K., Kim, B., Lee, K., 2023. Polypropylene nanoplastic exposure leads to lung inflammation through p38-mediated NF-κB pathway due to mitochondrial damage. Part. Fibre Toxicol. 17438977. 20 (1). Available from: https://particleandfibretoxicology.biomedcentral.com. doi:10.1186/s12989-022-00512-8.

Wu, B., Wu, X., Liu, S., Wang, Z., Chen, L., 2019. Size-dependent effects of polystyrene microplastics on cytotoxicity and efflux pump inhibition in human Caco-2 cells. Chemosphere. 18791298. 221, 333–341. Available from: https://www.elsevier.com/locate/chemosphere. doi:10.1016/j.chemosphere.2019.01.056.

Wu, Q., Liu, C., Liu, D., Wang, Y., Qi, H., Liu, X., Zhang, Y., Chen, H., Zeng, Y., Li, J., 2024. Polystyrene nanoplastics-induced lung apoptosis and ferroptosis via ROS-dependent endoplasmic reticulum stress. Sci. Total Environ. 18791026. 912. Available from: https://www.sciencedirect.com/science/journal/00489697. doi:10.1016/j.scitotenv.2023.169260.

Yacobi, N.R., Malmstadt, N., Fazlollahi, F., DeMaio, L., Marchelletta, R., Hamm-Alvarez, S.F., Borok, Z., Kim, K.J., Crandall, E.D., 2010. Mechanisms of alveolar epithelial translocation of a defined population of nanoparticles. Am. J. Respir. Cell Mol. Biol. 10441549. 42 (5), 604–614. Available from: http://ajrcmb.atsjournals.org/cgi/reprint/42/5/604. doi:10.1165/rcmb.2009-0138OC.

Yamashita, K., Yoshioka, Y., Higashisaka, K., Mimura, K., Morishita, Y., Nozaki, M., Yoshida, T., Ogura, T., Nabeshi, H., Nagano, K., Abe, Y., Kamada, H., Monobe, Y., Imazawa, T., Aoshima, H., Shishido, K., Kawai, Y., Mayumi, T., Tsunoda, S.I., Itoh, N., Yoshikawa, T., Yanagihara, I., Saito, S., Tsutsumi, Y., 2011. Silica and titanium dioxide nanoparticles cause pregnancy complications in mice. Nat. Nanotechnol. 17483395. 6 (5), 321–328. Available from: http://www.nature.com/nnano/index.html. doi:10.1038/nnano.2011.41.

Yang, M., Wang, W.X., 2022. Differential cascading cellular and subcellular toxicity induced by two sizes of nanoplastics. Sci. Total Environ. 18791026. 829. Available from: http://www.elsevier.com/locate/scitotenv. doi:10.1016/j.scitotenv.2022.154593.

Yang, Y., Wu, Q., Wang, D., 2021. Neuronal Gα subunits required for the control of response to polystyrene nanoparticles in the range of μg/L in C. elegans. Ecotoxicol. Environ. Saf. 10902414. 225. Available from: http://www.elsevier.com/inca/publications/store/6/2/2/8/1/9/index.htt. doi:10.1016/j.ecoenv.2021.112732.

Zantis, L.J., Bosker, T., Lawler, F., Nelms, S.E., O'Rorke, R., Constantine, R., Sewell, M., Carroll, E.L., 2022. Assessing microplastic exposure of large marine filter-feeders. Sci. Total Environ. 00489697. 818, 151815. doi:10.1016/j.scitotenv.2021.151815.

Zhang, D., Ji, L., Yang, Y., Weng, J., Ma, Y., Liu, L., Ma, W., 2024. Ceria nanoparticle systems alleviate degenerative changes in mouse postovulatory aging oocytes by reducing oxidative stress and improving mitochondrial functions. ACS Nano. 1936086X. 18 (21), 13618–13634. Available from: http://pubs.acs.org/journal/ancac3. doi:10.1021/acsnano.4c00383.

Zhang, Y., Yin, K., Wang, D., Wang, Y., Lu, H., Zhao, H., Xing, M., 2022. Polystyrene microplastics-induced cardiotoxicity in chickens via the ROS-driven NF-κB-NLRP3-GSDMD and AMPK-PGC-1α axes. Sci. Total Environ. 18791026. 840. Available from: https://www.elsevier.com/locate/scitotenv. doi:10.1016/j.scitotenv.2022.156727.

Zhao, Q., Fang, Z., Wang, P., Qian, Z., Yang, Y., Ran, L., Zheng, J., Tang, Y., Cui, X., Li, Y.Y., Zhang, Z., Jiang, H., 2025. Polylactic acid micro/nanoplastic exposure induces male reproductive toxicity by disrupting spermatogenesis and mitochondrial dysfunction in mice. ACS Nano. 1936086X. 19 (5), 5589–5603. Available from: http://pubs.acs.org/journal/ancac3. doi:10.1021/acsnano.4c15112.

Zhao, Y., Chen, J., Wang, R., Pu, X., Wang, D., 2023. A review of transgenerational and multigenerational toxicology in the in vivo model animal *Caenorhabditis elegans*. J. Appl. Toxicol. 10991263. 43 (1), 122–145. Available from: http://onlinelibrary.wiley.com/journal/10.1002/(ISSN)1099-1263. doi:10.1002/jat.4360.

Zou, W., Xia, M., Jiang, K., Cao, Z., Zhang, X., Hu, X., 2020. Photo-oxidative degradation mitigated the developmental toxicity of polyamide microplastics to zebrafish larvae by modulating macrophage-triggered proinflammatory responses and apoptosis. Environ. Sci. Technol. 15205851. 54 (21), 13888–13898. Available from: http://pubs.acs.org/journal/esthag. doi:10.1021/acs.est.0c05399.

CHAPTER

13

Transgenerational impacts of microplastics

Dimple Dimple[a], Ruby Ruby[a], Poonam Mundlia[a], Rajesh Kumar[b], Ranbir Chander Sobti[c], and Arup Giri[a]

[a]Department of Zoology, Baba Mastnath University, Asthal Bohar, Rohtak, Haryana, India
[b]Department of Biosciences, Himachal Pradesh University, Shimla, India
[c]Department of Biotechnology, Punjab University, Chandigarh, India

13.1 Introduction

Over the past few decades, the global production and consumption of plastics have increased dramatically, leading to widespread pollution that poses a significant threat to ecosystems and human health. In 2016 approximately 19–23 million metric tons of plastic waste, or 11% of total plastic waste, entered aquatic environments. Without effective intervention, annual emissions could rise to 53 million metric tons by 2030, even under the current ambitious governmental commitments (Borrelle et al., 2020; Junaid et al., 2023).

Microplastics (MPs) have emerged as pollutants of significant concern due to their ubiquity and ecological risks. They are typically defined as plastic particles 1 μm–5 mm in size, formed through the degradation of larger plastics. Further fragmentation produces nanoplastics (NPs), generally ranging from 1 to 1000 nm (Arif et al., 2024). Owing to their small size and persistence, MPs and NPs are readily ingested by aquatic organisms, leading to bioaccumulation and potential trophic transfer.

Once ingested, MPs can translocate across epithelial surfaces. These particles often contain chemical additives used to enhance their physicochemical properties, which may exert synergistic or additive toxic effects (Song et al., 2022). Additives can leach from plastics into the surrounding environment or within the bodies of aquatic organisms, including crustaceans, amphibians, and fish (Szkudlarek et al., 2024).

Collectively termed micro- and nanoplastics (MNPs), these particles are classified as primary, manufactured at microscopic sizes, or secondary, originating from the degradation of larger plastics. Due to their minute dimensions, NPs can cross biological barriers such as cell membranes and mucous layers, with permeability influenced by factors, including

Microplastics in Animal and Human Physiology
https://doi.org/10.1016/B978-0-443-45635-0.00009-4

particle size, surface charge, polymer type, and chemical interactions (Allouzi et al., 2021; Dube and Okuthe, 2023). Owing to their hydrophobicity and high surface area, MNPs readily adsorb organic pollutants and inorganic contaminants (Rafa et al., 2024).

Recent research has highlighted the transgenerational impacts of MPs, which can adversely affect offspring across diverse species by inducing reproductive and developmental abnormalities inherited from exposed parents (Junaid et al., 2023). MPs negatively affect survival, development, reproduction, and feeding in aquatic organisms (Junaid et al., 2023; Wang et al., 2019). The toxic effects of MNPs, including oxidative stress and tissue damage to inflammation, reproductive and growth impairments, immune and metabolic dysfunction, and gut microbiota disruption, are well-documented across freshwater and marine species, mammals, and terrestrial animals (Basak et al., 2025; Castro-Castellon et al., 2022; Junaid et al., 2023). Increasing evidence also shows transgenerational toxicity, with parental exposure enabling MNP transfer and bioaccumulation that harms offspring (Lee et al., 2021; Persinger and Ward, 2025; Wang et al., 2019).

These effects are commonly linked to the maternal transfer of MPs or associated chemicals, resulting in reproductive toxicity, endocrine disruption, oxidative stress, and metabolic disturbances (Mahmud et al., 2024; Song et al., 2022). Many studies focus solely on physiological and ecological effects, but MPs also affect molecular processes such as gene regulation. Determining correlations between MP exposure and heritable, non-DNA sequence-altering, or epigenetic changes in gene regulation allows us to better understand potential mechanisms behind phenotypic and population-level changes in species (Wade et al., 2025)

MNPs and their co-contaminants can be transmitted through gametes, accumulate in embryonic lipids, and impair early growth and reproduction (Pitt et al., 2018; Santhanam et al., 2024). Exposed offspring may experience reduced survival, altered gene expression, abnormal development, and disrupted behaviors such as impaired locomotion and feeding (Bhadsavle and Golding, 2022; Junaid et al., 2023). These effects can persist for multiple generations (Martins and Guilhermino, 2018), highlighting the enduring ecological threat posed by MNPs and their complex multigenerational impacts.

Studying transgenerational effects is crucial for revealing long-term risks, identifying vulnerable life stages, and improving predictions of population-level consequences, ensuring that the lasting impacts of MP exposure are not underestimated.

13.2 Bioaccumulation and biotranslocation of MPs across generations

MPs persist in the environment and move through biological systems, accumulating in organisms over time through contaminated food and water (Thacharodi et al., 2024). Numerous studies have documented their presence in various tissues, including the digestive tract, liver (Deng et al., 2021), kidneys, gonads (Wang et al., 2019), and even the brain (Alamgir and Shan, 2025). A closely related and essential phenomenon, known as biotranslocation, describes the movement of MP between different tissues within an organism or across trophic levels in food webs.

13.2.1 Species-specific accumulation patterns

The pattern of MP accumulation is dynamic across species, influenced by their differential retention and uptake capacities. Each species exhibits selective ingestion rates, and numerous studies have demonstrated species-specific variations in MP exposure and its effects. A study exposed three nematode species, *Caenorhabditis elegans, Acrobeloides nanus,* and *Plectus acuminatus,* to 1.0-μm polystyrene beads. All ingested the beads, but variations in their effects were observed in *C. elegans*, which showed reduced carrying capacity, while *A. nanus* populations grew faster. Long-term tests revealed subtler impacts than short-term exposure. These shifts in nematode dynamics may influence benthic food web structure (Mueller et al., 2020). Species-specific effects of MP, such as accelerated growth in one species and reduced carrying capacity in another, may alter species composition and disrupt benthic food web dynamics (Majdi and Traunspurger, 2015; Mueller et al., 2020). Another study reported that different fish species, Largemouth bass, Grass carp, and Jian carp, exhibit distinct physiological responses to MP exposure. These species-specific accumulation patterns and stress responses were attributed to variations in feeding behavior and gut morphology (Zheng et al., 2024).

13.2.2 Tissue-specific distribution and persistence

The chemical composition, shape, and surface properties of MPs likely influence their biological interactions and tissue-specific distribution. A study by Cui revealed distinct MP accumulation patterns between blood and atherosclerotic carotid plaques, with associations to circulating lipid biomarkers, highlighting polymer-specific behaviors. These patterns suggest that MP accumulation is governed by their physicochemical characteristics, such as surface charge, which, for instance, affects cellular uptake and tissue interactions (Cui et al., 2025). In a separate study, Deng investigated MP distribution and health risks in mice, focusing on the liver, kidney, and gut. The findings showed that particle size plays a key role: larger MPs were evenly distributed across tissues, while smaller MPs preferentially accumulated in the gut. Overall, MPs were found to accumulate in all three tissues, with a size-dependent distribution, and to induce alterations in biochemical markers and metabolomic profiles (Deng et al., 2017).

13.2.3 Influence of particle characteristics on bioavailability

Particle characteristics such as size, shape, polymer type, and surface properties (Fig. 13.1) significantly influence MPs' bioavailability across species and ecosystems. Smaller, irregular, and chemically active particles tend to be more readily ingested and retained.

13.2.3.1 *Size-dependent translocation and retention*

Smaller MP particles pose greater biological risks due to their enhanced mobility and retention across tissues. Their size increases bioavailability, making them more likely to be absorbed and retained within organisms. In a study on *Calanus finmarchicus*, MP translocation and retention were size-dependent, with smaller particles ingested more frequently than larger ones, indicating higher bioavailability of smaller MPs in this species (Vroom et al., 2017). Their uptake also correlates with increased cytotoxicity, as smaller particles tend to

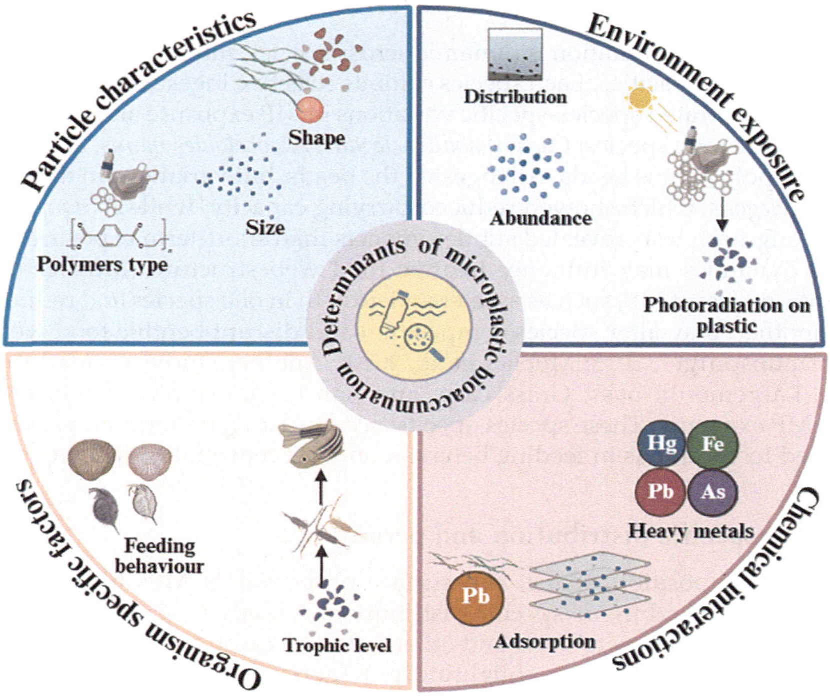

FIG. 13.1 Influencing factors of microplastic accumulation. Representation of key aspects of microplastic buildup, including environmental exposure (abundance, distribution), particle characteristics (size, shape, polymer type), organism-specific factors (feeding behavior), and chemical interactions (adsorbed pollutants, additive leaching, interactions with heavy metals).

induce stronger oxidative stress and inflammatory responses. Together, these findings underscore the importance of particle size in determining both the distribution and potential harm of MPs in biological systems (Mahmud et al., 2024).

13.2.3.2 *Surface chemistry and adsorption of environmental contaminants*

Tang's study showed that MPs increase in density due to biofouling, causing them to sink and settle. Smaller, film-type MPs settle faster due to their larger surface areas and higher biofouling rates. The plastisphere enhances metal and pollutant adsorption, shifts the mechanism from intraparticle to film diffusion, increases surface roughness, reduces pore size, and alters surface charge (Tang and Li, 2025). Metal ion adsorption onto polymer surfaces is mainly independent of the polymer's bulk composition and is primarily determined by surface physicochemical properties. Surface modifications of MPs through weathering, aging, contaminant adsorption, or biofilm formation enhance their ability to bind and precipitate metal ions. The hydrophobic nature of polymer surfaces confers a strong affinity for diverse metals, with adsorption influenced by surface chemistry, particle size, and exposure time (Lee et al., 2021) (Fig. 13.1).

13.2.3.3 *Photoaging and nanoparticle transformation effects*

Photoaging is a significant factor contributing to the rapid increase of plastic waste, particularly MPs in the environment (Jansen et al., 2024) (Fig. 13.1). A study demonstrated that photoaging affects plastic particles of different sizes, including small MPs (1–50 μm) and NPs (<1 μm). This process accelerates MP formation and increases surface roughness, introducing oxygen-containing functional groups that alter surface wettability. In addition, the presence of benzene rings in polymer structures may suppress MP generation while promoting a transition of the plastic surface from hydrophobic to hydrophilic during photoaging (Huang and Wang, 2023). Photooxidation of MNPs can influence pollutant biotransformation by altering MNPs' adsorption capacity and the microbial bioavailability of pollutants; secondly, by enhancing cometabolic transformation via labile dissolved organic carbon released from oxidized MNPs (Xu et al., 2024).

13.3 Physiological alterations induced by MPs

MPs induce physiological alterations through mechanisms such as oxidative stress, inflammation, and disruption of cellular and tissue functions. This can lead to a range of health problems, including hormonal imbalances, metabolic changes, DNA damage, and cellular injury (Mahmud et al., 2024). Numerous experimental studies have reported that MP exposure induces physiological alterations and disrupts growth and development across generations, as summarized in Table 13.1.

13.3.1 Growth and developmental perturbations

MP exposure can hinder growth, delay development, and disrupt physiological systems in both aquatic and terrestrial organisms. These effects are driven by mechanisms such as oxidative stress, hormonal imbalances, and interference with critical developmental signaling pathways (Subaramaniyam et al., 2023) (Fig. 13.2). For instance, Xiao and colleagues demonstrated in tadpoles that MPs induce oxidative stress, metabolic dysfunction, immune responses, and apoptosis in the liver by perturbing the signaling pathways (Xiao et al., 2025). Recent research has further shown that MPs exert evolutionary and ecological impacts across species, affecting development, reproduction, and behavior, and contributing to genotoxicity and stress. Moreover, MP exposure can produce transgenerational effects on development (Jimenez-Guri et al., 2021; Wade et al., 2025).

13.3.1.1 *Inflammatory response*

Micro- and nanoplastics (M-NP) exposure can trigger persistent inflammatory responses across generations, driven by oxidative stress, immune dysregulation, and epigenetic alterations (Fig. 13.2). M-NPs induce oxidative stress, inflammation, and abnormalities in glycolipid metabolism, contributing to metabolic and cardiovascular disorders, including diabetes, obesity, and atherosclerosis (Zheng et al., 2024). They exacerbate obesity and diabetes progression, damage vascular endothelial cells, and increase reactive oxygen species (ROS) production in cells, organoids, and animal models. Elevated ROS levels damage DNA,

TABLE 13.1 Summary of the effects of various compounds on model organisms across generations. The table presents the compound name, the model organism used in testing, the exposure duration, the generation examined, and the observed physiological effects.

Compound	Model organisms	Exposure duration	Generation	Reported effects	Reference
Polystyrene(10 μm)	*Oryzias melastigma*	60 days	2 generations	The findings showed sex-specific disruptions in the reproductive endocrine system, and prenatal exposure adversely affected early offspring development.	(Wang et al., 2019)
Polystyrene(1 μm)	*Caenorhabditis elegans*	96 h for each generation	5 generations	It showed reduced physiological responses, including diminished exploratory behavior (body bending) in nematodes, as well as decreased reproduction.	(Soares et al., 2023)
Polystyrene(1 μm)	*Moina macrocopa*	14 days	3 generations	The study revealed lower growth rates and longer intermolt durations than those observed in the first generation in each group.	(Kim and Rhee, 2021)
Low-density polyethylene	*Eisenia fetida*	98 days	2 generations	There were significant perturbations in cell cycle regulation, neural activity–associated pathways, and amino acid metabolism, underscoring the offspring's hyperactivated antioxidant defense response.	(Zhao et al., 2024)
Polyethylene	*Paramisgurnus dabryanus*	30 days	2 generations	Findings indicated that PE-MP exposure affected antioxidant system markers in the brain, liver, and gonads. In addition, PE-MPs accumulated in the gonads, interfered with HPG-axis gene transcription, altered sex hormone concentrations, and promoted cell apoptosis and pathological changes in gonadal tissue.	(Xia et al., 2023)
Polystyrene(1–10 μm)	*Neomysis awatschensis*	4 weeks	7 generations	The findings showed that MPs inhibited feeding rates and induced age-specific oxidative stress. Growth parameters were significantly impacted, with reduced 20-hydroxyecdysone (20E) levels and prolonged intermolt durations.	(Lee et al., 2021)
Polystyrene	Mice	6 weeks	–	The findings showed that changes in the hepatic transcriptome and serum metabolites indicated a potential risk of metabolic disorders. Subsequently, increased hepatic lipid accumulation was observed in adult F1 mice.	(Luo et al., 2019)
Microplastic	*Pimephales promelas*	30 days	2 generations	The results showed that parental MP exposure produces subtle, nonlethal effects on embryos and larvae, with potential consequences for individual fitness.	(Persinger and Ward, 2025)
Polyethylene(1–45 μm)	*Chironomus tepperi*	–	2 generations	Findings indicate that the parental generation of *C. tepperi* experienced marked reductions in survival, growth, and emergence.	(Lu et al., 2023)

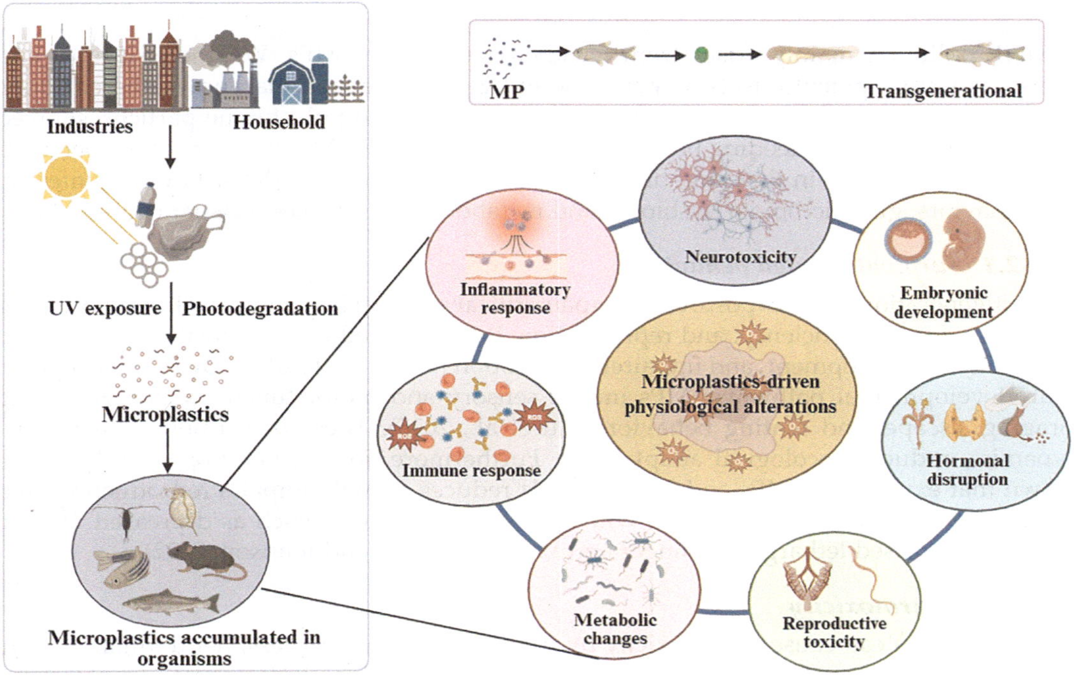

FIG. 13.2 Physiological alterations induced by microplastics. An overview of physical, chemical, and environmental processes, including photooxidation, which breaks down plastics into microplastics that disperse through ecosystems and accumulate in organisms. These accumulated particles can trigger a range of physiological effects, including neurotoxicity, immune responses, metabolic disturbances, reproductive toxicity, inflammatory responses, impaired embryonic development, and hormonal disruption.

proteins, and lipids. At the same time, direct or indirect interactions with immune cells stimulate mechanisms of proinflammatory cytokine release, which are central to the pathogenesis of various disorders, including diabetes, obesity, atherosclerosis, neurodegenerative, cardiovascular, and other age-related diseases (Mahmud et al., 2024).

13.3.1.2 Immunologic function

MP exposure can trigger immune responses that may be heritable, potentially contributing to inflammation and immune dysregulation. A study on *Daphnia magna* found increased immune cell counts under MP stress (Sadler et al., 2019). Various plastic types and sizes have been shown to alter immune gene expression in aquatic organisms such as *Mytilus galloprovincialis*, leading to stress and increased immune protein production (Capolupo et al., 2018). A key toxicological pathway involves ROS generation induced by MP exposure, which promotes the release of danger-associated molecular patterns, disrupts toll-like receptor signaling, and induces cytokine-mediated inflammation (Yang et al., 2022).

13.3.2 Behavioral modifications

MNP pollution mainly arises from weathered secondary MPs. Weathered MNPs and fibers exert more potent effects on behavior and gene expression than unweathered forms (Hutton et al., 2024). A study on *D. magna* exposed to 52-nm polystyrene particles showed changes in activity and feeding behavior (Lovern et al., 2007). MP exposure alters behavior and immune markers in liver and brain tissues. Mice exposed to MNPs showed dementia-like behaviors, gut microbiota dysbiosis, and metabolic disturbances (Gaspar et al., 2023).

13.3.2.1 *Survival-related behavior*

Multigenerational MP exposure can impair critical survival behaviors, including predator avoidance, feeding efficiency, and reproduction. It leads to reduced survival, altered immunity, delayed development, and impaired locomotion (Wade et al., 2025). By interfering with neurodevelopmental pathways, MPs impair sensory and motor functions, compromising foraging, escape, and mating behaviors (Lu et al., 2025). These disruptions affect social dynamics, reducing ecological adaptability. Furthermore, multigenerational studies have shown that exposure to MPs and heavy metals reduces growth, impairs reproduction, and alters locomotion in aquatic species, with behavioral endpoints such as decreased feeding rates and increased lethargy (Persinger and Ward, 2025; Uy and Johnson, 2022).

13.3.2.2 *Neurotoxicity*

MP exposure can cause neurotoxicity by triggering oxidative stress, neuroinflammation, and epigenetic disruption, impairing cognition, behavior, and neural development (Fig. 13.2). It enters the nervous system through ingestion, inhalation, or skin contact, primarily via oxidative stress, inflammation, and disrupted calcium signaling. MNPs reach the brain via olfactory pathways or by crossing the blood–brain barrier. Neurotoxic effects include altered neurotransmitter levels, synaptic dysfunction, anxiety, and reduced learning ability (Vojnits et al., 2025). *In vitro* studies show that MNPs induce apoptosis and mitochondrial damage in neural cells, while *in vivo* studies demonstrate impaired locomotion, memory loss, circadian disruption, and neurotransmitter imbalance (Araújo et al., 2025).

13.3.3 Reproductive strategy adjustments and trade-offs

MPs induce reproductive and endocrine toxicity. These effects are closely linked to MPs' physical and chemical characteristics, their bioaccumulation potential, and the molecular mechanisms they trigger in exposed organisms. They disrupt hormonal balance, oxidative stress regulation, and apoptotic pathways, leading to dysfunction of the gonadal hypothalamic–pituitary–gonadal (HPG) axis, impaired steroidogenesis, and gonadal damage (Hasan et al., 2025). Recent evidence by Wade et al. (2025) further indicates that MP exposure effects can be transmitted to offspring, revealing potential transgenerational reproductive risks (Wade et al., 2025).

13.3.3.1 *Embryonic development*

Recent investigations have demonstrated that MP exposure during parental or early developmental stages can elicit sublethal yet biologically significant effects on aquatic embryos and larvae. In fathead minnows (*Pimephales promelas*), adult exposure to virgin MPs or MPs combined with environmentally relevant concentrations of 17α-ethinyl estradiol over 30 days resulted in subtle developmental impairments in progeny, potentially compromising individual fitness (Persinger and Ward, 2025). A study on *D. magna* found that pristine polymer microspheres impaired growth and reproduction, leading to the collapse of the F1 generation, with MP residues persisting up to the F3 generation (Martins and Guilhermino, 2018). In *Oryzias melastigma* (marine medaka), polystyrene MPs delayed embryonic development, reduced cardiac activity, and decreased hatching success and larval length (Wang et al., 2019). Furthermore, higher depositions exacerbate developmental toxicity by suppressing growth and hatching rates, while lower doses showed minimal accumulation (Dubey et al., 2022).

13.3.3.2 *Fecundity and offspring development*

MNPs act as carriers for xenobiotics, bypassing physiological defenses and exerting direct cellular toxicity (Zurub et al., 2023). Fertility, which depends on gonadal integrity and gamete quality, is highly susceptible to MNP exposure. In aquatic organisms, reproductive toxicity is evident through reduced fecundity, fewer eggs or clutches, prolonged spawning intervals, and lower proportions of gravid females (Jewett et al., 2022). Martins and Guilhermino demonstrated that a 21-day exposure of *D. magna* to MPs resulted in severe impacts on growth and reproduction, leading to increased parental mortality, the extinction of the exposed F1 generation, and reduced reproductive output and population growth rate (Martins and Guilhermino, 2018). Collectively, these findings indicate that MNPs can induce transgenerational reproductive impairments and compromise long-term population sustainability (Ferrante et al., 2022).

13.3.3.3 *Sexual maturity*

MPs and NPs can translocate from the gut to reproductive organs, though only smaller NPs bypass the blood–gonad barrier. NPs have been found in human placenta and reproductive tissues, disrupting ovarian function, spermatogenesis, and sperm quality. In sea urchins, polystyrene particles accumulated in gonads, with smaller MPs showing higher deposition (Ferrante et al., 2022).

13.3.3.4 *Hormonal disruption*

MPs exert detrimental effects on reproductive health primarily through endocrine disruption and hormonal imbalance. They can bioaccumulate in gonadal tissues, induce oxidative stress, and trigger apoptosis in reproductive cells (Yong et al., 2020). MP exposure results in reduced fertility, impaired gametogenesis, and transgenerational reproductive dysfunction across various aquatic taxa. Moreover, MPs act as vectors for endocrine-disrupting chemicals, amplifying their toxic impacts on organisms and aquatic biodiversity. Emerging evidence implicates several key molecular pathways, such as mitogen-activated protein kinase (MAPK) , phosphatidylinositol 3-kinase -protein kinase B (PI3K-AKT), mechanistic target of rapamycin (mTOR), nuclear factor kappa-light-chain-enhancer of activated B cells (NF-κB), peroxisome Proliferator-Activated Receptor (PPAR), and the NLR family pyrin domain containing 3 (NLRP3) inflammasome, in mediating MP-induced reproductive toxicity (Hasan et al., 2025). By disrupting the HPG axis, MPs impair steroidogenesis and gametogenesis, leading to diminished fertility and multigenerational consequences (Shi et al., 2023).

13.4 Chronic and transgenerational effects in response to environmental concentrations

13.4.1 Genotoxic and mutagenic risks across generations

13.4.1.1 Germ cell mutations and heritable defects

Chronic exposure of organisms to environmentally realistic concentrations of MPs raises serious concerns not only for direct cytotoxicity but also for the induction of heritable genomic damage with long-term effects across generations (Bridgeman et al., 2025). These particles may be internalized by organisms and, under certain conditions, translocated to reproductive tissues (Ali et al., 2024). For instance, maternal exposure of Kunming mice to MPs has been shown to reduce oocyte maturation and fertilization rates, while inducing DNA damage in oocytes, including spindle disruption and chromosomal morphology defects (Zhang et al., 2023).

Germ cell mutation can occur through the direct interaction of particles or associated chemicals with DNA, chromosomes, and related proteins, or via indirect routes, such as ROS production from oxidative stress or inflammation (Tang, 2025). Once germ cells acquire mutations or chromosomal aberrations, these can be transmitted to subsequent generations, resulting in heritable defects such as reduced fertility, embryonic malformation, or developmental abnormalities. Parental exposure to MPs in mammals has also been linked to metabolic, reproductive, and developmental disorders in the offspring (He and Yin, 2024). Although direct evidence for germline mutagenesis by MPs remains limited, findings from aquatic and invertebrate model species indicate that exposure of the F0 generation can trigger adverse outcomes in F1, F2, and even F3 generations. For example, in the freshwater crustacean *D. magna*, chronic exposure of the F0 population to MPs resulted in persistently reduced growth, fecundity, and population growth rates across F1–F3 generations, even when the subsequent generations were reared in clean medium (Martins and Guilhermino, 2018). Such evidence suggests that MPs can induce heritable molecular and physiological impairments that may originate from germline DNA damage or epigenetic dysregulation.

13.4.1.2 Evidence from model species

Many model species studies provide clear evidence that MP exposure can induce genotoxic and mutagenic effects that persist across generations (Table 13.2). In *D. magna*, exposure to polyethylene MP fragments led to reduced somatic growth, decreased reproduction, and declining population growth by the third generation, though without detectable shifts in DNA methylation (Song et al., 2022). Soil invertebrate studies showed similar patterns; for instance, a study revealed that *Eisenia fetida*, when exposed to polystyrene or polyethylene MPs, exhibited significant DNA damage in F0 individuals, reduced reproductive output, and the continuation of these impairments into the F1 generation (Sobhani et al., 2021, 2022). In the nematode *C. elegans*, submicron polystyrene particles triggered DNA damage, germline apoptosis, and severe reproductive toxicity (Chen et al., 2022).

13.4.2 Epigenetic mechanisms in transgenerational toxicity

Epigenetic mechanisms, such as DNA methylation, histone modifications, and noncoding RNA regulation, offer plausible routes by which MPs' exposure in one generation may alter

TABLE 13.2 Experimental evidence of genotoxic and mutagenic transgenerational effects of microplastic exposure in model organisms.

Compound	Model species	Key findings	References
Polyethylene microplastics	*Daphnia magna*	• Decreased somatic growth and reproduction in the third generation • Decreased population growth • No significant variation in DNA methylation across four generations	(Song et al., 2022)
Polystyrene microplastics	*Eisenia fetida*	• DNA damage in the F0 and F1 generations induced genotoxicity • Decreased reproduction in both generations	(Sobhani et al., 2021)
Polyethylene microplastics	*Eisenia fetida*	• Considerable DNA damage caused genotoxicity in the F0 generation • Reduced reproduction in both F0 and F1 generations	(Sobhani et al., 2022)
Polystyrene microplastics	*Caenorhabditis elegans*	• Induced DNA damage and germline apoptosis • Severe reproductive toxicity	(Chen et al., 2022)
Polyethylene microplastics	Kunming mice	• Increased reactive oxygen species (ROS) levels in oocytes and embryos • DNA damage in oocytes • Induced oxidative stress, reproductive toxicity, and apoptosis	(Zhang et al., 2023b)

gene regulation and phenotype in subsequent generations without changing the DNA sequence. Recent investigations highlight that MPs can indeed trigger such heritable epigenetic and transcriptomic alterations, potentially influencing stress responses, metabolism, and reproduction (Poma et al., 2023).

In a two-generation study with fathead minnows (*P. promelas*), exposure of the F0 generation to polyethylene MPs led to differentially methylated regions in the unexposed F1 offspring, with genes in these regions enriched in endocrine and metabolic pathways (Wade et al., 2025). Similarly, in *D. magna*, exposure to MP fibers resulted in transgenerational impairments in reproduction and molting frequency across three generations, alongside gene-specific DNA methylation changes linked to developmental and metabolic processes (Shim et al., 2025). These findings suggest that MP exposure may leave a molecular epigenetic memory transmitted through the germline.

Mechanistically, MP exposure can induce oxidative stress and inflammation in parental tissues, which may alter germ cell epigenetic marks directly (via particle or chemical interaction with germ cells) or indirectly (through systemic physiological disruption). Perturbation of redox-sensitive transcription factors, such as Nrf2, in the parental (F0) generation can disrupt cellular redox homeostasis, potentially influencing germ cell methylation patterns and thereby altering gene expression in the offspring (Zhang et al., 2024). Furthermore, maternal exposure to polyethylene MP fragments containing benzophenone-3 in *D. magna* demonstrated transgenerational reductions in somatic growth and reproduction up to the F3 generation, with indications of cumulative toxicity from combined particle-additive exposure (Song et al., 2022). While no global DNA methylation changes were detected, the study highlighted the need for gene-specific analyses to elucidate epigenetic inheritance

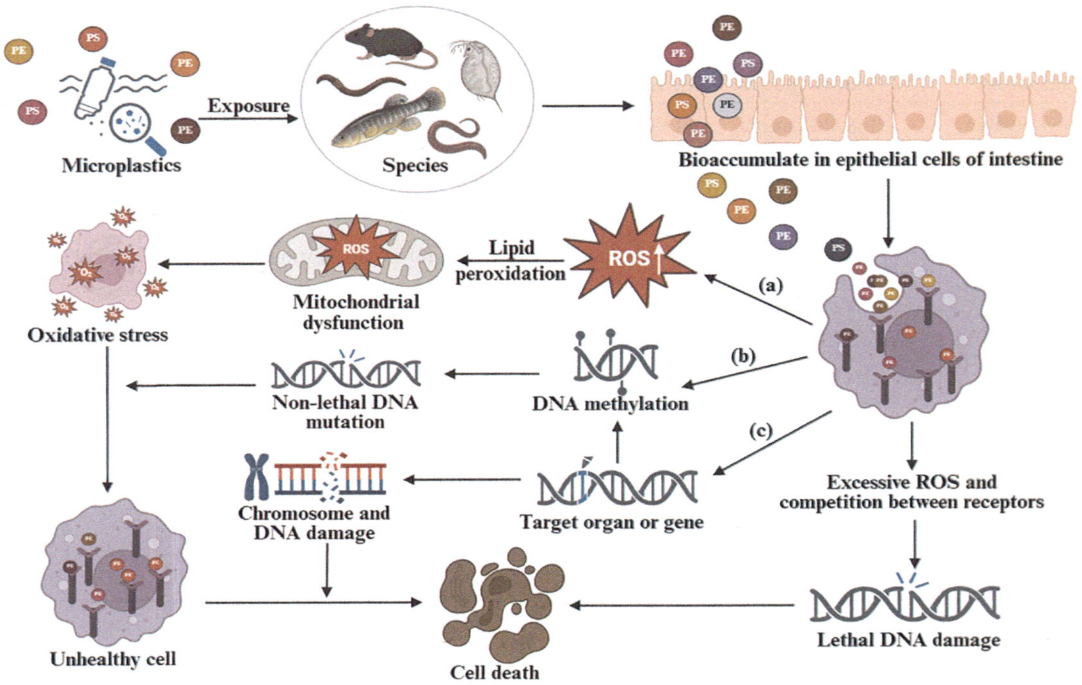

FIG. 13.3 Mechanisms underlying transgenerational effects of microplastics. After uptake, microplastics (MPs) and their associated chemicals activate three major pathways that can transmit effects across generations. (A) Oxidative stress: Excess reactive oxygen species (ROS) cause lipid peroxidation, mitochondrial dysfunction, and oxidative damage to germ cells, reducing gamete quality and enabling the inheritance of oxidative lesions. (B) Epigenetic disruption: MPs alter DNA methylation, histone marks, and noncoding RNAs in germ cells, reshaping gene expression and contributing to heritable metabolic and reproductive abnormalities. (C) Genotoxicity: MPs induce DNA strand breaks, chromosomal damage, and apoptosis; sublethal genomic alterations in germ cells can persist across generations.

mechanisms. Overall, these findings suggest that MPs act as epigenetic disruptors, inducing heritable molecular alterations that persist beyond direct exposure and contribute to reproductive decline across generations (Fig. 13.3).

13.5 Metabolic and proteomic responses underlying transgenerational effects

13.5.1 Oxidative stress and antioxidant defense systems

MPs trigger oxidative stress by elevating ROS production, which damages lipids, proteins, and DNA, ultimately disrupting cellular stability (Kadac-Czapska et al., 2024). Many organisms respond by upregulating antioxidant defenses, such as superoxide dismutase (SOD), catalase (CAT), and the glutathione system, but long-term exposure often exceeds these protective capacities. Importantly, oxidative alterations do not remain confined to directly exposed organisms. Cross-generational studies have shown inherited impairments in antioxidant signaling, including reduced Nrf2 pathway activation and increased basal ROS levels in unexposed offspring (Zhang et al., 2024). Similar patterns across taxa suggest that oxidative imbalance is a key mechanism by which MPs exert long-term, heritable physiological stress (Junaid et al., 2023).

13.5.2 Transcriptomic and proteomic profiling across generations

High-throughput molecular analyses showed that MP exposure affects multiple regulatory pathways, altering genes and proteins linked to detoxification, mitochondrial function, immune activity, and cellular stress responses (Poma et al., 2023). However, these disruptions are not limited to the exposed generation. Mounting evidence showed that they can be transmitted to progeny. In *Drosophila melanogaster*, parental MP exposure resulted in altered developmental gene expression and impaired stress response pathways in unexposed offspring, indicating molecular inheritance (Jimenez-Guri et al., 2021). Similar findings in fish demonstrated that MP-associated epigenomic alterations, such as changes in methylation and chromatin structure, can lead to persistent shifts in gene expression across generations (Wade et al., 2025). These results suggest that MPs can act as environmental epimutagens, with lasting effects on transcriptomic and proteomic profiles.

13.5.3 Metabolic disruption and energy allocation shifts

MPs interfere with nutrient absorption, digestive efficiency, and microbial balance, collectively imposing metabolic alterations on exposed organisms (Kadac-Czapska et al., 2024). To cope with increased oxidative and cellular stress, organisms must redirect energy toward defense and repair pathways, often at the expense of growth, reproduction, and immune function. Transgenerational studies revealed that these metabolic strains persist, as offspring of MP-exposed parents frequently showed altered carbohydrate and lipid metabolism, reduced growth rates, and impaired energy allocation efficiency (Junaid et al., 2023).

13.6 Human health and ecological sustainability concerns

MPs have transitioned from being perceived as primarily marine pollutants to a pervasive global contaminant capable of infiltrating terrestrial, freshwater, atmospheric, and biological systems. Their durability, high surface reactivity, ability to adsorb hazardous chemicals, and capacity to bioaccumulate across trophic levels highlight them as emerging environmental toxicants (Fig. 13.4). Increasing evidence suggests that MPs and NPs can cross biological barriers, induce cellular toxicity, influence population dynamics, and alter ecosystem functioning, thereby posing significant challenges to both human health and ecological sustainability (Ali et al., 2025).

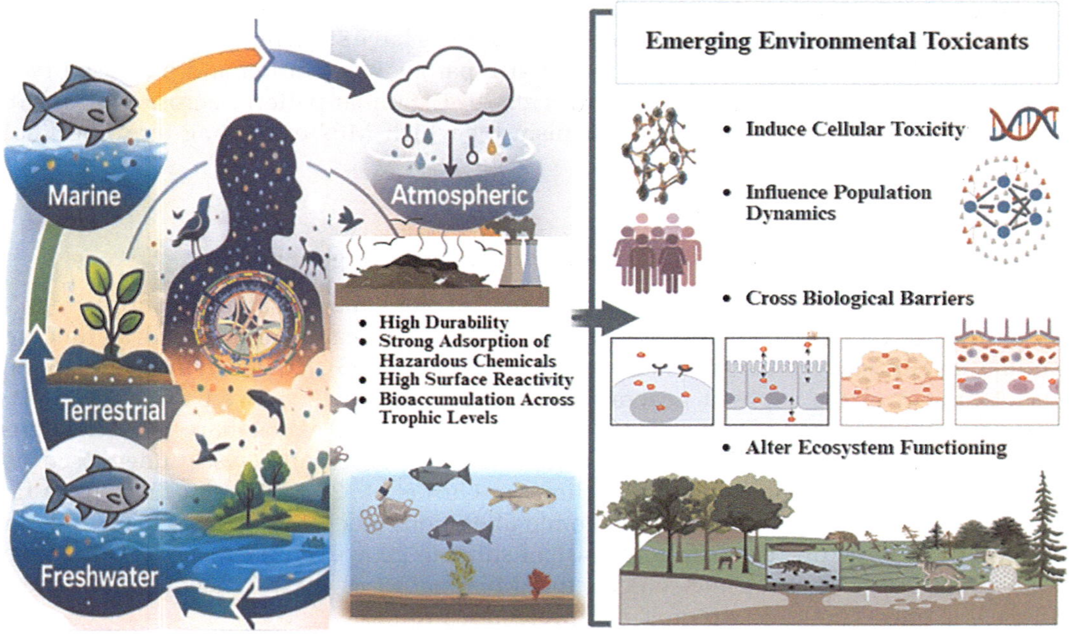

FIG. 13.4 Potential routes to human microplastic exposure. Microplastics increasingly contaminate marine, terrestrial, freshwater, and atmospheric systems, where their durability and chemical reactivity enable widespread movement and bioaccumulation. By entering organisms and crossing biological barriers, they induce cellular toxicity and ultimately disrupt ecosystem functioning.

13.6.1 Potential human exposure pathways via food chains

Humans are exposed to MPs primarily through ingestion and inhalation (Yang et al., 2023), with dietary intake being a major route via seafood (Borriello et al., 2023), drinking water (Udovicki et al., 2022), salt (Kibria, 2024), produce, and processed foods (Udovicki et al., 2022). MPs have now been detected in human blood and lung tissue, demonstrating their capacity for systemic distribution. Once ingested, they can induce oxidative stress, inflammation, metabolic alterations, and dysbiosis of the gut microbiota (Wang et al., 2019). MPs' translocation capacity significantly heightens the potential for endocrine disturbances, immune dysregulation, and reproductive impairments, underscoring their emerging relevance as human health toxicants (Ullah et al., 2023). In addition, MPs can act as carriers for cooccurring contaminants such as antibiotics and persistent organic pollutants, further intensifying their toxicological significance.

13.6.2 Implications for ecosystem services and biodiversity

MP contamination disrupts ecosystem functioning by altering community composition, trophic interactions, and biogeochemical processes across aquatic and terrestrial habitats. Primary producers, such as phytoplankton and microalgae, experience reduced photosynthetic efficiency, chlorophyll degradation, and inhibited carbon fixation when exposed to suspended MPs, resulting in lowered aquatic primary productivity and oxygen generation (Dave and Sharaff, 2024). These impairments cascade upward through the food web, affecting herbivore feeding dynamics, predator–prey balance, and nutrient flow, ultimately weakening ecosystem resilience and energy transfer efficiency (Ilyas et al., 2025). In benthic environments, MPs interfere with nutrient mineralization, sediment mixing, and microbial enzymatic activity, destabilizing nitrogen and carbon cycling and reducing the capacity of ecosystems to support biodiversity and essential services such as fisheries, water purification, and soil fertility (Yang et al., 2025).

13.6.2.1 *Community structure shifts and loss of keystone species*

Keystone species, such as zooplankton, coral–algal symbionts, and soil invertebrates, are susceptible to MP stress because they occupy foundational roles in energy transfer and nutrient cycling. In aquatic food webs, chronic ingestion of MPs reduces zooplankton feeding efficiency, reproductive output, lipid reserves, and population growth, thereby diminishing prey availability for fish and disrupting upper-trophic energy flow (Wang et al., 2019).

In coral reef systems, MPs adhere to coral surfaces, impairing the photosynthetic performance and antioxidant balance of symbiotic microalgae, thereby weakening coral calcification and stress tolerance (Wang et al., 2019). Soil invertebrates, such as earthworms and collembolans, show reduced burrowing, disrupted gut microbiota, and lower nutrient-mineralization capacity when exposed to MPs. These changes directly reduce soil fertility and, in turn, weaken overall terrestrial ecosystem productivity.

13.6.2.2 *Disruption of nutrient cycling and ecosystem stability*

MPs disrupt nitrogen and carbon cycling by altering microbial community structure and reducing key soil and sediment enzyme activities, thereby decreasing biogeochemical efficiency in both land and water systems (Pan et al., 2024). These microbial impairments weaken ecosystem stability, slow natural recovery processes, and raise vulnerability to environmental stresses. Sensitive habitats, such as coral reefs, mangroves, and wetlands, also show reduced resilience under the combined effects of MP pollution and climate-related pressures, increasing their risk of long-term ecological decline (Shi et al., 2023).

13.6.3 Emerging challenges in MP toxicology

Although MP toxicity is increasingly documented, essential knowledge gaps persist regarding its long-term, low-dose, and transgenerational effects. Variability in polymer types, environmental weathering, and biofouling results in dynamic surface properties that complicate exposure and hazard assessment. Detection challenges are even greater for NPs, which can cross cellular and physiological barriers but remain difficult to quantify in tissues and environmental matrices (Ayejoto et al., 2024). Furthermore, MPs frequently act as carriers of cocontaminants, including antibiotics, per- and polyfluoroalkyl substances (PFAS), and heavy metals, leading to synergistic toxicity that cannot be predicted from single-stressor tests. Growing evidence of placental translocation, microbiome disruption, and heritable epigenetic modifications further underscores the need for multi-generational and long-term monitoring to fully understand population-level health risks (Nguyen et al., 2025).

13.6.3.1 *Complexity of polymer diversity and degradation products*

MPs originate from a variety of polymers (e.g., polyethylene, polypropylene, polystyrene), each exhibiting distinct surface chemistry, degradation rates, and interaction potentials within biological systems. During weathering, these polymers release additives such as bisphenols, phthalates, and brominated flame retardants, many of which act as endocrine-disrupting compounds and can disrupt hormone signaling, reproductive development, and metabolic regulation in exposed organisms (Niu et al., 2023). Environmental fragmentation further produces NPs that exhibit increased surface reactivity and an enhanced ability to cross cellular and physiological barriers, thereby intensifying oxidative stress, inflammation, and genotoxicity. In addition, the chemical alteration of plastics under UV radiation, heat, and microbial colonization leads to the formation of secondary degradation products, whose toxicological profiles are often more bioactive and less well understood than those of the parent polymers, complicating exposure assessment and risk evaluation (Li and Xiao, 2023).

13.6.3.2 *Combined effects with climate change stressors*

Climate-linked increases in temperature, salinity, and ocean acidification intensify the toxicity of MPs by enhancing polymer weathering, microbial colonization, and contaminant adsorption. These interactive stressors amplify adverse effects on metabolism, behavior, and survival in aquatic organisms. These simultaneous stressors exert synergistic and more severe impacts on aquatic biota, impairing metabolic efficiency, disrupting ion regulation, altering behavioral responses, and substantially reducing growth, reproduction, and overall survival. Furthermore, increasing evidence demonstrates that warming and acidification conditions markedly amplify MP-mediated oxidative stress, inflammatory damage, and developmental abnormalities, thereby heightening organismal vulnerability in multistressor environments (Li et al., 2024).

13.6.4 Long-term evolutionary consequences

Prolonged MP exposure imposes novel selection pressures that can gradually shape the evolutionary trajectories of microorganisms, plants, and animals. MPs and associated chemical additives may induce heritable genetic mutations, alter epigenetic markers, and affect gene expression pathways, leading to changes in stress tolerance, reproductive timing, and developmental processes (Tang, 2025). For instance, microbial communities in plastic-rich environments show shifts toward plastic-degrading taxa, while some aquatic species exhibit altered life-history traits to compensate for chronic physiological stress (Haney and Rochman, 2025).

13.6.5 Policy and risk assessment frameworks

Global measures to address MP pollution remain limited, prompting the 2022 United Nations Environment Assembly to initiate negotiations for a Global Plastics Treaty focused on lifecycle-based plastic reduction, improved waste management, and standardized monitoring. In India, the Plastic Waste Management Rules, 2016, strengthened through amendments from 2021 to 2024, promote the phase-out of single-use plastics and mandate extended producer responsibility (EPR) for plastic paçkaging, which is monitored through the Central Pollution Control Board Extended Producer Responsibility (CPCB EPR) digital portal. However, no MP-specific standards currently exist for drinking water, soil, or food, creating a gap between regulation and emerging health evidence. Recent studies reporting MPs in Indian rivers, coastal waters, agricultural soils, and seafood indicate active land–sea transfer pathways and potential dietary exposure risks (Kumar et al., 2025).

13.7 Conclusions

The pervasive presence of MNPs in natural and engineered environments has transformed plastic pollution from a visible waste management concern into a complex, long-term ecological and biological crisis. As discussed throughout this chapter, the impacts of MNPs extend far beyond immediate physiological stress, encompassing molecular, developmental, reproductive, and behavioral disruptions that may propagate across generations. The evidence synthesized here clearly indicates that transgenerational toxicity is not an isolated phenomenon but an emerging and consequential dimension of MP pollution.

MNPs exhibit high bioavailability due to their minute size, hydrophobicity, and large surface area, enabling them to penetrate biological membranes, translocate across tissues, and accumulate in reproductive organs and developing embryos. Their ability to act as vectors for chemical additives and cocontaminants further amplifies toxicological risks. The combined effects of oxidative stress, endocrine disruption, immune dysregulation, metabolic imbalance, and inflammatory responses compromise organismal fitness and reproductive success. Importantly, these adverse outcomes are not confined to directly exposed individuals; parental exposure can mediate effects in offspring through maternal transfer, altered gamete quality, and epigenetic modifications.

Transgenerational consequences—including impaired embryonic development, reduced survival, altered growth trajectories, behavioral abnormalities, and persistent gene expression changes—underscore the depth of ecological vulnerability associated with MNP exposure. The growing body of literature documenting epigenetic inheritance mechanisms, such as DNA methylation changes, histone modifications, and noncoding RNA regulation, suggests that MPs may influence phenotypes without altering DNA sequences. Such heritable modifications have profound implications for adaptation, resilience, and evolutionary trajectories of species inhabiting contaminated environments.

At the ecosystem level, the cumulative and multigenerational effects of MNPs threaten biodiversity, alter community composition, and destabilize trophic interactions. Disruptions in keystone species or early life stages can cascade through food webs, affecting nutrient cycling and biogeochemical processes. Given the potential for trophic transfer and biomagnification, these ecological disturbances may ultimately intersect with human health concerns, particularly in populations reliant on aquatic and terrestrial food systems.

Despite significant advances, critical knowledge gaps remain. Standardized methodologies for multigenerational exposure studies are lacking, and the variability in polymer types, particle sizes, weathering states, and experimental conditions complicates cross-study comparisons. The interactive effects of MNPs with climate change stressors—such as rising temperatures, ocean acidification, and hypoxia—further challenge our predictive capacity. Moreover, the long-term ecological thresholds at which sublethal transgenerational effects translate into population declines are still poorly defined.

Addressing these challenges requires an integrative and interdisciplinary research framework that bridges toxicology, molecular biology, ecology, evolutionary science, and environmental policy. Long-term, multigenerational studies under environmentally realistic exposure conditions are essential to capture the cumulative burden of MNP contamination. Greater emphasis should be placed on mechanistic investigations into epigenetic inheritance and transgenerational plasticity to clarify whether the observed changes represent adaptive responses or maladaptive consequences.

From a policy perspective, current regulatory frameworks remain insufficient to address the complexity of MP pollution. Effective mitigation strategies must incorporate source reduction, improved waste management, polymer redesign, and stricter control of additive chemicals. Monitoring programs should include reproductive and developmental endpoints and consider multigenerational risk assessment rather than focusing solely on acute toxicity. International cooperation is crucial, as MP contamination transcends political boundaries and disperses globally through atmospheric and oceanic transport.

Ultimately, the transgenerational impacts of MPs challenge traditional paradigms of environmental risk assessment, which often underestimate delayed and inherited effects. The persistence of MNPs in ecosystems, coupled with their capacity to influence biological systems across generations, underscores the urgency of coordinated scientific, regulatory, and societal action. Protecting biodiversity and ensuring ecological sustainability will depend on recognizing that the consequences of plastic pollution are not limited to the present generation but may shape the health, resilience, and evolutionary potential of future life on Earth.

MP pollution represents a silent yet enduring stressor with far-reaching, multigenerational implications. Only through comprehensive research, informed policymaking, and proactive global mitigation efforts can we prevent the long-term ecological and biological legacy of plastics from becoming irreversible.

AI disclosure

During the preparation of this work, the author(s) used ChatGPT to compile rough content and design diagrams. However, the content was cross-checked, verified, rephrased, and aligned with scholarly literature. Only ideas were taken from AI sources for the drawing of diagrams. After using this tool/service, the author(s) reviewed and edited the content as needed and take(s) full responsibility for the content of the publication.

References

Alamgir, W., Shan, H., 2025. Bioaccumulation of microplastics: from environmental pollutants to multisystemic toxicity in humans. Life Sci. 2708-2970. 6 (2), 02. doi:10.37185/lns.1.1.946.

Ali, M., Chang, X., Kun, L., 2025. Micro- and nanoplastics as emerging threats to both terrestrial and aquatic animals: a comprehensive review. Vet. Sci. 2306-7381. 12 (8), 688. doi:10.3390/vetsci12080688.

Ali, N., Katsouli, J., Marczylo, E.L., Gant, T.W., Wright, S., de la Serna, J.B., 2024. The potential impacts of micro-and -nano plastics on various organ systems in humans. eBioMedicine. 23523964. 99. Available from: https://www.sciencedirect.com/science/journal/23523964. doi:10.1016/j.ebiom.2023.104901.

Allouzi, M.M.A., Tang, D.Y.Y., Chew, K.W., Rinklebe, J., Bolan, N., Allouzi, S.M.A., Show, P.L., 2021. Micro (nano) plastic pollution: the ecological influence on soil-plant system and human health. Sci. Total Environ. 18791026. 788. Available from: https://www.elsevier.com/locate/scitotenv. doi:10.1016/j.scitotenv.2021.147815.

Araújo, A.M., Mota, C., Ramos, H., Faria, M.A., Carvalho, M., Ferreira, I.M.P.L.V.O., 2025. The neurotoxic threat of micro- and nanoplastics: evidence from in vitro and in vivo models. Arch. Toxicol. 14320738. 99 (9), 3505–3525. Available from: https://www.springer.com/journal/204. doi:10.1007/s00204-025-04091-3.

Arif, Y., Mir, A.R., Zieliński, P., Hayat, S., Bajguz, A., 2024. Microplastics and nanoplastics: source, behavior, remediation, and multi-level environmental impact. J. Environ. Manag. 03014797. 356, 120618. doi:10.1016/j.jenvman.2024.120618.

Ayejoto, D.A., Egbueri, J.C., Onuba, L.N., Agbasi, J.C., Nweke, N.D., 2024. Toxicological Effects of Ingested Microplastics on Human Health. Springer Science and Business Media LLC, pp. 427–461. doi:10.1007/978-3-031-64253-1_21.

Basak, S., Dixit, A.K., Dey, R.K., Babu, G., 2025. MNPs in marine environment: sources, distribution, trophic transfer, toxicity, fate and the remediating role of epiplastic syntrophic microbial consortia (Biofilms). Mar. Chem. 03044203. 273. Available from: https://www.sciencedirect.com/science/journal/03044203. doi:10.1016/j.marchem.2025.104553.

Bhadsavle, S.S., Golding, M.C., 2022. Paternal epigenetic influences on placental health and their impacts on offspring development and disease. Front. Genet. 16648021. 13. Available from: https://www.frontiersin.org/journals/genetics#. doi:10.3389/fgene.2022.1068408.

Borrelle, S.B., Ringma, J., Law, K.L., Monnahan, C.C., Lebreton, L., McGivern, A., Murphy, E., Jambeck, J., Leonard, G.H., Hilleary, M.A., Eriksen, M., Possingham, H.P., De Frond, H., Gerber, L.R., Polidoro, B., Tahir, A., Bernard, M., Mallos, N., Barnes, M., Rochman, C.M., 2020. Predicted growth in plastic waste exceeds efforts to mitigate plastic pollution. Science. 10959203. 369 (6509), 1515–1518. Available from: https://www.sciencemag.org/lookup/doi/10.1126/science.aba3656. doi:10.1126/SCIENCE.ABA3656.

Borriello, L., Scivicco, M., Cacciola, N.A., Esposito, F., Severino, L., Cirillo, T., 2023. Microplastics, a global issue: human exposure through environmental and dietary sources. Foods. 23048158. 12 (18). Available from: https://www.mdpi.com/journal/foods. doi:10.3390/foods12183396.

Bridgeman, L., Cimbalo, A., López-Rodríguez, D., Pamies, D., Frangiamone, M., 2025. Exploring toxicological pathways of microplastics and nanoplastics: insights from animal and cellular models. J. Hazard. Mater. 03043894. 490, 137795. doi:10.1016/j.jhazmat.2025.137795.

Capolupo, M., Franzellitti, S., Valbonesi, P., Lanzas, C.S., Fabbri, E., 2018. Uptake and transcriptional effects of polystyrene microplastics in larval stages of the Mediterranean mussel *Mytilus galloprovincialis*. Environ. Pollut. 18736424. 241, 1038–1047. Available from: https://www.elsevier.com/inca/publications/store/4/0/5/8/5/6. doi:10.1016/j.envpol.2018.06.035.

Castro-Castellon, A.T., Horton, A.A., Hughes, J.M.R., Rampley, C., Jeffers, E.S., Bussi, G., Whitehead, P., 2022. Ecotoxicity of microplastics to freshwater biota: considering exposure and hazard across trophic levels. Sci. Total Environ. 18791026. 816. Available from: https://www.elsevier.com/locate/scitotenv. doi:10.1016/j.scitotenv.2021.151638.

Chen, H., Yang, Y., Wang, C., Hua, X., Hui, L., Xie, D., Xiang, M., Yunjiang, Y., 2022. Reproductive toxicity of UV-photodegraded polystyrene microplastics induced by DNA damage-dependent cell apoptosis in *Caenorhabditis elegans*. Sci. Total Environ. 00489697. 811, 152350. doi:10.1016/j.scitotenv.2021.152350.

Cui, C., Guo, Z., Liu, Y., Han, N., Song, J., Chen, Y., Zheng, Y., Sheng, C., Balmer, L., Haibin, L., Liu, R., Liu, L., Du, J., Zhiyuan, W., 2025. Tissue-specific distribution of microplastics in human blood and carotid plaques: a paired sample analysis. Environ. Int. 01604120. 203, 109743. doi:10.1016/j.envint.2025.109743.

Dave, G., Sharaff, M., 2024. Effects of microplastics on coastal planktonic community. In: Microplastics: Origins, Risks, and Mitigation. Elsevier, India, pp. 259–277. Available from: https://www.sciencedirect.com/book/9780443298042. doi:10.1016/B978-0-443-29804-2.00010-X.

Deng, Y., Yan, Z., Shen, R., Huang, Y., Ren, H., Zhang, Y., 2021. Enhanced reproductive toxicities induced by phthalates contaminated microplastics in male mice (*Mus musculus*). J. Hazard. Mater. 03043894. 406, 124644. doi:10.1016/j.jhazmat.2020.124644.

Deng, Y., Zhang, Y., Lemos, B., Ren, H., 2017. Tissue accumulation of microplastics in mice and biomarker responses suggest widespread health risks of exposure. Sci. Rep. 2045-2322. 7 (1). doi:10.1038/srep46687.

Dube, E., Okuthe, G.E., 2023. Plastics and micro/nano-plastics (MNPs) in the environment: occurrence, impact, and toxicity. Int. J. Environ. Res. Public Health. 1660-4601. 20 (17), 6667. doi:10.3390/ijerph20176667.

Dubey, I., Khan, S., Kushwaha, S., 2022. Developmental and reproductive toxic effects of exposure to microplastics: a review of associated signaling pathways. Front. Toxicol. 2673-3080. 4. doi:10.3389/ftox.2022.901798.

Ferrante, M.C., Monnolo, A., Piano, F.D., Raso, G.M., Meli, R., 2022. The pressing issue of micro- and nanoplastic contamination: profiling the reproductive alterations mediated by oxidative stress. Antioxidants. 20763921. 11 (2). Available from: https://www.mdpi.com/2076-3921/11/2/193/pdf. doi:10.3390/antiox11020193.

Gaspar, L., Bartman, S., Coppotelli, G., Ross, J.M., 2023. Acute exposure to microplastics induced changes in behavior and inflammation in young and old mice. Int. J. Mol. Sci. 1422-0067. 24 (15), 12308. doi:10.3390/ijms241512308.

Haney, J., Rochman, C.M., 2025. Plastic pollution has the potential to alter ecological and evolutionary processes in aquatic ecosystems. Nat. Ecol. Evol. 2397334X. 9 (5), 762–768. Available from: https://www.nature.com/natecolevol/. doi:10.1038/s41559-025-02678-8.

Hasan, A.K.M.M., Hossain, M.F., Uddin, M., Khan, M.T., Saif, U.M., Hamed, M., Martyniuk, C.J., Chivers, D.P., 2025. Mechanistic insights into microplastic-induced reproductive toxicity in aquatic organisms: a comprehensive review. Aquat. Toxicol. 18791514. 286. Available from: https://www.sciencedirect.com/science/journal/0166445X. doi:10.1016/j.aquatox.2025.107478.

Yuchong, H., Yin, R., 2024. The reproductive and transgenerational toxicity of microplastics and nanoplastics: a threat to mammalian fertility in both sexes. J. Appl. Toxicol. 0260-437X. 44 (1), 66–85. doi:10.1002/jat.4510.

Huang, Z., Wang, H., 2023. Study on the impact of photoaging on the generation of very small microplastics (MPs) and nanoplastics (NPs) and the wettability of plastic surface. Environ. Sci. Pollut. Res. 1614-7499. 30 (40), 92963–92982. doi:10.1007/s11356-023-29012-8.

Hutton, S.J., Kashiwabara, L., Anderson, E., Siddiqui, S., Harper, B., Harper, S., Brander, S.M., 2024. Behavioral and molecular effects of micro and nanoplastics across three plastic types in fish: weathered microfibers induce a similar response to nanosized particles. Front. Toxicol. 26733080. 6. Available from: https://www.frontiersin.org/journals/toxicology/about. doi:10.3389/ftox.2024.1490223.

Ilyas, M., Duarte, C.M., Xu, E.G., Xu, G., Yang, J., 2025. Ecological effects of micro/nanoplastics on plant-associated food webs. Trends Plant Sci. 13601385. 30 (5), 526–538. Available from: https://www.sciencedirect.com/science/journal/13601385. doi:10.1016/j.tplants.2024.11.018.

Jansen, M.A.K., Andrady, A.L., Bornman, J.F., Aucamp, P.J., Bais, A.F., Banaszak, A.T., Barnes, P.W., Bernhard, G.H., Bruckman, L.S., Busquets, R., Häder, D.P., Hanson, M.L., Heikkilä, A.M., Hylander, S., Lucas, R.M., Mackenzie, R., Madronich, S., Neale, P.J., Neale, R.E., Olsen, C.M., Ossola, R., Pandey, K.K., Petropavlovskikh, I., Revell, L.E., Robinson, S.A., Robson, T.M., Rose, K.C., Solomon, K.R., Andersen, M.P.S., Sulzberger, B., Wallington, T.J., Wang, Q. W., Wängberg, S.Å., White, C.C., Young, A.R., Zepp, R.G., Zhu, L., 2024. Plastics in the environment in the context of UV radiation, climate change and the montreal protocol: UNEP environmental effects assessment panel, update 2023. Photochem. Photobiol. Sci. 14749092. 23 (4), 629–650. Available from: https://www.springer.com/journal/43630. doi:10.1007/s43630-024-00552-3.

Jewett, E., Arnott, G., Connolly, L., Vasudevan, N., Kevei, E., 2022. Microplastics and their impact on reproduction—can we learn from the C. elegans model? Front. Toxicol. 2673-3080. 4. doi:10.3389/ftox.2022.748912.

Jimenez-Guri, E., Roberts, K.E., García, F.C., Tourmente, M., Longdon, B., Godley, B.J., 2021. Transgenerational effects on development following microplastic exposure in *Drosophila melanogaster*. PeerJ. 21678359. 9. Available from: https://peerj.com/articles/11369/. doi:10.7717/peerj.11369.

Junaid, M., Liu, S., Chen, G., Liao, H., Wang, J., 2023. Transgenerational impacts of micro(nano)plastics in the aquatic and terrestrial environment. J. Hazard. Mater. 03043894. 443 (Pt B), 130274. doi:10.1016/j.jhazmat.2022.130274.

Kadac-Czapska, K., Ośko, J., Knez, E., Grembecka, M., 2024. Microplastics and oxidative stress—current problems and prospects. Antioxidants. 2076-3921. 13 (5), 579. doi:10.3390/antiox13050579.

Kibria, G., 2024. Exposure routes of microplastics (MPs) to humans and possible risks of MPs to human health from food and the environment: a short review. J. Food Saf. Hyg. 2476–3241. doi:10.18502/jfsh.v10i1.16440.

Kim, J., Rhee, J.S., 2021. Biochemical and physiological responses of the water flea *Moina macrocopa* to microplastics: a multigenerational study. Mol. Cell. Toxicol. 20928467. 17 (4), 523–532. Available from: http://www.springerlink.com/content/1738-642X. doi:10.1007/s13273-021-00162-5.

Kumar, S.B., Pachiyappan, J.K., Chandrasekar, S., Karri, V.V.S.R., Krishnamurthy, P.T., Natarajan, J., Palaniami, T., Kuppusamy, G., 2025. Microplastics and health hazards: gastrointestinal risk assessment across a multi-species perspective. Asian J. Atmos. Environ. 22871160. 19 (1). Available from: https://www.springer.com/journal/44273. doi:10.1007/s44273-025-00068-x.

Lee, A., Mondon, J., Merenda, A., Dumée, L.F., Callahan, D.L., 2021. Surface adsorption of metallic species onto microplastics with long-term exposure to the natural marine environment. Sci. Total Environ. 18791026. 780. Available from: https://www.elsevier.com/locate/scitotenv. doi:10.1016/j.scitotenv.2021.146613.

Li, K., Du, L., Qin, C., Bolan, N., Wang, H., Wang, H., 2024. Microplastic pollution as an environmental risk exacerbating the greenhouse effect and climate change: a review. Carbon Res. 27316696. 3 (1). Available from: https://www.springer.com/journal/44246. doi:10.1007/s44246-023-00097-7.

Wanlin, L., Xiao, Y., 2023. Microplastics increase soil microbial network complexity and trigger diversity-driven community assembly. Environ. Pollut. 02697491. 333, 122095. doi:10.1016/j.envpol.2023.122095.

Lovern, S.B., Strickler, J.R., Klaper, R., 2007. Behavioral and physiological changes in *Daphnia magna* when exposed to nanoparticle suspensions (titanium dioxide, nano-C60, and C 60HxC70Hx). Environ. Sci. Technol. 0013936X. 41 (12), 4465–4470. doi:10.1021/es062146p.

Lu, H.C., Kumar, A., Melvin, S.D., Ziajahromi, S., Neale, P.A., Leusch, F.D.L., 2023. Metabolomic responses in freshwater benthic invertebrate, *Chironomus tepperi*, exposed to polyethylene microplastics: a two-generational investigation. J. Hazard. Mater. 18733336. 459. Available from: https://www.elsevier.com/locate/jhazmat. doi:10.1016/j.jhazmat.2023.132097.

Lu, K., Que, Y., Wang, L., Wang, Y., Qiu, J., Jia, Y., Ding, C., Wang, D., Cheng, W., Zhang, Y., 2025. Environmental exposure pathways of microplastics and their toxic effects on ecosystems and the nervous system. Front. Toxicol. 2673-3080. 7. doi:10.3389/ftox.2025.1649282.

Luo, T., Wang, C., Pan, Z., Jin, C., Zhengwei, F., Jin, Y., 2019. Maternal polystyrene microplastic exposure during gestation and lactation altered metabolic homeostasis in the dams and their F1 and F2 offspring. Environ. Sci. Technol. 0013-936X. 53 (18), 10978–10992.,doi:10.1021/acs.est.9b03191.

Mahmud, F., Sarker, D.B., Jocelyn, J.A., Sang, Q.X.A., 2024. Molecular and cellular effects of microplastics and nanoplastics: focus on inflammation and senescence. Cells. 20734409. 13 (21). Available from: https://www.mdpi.com/journal/cells. doi:10.3390/cells13211788.

Majdi, N., Traunspurger, W., 2015. Free-living nematodes in the freshwater food web: a review. J. Nematol. 0022300X. 47 (1), 28–44. Available from http://journals.fcla.edu/jon/article/view/84525/81566.

Martins, A., Guilhermino, L., 2018. Transgenerational effects and recovery of microplastics exposure in model populations of the freshwater cladoceran *Daphnia magna* Straus. Sci. Total Environ. 00489697. 631-632, 421–428. doi:10.1016/j.scitotenv.2018.03.054.

Mueller, M.-T., Fueser, H., Höss, S., Traunspurger, W., 2020. Species-specific effects of long-term microplastic exposure on the population growth of nematodes, with a focus on microplastic ingestion. Ecol. Indic. 1470160X. 118, 106698. doi:10.1016/j.ecolind.2020.106698.

Nguyen, M.K., Lin, C., Nguyen, D.D., 2025. Microplastic pollution associated with probabilistic human health risks: potential hazards, critical factors, challenges, and limitations. Mar. Pollut. Bull. 18793363. 221. Available from: https://www.sciencedirect.com/science/journal/0025326X. doi:10.1016/j.marpolbul.2025.118528.

Niu, L., Chen, Y., Li, Y., Wang, Y., Shen, J., Wang, L., Zhang, W., Zhang, H., Zhao, B., 2023. Diversity, abundance and distribution characteristics of potential polyethylene and polypropylene microplastic degradation bacterial communities in the urban river. Water Res. 00431354. 232, 119704. doi:10.1016/j.watres.2023.119704.

Pan, X., Lin, L., Cao, X., Jing, Z., Dong, L., Zhai, W., 2024. Response of microbial communities and biogeochemical cycling functions to sediment physicochemical properties and microplastic pollution under damming and water diversion projects. Sci. Total Environ. 00489697. 940, 173209. doi:10.1016/j.scitotenv.2024.173209.

Persinger, M., Ward, J., 2025. Evaluation of cross-generational exposure to microplastics and co-occurring contaminants on embryonic and larval behavior in fathead minnows, *Pimephales promelas*. PeerJ. 2167-8359. 13, e19927. doi:10.7717/peerj.19927.

Pitt, J.A., Kozal, J.S., Jayasundara, N., Massarsky, A., Trevisan, R., Geitner, N., Wiesner, M., Levin, E.D., Di Giulio, R. T., 2018. Uptake, tissue distribution, and toxicity of polystyrene nanoparticles in developing zebrafish (*Danio rerio*). Aquat. Toxicol. 18791514. 194, 185–194. Available from: http://www.elsevier.com/wps/find/journaldescription.cws_home/505509/description#description. doi:10.1016/j.aquatox.2017.11.017.

Poma, A.M.G., Morciano, P., Aloisi, M., 2023. Beyond genetics: can micro and nanoplastics induce epigenetic and gene-expression modifications? Front. Epigenet. Epigenomics. 2813-706X. 1. doi:10.3389/freae.2023.1241583.

Rafa, N., Ahmed, B., Zohora, F., Bakya, J., Ahmed, S., Ahmed, S.F., Mofijur, M., Chowdhury, A.A., Almomani, F., 2024. Microplastics as carriers of toxic pollutants: source, transport, and toxicological effects. Environ. Pollut. 18736424. 343. Available from: https://www.sciencedirect.com/science/journal/02697491. doi:10.1016/j.envpol.2023.123190.

Sadler, D.E., Brunner, F.S., Plaistow, S.J., 2019. Temperature and clone-dependent effects of microplastics on immunity and life history in *Daphnia magna*. Environ. Pollut. 18736424. 255. Available from: https://www.journals.elsevier.com/environmental-pollution. doi:10.1016/j.envpol.2019.113178.

Santhanam, S.D., Ramamurthy, K., Priya, P.S., Sudhakaran, G., Guru, A., Arockiaraj, J., 2024. A combinational threat of micro- and nano-plastics (MNPs) as potential emerging vectors for per- and polyfluoroalkyl substances (PFAS) to human health. Environ. Monit. Assess. 15732959. 196 (12). Available from: https://www.springer.com/journal/10661. doi:10.1007/s10661-024-13292-9.

Shi, J., Wang, Z., Peng, Y., Zhang, Z., Fan, Z., Wang, J., Wang, X., 2023. Microbes drive metabolism, community diversity, and interactions in response to microplastic-induced nutrient imbalance. Sci. Total Environ. 00489697. 877, 162885. doi:10.1016/j.scitotenv.2023.162885.

Shim, B.S., Yoon, H.M., An, J.H., Chen, Q., Kim, G.J., Lee, J.S., Park, H.G., Lee, Y.H., 2025. Microplastic fiber-induced transgenerational epigenetic disruption impairs fitness in *Daphnia magna*. Aquat. Toxicol. 18791514. 289. Available from: https://www.sciencedirect.com/science/journal/0166445X. doi:10.1016/j.aquatox.2025.107579.

Soares, G.C., Müller, L., Josende, M.E., Ventura-Lima, J., 2023. Biochemical and physiological effects of multigenerational exposure to spheric polystyrene microplastics in *Caenorhabditis elegans*. Environ. Sci. Pollut. Res. 16147499. 30 (26), 69307–69320. Available from: https://www.springer.com/journal/11356. doi:10.1007/s11356-023-27162-3.

Sobhani, Z., Panneerselvan, L., Fang, C., Naidu, R., Megharaj, M., 2021. Chronic and transgenerational effects of polystyrene microplastics at environmentally relevant concentrations in earthworms (*Eisenia fetida*). Environ. Toxicol. Chem. 0730-7268. 40 (8), 2240–2246. doi:10.1002/etc.5072.

Sobhani, Z., Panneerselvan, L., Fang, C., Naidu, R., Megharaj, M., 2022. Chronic and transgenerational effects of polyethylene microplastics at environmentally relevant concentrations in earthworms. Environ. Technol. Innov. 23521864. 25, 102226. doi:10.1016/j.eti.2021.102226.

Song, J., Kim, C., Joorim, N., Sivri, N., Samanta, P., Jung, J., 2022. Transgenerational effects of polyethylene microplastic fragments containing benzophenone-3 additive in *Daphnia magna*. J. Hazard. Mater. 03043894. 436, 129225. doi:10.1016/j.jhazmat.2022.129225.

Subaramaniyam, U., Allimuthu, R.S., Vappu, S., Ramalingam, D., Balan, R., Paital, B., Panda, N., Rath, P.K., Ramalingam, N., Sahoo, D.K., 2023. Effects of microplastics, pesticides and nano-materials on fish health, oxidative stress and antioxidant defense mechanism. Front. Physiol. 1664042X. 14. Available from: http://www.frontiersin.org/Physiology/archive/. doi:10.3389/fphys.2023.1217666.

Szkudlarek, M., Najbar, B., Jankowiak, Ł., 2024. Variation in microplastic characteristics among amphibian larvae: a comparative study across different species and the influence of human activity. Sci. Rep. 2045-2322. 14 (1). doi:10.1038/s41598-024-61432-5.

Tang, K.H.D., 2025. Genotoxicity of microplastics on living organisms: effects on chromosomes, DNA and gene expression. Environments. 20763298. 12 (1). Available from: https://www.mdpi.com/journal/environments. doi:10.3390/environments12010010.

Tang, K.H.D., Li, R., 2025. The effects of plastisphere on the physicochemical properties of microplastics. Bioprocess Biosyst. Eng. 16157605. 48 (1), 1–15. Available from: https://www.springer.com/journal/449. doi:10.1007/s00449-024-03059-4.

Thacharodi, A., Meenatchi, R., Hassan, S., Hussain, N., Bhat, M.A., Arockiaraj, J., Ngo, H.H., Le, Q.H., Pugazhendhi, A., 2024. Microplastics in the environment: a critical overview on its fate, toxicity, implications, management, and bioremediation strategies. J. Environ. Manag. 10958630. 349. Available from: https://www.sciencedirect.com/journal/journal-of-environmental-management. doi:10.1016/j.jenvman.2023.119433.

Udovicki, B., Andjelkovic, M., Cirkovic-Velickovic, T., Rajkovic, A., 2022. Microplastics in food: scoping review on health effects, occurrence, and human exposure. Int. J. Food Contam. 21962804. 9 (1). Available from: https://foodcontaminationjournal.biomedcentral.com/. doi:10.1186/s40550-022-00093-6.

Ullah, S., Ahmad, S., Guo, X., Ullah, S., Ullah, S., Nabi, G., Wanghe, K., 2023. A review of the endocrine disrupting effects of micro and nano plastic and their associated chemicals in mammals. Front. Endocrinol. 1664-2392. 13. doi:10.3389/fendo.2022.1084236.

Uy, C.A., Johnson, D.W., 2022. Effects of microplastics on the feeding rates of larvae of a coastal fish: direct consumption, trophic transfer, and effects on growth and survival. Mar. Biol. 14321793. 169 (2). Available from: https://www.link.springer.de/link/service/journals/00227/index.htm. doi:10.1007/s00227-021-04010-x.

Vojnits, K., de León, A., Gibon, J., Barker, P., Mahmoudi, M., Pakpour, S., 2025. A systematic review of the potential neurotoxicity of micro-and nanoplastics: the known and unknown. Part. Fibre Toxicol. 17438977. 22 (1). Available from: https://particleandfibretoxicology.biomedcentral.com. doi:10.1186/s12989-025-00647-4.

Vroom, R.J.E., Koelmans, A.A., Besseling, E., Halsband, C., 2017. Aging of microplastics promotes their ingestion by marine zooplankton. Environ. Pollut. 18736424. 231, 987–996. Available from: https://www.elsevier.com/inca/publications/store/4/0/5/8/5/6. doi:10.1016/j.envpol.2017.08.088.

Wade, M.J., Bucci, K., Rochman, C.M., Meek, M.H., 2025. Microplastic exposure is associated with epigenomic effects in the model organism *Pimephales promelas* (fathead minnow). J. Hered. 14657333. 116 (2), 113–125. Available from: http://jhered.oxfordjournals.org/. doi:10.1093/jhered/esae027.

Wang, J., Yuejiao, L., Lin, L., Zheng, M., Zhang, X., Tian, H., Wang, W., Shaoguo, R., 2019. Polystyrene microplastics cause tissue damages, sex-specific reproductive disruption and transgenerational effects in marine medaka (*Oryzias melastigma*). Environ. Pollut. 02697491. 254 (Pt B), 113024. doi:10.1016/j.envpol.2019.113024.

Xia, X., Guo, W., Xiaoyu, M., Liang, N., Duan, X., Zhang, P., Zhang, Y., Chang, Z., Zhang, X., 2023. Reproductive toxicity and cross-generational effect of polyethylene microplastics in *Paramisgurnus dabryanus*. Chemosphere. 00456535. 313, 137440. doi:10.1016/j.chemosphere.2022.137440.

Xiao, S., Chen, H., Gao, X., He, X., Jin, R., Wei, Y., Li, S., Xie, L., Zhang, Y., 2025. Effects of exposure to different types of microplastics on the growth and development of *Rana zhenhaiensis* Tadpoles. Toxics. 2305-6304. 13 (3), 165. doi:10.3390/toxics13030165.

Xu, Y., Ou, Q., van der Hoek, J.P., Liu, G., Lompe, K.M., 2024. Photo-oxidation of micro- and nanoplastics: physical, chemical, and biological effects in environments. Environ. Sci. Technol. 15205851. 58 (2), 991–1009. Available from: http://pubs.acs.org/journal/esthag. doi:10.1021/acs.est.3c07035.

Yang, C., Yang, R., Feng, Y., Wang, Y., Zou, Q., Song, J., Duan, J., Hongyue, L., Gao, X., Chen, M., Zhang, H., 2025. Microplastics affect organic nitrogen in sediment: the response of organic nitrogen mineralization to microbes and benthic animals. J. Hazard. Mater. 03043894. 485, 136926. doi:10.1016/j.jhazmat.2024.136926.

Yang, W., Jannatun, N., Zeng, Y., Liu, T., Zhang, G., Chen, C., Li, Y., 2022. Impacts of microplastics on immunity. Front. Toxicol. 2673-3080. 4. doi:10.3389/ftox.2022.956885.

Yang, Z., Wang, M., Feng, Z., Wang, Z., Lv, M., Chang, J., Chen, L., Wang, C., 2023. Human microplastics exposure and potential health risks to target organs by different routes: a review. Curr. Pollut. Rep. 2198-6592. 9 (3), 468–485. doi:10.1007/s40726-023-00273-8.

Yong, C.Q.Y., Valiyaveettil, S., Tang, B.L., 2020. Toxicity of microplastics and nanoplastics in mammalian systems. Int. J. Environ. Res. Public Health. 16604601. 17 (5). Available from: https://www.mdpi.com/1660-4601/17/5/1509/pdf. doi:10.3390/ijerph17051509.

Zhang, S., Haodi, W., Hou, J., 2023a. Progress on the effects of microplastics on aquatic crustaceans: a review. Int. J. Mol. Sci. 1422-0067. 24 (6), 5523. doi:10.3390/ijms24065523.

Zhang, X., Chen, X., Gao, L., Zhang, H.T., Li, J., Ye, Y., Zhu, Q.L., Zheng, J.L., Yan, X., 2024. Transgenerational effects of microplastics on Nrf2 signaling, GH/IGF, and HPI axis in marine medaka *Oryzias melastigma* under different salinities. Sci. Total Environ. 18791026. 906. Available from: https://www.elsevier.com/locate/scitotenv. doi:10.1016/j.scitotenv.2023.167170.

Zhang, Y., Wang, X., Zhao, Y., Zhao, J., Yu, T., Yao, Y., Zhao, R., Yu, R., Liu, J., Su, J., 2023b. Reproductive toxicity of microplastics in female mice and their offspring from induction of oxidative stress. Environ. Pollut. 02697491. 327, 121482. doi:10.1016/j.envpol.2023.121482.

Zhao, Y., Jia, H., Deng, H., Chengjun, G., Luo, H., Zhang, Y., 2024. Cross-generational exposure to low-density polyethylene microplastics induced hyperactive responses in *Eisenia fetida* offsprings. Environ. Sci. Technol. 0013-936X. 58 (50), 21918–21929. doi:10.1021/acs.est.4c05208.

Zheng, P.C., Li, R., Lai, K.P., Zhang, X.X., 2024. Biological exposure to microplastics and nanoplastics and plastic additives: impairment of glycolipid metabolism and adverse effects on metabolic diseases. Environ. Sci. Pollut. Res. 16147499. 31 (51), 60778–60791. Available from: https://www.springer.com/journal/11356. doi:10.1007/s11356-024-35220-7.

Zurub, R.E., Cariaco, Y., Wade, M.G., Bainbridge, S.A., 2023. Microplastics exposure: implications for human fertility, pregnancy and child health. Front. Endocrinol. 16642392. 14. Available from: https://www.frontiersin.org/journals/endocrinology#. doi:10.3389/fendo.2023.1330396.

CHAPTER

14

Microplastics in animal and human physiology: Toxicity, mechanisms, and mitigation

Giuseppe Piccione and Federica Arrigo
Department of Veterinary Science, University of Messina, Messina, Italy

14.1 Introduction

In recent decades, the problem of plastics has become increasingly important. Plastics are synthetic polymers combined with other materials to acquire additional characteristics, such as ductility and elasticity. Microplastics form when plastic debris undergoes mechanical abrasion and degradation. Microplastics are generally divided into primary and secondary sources, which can help explain how they reach various ecosystems and living beings. In general, primary microplastics are characterized by a size of <5 mm, whereas the breakdown of larger objects generates secondary microplastics. In addition to having a devastating effect on the environment, this has also affected the food chain and, subsequently, animals (from fish to large mammals) and even humans. Moreover, they also carry other pollutants present in the environment because their characteristics enable them to absorb water and xenobiotics (Rios et al., 2007; Teuten et al., 2007).

Contrary to popular belief, invertebrates are not the only organisms affected by plastic. In fact, in 2016 the United Nations revealed that, through various processes, approximately 880 species have been contaminated by plastic, with the most common methods believed to be ingestion and entanglement (Murray and Cowie, 2011). The most significant concerns regarding plastic ingestion include the formation of reactive oxygen species, oxidative stress, and inflammation, with the respiratory system and intestines being the most affected organs. The species most affected are certainly invertebrates, fish, and birds, but plastic can now reach other species through the food web (Haave et al., 2021). Domestic animals, such as dogs and cats, are further affected by, for example, microplastics that have been found in various parts of the body, including the lungs, ileum, liver, and kidneys, and even in feces, where they were subsequently detected (Zhang et al., 2019). Furthermore, owing to continuous exposure—for

Microplastics in Animal and Human Physiology
https://doi.org/10.1016/B978-0-443-45635-0.00001-X

example, through food or contact with environmental particles—to humans, microplastics have been detected and appear to have caused intestinal and hepatic hepatotoxicity (Afreen et al., 2023; Yin et al., 2021).

14.2 Presence and effects of microplastics in different ecosystems

The increasing use of plastic products and the release of these fragments into the environment have become increasingly urgent problems (Amereh et al., 2020). Plastic production reached 335 million tons in 2020, then 367 million tons, and is unfortunately set to rise further. In fact, the COVID-19 pandemic, the medical emergency caused by the new SARS-CoV-2 coronavirus, has worsened this current global plastic emergency (Tiwari et al., 2023; Devereux et al., 2023). The arrival of the pandemic has led to specific requirements, such as the use of protective equipment—gloves and masks—to prevent the spread of the virus. The excessive use of personal hygiene products (hand sanitizer), COVID test kits, and vaccine administration kits has increased plastic use. According to the data collected, 65 billion gloves and 129 billion masks were used each month (Elachola et al., 2020). Plastic waste, both large and small, enters the environment as a result of a series of events involving photooxidation, atmospheric agents, and biological and mechanical degradation (Ammendolia et al., 2021). The accumulation of various plastic fragments is generally due to their stable chemical properties, which interact with other adsorbed pollutants, causing particular toxicity that affects organisms. Considering the degradation methods of microplastics, several methods have been identified, but unfortunately, their distribution is mainly linked to human activities. The ecosystems in which plastic waste and microplastics are most commonly found are soil/land, water, and air, and the methods by which they enter these ecosystems vary.

In soil, plastic packaging is usually disposed of directly, and sources of microplastics can include agricultural practices such as the use of sludge and wastewater irrigation, as well as other human activities such as road runoff and atmospheric transport (Defu et al., 2018). In this case, microbial activity, UV radiation, physical abrasion, technical oxidation, and soil colloids remain in the soil, polluting it. Each plastic fragment can be different due to its chemical and crystalline structure. Its degradation involves the release of various chemicals, including oligomers, stabilizers, and pigments (Zhou et al., 2019). The migration and degradation of microplastics are continuous processes, meaning that they can settle deeper into the soil or spread into the surrounding environment. This can occur through processes such as leaching and soil infiltration.

When microplastics settle in the soil, they can exert various effects, influencing soil characteristics such as mass density and water dynamics. The most significant effects are experienced by soil animals that interact with and are exposed to microplastics. In particular, most research has focused on earthworms, nematodes, and enchytraeid worms. The effects include accumulation in the gut following ingestion and subsequent breakdown into smaller sections. Among the various effects is the disruption of vital relationships between soil and water, which affects their structure and microbial function (Defu et al., 2018). In addition to microplastic accumulation in soil, there is considerable concern about microplastic accumulation in water. Microplastics have been found in oceans and seas and even in drinking water. Data suggest that there are approximately 250,000 tons of microplastics and floating plastics in the ocean. In addition to the accumulation of these substances, their ability to promote bacterial and viral accumulation on their surfaces also poses a danger. Depending on size and type, the biological impact concerns the effects on organisms. Most microplastics, in fact, come into contact with organisms through ingestion. Many marine species, including crustaceans, fish, marine mammals, and turtles, seem to ingest microplastics. Microplastics accumulate in digestive systems, and because organisms lack enzymes capable of breaking down plastic, this accumulation can block the digestive system and even cause death.

14.2.1 Determination in marine and freshwater environments

The main factors contributing to the accumulation of microplastics in marine and freshwater environments are wind-driven currents, plant discharge, and atmospheric deposition. One example is rivers in large cities, such as the Thames in the United Kingdom, where a maximum abundance of 660 units per kg^{-1} of large particles has been identified in sediments downstream of drainage outlets (Hidalgo-Ruz et al., 2012; Stock et al., 2019). This can have a particular impact on isolated freshwater systems, where microplastics tend to accumulate because of their long residence time in the water (Free et al., 2014). Given that these areas differ greatly, microplastic concentrations vary widely. For example, microplastic concentrations increase in densely populated areas and in estuaries receiving inflows from coastal and marine environments. In this context, degradation patterns are fundamental to consider, as the shape, size, and texture of microplastic fragments interact with factors that influence their presence in the environment and their transport (Mai et al., 2018).

The high concentrations of primary microplastics, such as polyethylene, polypropylene, and polystyrene (PS), are due to industrial and human activities. In the case of surface water, for example, in some rivers in China and the Seine River, the dominant particles consisted mainly of fibers linked to numerous activities, such as the production of textiles, clothing, blankets, and furniture, as well as sports equipment such as backpacks, ropes, and fishing nets (Kharraz et al., 2024). Compared with contact with organisms, contact with microplastics occurs through breathing and ingestion. However, microplastics have been detected at various levels, including genes, cells, and tissues. As in other ecosystems, microplastics tend to bind with other xenobiotics, thereby increasing their toxicity. Benthic organisms and wild freshwater mussels carefully filter microplastics through sediment, and the effects are deleterious, such as genotoxicity, neurotoxicity, and impacts on the immune system; this could be a problem if their fate is the food chain (Besseling et al., 2017; Avio et al., 2015) (Fig. 14.1). Considering the digestive gland as the target organ, histological changes were observed in the digestive glands of mussels exposed to polyethylene. *Mytilus edulis* is among the species used as a model organism to assess the effects of various xenobiotics, including microplastics. The most notable effects are negative impacts on feeding and energy balance (Fig. 14.1). The same occurs in oysters, where reproductive parameters and feeding behavior are affected (Phuong et al., 2018).

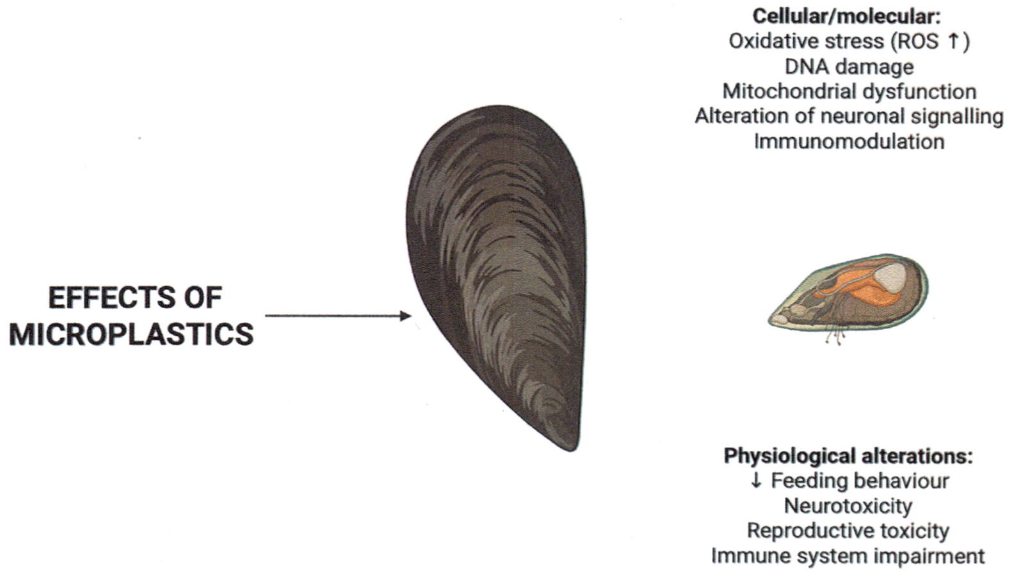

FIG. 14.1 Schematic representation of microplastic-induced effects in an aquatic organism, showing the progression from microplastic exposure to cellular and molecular disturbances, and subsequent physiological impairments at the organism level.

This figure schematically illustrates how exposure to microplastics can lead to a cascade of biological effects in aquatic organisms. After uptake, microplastics interact with biological tissues, inducing molecular and cellular alterations, including oxidative stress, DNA damage, mitochondrial dysfunction, altered neuronal signaling, and immune modulation. These early responses may subsequently result in broader physiological consequences, including impaired feeding behavior, neurotoxicity, reproductive toxicity, and immune system dysfunction.

14.2.2 Bioaccumulation of microplastics in food chains

The excessive use of plastic has undoubtedly led to the presence of microplastics in the food chain and in beverages. Sediment is one of the first places where microplastics bioaccumulate, and samples at other concentrations have also been found in various types of plants, especially spheres, amorphous structures, and fibers.

Organisms come into contact with microplastics through direct or indirect methods. Direct contact occurs when organisms ingest plastic that they mistake for food. The method is indirect; however, it occurs when organisms eat prey that has previously ingested microplastics. Importantly, microplastics often release toxic substances and, upon ingestion, can cause metabolic disorders, oxidative stress, and altered gene expression. Microplastics, including those consumed by zooplankton, have been identified at low trophic levels in marine environments. The most serious accumulation of microplastics is found in crustaceans, seafood, and other organisms intended for human consumption. In this context, several studies have reported MS data for 20 brands of sardines and oysters (Karami et al., 2018).

TABLE 14.1 Concentrations of microplastic particles detected in various food samples according to various studies.

Sample	Concentrations identified	References
Mussels	3.2 ± 0.9 particles/mussels	(Catarino et al., 2018)
Beer	2–79 particlesL^{-1}	(Lachenmeier et al., 2015)
Milk	6500 particlesm^{-3}	(Kutralam-Muniasamy et al., 2020)
Honey	166 ± 147 particlesKg^{-1}	(Liebezeit and Liebezeit, 2013)
Sugar	217 ± 123 particlesKg^{-1}	(Liebezeit and Liebezeit, 2013)
Fruit	98,325–302,250 particlesg^{-1}	(Oliveri Conti et al., 2020)
Vegetables	72,175–130,500 particlesg^{-1}	(Oliveri Conti et al., 2020)

Furthermore, MS is not limited to crustaceans but also extends to other foods and beverages consumed daily, such as honey and sugar. For several years, plastic has been detected in water, with an average of 10.4 particles per liter in bottled water. In addition to bottled water, microplastics have also been found in tea bags, with the most significant fragments being nylon or polyethylene terephthalate fibers (Cverenkárová et al., 2021). Although the types of contact may vary, food can come into contact with plastic through various types of plastic. This occurs, for example, when using plastic bottles or other packaging materials to transport food.

Regardless of the brands used across studies, under mechanical stress, some bottles contain higher concentrations of microplastics than others do (Winkler et al., 2019). In particular, the containers that release the most microplastics into food are those used for takeaway food (Winkler et al., 2019). Microplastics can contaminate contents during the production process and then be ingested (Hernandez et al., 2019). Their detection in human feces has confirmed the presence of microplastics in food chains. Notably, approximately 20 MP particles per 10 g of feces were counted (Table 14.1).

14.2.3 Effects of microplastics on animal tissues

Despite various attempts to reduce the spread of microplastics and their potential effects, high concentrations of microplastics have a significant impact on the tissues of various animal species. These effects depend, in particular, on the distribution, accumulation, and excretion of Microplastics . The risks to invertebrates and fish are certainly greater than those to other species because they are in direct contact with microplastics. Marine and freshwater benthic invertebrates, such as sea urchins and bivalves, exhibit decreases in feeding activity and fertility. For example, the common mussel *M. edulis* feeds on suspended particles, and microplastics (as well as mesoplastics) are the right size to be identified as optimal prey. For these reasons, microplastic particles ranging from 2 to 16 μm have been identified. Although it depends on the size and type of microplastics, as well as their presence in the sediment, it is necessary to pay close attention to this organism. *M. edulis* is considered a model organism and has various functions in ecotoxicology. In this case, it occupies an excellent basal position in the food web and is easily accessible. Most studies (Wright et al., 2013; von Moos et al., 2012) have confirmed that some MP particles enter the digestive tract. In particular, microspheres are likely to become trapped in the intestinal cavity and digestive tubules. The interference of MP fragments, which are often industrial high-density polyethylene particles, causes the formation of granulocytes and a decrease in lysosomal stability after approximately 3–6 h. In fish, the accumulation and associated problems caused by MPs vary depending on MP size and type. Another widely recognized model organism is *Danio rerio*. Experimental studies have shown that PS particles ranging from 25 to 70 nm can penetrate the egg membrane and accumulate in the yolk sac (Wright et al., 2013). The most significant effect is on fish embryos, in which case the particles settle in other tissues, such as the intestine, liver, pancreas, brain, and heart, making them even more dangerous. Like many other species, mammals, particularly some domestic animals, such as dogs and cats, come into contact with microplastics through their diet. Even the smallest MP can move from the lumen into intra- and intercellular spaces and spread throughout the body. One study identified mesoplastics in 35 animals, which were found throughout organs, including the kidneys, lungs, and ileum, and even in blood clots. This confirms how microplastics spread and how they can affect the physiology of even higher organisms that do not come into close contact with them, such as bivalves and fish.

14.3 Identification of microplastics in animal tissue

The information provided in the previous paragraphs helps us understand that microplastics are everywhere due to the global production of plastic material, which has increased significantly over the last six decades. As discussed previously, microplastics can accumulate in ecosystems and tissues, causing various forms of damage. Their identification depends on the research question, the available resources, and, ultimately, the organism. This procedure begins with animal sampling, which must be carried out without contamination. Most invertebrates or fish species, for example, are collected via trawl nets (Phuong et al., 2018). In addition to avoiding contamination, eliminating the possibility of distortion during sampling analyses or mitigation protocols is important. For the identification of microplastics, ensuring proper storage is essential. Importantly, the storage conditions do not affect the chemical composition, structure, or analytical properties of microplastics. The most commonly used methods are 4% formaldehyde, 70% ethanol, drying, and freezing. Importantly, the methods used depend on the type of organism being treated. After a sample has been adequately preserved, fluorescence or various staining methods are commonly used to detect microplastics. This technique is frequently used across various tissues, such as the gills, the intestinal tract, and the liver. One of the most commonly used dyes is "Nile Red (NR)," which is effective in highlighting certain microplastics, such as those made of polyethylene and polycarbonate (Lulu et al., 2021). There are several techniques used to identify and isolate microplastics, the main ones being dissection, purification, and digestion. One of the most commonly used techniques is dissection, especially for specific sections of tissue from large animals, such as the stomach and intestines. After being spread out, the tissue was inspected visually.

Following this procedure, especially if a sample is used multiple times, it can be fixed with formalin or frozen at −80°C. To avoid confusing fragments with food particles or algae, other methods, such as sieving and filtration, must be used. In particular, these are used to divide the material of interest by size or to separate it from a liquid matrix. Finally, the purification method described by Zarfl (2019) was used. Purification, on the contrary, aims to remove microplastics in specific regions, such as the digestive tract or gills. This is done to remove and extract surface microplastics from organisms. This procedure is often carried out using a process that allows potent oxidizing agents to be applied to microplastics (Lulu et al., 2021).

14.3.1 Interaction with biological matrices

The presence and ingestion of microplastics by organisms and ecosystems make this issue a serious concern. Their size, shape, and particular variability, which allow them to come into contact with various matrices, contribute to the complexity of the issue (Wenyi et al., 2022). The danger posed by microplastics binding to matrices is genuinely significant, and the conditions they encounter reduce the possibility of degradation and instead allow them to take root. Several studies have investigated the interactions of microplastics with biological matrices. The most common among those studied are animal tissues and sediments, which have already been discussed, as well as corals, sponges, ascidians, and sea cucumbers. These procedures may typically involve changes in the concentration of the reagent or, for example, the temperature, type of microplastic, or matrix. In general, the primary separation methods involve acid (nitric acid) or alkaline (potassium hydroxide) reagents (Malafaia et al., 2022). The first procedure is chemical digestion, typically performed with HNO_3or KOH. In the case of coral skeletons, for example, studies have shown that $HNO_3$is effective for up to 3 h. The aim was to preserve the chemical and physical integrity of the fibers and fragments. In this context, care must be taken, as the use of particularly acidic reagents can lead to degradation, as seen with polyvinyl chloride. Other digestion methods that are considered milder and more delicate include KOH and NaOH (Prata et al., 2021; Ruggero et al., 2020). After microplastics are separated from the matrix via digestion, the next step is filtration, which removes other substances and impurities and selects only the microplastics suitable for analysis. The filters used are made of materials such as nitrocellulose and have porosities selected to match the sample size. After the microplastics are separated and isolated, various analyses can be performed via optical, electronic, and scanning microscopy. Among the most specific techniques to be used are transmission infrared spectroscopy [Fourier transform infrared (FTIR)] and Raman spectroscopy (Lin et al., 2025).

14.3.2 Levels of detection and sensitivity

Despite the growing recognition of microplastics in the environment, much remains to be done regarding their effects and how to reduce or address them. The greatest problem in this regard is the lack of limits on determination, leading to inaccurate data and a lack of standardization. All the techniques explained above, which are discussed in more detail in the following paragraphs, can be considered valid but must be well organized. In addition to paying close attention to sample contamination, it is important to assess where sampling occurs. For example, the determination and detection limits change depending on whether the samples are in water or sediment. Once the best method for separating the matrix has been determined, the most significant difficulty lies in the laboratory sample analysis phase. The most frequent stages, and in most cases, regardless of the previous steps, are separation by density, sieving (not to be confused with the previous stages), and visual recognition. In separation by density, the densities of the various polymers are considered, particularly those of the raw plastic fragments. The weights of these fragments, ranging from 0.8 to 1.4 $g \times cm^{-3}$, must be considered. Propylene and PS have been reported to have limits of 0.85 to 0.94 $g \times cm^{-3}$ or <0.05 to 1.00 $g \times cm^{-3}$ (Ye and Andrady, 1991). The density of the environment in which the fragments are collected is used to separate microplastics from other particles. This is done through a mixing procedure that causes the heavier material to settle to the bottom and the lighter particles to remain in suspension. The duration of the mixing phase depends on both the material that will settle to the bottom and the supernatant. This can vary from 30 s to 2 h, after which it may be necessary to wait for the displaced particles to stabilize, from 2 min to 6 h (Browne et al., 2010).

The additional filtration procedure primarily separates the supernatant from the sediment (Costa et al., 2021). The particles are passed through filters of different sizes, ranging from 1 to 2 μm, but the most commonly used filters are 1–2 μm. To avoid contamination, one possible solution is to use vacuum-assisted, sealed filters. The separation of microplastic particles can be complex, especially when the number of fragments is small and visual separation is insufficient (Harrison et al., 2012). Techniques such as FTIR spectroscopy can be useful for this purpose. During this procedure, microplastics are separated according to specific criteria, such as the absence of cellular and/or organic structures, and the fibers differ in size. If they are white or transparent, they are examined via a fluorescence microscope.

14.4 Techniques for detecting microplastics in animal tissues

Different types of microplastics have been identified practically everywhere, but they vary in shape and composition depending on where they are found; moreover, their composition and structure may change once they degrade. The techniques used to identify them are therefore different.

14.4.1 Labeling and staining technique

The visual inspection phase, as discussed, requires the use of special dyes, although their application may be optional. Despite the various procedures aimed at removing impurities from MP fragments, some particles (organic and inorganic) may remain. The best type of dye for this purpose is a fluorescent dye. The use of fluorescent dyes allows the identification of only MP particles without highlighting other structures or additional materials. This enables them to be easily identified visually (Ribeiro et al., 2024). The fluorescence method involves the emission of light from a substance that has absorbed light or other radiation. There are different fluorescence methods, depending on whether the staining is performed on filter paper or in solution. In the former procedure, after the samples are collected and separated, the dye is added directly to filter paper, a process that takes 2–4 h. The other method, which appears to be more complex, involves staining the MP in solution. For example, NR is among the most commonly used dyes for fluorescence microscopy of microplastics (Ribeiro et al., 2024). The first step is to prepare the dye solution by dissolving the NR in a solvent. The dye concentration used was approximately 10–20 mg/L. Next, the MPs are dissolved directly in the dye, and these solutions are heated and/or cooled to enhance fluorescence. The use of this dye usually results in red/purple staining (Cook et al., 2020).

The NR dye is recognized as one of the most widely used dyes. It is a lipophilic dye that exhibits fluorescence, which can vary depending on the polarity of the polymers. However, the use of this dye has limitations, for example, regarding incubation time, solvent type, and temperature. Temperature, for example, is considered particularly important, as one of the limitations of NR is undoubtedly leaching. This becomes even more significant in cases such as microplastics, tissues, and organs. In fact, the dye could become available to organisms. To overcome this limitation, the most commonly used approach is thermal expansion. The dye molecules are then heated to 50–70°C or 100°C for different staining times, allowing the structure to become denser.

14.4.2 Polarization microscopy

After staining, additional investigations are often necessary for more accurate identification of microplastics. One technique that does not necessarily rely on spectroscopy is polarized light optical microscopy. This approach can be very useful because many plastics may appear bright and characterized by colored patterns due to birefringence or contrast with other materials (Labbe et al., 2020). According to some studies, polarized microscopy can help identify MPs, for example, in sediments, water samples, or other surfaces (Labbe et al., 2020). Sample preparation for subsequent polarized microscopy involves the use of bleaching solutions and oxidizing agents. This type of microscopy uses crossed polarizers that allow polarized light to pass through. Regarding the polarization type, an 8×8 polarized film can be used on the main base of the microscope, with a second film placed on the upper part. For this reason, the use of polarized light microscopy, together with NR staining and fluorescence, has proven particularly useful for confirming the identity of MPs.

14.4.3 Spectroscopy techniques: Raman and FTIR

As we have said, there are numerous limitations in detecting microplastics, and in recent years, various techniques have been refined to enable effective recognition. This situation has only worsened, as microplastics are increasingly difficult to detect due to their small size. Interestingly, these techniques can be used together, but none of them can provide a complete profile on their own. In fact, spectroscopy techniques are often used in conjunction with microscopy techniques. Raman spectroscopy is emerging as one of the most important tools for recognizing fragments as small as 1 μm, which remains a significant challenge. Raman spectroscopy is a vibrational technique that uses inelastic light to provide molecular details. The molecular details subsequently yield a Raman spectrum featuring prominent vibrations.

The choice of laser can vary from one instrument to another, depending on the type of particles that can be detected. This affects the sample's signal strength and fluorescence. The use of these techniques enables recognition and analysis with greater accuracy than other techniques do. Simple Raman spectroscopy has been used for numerous detections to study a wide range of polymer properties across diverse environments, from marine MPs to animal and human tissues. Micro-Raman spectroscopy, on the contrary, is an even more specific technique than Raman for several reasons, including its ability to investigate tiny fragments down to 0.5–1 μm. This is considered a nondestructive technique and can be used with different tissues and/or ecosystems. In recent years, microplastics have been refined for use in cells and tissues, and several studies have reported their detection in female reproductive organs, the intestine, and other internal organs. FTIR spectroscopy is a different technique that uses infrared light absorption and, through its vibrational properties, can be used to determine structures that are difficult to identify via other methods. This yields an absorption spectrum, allowing the identification of the constituent functional groups. Compared with Raman, FTIR is more useful for applications such as detecting microplastics in biological tissues.

14.5 Limit detection

The limits of intervention and determination are considered widely different because microplastics and plastics in general have now been found everywhere. The first problem is undoubtedly related to sample collection. In fact, regardless of the organism and the organ or tissue from which it originates, the risk of contamination is very high. In addition, microplastic contamination is subject to continuous leaching, and the material can undergo structural changes due to numerous atmospheric agents (Cook et al., 2020). Furthermore, when a section is analyzed via the abovementioned processes, the fragments differ in both size and material type. For this reason, choosing the most suitable extraction method and subsequent determination can be complex. Although sampling methods and their determination have developed considerably, there are still limitations in detecting fragments between 1 and 10 μm, which can sometimes be difficult to detect. All the techniques mentioned differ and are used differently depending on the type of tissue present. Among the most important limitations, it is well known that there is no standardization across the various techniques and that the approach varies depending on the type of study and the material under consideration. For this reason, no single technique is effective on its own; rather, a combination of techniques is necessary.

14.6 Future perspective

The previous paragraph highlights the specific difficulties encountered in investigations into microplastics. Although this problem is now well known and has been discussed for some time, as mentioned, there are still limitations in determining it. This is undoubtedly reflected in the effects of very high levels of microplastics on ecosystems, animals, and humans. It will be necessary to focus, for example, on MPs' effects on tissues and cells, including any effects on DNA. The most significant problem arises with microplastics, which tend to accumulate in ecosystems and organisms through the food chain. The harmful effects they cause remain largely unknown, as it is difficult to track them comprehensively. Therefore in addition to standardizing the ability to identify and determine MPs, it is necessary to reduce the use of materials other than plastic that can mitigate their consumption.

14.7 Multiomics integration

Given the potential limitations and prospects that need to be investigated, a multiomics approach has been proposed in recent years. In particular, one study used transcriptomics to investigate changes in the gene expression profile of fish liver and the effects on the microbiota following exposure to microplastics. This study revealed deleterious effects of MPs on the body, as evidenced by abnormalities in phenylalanine metabolism in carp. Further studies investigating the effects of MPs on different tissues, including intestinal tissue, revealed metabolic and inflammatory disruptions through transcriptomic, proteomic, and metabolomic analyses.

14.8 Conclusions

In conclusion, although microplastic emergence is becoming increasingly important, the data indicate that it is also becoming increasingly dangerous. Unfortunately, in recent decades, the focus has mainly been on the environmental damage caused by the widespread presence of microplastics in all ecosystems, but microplastics have now reached living beings. It is now a global pollution problem that has affected all species, including humans, and has damaged cells and tissues. To improve identification and detection, it is necessary to standardize the processes across the different stages. At present, there is no single technique or process that can reliably identify a sample. Although improving and standardizing techniques is important, government action is still needed to reduce the concentrations of microplastics that spread throughout the environment.

References

Afreen, V., Hashmi, K., Nasir, R., Saleem, A., Khan, M.I., Akhtar, M.F., 2023. Adverse health effects and mechanisms of microplastics on female reproductive system: a descriptive review. Environ. Sci. Pollut. Res. 16147499. 30 (31), 76283–76296. Available from: https://www.springer.com/journal/11356. doi:10.1007/s11356-023-27930-1.

Amereh, F., Babaei, M., Eslami, A., Fazelipour, S., Rafiee, M., 2020. The emerging risk of exposure to nano(micro) plastics on endocrine disturbance and reproductive toxicity: from a hypothetical scenario to a global public health challenge. Environ. Pollut. 02697491. 261, 114158. doi:10.1016/j.envpol.2020.114158.

Ammendolia, J., Saturno, J., Brooks, A.L., Jacobs, S., Jambeck, J.R., 2021. An emerging source of plastic pollution: environmental presence of plastic personal protective equipment (PPE) debris related to COVID-19 in a metropolitan city. Environ. Pollut. 18736424. 269. Available from: https://www.journals.elsevier.com/environmental-pollution. doi:10.1016/j.envpol.2020.116160.

Avio, C.G., Gorbi, S., Milan, M., Benedetti, M., Fattorini, D., D'Errico, G., Pauletto, M., Bargelloni, L., Regoli, F., 2015. Pollutants bioavailability and toxicological risk from microplastics to marine mussels. Environ. Pollut. 18736424. 198, 211–222. Available from: https://www.journals.elsevier.com/environmental-pollution. doi:10.1016/j.envpol.2014.12.021.

Besseling, E., Quik, J.T.K., Sun, M., Koelmans, A.A., 2017. Fate of nano- and microplastic in freshwater systems: a modeling study. Environ. Pollut. 18736424. 220, 540–548. Available from: https://www.journals.elsevier.com/environmental-pollution. doi:10.1016/j.envpol.2016.10.001.

Browne, M.A., Galloway, T.S., Thompson, R.C., 2010. Spatial patterns of plastic debris along estuarine shorelines. Environ. Sci. Technol. 15205851. 44 (9), 3404–3409. doi:10.1021/es903784e.

Catarino, A.I., Macchia, V., Sanderson, W.G., Thompson, R.C., Henry, T.B., 2018. Low levels of microplastics (MP) in wild mussels indicate that MP ingestion by humans is minimal compared to exposure via household fibres fallout during a meal. Environ. Pollut. 18736424. 237, 675–684. Available from: www.elsevier.com/inca/publications/store/4/0/5/8/5/6. doi:10.1016/j.envpol.2018.02.069.

Conti, G.O., Ferrante, M., Banni, M., Favara, C., Nicolosi, I., Cristaldi, A., Fiore, M., Zuccarello, P., 2020. Micro- and nano-plastics in edible fruit and vegetables. The first diet risks assessment for the general population. Environ. Res. 00139351. 187, 109677. doi:10.1016/j.envres.2020.109677.

Cook, S., Chan, H.L., Abolfathi, S., Bending, G.D., Schäfer, H., Pearson, J.M., 2020. Longitudinal dispersion of microplastics in aquatic flows using fluorometric techniques. Water Res. 18792448. 170. Available from: www.elsevier.com/locate/watres. doi:10.1016/j.watres.2019.115337.

Costa, C.Q.V., Cruz, J., Martins, J., Teodósio, M.A.A., Jockusch, S., Ramamurthy, V., Silva, J.P.D., 2021. Fluorescence sensing of microplastics on surfaces. Environ. Chem. Lett. 16103661. 19 (2), 1797–1802. Available from: https://www.springer.com/journal/10311. doi:10.1007/s10311-020-01136-0.

Cverenkárová, K., Valachovičová, M., Mackuľak, T., Žemlička, L., Bírošová, L., 2021. Microplastics in the food chain. Life. 2075-1729. 11 (12), 1349. doi:10.3390/life11121349.

Defu, H., Luo, Y., Shibo, L., Liu, M., Song, Y., Lei, L., 2018. Microplastics in soils: analytical methods, pollution characteristics and ecological risks. TrAC Trends Anal. Chem. 01659936. 109, 163–172. doi:10.1016/j.trac.2018.10.006.

Devereux, R., Ayati, B., Westhead, E.K., Jayaratne, R., Newport, D., 2023. Impact of the Covid-19 pandemic on microplastic abundance along the River Thames. Mar. Pollut. Bull. 0025326X. 189, 114763. doi:10.1016/j.marpolbul.2023.114763.

Elachola, H., Ebrahim, S.H., Gozzer, E., 2020. COVID-19: facemask use prevalence in international airports in Asia, Europe and the Americas, March 2020. Travel Med. Infect. Dis. 18730442. 35. Available from: http://www.elsevier.com/wps/find/journaldescription.cws_home/643125/description#description. doi:10.1016/j.tmaid.2020.101637.

Free, C.M., Jensen, O.P., Mason, S.A., Eriksen, M., Williamson, N.J., Boldgiv, B., 2014. High-levels of microplastic pollution in a large, remote, mountain lake. Mar. Pollut. Bull. 18793363. 85 (1), 156–163. Available from: www.elsevier.com/locate/marpolbul. doi:10.1016/j.marpolbul.2014.06.001.

Haave, M., Gomiero, A., Schönheit, J., Nilsen, H., Olsen, A.B., 2021. Documentation of Microplastics in Tissues of Wild Coastal Animals. Front. Environ. Sci. 2296665X. 9. Available from: journal.frontiersin.org/journal/environmental-science. doi:10.3389/fenvs.2021.575058.

Harrison, J.P., Ojeda, J.J., Romero-González, M.E., 2012. The applicability of reflectance micro-Fourier-transform infrared spectroscopy for the detection of synthetic microplastics in marine sediments. Sci. Total Environ. 18791026. 416, 455–463. doi:10.1016/j.scitotenv.2011.11.078.

Hernandez, L.M., Xu, E.G., Larsson, H.C.E., Tahara, R., Maisuria, V.B., Tufenkji, N., 2019. Plastic Teabags Release Billions of Microparticles and Nanoparticles into Tea. Environ. Sci. Technol. 15205851. 53 (21), 12300–12310. Available from: http://pubs.acs.org/journal/esthag. doi:10.1021/acs.est.9b02540.

Hidalgo-Ruz, V., Gutow, L., Thompson, R.C., Thiel, M., 2012. Microplastics in the marine environment: a review of the methods used for identification and quantification. Environ. Sci. Technol. 15205851. 46 (6), 3060–3075. doi:10.1021/es2031505.

Karami, A., Golieskardi, A., Choo, C.K., Larat, V., Karbalaei, S., Salamatinia, B., 2018. Microplastic and mesoplastic contamination in canned sardines and sprats. Sci. Total Environ. 18791026. 612, 1380–1386. Available from: www.elsevier.com/locate/scitotenv. doi:10.1016/j.scitotenv.2017.09.005.

Kharraz, J.A., Jia, M., Farid, M.U., Khanzada, N.K., Hilal, N., Hasan, S.W., An, A.K., 2024. Determination of microplastic pollution in marine ecosystems and its effective removal using an advanced nanobubble flotation technique. J. Water Process Eng. 22147144. 57. Available from: https://www.sciencedirect.com/science/journal/22147144. doi:10.1016/j.jwpe.2023.104637.

Kutralam-Muniasamy, G., Fermín Pérez-Guevara, I.E.-M., Shruti, V.C., 2020. Branded milks – are they immune from microplastics contamination? Sci. Total Environ. 00489697. 714, 136823. doi:10.1016/j.scitotenv.2020.136823.

Labbe, A.B., Bagshaw, C.R., Uttal, L., 2020. Inexpensive adaptations of basic microscopes for the identification of microplastic contamination using polarization and Nile red fluorescence detection. J. Chem. Educ. 19381328. 97 (11), 4026–4032. Available from: http://pubs.acs.org/loi/jceda8. doi:10.1021/acs.jchemed.0c00518.

Lachenmeier, D.W., Kocareva, J., Noack, D., Kuballa, T., 2015. Microplastic identification in German beer-an artefact of laboratory contamination? Deutsche Lebensmittel-Rundschau 00120413. 111 (10), 437–440. Available from: https://content.behrs-online.de/wissen/document-view/nl_deutsche-lebensmittel-rundschau_2018.

Liebezeit, G., Liebezeit, E., 2013. Non-pollen particulates in honey and sugar. Food Addit. Contam. Part A. 1944-0049. 30 (12), 2136–2140. doi:10.1080/19440049.2013.843025.

Lin, J.T., Chung, Y.P., Lee, Y.Y., Wu, T.L., Huynh, T.T.T., Nguyen, P.T., Lu, M.J., Huang, B.W., Sriram, B., Wang, S.F., Chang-Chien, G.P., Kogularasu, S., Lin, W.C., 2025. Emerging analytical frontiers in microplastic detection: from spectroscopy to smart sensor technologies. Talanta Open. 26668319. 12. Available from: https://www.sciencedirect.com/science/journal/26668319. doi:10.1016/j.talo.2025.100514.

Lulu, L., Yan, X., Feng, L., Jiang, S., Zifan, L., Xie, H., Sun, S., Chen, J., Chengyong, L., 2021. Challenge for the detection of microplastics in the environment. Water Environ. Res. 1061-4303. 93 (1), 5–15. doi:10.1002/wer.1281.

Mai, L., Bao, L.J., Shi, L., Wong, C.S., Zeng, E.Y., 2018. A review of methods for measuring microplastics in aquatic environments. Environ. Sci. Pollut. Res. 16147499. 25 (12), 11319–11332. Available from: http://www.springerlink.com/content/0944-1344. doi:10.1007/s11356-018-1692-0.

Malafaia, G., da Luz, T.M., Araújo, A.P.D.C., Ahmed, M.A.I., Rocha-Santos, T., Barceló, D., 2022. Novel methodology for identification and quantification of microplastics in biological samples. Environ. Pollut. 18736424. 292. Available from: https://www.journals.elsevier.com/environmental-pollution. doi:10.1016/j.envpol.2021.118466.

Murray, F., Cowie, P.R., 2011. Plastic contamination in the decapod crustacean Nephrops norvegicus (Linnaeus, 1758). Mar. Pollut. Bull. 18793363. 62 (6), 1207–1217. doi:10.1016/j.marpolbul.2011.03.032.

Phuong, N.N., Zalouk-Vergnoux, A., Kamari, A., Mouneyrac, C., Amiard, F., Poirier, L., Lagarde, F., 2018. Quantification and characterization of microplastics in blue mussels (Mytilus edulis): protocol setup and preliminary data on the contamination of the French Atlantic coast. Environ. Sci. Pollut. Res. 16147499. 25 (7), 6135–6144. Available from: http://www.springerlink.com/content/0944-1344. doi:10.1007/s11356-017-8862-3.

Prata, J.C., Sequeira, I.F., Monteiro, S.S., Silva, A.L.P., da Costada Costa, J.P., Dias-Pereira, P., Fernandes, A.J.S., da Costada Costa, F.M., Duarte, A.C., Rocha-Santos, T., 2021. Preparation of biological samples for microplastic identification by Nile Red. Sci. Total Environ. 18791026. 783. Available from: www.elsevier.com/locate/scitotenv. doi:10.1016/j.scitotenv.2021.147065.

Ribeiro, F., Duarte, A.C., da Costa, J.P., 2024. Staining methodologies for microplastics screening. TrAC Trends Anal. Chem. 18793142. 172. Available from: https://www.sciencedirect.com/science/journal/01659936. doi:10.1016/j.trac.2024.117555.

Rios, L.M., Moore, C., Jones, P.R., 2007. Persistent organic pollutants carried by synthetic polymers in the ocean environment. Mar. Pollut. Bull. 0025326X. 54 (8), 1230–1237. doi:10.1016/j.marpolbul.2007.03.022.

Ruggero, F., Gori, R., Lubello, C., 2020. Methodologies for microplastics recovery and identification in heterogeneous solid matrices: a review. J. Polym. Environ. 1566-2543. 28 (3), 739–748. doi:10.1007/s10924-019-01644-3.

Stock, F., Kochleus, C., Bänsch-Baltruschat, B., Brennholt, N., Reifferscheid, G., 2019. Sampling techniques and preparation methods for microplastic analyses in the aquatic environment – a review. TrAC Trends Anal. Chem. 01659936. 113, 84–92. doi:10.1016/j.trac.2019.01.014.

Teuten, E.L., Rowland, S.J., Galloway, T.S., Thompson, R.C., 2007. Potential for plastics to transport hydrophobic contaminants. Environ. Sci. Technol. 0013936X. 41 (22), 7759–7764. doi:10.1021/es071737s.

Tiwari, B.R., Lecka, J., Pulicharla, R., Brar, S.K., 2023. Microplastic pollution and associated health hazards: impact of COVID-19 pandemic. Curr. Opin. Environ. Sci. Health. 24685844. 34. Available from: www.journals.elsevier.com/current-opinion-in-environmental-science-and-health. doi:10.1016/j.coesh.2023.100480.

von Moos, N., Burkhardt-Holm, P., Köhler, A., 2012. Uptake and effects of microplastics on cells and tissue of the blue mussel Mytilus Edulis l. after an experimental exposure. Environ. Sci. Technol. 0013-936X. 46 (20), 11327–11335. doi:10.1021/es302332w.

Wenyi, Y., Chen, J., Zhang, S., Zhao, Y., Fang, M., Deng, Y., Zhang, Y., 2022. Extraction of biodegradable microplastics from tissues of aquatic organisms. Sci. Total Environ. 00489697. 838 (Pt 4), 156396. doi:10.1016/j.scitotenv.2022.156396.

Winkler, A., Santo, N., Ortenzi, M.A., Bolzoni, E., Bacchetta, R., Tremolada, P., 2019. Does mechanical stress cause microplastic release from plastic water bottles? Water Res. 00431354. 166, 115082. doi:10.1016/j.watres.2019.115082.

Wright, S.L., Thompson, R.C., Galloway, T.S., 2013. The physical impacts of microplastics on marine organisms: a review. Environ. Pollut. 02697491. 178, 483–492. doi:10.1016/j.envpol.2013.02.031.

Ye, S., Andrady, A.L., 1991. Fouling of floating plastic debris under Biscayne Bay exposure conditions. Mar. Pollut. Bull. 0025326X. 22 (12), 608–613. doi:10.1016/0025-326X(91)90249-R.

Yin, K., Wang, Y., Zhao, H., Wang, D., Guo, M., Mengyao, M., Liu, Y., Nie, X., Baoying, L., Jingyan, L., Xing, M., 2021. A comparative review of microplastics and nanoplastics: toxicity hazards on digestive, reproductive and nervous system. Sci. Total Environ. 00489697. 774, 145758. doi:10.1016/j.scitotenv.2021.145758.

Zarfl, C., 2019. Promising techniques and open challenges for microplastic identification and quantification in environmental matrices. Anal. Bioanal. Chem. 1618-2642. 411 (17), 3743–3756. doi:10.1007/s00216-019-01763-9.

Zhang, J., Wang, L., Kannan, K., 2019. Polyethylene terephthalate and polycarbonate microplastics in pet food and feces from the united states. Environ. Sci. Technol. 0013-936X. 53 (20), 12035–12042. doi:10.1021/acs.est.9b03912.

Zhou, Y., Liu, X., Wang, J., 2019. Characterization of microplastics and the association of heavy metals with microplastics in suburban soil of central China. Sci. Total Environ. 00489697. 694, 133798. doi:10.1016/j.scitotenv.2019.133798.

CHAPTER

15

Mitigation strategies for microplastic-induced physiological impacts in animals: Approaches and future directions

Toha Tasnem, Salvina Chowdhury Neha, and Rabiul Islam

School of Pharmacy, BRAC University, Dhaka, Bangladesh

15.1 Introduction

Microplastics (MPs) are plastic particles measuring <5 mm. The long-chain molecular structure and high molecular weight of MPs are some of the distinctive physicochemical traits that make them resilient to biodegradation. They are hydrophobic in nature (Bhattacharjee et al., 2025). MPs, on the basis of their origin, are broadly classified into primary and secondary MPs. MPs that are intentionally produced and infused into consumer products are termed primary MPs. Conversely, secondary MPs originate from larger plastics through degradation (Lackner and Branka, 2024). Various abiotic factors, including photooxidation, mechanical degradation, weathering, and biological degradation, contribute to plastic degradation (Mao et al., 2022). In addition, MPs are classified into six types on the basis of their chemical composition: polyvinyl chloride (PVC), polystyrene (PS), polyethylene terephthalate (PET), polyethylene (PE), and polyurethane (Osman et al., 2023). Currently, MPs are widely dispersed throughout the environment, and traces have been detected in oceans, freshwater, the atmosphere, soil, and even human placentas (Liu et al., 2022). Since the 1950s, a total of 8000 million plastic items have been manufactured, of which only 9% are recycled, 12% are incinerated, and the remaining 79% accumulate in the environment. The net plastic production in 2017 was 348 million tons, and it is estimated to reach 33 billion tons by 2050 (First Sentier MUFG. Sustainable Investment Institute, n.d.). This indicates a tremendous increase in the generation and deposition of MPs in the environment over time. One study suggested that MPs will account for 13.2% of the global burden of plastic pollution by 2060 (Sharma et al., 2021).

Microplastics in Animal and Human Physiology
https://doi.org/10.1016/B978-0-443-45635-0.00002-1

The most significant pathway for the influx of primary MPs into the ecosystem is road runoff, accounting for 66%, followed by wastewater treatment plants (WWTPs), which contribute 25% of the total MPs. In contrast, eolian transport and marine activities contribute smaller portions, including 7% and 2%, respectively (First Sentier MUFG. Sustainable Investment Institute, n.d.). Both terrestrial and ocean-based MPs eventually enter the marine environment, causing substantial ecological harm (Arat, 2024). Strikingly, approximately 80%–90% of MPs in aquatic environments arise from land-based activities, whereas the remaining 10%–20% are attributed to ocean-based sources (Osman et al., 2023). Severe climatic events, including storms, hurricanes, high winds, heavy rains, and tsunamis, sweep land-based plastics into marine habitats. Furthermore, airborne MPs disperse over long distances and circulate at the marine boundary, allowing bidirectional exchange between the atmosphere and the ocean (Arat, 2024).

A variety of sources account for the prevalence of primary MPs. First, synthetic textiles are made by combining synthetic fibers (nylon, polyester, acetate, and polypropylene) and natural fibers (cotton, wool, etc.). Therefore primary MPs are washed off during washing and end up in wastewater (Lackner and Branka, 2024), accounting for more than one-third of the MPs that infiltrate the marine environment. Vehicle tire abrasion, driven by factors such as tire composition, speed, braking, and road texture, is the second-largest source of primary MPs, accounting for 28% of the global MPs entering the marine environment. The erosion and surface wear of road markings involve polymer binders that eventually lead to 7% of the yearly entry of primary MPs into the environment (First Sentier MUFG. Sustainable Investment Institute, n.d.).

Furthermore, various types of marine coatings contain substances such as epoxy coatings, polyurethane, vinyl, and lacquers, as well as additional substances such as metals, which are applied to the exterior of industrial and personal boats, such as yachts and jet skis, to protect them from UV rays, surface degradation, and biofilm accumulation. However, degradation, abrasion, sanding, and uncontrolled dumping of such substances are also considered sources of primary MPs, accounting for 3.7% of the total MPs in the marine environment. In addition, personal care products, such as exfoliants and leave-on cosmetics, contain microbeads and various types of solid, insoluble plastics, typically PE and polyurethane. This addition improved the formulation by increasing the viscosity, thickness, binding properties, color, opacity, dispersion, and stability of the products. Nutrient pills, which contain nutrients encased in a thin polymer coating, are widely used in agricultural fields for controlled, sustained nutrient release to increase yield. The overuse of chemical fertilizers and plastic-film coatings in agricultural fields can infuse MPs into the soil environment, which eventually leach into the marine environment. In addition to these sources, MPs are also introduced into the environment through city dust, such as the weathering of dry paint containing microfibers from building walls and the wearing of turf composed of polymeric infill (First Sentier MUFG. Sustainable Investment Institute, n.d.). Finally, studies suggest that advanced membranes, such as the reverse osmosis (RO) membranes used in wastewater purification, allow MPs to pass through, thereby failing to meet their intended purpose (Arat, 2024).

On the contrary, secondary MPs refer to the type of materials formed from the degradation of larger plastics into smaller ones due to exposure to UV rays, wind, etc. (Lackner and Branka, 2024). These materials include fishing gear, plastic bottles, plastic bags, plastic food containers, cutlery, and other plastic packaging materials (Osman et al., 2023). The sources of MPs entering the ecosystem are illustrated in Fig. 15.1.

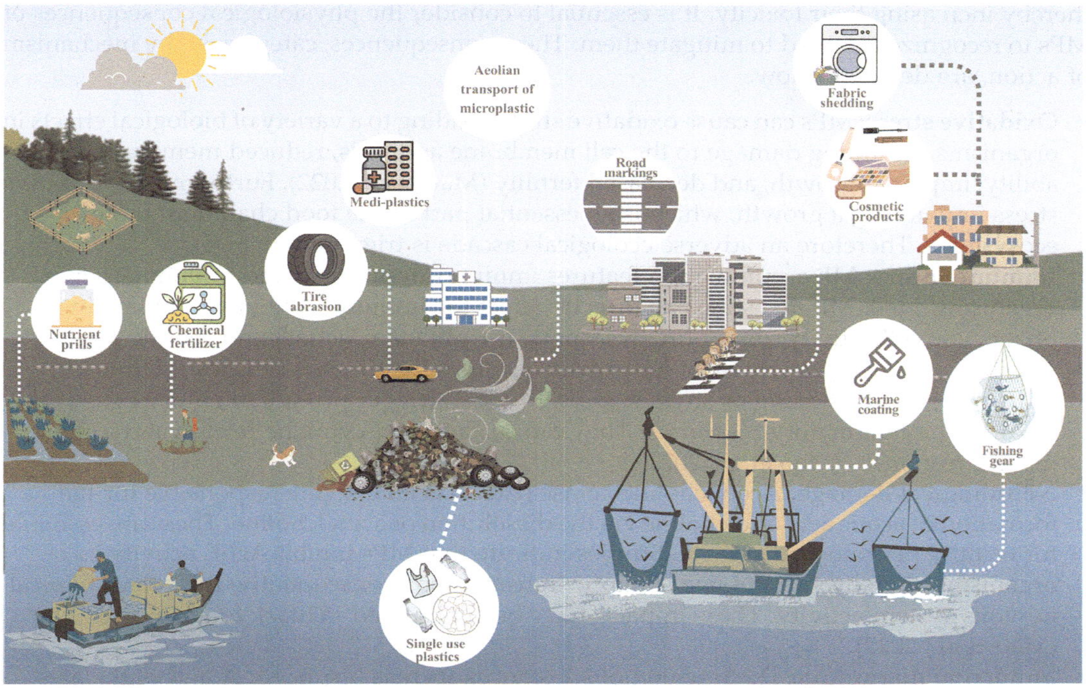

FIG. 15.1 Primary and secondary sources of microplastics entering the ecosystem via land, water, and air. Land sources include nutrient prills, chemical fertilizers, tire abrasion, media plastics, road markings, cosmetic products, and fabric shedding. Water sources include marine coatings and fishing gear. Single-use plastic is dumped into water bodies, where it can infiltrate both land and water. Air sources of MPs include the eolian transport of microplastics.

Several factors make plastics a threat to the environment and human health. The physiological impacts of MPs on animals arise from two main effects:

15.1.1 Self-toxic effects

These effects include reproductive issues, endocrine disruption, inflammation, intestinal blockage, lacerations, growth disruption, impaired reproduction, and significant structural changes in the intestinal flora resulting from the direct intake of MPs (Liu et al., 2022). Moreover, significant adverse environmental effects are observed when additives, such as plasticizers, antioxidants, biocides, flame retardants, pigments, and slip agents, are incorporated into plastic production and leach off under unfavorable conditions.

15.1.2 Load toxicity effects

Because MP particles have relatively large surface areas due to their small particle size, they exhibit very high adsorption capacities. This enables MPs to absorb various harmful particles from the environment (Liu et al., 2022). These toxins are taken up by organisms, thereby increasing their toxicity. It is essential to consider the physiological consequences of MPs to recognize the need to mitigate them. These consequences, categorized by mechanism of action, are detailed below.

- **Oxidative stress:** MPs can cause oxidative stress, leading to a variety of biological effects in organisms, including damage to the cell membrane and cells, reduced membrane permeability, impaired growth, and decreased fertility (Mao et al., 2022). Furthermore, oxidative stress impairs algal growth, which is an essential part of the food chain in marine ecosystems. Therefore an adverse ecological cascade is triggered.
- **Immune injury:** MP accumulation features immunosuppression and inflammation in organisms. First, MP ingestion can directly affect the immune system by minimizing phagocytic cell activity, disrupting lysosomal membrane function, and reducing cell viability, allowing the release of hydrolytic enzymes that initiate self-destruction. Moreover, by functioning as stress signals, MPs can trigger proinflammatory cytokines, leading to inflammatory responses. They can further alter cytokine release and inflammatory responses (Mao et al., 2022).
- **Neurological damage:** Acetylcholinesterase (AchE) is an enzyme responsible for the formation of acetic acid and choline via the dissolution of acetylcholine. These are essential for healthy neurological activity. Studies indicate that MPs inhibit AchE activity in organisms, thereby suppressing synaptic transmission. Consequently, this affects typical nervous system activity. These impacts have been observed in black crucian carp (Mao et al., 2022).
- **Endocrine disruption:** The leaching of substances such as bisphenol A and phthalates from MPs has been shown to induce endocrine disruption in organisms, including altered metamorphosis and impaired thyroid activity in amphibians, hampering normal reproduction and development. In fish, phthalate effects on signaling pathways that target nuclear hormone receptors may initiate endocrine disruption (Mao et al., 2022).

In addition, MPs have been proven to be associated with DNA damage, which can influence cancer development. This is due to their increased cytotoxicity toward tissues and cells, which might cause DNA damage (Osman et al., 2023). On the contrary, the impacts of plasticizers on reproduction, the induction of genetic aberrations in invertebrates, and impaired development have also been demonstrated (First Sentier MUFG. Sustainable Investment Institute, n.d.). Moreover, some studies highlight the link between MP exposure and changes in the gut microbiota, increasing the risk of hepatic disease and its potential to impair fetal growth by crossing the placental barrier (Lackner and Branka, 2024). Fig. 15.2 schematically represents the physiological impacts-

This chapter aims to explain mitigation strategies for MPs, which can have severe physiological impacts on animals and vegetation. Specifically, it describes the primary adverse outcomes, including inflammation, impaired reproduction, oxidative stress, reproductive impairments, and potentially fatal consequences, thereby providing a comprehensive overview of the biological risks associated with MPs. Furthermore, this chapter aims to mitigate the impact of MPs by encompassing biological interventions, technological innovations, advanced waste management practices, and policy-driven frameworks by articulating future research priorities, including the development of standardized methodologies, the

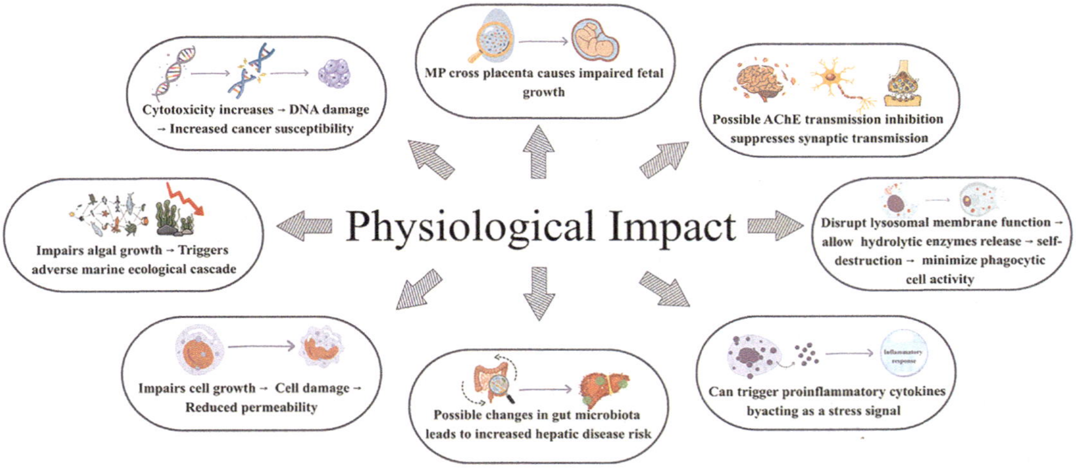

FIG. 15.2 Schematic representation of the physiological impacts of microplastics on animals. Microplastic causes impaired fetal growth, the suppression of synaptic transmission, minimized phagocytic cell activity, the triggering of proinflammatory cytokines, increased hepatic disease risk, cell damage, and reduced permeability, triggering a marine ecological cascade due to impaired algal growth by the MPs and increasing cancer susceptibility.

identification of early-detection biomarkers, and the promotion of interdisciplinary collaboration. The chapter establishes a forward-looking agenda intended to guide scientific inquiry and policy development over the coming decade.

15.2 Mitigation strategies for MPs

The effective removal of MPs has become essential because of persistent ecological and health threats across ecosystems. Extensive in-depth analyses are being carried out to address this problem. Various studies indicate that wetlands are among the largest environmental sinks for MPs. Over the years, a set of mitigation strategies has been developed, including biological, conventional, and innovative technological advancements (Osman et al., 2023). A metaanalysis on the alleviation of MPs revealed that removal efficiencies ranging from 74% to 100% can be achieved via multiple biological, physical, and chemical approaches (Liu et al., 2022).

15.2.1 Developing biological advancements

Biological advancements are emerging as indispensable strategies for mitigating MP pollution, particularly in soil (Hettiarachchi and Meegoda, 2023). Researchers have explored biomimetic solutions to attenuate MPs in soil. Among these methods, bioremediation and phytoremediation stand out for degrading MPs incorporated into the soil. These processes involve the natural fauna present in the soil and plants in degrading MPs (Sajjad et al., 2022).

Therefore these approaches offer a long-term, sustainable, and eco-friendly pathway by integrating natural systems into waste management and mitigation strategies (Hechmi et al., 2024).

15.2.1.1 *Microbe treatment to degrade MPs*

15.2.1.1.1 Bioremediation

Bioremediation is a type of biodegradation in which MPs are broken down by soil fauna through hydrolysis (Hettiarachchi and Meegoda, 2023). The soil contains microorganisms, such as fungi, bacteria, earthworms, and snails, that ingest MPs and secrete enzymes, including hydrolases, laccases, peroxidases, oxidoreductases, and esterases (Hechmi et al., 2024). These enzymes degrade long polymer chains into monomers via hydrolysis, thereby mineralizing the monomers into carbon dioxide and water (Hettiarachchi and Meegoda, 2023). The biodegradation of plastics is influenced by numerous factors that are not yet fully understood. However, the structure of MPs and the living conditions of degrading microbes, which are necessary for their proliferation and metabolic activity, can affect the efficacy of plastic biodegradation. Together, oxygen, a favorable pH, and temperature are required for optimal polymer degradation. The amount of sunlight and nutrients present is directly linked to microbial growth, which, in turn, influences the degradation process. In addition, abiotic degradation of the MP surface can also be triggered by sunlight exposure (Ebrahimbabaie et al., 2022).

Among microbes, fungi can degrade MPs more effectively because they release more enzymes than bacteria do. For example, studies have shown that fungi can effectively decompose lignin. As PE has a similar chemical structure to lignin, its breakdown is associated with fungal ligninolytic activity, with manganese peroxidase as the primary enzyme. Although other microbes can degrade only up to 15% of MPs (Ebrahimbabaie et al., 2022), bacteria, particularly *Pseudomonas* spp., and mealworms, particularly *Tenebrio molitor* Linnaeus, have exhibited remarkable efficiency in reducing the MP mass in soil. *T. molitor* Linnaeus competently degrades 40.01% of PE within 24 h. Nevertheless, the efficacy rate remains insufficient to implement this technique on an industrial scale (Hechmi et al., 2024).

To overcome this limitation, several combinations have been tested to achieve better results. The combination of bacteria and algae synergistically accelerated MP decay, achieving up to 96% efficacy (Shuo et al., 2023). Second, to increase the biodegradation of MPs, genetically modified microbes, which are also affordable, are introduced. However, for large-scale use, meticulous testing is needed to identify potential environmental impacts (Bhattacharjee et al., 2025). In addition, the culture includes carbon sources such as glucose and malt. Therefore abundant microbial growth is evident, leading to an increased number of microbes capable of breaking down MPs (Shuo et al., 2023). Finally, incorporating bioreactors creates an optimal environment for microorganism growth, thereby increasing biodegradation (Hechmi et al., 2024).

15.2.1.1.2 Phytoremediation

Phytoremediation is another type of plastic degradation and management process, particularly with the help of vegetation and its symbiotic microorganisms. This process can be further classified into various techniques for managing MPs in soil. Phytoextraction, also known as phytoaccumulation, is a technique that enables MPs to accumulate in soil or plants. Plant roots take up MPs and translocate them to above-ground tissues. For example, studies have shown the presence of PS microbeads in the root cap of lettuce plants.

On the contrary, phytostabilization focuses on immobilizing MPs within the soil system. The rhizospheres of plants, which contain microbes, facilitate the precipitation of MPs in the surrounding area, while some MPs are adsorbed onto the plant's roots. This reduces mobility and lowers the risk of their transfer to the food chain and water bodies. However, this mechanism of action has been observed in hydrophytes, but no prominent information is available for terrestrial plants. Finally, phytofiltration or rhizofiltration follows a technique similar to phytoaccumulation. This method has shown positive results in eliminating MPs from the soil. It involves the absorption and adsorption of MPs by plant roots or seedlings, which are subsequently extracted from the soil, thereby removing MPs along with them. Coupled with this process, rhizoremediation leverages rhizosphere microbes that amplify plant growth and help detoxify the environment. Rhizosphere microbes, including Cyanobacteria, Proteobacteria, Actinobacteria, and fungi such as Ascomycota and Basidiomycota, release root mucilage, which is rich in organic acids and amino acids, to coat and trap MPs, thereby preventing their uptake by plants. In addition, MPs are utilized as a carbon source to increase their biodegradation. The effectiveness of this approach is strongly influenced by environmental factors such as pH, soil moisture, the molecular weight of the plastic, and other inherent polymer characteristics, which determine microbial survival, hydrolytic activity, and the overall degradation rate of MPs (Hechmi et al., 2024). Table 15.1 summarizes the different techniques, their mechanisms of action, factors affecting efficacy, and limitations.

15.2.2 Developing technological advancements

Technological advancements involve removing MPs from the water matrix via physical and chemical methods, which are further classified into two major categories: conventional and innovative treatment techniques. Conventional techniques include coagulation, adsorption, rapid filter separation, and membrane bioreactor (MBR) methods. Innovative treatment techniques include electrocoagulation, magnetic separation, and advanced oxidative processes. The efficacy, general mechanisms, advantages over other strategies, and drawbacks of each method are summarized.

15.2.2.1 Conventional treatment technique

15.2.2.1.1 Coagulation

Coagulation is a widely used procedure that employs various synthetic or natural coagulants to destabilize and clump particles in suspension or solution, which can then be removed via sedimentation (Osman et al., 2023). Synthetic coagulants, e.g., polyaluminum chloride (PAC) and polyacrylamide (PAM), are used because of their high efficiency and ease of use. However, their use has several disadvantages, which pose severe environmental threats. Some drawbacks include increased sludge volume, leading to residual toxicity; additional operational costs; biostability issues; and potential health risks associated with elevated aquatic aluminum levels (Osman et al., 2023; Ramakrishnan and Sathiyamoorthy, 2025). In addition,

TABLE 15.1 Mechanisms, factors affecting performance, and limitations of various advanced biological methods for mitigating microplastics in the ecosystem.

Method	Basic mechanism	Factors affecting efficacy	Limitations
Bioremediation	Microorganisms in the soil that ingest MPs and secrete enzymes that degrade the long polymer chains of plastics into monomers via hydrolysis, thereby mineralizing the monomers into carbon dioxide and water through digestion.	MPs structure, oxygen, a favorable pH, temperature, sunlight exposure, and nutrient amount for the microorganisms to proliferate and perform metabolic activity.	• The efficacy rate is only (15%–40%), which is insufficient to implement for industrial-scale application.
Phytoremediation	It includes three mechanisms: Phytoaccumulation enables MP to be incarcerated, where plant roots uptake MP and translocate it to the above-ground tissue. Phytostabilization immobilizes the MP within the soil, and rhizoremediation facilitates the precipitation of MP to the roots or seedlings, which are later isolated. Coupled with this, the rhizosphere microbes amplify microbial growth for increased release of root mucilage to coat and trap the MP, preventing it from being taken up by the plants.	Environmental factors like pH, soil moisture, molecular weight of the plastic, and other innate characteristics of the polymer, which determine the survival of microbes, hydrolytic activity, and the overall degradation rate of MPs.	• Predominantly relies on biological processes, and climate conditions make it a slower method.

the use of coagulation additives to increase removal efficiency might not be cost-effective for the process (Osman et al., 2023). Consequently, switching to natural coagulants can help overcome the limitations of synthetic coagulants (Ramakrishnan and Sathiyamoorthy, 2025). In recent years, there has been a significant increase in the use of plant-based natural coagulants. Natural coagulants can be derived from plant-based and microbial sources. Plant-based coagulants, such as *Moringa oleifera* seeds, which contain cationic proteins as the active ingredient, offer two key advantages: they are biodegradable and reduce the sludge volume. With high coagulation efficiency (85%–95% turbidity removal) and 70%–90% pathogen reduction, *M. oleifera* seeds have been employed as coagulants across different water matrices (Agarwal et al., 2024). Moreover, plant-derived cactus mucilage, which contains polysaccharides, serves as a coagulant by interparticle bridging and adsorption. The advantage of this cactus mucilage is its affordability and local availability. In addition, the use of animal-derived coagulants, such as keratin (derived from poultry feathers), achieves coagulation through particle entrapment and charge neutralization, thereby assisting in waste valorization (Badawi et al., 2025). Furthermore, in complex natural water sources, commonly used coagulants often have reduced removal efficiency. This occurs because of the presence of dissolved organic compounds and inconsistent ionic compositions. It has been reported that natural coagulants, specifically *Strychnos potatorum*, retain their removal efficiency even in such circumstances. All of these findings indicate the potential of these natural coagulants as eco-friendly substitutes for synthetic coagulants in MP mitigation. Furthermore, these natural coagulants are biodegradable, locally available, and less toxic, which adds to their advantages for mitigating MP pollution, especially in areas with limited resources and limited infrastructure (Ramakrishnan and Sathiyamoorthy, 2025). Even with the widespread application of coagulation techniques to mitigate MP pollution, their limitations in treating multiple pollutants simultaneously persist. On the contrary, given the varying compositions of wastewater, different coagulants are needed, and rigorous process parameter optimization is required to address different contaminants. These factors, taken together, can increase the process's costs, making it economically unfeasible (Osman et al., 2023).

15.2.2.1.2 Adsorption

Adsorption is a straightforward process, making it a widely recognized technique for MP treatment. It requires much less time, and its capacity to recycle adsorbents makes it stand out among the other developed techniques (Liu and Khor, 2024). Its mechanism relies on the adsorption of small particles, contaminants, atoms, ions, or molecules through surface adherence of the adsorbent via physical interactions such as van der Waals forces, electrostatic forces, hydrogen bonding, or other chemical interactions. Specifically, for wastewater plastic treatment, this technique is generally analyzed by emphasizing equilibrium behavior and kinetics, often described by pseudo-first- and pseudo-second-order models (Chabi et al., 2024). Pore material and pore size play significant roles in adsorption mechanisms (Xing et al., 2023). Examples of adsorbents include biochar, coffee grounds, double hydroxide layers, chitin, graphene oxide, and magnetic particles (Liu and Khor, 2024). Among them, activated charcoal, or biochar, is commonly used as an adsorbent because of its microporous structure and large surface area. It optimally isolates PS nanoparticles from drinking water, primarily via electrostatic interactions. Studies have been conducted to enhance the efficacy of activated charcoal. The use of activators such as zinc chloride ($ZnCl_2$) with activated charcoal enhances the overall performance of the adsorbent. Zinc chloride promotes structural perforation, enhancing the surface area and facilitating pollutant adsorption (Xing et al., 2023). Even though it shows up to 100% efficacy in removing MPs via double-hydroxide layers or activated charcoal, the nonselective behavior of the adsorbents reduces their potency. They are unable to distinguish between necessary particles, such as natural organic matter, metal ions, and MPs. Therefore to eliminate MPs via the adsorption technique on a large, durable scale, further research is needed (Osman et al., 2023; Tiwari et al., 2020). Owing to its limitations, the adsorption method is combined with other techniques, such as regenerated silk fibroin, to achieve better results.

15.2.2.1.3 Rapid sand filtration

Rapid sand filtration has conventionally been used to remove contaminants such as viruses and dispersed clay particles (Osman et al., 2023). It primarily targets MPs in tap water or drinking water, which contain polypropylene (PP), PE, polyamide (PA), polyester, and PVC. With features such as reduced operational space, stability based on water quality, and rapid flow, RSF is a well-known and promising method for removing MPs in wastewater treatment (Osman et al., 2023). Studies have shown that RSF performance is markedly affected by MP particle size and filter media characteristics, including the appropriate silica particle size (ES), flow rate, and backwashing rate. Studies have revealed that up to 91.3% of MPs < 400 μm can be removed with an ES value of 0.4 mm via double-media RSF. For MPs > 400 μm, the efficiency is 95.8% (Wulandari et al., 2024). Further experimental results have shown that RSF is an effective conventional treatment technique for removing MPs <10 μm in size, with an efficiency of 84%–98% (Chabi et al., 2024).

The RSF is designed to physically eliminate contaminants by using filter media for short periods. The removal mechanisms include transportation, mechanical strain, and collision (Wulandari et al., 2024), which involve interception, trapping, entanglement, and adsorption via physical and chemical interactions. During interception, MPs become trapped between grains of sand, preventing them from moving further down through the filter. Trapping occurs when the filter material has tiny pores, and the water moves slowly, allowing particles to settle within them. Moreover, entanglement occurs when deeper gaps in the sand create obstacles that prevent the MPs from being carried away by the water flow. All these processes occur under mechanical strain. Finally, owing to collisions (Wulandari et al., 2024), adhesion occurs when smaller MPs adhere to the surface of silicate sand via van der Waals forces. This mechanism is known as adsorption (Wulandari et al., 2024).

Furthermore, synergistic coagulation–flocculation with activated charcoal (biochar) significantly enhances the MP retention efficiency. Coagulants such as PAC and PAM at pH 8 neutralize the hydrophobicity of smaller microbeads, bringing them together; flocculation then causes them to aggregate into larger clumps. Hence, it is easier to restrain the filter, increasing the efficacy above 97%. On the contrary, activated charcoal has a large surface area and a porous structure, which enhances its ability to adsorb MPs. Biochar is a carbon-rich material produced by the pyrolysis of biomass. It is widely used in agriculture to improve soil quality and reduce extraction costs. Corn straw biochar manufactured at 500°C (C500) layered with RSF exhibited 100% efficacy in removing 10-μm MPs from wastewater, which was attributed to its honeycomb-like porous structure (Garfansa et al., 2025). It follows three mechanisms: stuck, trapped, and entangled. Compared with activated charcoal, the low manufacturing cost and high efficiency of corn straw biochar make it a robust technique for removing microbeads.

Despite its significant performance, some limitations persist. MPs saturate the filter media, thereby reducing the efficacy of further retention. However, this can be alleviated by backwashing filter media, which desorb up to 97% of the trapped MPs. Therefore the filtration capacity can be restored. Nonetheless, the effectiveness of this technique is constrained by the absence of expensive, harmful synthetic coagulants and flocculating agents, thereby increasing overall operational costs (Osman et al., 2023). For this reason, detailed kinetic modeling, long-term operational studies, and comparisons across polymer types remain necessary to increase scalability and real-world applicability.

15.2.2.1.4 Membrane bioreactor technology

Another effective way to separate MPs from wastewater is MBR technology. Recently, the application of MBR technology in MP mitigation has increased (Osman et al., 2023). MBR uses biological catalysts, such as bacteria and enzymes, for biological cleaning and is integrated with a membrane-based separation process, typically microfiltration or ultrafiltration. The main objective of this system in MP-containing wastewater treatment is to reduce solution complexity by degrading organic matter, thereby facilitating MP removal. MBR technology has been used in WWTPs, achieving 99% removal of MPs. This MBR was used in an actual treatment plant in Finland, which comprised a submerged membrane unit along with an ultrafiltration process. Compared with conventional activated sludge processes, substantial improvements in MP removal rates were observed. This approach also results in cleaner, higher-quality effluent and the potential to simplify the treatment process by effectively reducing the number of steps. In related research, the MBR demonstrated an MP removal rate of 99.69%, indicating that it is more efficient than the sand filtration system employed in a WWTP (Michielssen et al., 2016; Nguyen et al., 2025). MBR can achieve a removal efficiency of up to 99.9% from primary wastewater while removing MPs of a wide range of sizes and shapes, exceeding those of other advanced and established treatment methods (Nguyen et al., 2025; Talvitie et al., 2017a). Certain controlling factors in MBR technology include microorganisms, flow rates, membrane characteristics, and pollutant loads (Osman et al., 2023).

A two-step process can describe the mechanism of MP removal by an MBR. First, the process starts with the entry of a pretreated stream into the bioreactor. The organic matter present in the stream undergoes biodegradation, reducing its complexity. The resulting liquor is driven across a filtration system for the separation of MPs. The retentate stream contains the separated MPs. The results from a recent study comparing the efficiencies of MBR, RO, rapid sand filter, disc filter with various pore sizes (10 and 20 μm), and dissolved air flotation indicate that the final effluent from the MBR contained the lowest MP content. Moreover, incorporating MBR with other existing techniques might further increase its MP removal efficiency. One such possibility has been demonstrated by removing di(2-ethylhexyl) phthalate via an MBR in WWTPs, resulting in an 83% removal rate, with preliminary absorption, which was previously 70% when used independently. The complete removal of MPs was achieved by incorporating an MBR via a preliminary anaerobic process, followed by RO filtration. Notably, *Ideonella sakaiensis* is a bacterium that, as its primary energy and carbon source, can utilize PET. In other words, it degrades PET into ethylene glycol and terephthalic acid by producing enzymes. The resulting monomers are less hazardous and metabolizable. Consequently, incorporating MBR technology with *I. sakaiensis* will enable concurrent physical removal and biodegradation of MPs, representing a significant breakthrough (Talvitie et al., 2017a).

Finally, even highly efficient techniques such as MBR technology have their own drawbacks. Some noteworthy limitations that result in elevated operational costs include membrane fouling, frequent filter membrane replacement, and a requirement for a consistent, nutritious supply for microorganism growth (Osman et al., 2023). A previous study indicated that MPs ranging in size from 0.5 to 0.8 and 40 to 48 μm, at concentrations of 0.1 and 0.2 g L^{-1}, significantly reduce the effectiveness of membrane filters by 9%–54% (Michielssen et al., 2016) (Table 15.2).

TABLE 15.2 Comparative evaluation of different conventional microplastic contamination treatment approaches.

Method	Basic mechanism	Factors affecting efficacy	Limitations
Coagulation	Various synthetic or natural coagulants are used to destabilize and clump up the particles in a suspended or dissolved state. It is then removed via sedimentation.	Dissolved organic compounds, inconsistent ionic composition in wastewater.	• Treating numerous pollutants at once. • Increased operational costs due to rigorous process parameter optimization and the use of different coagulants based on varying compositions of wastewater.
Adsorption	Small particles, contaminants, atoms, ions, or molecules are adsorbed through surface adherence to the adsorbent by physical interactions like Van der Waals forces, electrostatic forces, hydrogen bonding, or other chemical interactions.	Pore material and pore size play a significant role in adsorbing mechanisms.	• Low potency due to the nonselective behavior of the adsorbents.
Rapid sand filtration	Transportation, mechanical straining, and collision are incorporated that involve the processes of interception, trapping, entanglement, and adsorption through physical and chemical interactions.	MP particle size, the characteristics of filter media such as the effective size of silica (ES), flow rate, and rate of backwashing.	• The synthetic coagulants and flocculating agents are not only harmful to the environment but also expensive.
Membrane bioreactor technology	Biological catalysts, such as bacteria and enzymes, are used for biological cleaning, integrated with a membrane-based separation process, usually incorporating microfiltration or ultrafiltration.	Microorganisms, flow rate, membrane characteristics, and pollutant load.	• Elevated operational costs link to membrane fouling, frequent filter membrane replacement, and the consistent supply of nutrients for microorganism growth. • Smaller MPs, ranging in size from 0.5 to 0.8 μm and 40 to 48 μm, at concentrations of 0.1 and 0.2 g L–1, significantly reduce the effectiveness of membrane filters by 9%–54%.

15.2.2.2 *Innovative treatment technique*

15.2.2.2.1 Electrocoagulation

The electrocoagulation technique uses an electrically induced process based on the principle of coagulation–flocculation (Senathirajah et al., 2023). In various studies on MP pollution, MP removal has been described as an economical and resource-efficient approach (Shuo et al., 2023). Electrocoagulation processes do not require the external coagulants used in traditional coagulation techniques, which makes them time- and cost-efficient and environmentally friendly (Shuo et al., 2023; Osman et al., 2023). This technique uses an electric current to release cations from metallic electrodes, which then generate flocs (Osman et al., 2023).

Various studies have shown that the MP removal efficiency of electrocoagulation exceeds 90% for municipal wastewater, laundry wastewater, and industrial freshwater. However, for landfill leachate, an MP removal efficiency of 75% was observed (Senathirajah et al., 2023).

The working principle of electrocoagulation involves an electrochemical reaction that neutralizes charges, followed by the coagulation and flocculation of the MPs. An electrocoagulation reactor includes an electrolytic cell and electrodes. Here, an electric current is passed through the electrolytic cell, which causes the oxidation of metal compounds at the anode, the dissolution of the anode, and the release of highly charged metal atoms (Senathirajah et al., 2023). Under these circumstances, as the anode oxidizes and dissolves rapidly, it needs to be replaced regularly (Osman et al., 2023). For both the anode and the cathode, inexpensive metals such as aluminum and iron are used (Senathirajah et al., 2023). Concurrently, hydroxide ions (OH^-) and hydrogen gas are produced by the cathodic ionization of water molecules. The positively charged metal ions from the anode combine with the negatively charged hydroxide ions from the cathode to form metal hydroxides, which are colloidal flocs that eventually act as coagulants. That is, positively charged hydrolysis products can attract and bind negatively charged MPs, destabilizing them and promoting aggregation. During this interaction, the positive metal ions are converted to noncrystalline or amorphous metal hydroxides. Simultaneously, the oxygen gas generated by oxidation at the anode and the hydrogen bubbles from reduction at the cathode attach to the flocs and help the less dense aggregates rise to the water surface, while the denser aggregates sink, forming sludge. A previous study revealed that electrocoagulation treatment induces visible morphological and topographical changes in polymers, such as cracking, flaking, and pits, which eventually facilitate further degradation of the MPs (Senathirajah et al., 2023).

Electrocoagulation has multiple advantages, including its adaptability in treating wastewater of versatile qualities, reduced sludge production, and treated water with fewer total dissolved solids. All these qualities position these materials as eco-friendly, feasible alternatives to conventional coagulation techniques (Osman et al., 2023). Despite its outstanding performance in removing MPs from wastewater, several variables, such as water quality, can affect its efficacy. Specifically, pH, temperature, electrolyte ionic strength, organic material concentration, and wastewater turbidity can influence the performance of this technique (Senathirajah et al., 2023).

From another perspective, incorporating biofilm technology with electrocoagulation can facilitate the mitigation of MPs. Combining these two techniques allows simultaneous aggregation and removal of MPs via electrocoagulation, with subsequent degradation via biofilm technology. Thus biofilms can be used as biocatalysts. Because biofilms contain microbial communities, electrocoagulation can help simulate an environment that enhances microbial activity within them. This leads to increased production of plastic-degrading enzymes, ultimately facilitating the degradation of MPs. Moreover, biofilms can facilitate electrocoagulation by reducing fouling and sludge formation through the biodegradation of organic substances (Bhattacharjee et al., 2025). Further advancements in the electrocoagulation technique may involve optimizing the reaction time, electrolyte concentration, electrode type, and other parameters to mitigate MPs more effectively (Bhattacharjee et al., 2025).

15.2.2.2.2 Magnetic separation

The removal of MPs via magnetic separation has been employed in recent years for freshwater and marine water samples (Osman et al., 2023). The principle of magnetic separation relies on the hydrophobic traits of MPs. It operates by binding MPs to magnetic materials as they are magnetized. The movements of magnetized MPs can be regulated via a magnetic field. PVC, PP, PET, PS, PA, and PE are magnetizable MPs that can be removed via magnetic separation. Magnetic separation uses magnetic seeds to agglomerate and remove MPs. Variation in efficacy and mechanism is observed across different magnetic materials. Commonly employed magnetic seeds include carbon nanotubes (CNTs), ferric nanoparticles (Fe NPs), microrobots, and micromotors. Fe NPs are cost-effective magnetic nanoparticles (MNPs) that, upon hydrophobization, exhibit high MP removal efficiency during magnetic separation. They have been shown to remove 96% of PE beads (<20 μm) and 88% of PS beads (< 20 μm) from artificial seawater, demonstrating their high efficiency (Grbic et al., 2019). On the contrary, superhydrophobic Fe NPs can achieve up to 100% removal of MPs, such as high-density polyethylene, at MP concentrations ranging from 0.2 to 1.0 MP/mL. Moreover, an increase in concentration reduces the efficiency, possibly due to a limited surface area. In addition, a 92.6% separation efficiency was achieved by employing magnetic coagulants comprising Fe_3O_4 and $Mg(OH)_2$ (Zhang et al., 2021). Magnetic CNTs are another highly efficient material for MP removal, along with a noteworthy advantage: they can be recycled by heat treatment (600°C) while retaining a separation efficiency of up to 80%, even after being used four times.

The magnetization of MPs via NPs and the directional transport of magnetized MPs are the two stages involved in the mechanism of MP removal in magnetic separation. In the first step, the MNPs or microrobots attach to the MPs and magnetize them. MNPs possess a high surface area and hydrophobic properties. MNPs bind to MP particles via electrostatic attraction because of these properties. Following this, the direction of the MNPs can be regulated by external magnetic fields, enabling them to approach the MP particles and eventually collect them, separating them from the water matrix. In addition, MPs can be decomposed via magnetic fields and light irradiation and then extracted and transported via catalytic and functionalized magnetic nanostructures. To overcome the limitations of weakened adsorption caused by excess pollutants in water, various surface-altering approaches have been developed. Technological advancements have enabled the entrapment and decomposition of MPs in water via magnetic microrobots. To facilitate further development, polymer-specific enzymes can be incorporated into MNPs to increase degradation, which is more sustainable and less toxic than degradation via photocatalysis or photo-Fenton-like processes. Even after significant advancements in magnetic separation for MP mitigation, its limited efficiency and selectivity in the presence of other pollutants, along with compromised separation due to the diverse sizes and shapes of MPs, persist (Osman et al., 2023).

15.2.2.2.3 Advanced oxidation processes

Advanced oxidation processes (AOPs) represent a sustainable and efficient green strategy for eliminating MPs, achieving 50%–99% removal, and alleviating environmental and health risks. They act through a two-step mechanism to degrade MPs in an aqueous system. The initial step involves the generation of highly reactive oxidants. In contrast, the second step involves persistent interactions with MPs, initiating oxidative reactions that break their polymer chains into smaller, more benign fragments, thereby reducing their bioavailability in water and mitigating their ecological and health impacts. The oxidizing agents commonly used are hydrogen peroxide (H_2O_2), ozone (O_3), persulfate ($S_2O_8^{2-}$), and peroxymonosulfate (HSO_5-), which form reactive oxygen species (ROS), including sulfate radicals (-SO_4^-), superoxide radicals (-O_2^-), singlet oxygen ($^1O_2$), and hydroxyl radicals (-OH), which play a central role in this process. In recent years, with this primary mechanism, various methods have been established to eliminate MPs. These include photocatalytic oxidation, electrochemical oxidation, Fenton oxidation, sonochemical oxidation, ozonation, and sulfate radical-based AOPs. Techniques such as photocatalysis, Fenton reactions, and ozonation effectively utilize these radicals to break down complex polymer structures (Kumar et al., 2024).

The photodegradation technique has shown exceptional proficiency in treating wastewater MPs. It consists of a semiconductor that absorbs UV light or visible light. When the light energy surpasses the bandgap of the semiconductor, electrons are excited from the valence band to the conduction band. This phenomenon creates positively charged holes, which subsequently participate in redox reactions to initiate the formation of free radicals, including ROS such as $^1O_2$, -OH, and -O_2^-. These free radicals, in turn, actively participate in degrading the long polymer chains of MPs into smaller fragments. In recent investigations, green-synthesized iron–zinc oxide nanocomposites have gained significant recognition as robust semiconductors for the successful degradation of PE MP. In addition, pH, temperature, photocatalyst, light intensity, and MP concentration are the controlling factors that affect the reaction. Photocatalytic degradation offers a primary benefit in harnessing UV or visible light as a clean energy source for optimal reactions with minimal ecological footprints and the generation of benign by-products (Kumar et al., 2024). However, one major limitation of this technique is its high energy consumption due to the continuous use of UV light (Osman et al., 2023).

Electrochemical oxidation is another sustainable and economical method for treating wastewater. It is a two-step process involving direct and indirect electrochemical oxidation (Osman et al., 2023). Indirect electrochemical oxidation initiates oxidation at the electrode by applying an electrical current without the use of any chemicals (Kumar et al., 2024; Osman et al., 2023). This process involves the formation of ROS (SO_4^{2-},-OH^-, -CO_3^{2-}, and O_2^- and reactive chlorine species) via water electrolysis at the surface of the anode. Simultaneously, direct electrochemical oxidation, also referred to as anodic oxidation, utilizes -OH to facilitate direct electron transfer at the anode surface and to oxidize organic molecules (Falco et al., 2024). This enables the continuous oxidation of MPs into smaller, harmless fragments, such as carbon dioxide and water vapor (Osman et al., 2023). Anodes can produce abundant -OH and additional secondary oxidizing agents. They are classified as active or inactive based on their ability to interact with hydroxyl radicals. Titanium/platinum, graphite, and dextran sulfate sodium salt are typical examples of active anodes. These have a high affinity for -OH groups and partially degrade contaminants.

On the contrary, a nonactive anode has a low affinity for -OH groups, thereby facilitating the complete degradation and mineralization of MPs. Examples include boron-doped diamond anodes (BDDs), lead oxides (PbO_2), and tin (IV) oxides (SnO_2) (Falco et al., 2024). Kiendrebeogo and others demonstrated an outstanding efficacy of 86.8% and an approximately 90% reduction in operational costs by eliminating nanoplastics via ROS on the BDD anode in 0.67 h (Marthe Kiendrebeogo et al., 2022). In addition, although degradation at 1 μm is very difficult, studies have shown that 65% of MPs can be degraded in 5 h when Na_2SO_4 (0.02 M) is used as the supporting electrolyte. In this case, the polymer chain breakdown was mainly due to the formation of reactive oxidizing radicals (Falco et al., 2024). Therefore the reaction time, surface area, choice of anode material, current intensity, and electrolyte concentration significantly impact removal (Osman et al., 2023). In contrast, this method results in high energy consumption and electrode corrosion in complex wastewater, requiring frequent replacement, thereby increasing operational costs and, in turn, limiting the effectiveness of the method. Second, although the mechanism does not generate any toxic by-products, depending on the composition of the water matrix, harmful substances, such as chlorinated compounds, may form. There is no such mechanism to control these toxic by-products (Kumar et al., 2024).

Fenton oxidation provides another practical method for water treatment. It is primarily used as a pretreatment to isolate the MPs from the aqueous matrix; it also contributes to the partial degradation of the MPs. Research has shown that the efficacy of this method increases as the MP size decreases because smaller MPs have greater surface areas and undergo surface oxidation by hydroxyl radicals more readily. The pH of the system decreases significantly from 9 to 4 after oxidation, creating an acidic environment. Various polar functional groups, including ketone carbonyl, hydroxyl, carboxyl, alcohol, and ester groups, are formed during the breakdown of MP polymer chains. This enhances hydrophilicity, allowing interaction with the aqueous medium and increasing its susceptibility to hydroxyl radical degradation. The structural characteristics of polyethylene terephthalate (PET) and expanded polystyrene (EPS) MPs are more likely to demonstrate this phenomenon (Ortiz et al., 2022). The Fenton oxidation method also involves two simultaneous degradation routes, specifically for PVC MPs. The initial route includes synchronous cathodic reduction dechlorination at the TiO_2/C cathode. PVC is reduced by electron transfer at the cathode, and this process is enhanced by applying heat, ultimately promoting dechlorination. The following route involves oxidative attack by hydroxyl radicals on reduced PVC, forming organic intermediates, particularly exfoliated hydrocarbons, which further break down into alcohols, esters, and carboxylic acids and ultimately mineralize to water and carbon dioxide. However, this technique causes additional environmental side effects. First, operating under acidic conditions (pH of approximately 3) raises ecological concerns when this method is applied on a large scale. Second, the presence of iron ions from the reactions can act as a secondary pollutant, introducing a new contaminant in drinking water. Furthermore, polymers of MPs with structural complexity are less susceptible to oxidative degradation. In addition, the hydroxyl radicals produced in this process are unable to degrade MPs completely, leading to the formation of toxic intermediates as by-products (Kumar et al., 2024).

Other advanced oxidation techniques, including sonochemical oxidation, ozonation, and sulfate radical-based advanced oxidation, have also been developed to eradicate MPs. Sonochemical oxidation reduces the polymer chain via high-frequency ultrasonic waves at 20–1 MHz. Ultrasonic waves are introduced into a solvent system containing MPs to induce cavitation-induced implosion via controlled pressure cycling. Microbubble bursting can lead to elevated temperatures of up to 5000 K and pressures over 1000 atm, resulting in the formation of numerous highly reactive radicals that, in turn, cause oxidative degradation. On the contrary, exposure of MPs to ozone clearly increases the surface tension, hydrophilicity, and solubility of PP, PET, and PE, thereby enhancing their adsorption characteristics. Studies have shown the addition of hydrogen peroxide (H_2O_2).

Nevertheless, the heterogeneity and complex structure of MPs hinder their complete degradation, leading to the formation of secondary pollutants in drinking water and limiting their efficiency. Sulfate-based oxidation involves the generation of ROS—primarily sulfate and hydroxyl radicals—using catalysts to degrade the polymer chain in an acidic environment. However, this process cannot fully mineralize MP fragments and can form toxic by-products. Although three of these mechanisms have shown promising results in eradicating MPs and mineralizing them into carbon dioxide and water, their high operational costs, extreme energy consumption, and formation of toxic by-products can pose significant environmental and physiological hazards, hindering their large-scale practical use. Moreover, ongoing research aims to better understand the factors affecting their efficacy and to introduce new elements to enhance the synergistic degradation of MPs (Kumar et al., 2024) (Table 15.3).

TABLE 15.3 Innovative techniques for treating microplastic contamination and its basic mechanisms, considering factors and limitations.

Method	Basic mechanism	Considering factors	Limitations
Electrocoagulation	It utilizes electric current to form cations released from metallic electrodes, which generate flocs. The working principle involves an electrochemical reaction, charge neutralization, followed by coagulation and flocculation of MPs.	pH, temperature, ionic strength of electrolytes, organic material concentration, as well as turbidity of the wastewater.	➧ Requires frequent replacement of electrodes due to rapid decay, increasing operational costs.
Magnetic separation	The principle of magnetic separation relies on the hydrophobic traits of MPs. MPs bind to magnetic materials following their magnetization. Subsequently, by employing a magnetic field, the movements of the magnetized MPs are regulated to isolate them from the medium.	Hydrophobicity and surface area of the magnetic material employed the strength of the magnetic field.	➧ Reduced efficiency and selectivity in the presence of other pollutants. ➧ Different sizes and shapes of MPs affect separation.
Photocatalytic degradation	It consists of a semiconductor that absorbs UV light or visible light. When the light energy surpasses the bandgap of the semiconductor, electrons are excited from the valence band to the conduction band, creating positively charged holes, which	pH, temperature, photocatalyst, light intensity, and the concentration of the MPs.	➧ It requires high energy due to the continuous use of UV light.

Continued

TABLE 15.3 Innovative techniques for treating microplastic contamination and its basic mechanisms, considering factors and limitations—cont'd

Method	Basic mechanism	Considering factors	Limitations
	subsequently participate in redox reactions to initiate the formation of free radicals, including ROS. These free radicals, in turn, actively participate in degrading the long polymer chains of MPs into smaller fragments.		
Electrochemical oxidation	It is a two-step process. Indirect Electrochemical Oxidation initiates the oxidation reaction at the electrode by applying electricity, without the use of any chemicals. It involves the formation of ROS via water electrolysis at the surface of the anode. Simultaneously, Direct Electrochemical Oxidation utilizes -OH to facilitate direct electron transfer within the surface of the anode and organic molecules. This enables the continuous breakdown of MPs through oxidation into smaller, harmless fragments like carbon dioxide and water vapor.	Reaction time, surface area, choice of anode materials, intensity of the current, and the concentration of electrolytes.	• Erosion of electrodes in complex wastewater requires regular replacement of anodes, leading to an elevation in operational costs. • High energy consumption. • Harmful substances, such as toxic chlorinated compounds, are formed. There is no mechanism to control these toxic by-products.
Fenton Oxidation	Poses two simultaneous degradation routes, specifically for PVC MPs. The initial route includes synchronous cathodic reduction dechlorination at the $TiO2/C$ cathode. PVC is reduced by electron transfer at the cathode, and this process is enhanced by applying heat, ultimately promoting dechlorination. The following route includes oxidative attack by hydroxyl radicals on the reduced PVC, forming organic intermediates, particularly exfoliated hydrocarbons, which further break down into alcohols, esters, and carboxylic acids, and ultimately mineralize into water and carbon dioxide.	MP size, surface area, and pH.	• A pH of around 3 is too acidic and harmful to the environment. • Iron ions formed from the reaction act as secondary pollutants in drinking water. • MPs with complex polymer structures are less susceptible to oxidative degradation. • Incomplete degradation of MPs by the hydroxyl group forms toxic intermediates as by-products.

Method	Basic mechanism	Considering factors	Limitations
Sonochemical oxidation	Sonochemical oxidation reduces the polymer chain using high-frequency ultrasonic waves of frequencies ranging from 20 kHz to 1 MHz. The ultrasonic waves are introduced into a solvent system containing MPs to form cavitation-induced implosions through controlled pressure cycling. The microbubble bursting leads to heightened temperatures of up to 5000 K and pressures exceeding 1000 atm, resulting in the formation of numerous highly reactive radicals that, in turn, cause oxidative degradation.	Frequency of ultrasonic waves, temperature, and MP concentration in the solvent.	• Requires a specific frequency and intensity of ultrasonic waves. • Hydrophobicity of MPs. • Produces toxic by-products and secondary pollutants during the reactions. • High consumption of energy, therefore less cost-effective in large-scale production.
Ozonation	MPs exposed to ozone elevate the surface tension, hydrophilicity, and solubility of PP, PET, and PE, thereby enhancing their adsorption characteristics.	Ongoing research is being conducted to understand the factors, but studies have shown that combining ozone with hydrogen peroxide improves its efficacy.	• MP with complex polymer structures and heterogeneity lowers its efficacy. • Produces toxic by-products and secondary pollutants during the reactions. • High consumption of energy, therefore less cost-effective in large-scale production.
Sulfate radical-based oxidation	During the formation of ROS using catalysts, primarily sulfate radicals and hydroxyl radicals degrade the polymer chain into nontoxic by-products.	Number of radicals, mainly sulfate and hydroxyl radicals, catalysts, pH, and temperature.	• Limited availability of required catalysts, therefore hampering large-scale production. • Over time, the efficiency of the process is reduced after reusing the catalysts several times. • Acidic environment with a pH of 3 is needed for the reaction, which is harmful to the environment and living beings. • Incomplete mineralization leads to the formation of toxic by-products, which act as secondary pollutants.

15.3 Management of MP waste

The alarming growth of landfills threatens waste abatement measures and, as a result, destroys the environment (Nafea et al., 2024). Hence, along with remediation, prevention measures must also be implemented. While various technological techniques eradicate existing MPs in the ecosystem, preventing their entry through comprehensive management and multidisciplinary tactics simultaneously is vital. As a significant number of MPs are generated from secondary MPs, overall plastic waste also helps reduce MP fouling (Nafea et al., 2024). Consequently, the integrated and interdisciplinary framework encompasses 4 R's: ecolabeling, imposing fines and bans, promoting awareness campaigns, incorporating policies and legislation, and leveraging bioplastics. This section further discusses these management strategies in detail.

15.3.1 Improving the sustainability of waste disposal operations

15.3.1.1 4-R hierarchy: Principle of sustainable waste management

An efficient and impactful method to support the management of MPs is the incorporation of the 4-R principle—reduce, reuse, recycle, and recovery—along with upcycling strategies (Nafea et al., 2024).

Reducing: This involves reducing total MP aggregation by scaling down the use of plastic bags and microbeads, which are extensively used daily across the globe. In return, green bags made of natural materials, such as jute or cloth, can be utilized. Other sustainable substitutes, including the use of bioplastics, curbing packaging waste, and imposing bans on specific uses of MPs, such as single-use plastics, also help reduce their accumulation in landfills and aquatic bodies (Nafea et al., 2024).

Reuse: Reuse directly contributes to reducing the overall plastic footprint. Substituting single-use plastics with reusable or refillable alternatives will not only eliminate plastic pollution but also be cost-effective and help build business and country economies. It has been shown that USD 10 billion could be achieved by switching only 20% of single-use plastics to refillable or reusable alternatives. In addition, a significant amount of plastic is generated by single-use plastic utilities; approximately 5 million tons of plastic are used in packaging in the UK. Therefore "reuse" is evidently essential for eradicating MP pollution (Nafea et al., 2024).

Recycling: Plastic recycling is the process of collecting plastic waste and converting it into a new, useful product (Merrington, 2011). It therefore reduces the outflow of plastics into the land, enhancing sustainability through efficient resource and energy management. Studies have shown that approximately 130 million kJ of energy can be conserved by recycling 1 ton of plastic litter (Garcia and Robertson, 2017; Nafea et al., 2024). However, even with significant global efforts, the recycling rate for plastic remains below 9%, primarily due to high costs and limited efficiency. The plastic waste is separated before processing. Although mechanical recycling is commonly performed, it reduces the quality of recycled plastic, ultimately limiting its potential for further recycling.

In contrast, chemical recycling has been developed recently and is not yet widely employed. It breaks the polymer into its basic molecular units and then reassembles them into their original form. Given the necessity of plastic recycling and its advantages in mitigating plastic waste and improving the environment, increasing global recycling rates is crucial (Nafea et al., 2024). In addition, community-level actions in collecting MPs from their respective areas have shown promising results in managing waste plastics. The introduction of such a policy, along with economic models, encouraged the public to reuse and recycle plastics, thereby increasing the recycling rate to approximately 40% (Bhattacharjee et al., 2025).

Recovery: The main objective of this point is to ensure maximum energy recovery from MP waste that cannot be recycled. Even though the process has very low efficiency, it can contribute to a slight reduction in MP deposition (Nafea et al., 2024).

Upcycling offers a promising solution to the challenges of conventional recycling. It transforms ubiquitous plastic debris into higher-value products, such as chemicals, fuels, and materials, by recovering energy, carbon, hydrogen, and macromolecular structures. This alternative pathway for plastic disposal significantly reduces plastic littering in landfills and surrounding ecosystems. Upcycling not only alleviates the environmental burden but also reduces the need for new plastic production and the inflow of harmful MPs. Consequently, the reintegration of plastic waste for new productive applications contributes to the conservation of resources and energy. This paves the way for long-term sustainability in waste management. The upcycling technique has recently emerged and is still not widely recognized, so its efficiency in managing plastics at a large scale remains unknown. Its efficacy in reducing plastic and MP contents in the environment depends on the quality and market demand for new products, as well as the operational materials and costs (Nafea et al., 2024).

Although industrial plastic waste is produced in amounts similar to municipal waste, the predominant focus is on mitigation strategies for municipal waste. Therefore it is crucial to shed light on the reduction and management strategies for industrial plastic waste, including the 4-R hierarchy (Nafea et al., 2024).

15.3.1.2 *Waste-to-fuel technique*

Among all the measures taken to manage MPs efficiently, converting MP waste into usable energy is among the most remarkable approaches. This strategy emphasizes the production of fuel, electricity, heat, and steam while overcoming several recycling limitations, thereby facilitating efficient MP management (Prata et al., 2019). For example, plastic waste that cannot be reused, recycled, or upcycled can be used as a feedstock for energy generation (Nafea et al., 2024). Processes such as gasification and pyrolysis in the presence of catalysts have been shown to effectively convert MP waste into fuels that are costly and possess physicochemical properties similar to those of gasoline (Prata et al., 2019). In addition, pyrolysis at approximately 500°C has been shown to achieve up to 80% conversion of plastic into liquid oil (FakhrHoseini et al., 2013; Liu et al., 2024).

Furthermore, compared with landfilling or conventional incineration, pyrolysis can fully recycle its products in a closed, anoxic environment with reduced hazardous gas emissions, making it an advantageous alternative (Liu et al., 2024). Another study provided evidence that functional organic molecules—glycolate, ethanol, acetate, etc.—can be produced during PET degradation via photocatalysis. Specifically, the waste-energy approach offers an alternative with an efficient, higher electricity-to-heat ratio than fossil fuels do, enhanced energy efficiency through recycling, the ability to treat contaminated and mixed water, moderately positive environmental effects, and no pretreatment requirements. Moreover, its retention of the linear economy framework, prolonged investments, and the possibility of harmful substance release constitute notable limitations (Prata et al., 2019). Therefore as this approach has the potential to significantly reduce MP waste, further research is needed to address its limitations.

15.3.2 Preventing infiltration of MPs into the ecosystem

First, regulated disposal is the initial step in isolating plastic scraps for further conventional processing. Therefore implementing it with a strict policy is inevitable. In addition, proper litter collection reduces secondary sources of MPs in the environment (Lackner and Branka, 2024).

Second, working on the inflow of primary MPs from natural and synthetic fiber shedding will significantly reduce MP generation. To curb this influx of MPs, policies are implemented through a maximum limit on the release of fiber from textiles to clothing manufacturers (Lackner and Branka, 2024). For example, the California State Water Resources Control Board has been ordered by the state of California's Assembly Bill (AB) No. 129 to develop methodologies that can evaluate microfiber filtration systems. In addition, MP filtration in washing systems attenuated MP infiltration into the water. Finally, individuals are encouraged to opt for cotton and wool clothing. Eliminating the use of microbead personal care products also plays a vital role in managing MPs (Osman et al., 2023). In addition, laws have been enacted in some regions requiring manufacturers to pay a specific amount for the filtration systems used in wastewater treatment to cover the high maintenance costs. This would also prompt the manufacturer to reduce its use of MPs in its products (First Sentier MUFG. Sustainable Investment Institute, n.d..

In agriculture and poultry farming, employing alternatives such as coated seeds without conventional plastics or nonplastic food storage can significantly reduce MP entry into animals and soil. The use of natural polymers, such as chitosan or PHA-based materials, is encouraged (Lackner and Branka, 2024).

Ecolabels foster a link between consumer choice and sustainability, encouraging them to make sustainable purchases. Ecolabels are given to products that are eco-friendly and pose no health hazards. This instills people's awareness of buying goods that are less harmful to the environment. Their use is prevalent in the Nordic region, which offers "Nordic Swan" ecolabeled disposables for food, indicating that the products are minimally harmful to both the environment and human health, as they primarily use recycled plastic rather than fossil carbon. Endorsing such an idea worldwide will promote sustainable consumption patterns and enable companies to adopt environmentally responsible practices.

Raising public awareness through campaigns and educational outreach that emphasize the dangers and risks posed by plastic waste is crucial to promoting the need for such measures. Using social media is an excellent tool to reach the public through advertisements, stories, and videos about the detrimental effects and their management strategies. In addition, campaigns conducted in schools and local communities have achieved notable success in various regions.

In addition, banning single-use plastic bags or other cutlery, such as spoons, plates, and glasses, illegal plastic disposal, and imposing heavy fines for breaking regulatory rules can serve as strong deterrents. Already, limited or absolute bans on plastic bags have been implemented in >30 countries across Asia, Europe, Africa, the Americas, and Oceania. In addition, Canada imposed a complete regulatory ban under the Canadian Environmental Protection Act, 2017, on nonprescription drugs and natural health products formulated with microbeads, which became effective in July 2019 (Cao et al., 2022).

Among them, extended producer responsibility (EPR) is increasingly being embraced in developed regions of the world, such as Europe, Japan, Canada, and South Korea. EPR is a strategic approach to waste management and is designed to strengthen the recycling system and curb the continuous deposition of plastic into landfills. This affirms that the company that produces and uses plastics assumes responsibility for plastic management. It involves collection, recycling, processing, and mineralizing. This study therefore helps regulate budget-friendly plastic waste management. Increased implementation of such a policy is needed worldwide (Cao et al., 2022).

Some countries have enforced incentive-based schemes to promote the efficient disposal and recycling of plastics. Retailers' take-back schemes for single-use bags in the USA and the government-sponsored buyback of fishing gear in South Korea are two examples that have achieved promising results. In addition, Basel Convention Amendments (2019) have been introduced, confirming proper waste disposal through strict site management (Bhattacharjee et al., 2025). These frameworks significantly reduce marine plastic pollution while driving the adoption of sustainable bioplastic alternatives (Cao et al., 2022).

15.4 Bioplastics in the management and mitigation of MPs

Bioplastics are biodegradable plastics (BPs) that are widely recognized as substitutes for nondegradable plastic polymers (Lackner and Branka, 2024). According to the American Society for Testing and Materials, bioplastics are biogenic materials that, under aerobic conditions and in the presence of microorganisms, degrade into smaller fragments and mineralize to form carbon dioxide, water, and organic matter. Examples of biobased plastics include polyhydroxybutyrate, polycaprolactone, polybutylene succinate, polyhydroxyalkanoate (PHA), polyglycolide, polylactic acid, polybutylene adipate-coterephthalate (PBAT), and starch-based blends. Therefore they act as a viable alternative to conventional plastics previously used in multiple settings (Sajjad et al., 2022). Completely degradable PHAs are already being employed in food and pharmaceutical packaging, as well as in agriculture and crop cultivation, as soil-covering films to prevent plastic contamination in soil and crops. The lightweight, high-resilience attributes of bioplastics enable their use in electronic devices, from smartphones and touchscreens to circuit boards and storage media. This also includes their use in the automotive industry as substitutes for manufacturing seat covers and airbags (Osman et al., 2023).

A composting environment is needed for bioplastics to degrade completely. However, these conditions are not met in the natural environment, resulting in only partial decomposition of BPs. This phenomenon leaves behind notable fragments of MPs in the ecosystem. Research has shown that PAEs are released as toxic by-products from the application and breakdown of MPs, thereby polluting the soil. Both BPs and conventional MPs adsorb hydrophobic organic compounds such as polycyclic aromatic hydrocarbons (PAHs), polybrominated diphenyl ethers (PBDEs), and polychlorinated biphenyls (PCBs) and exhibit similar toxicity. In one study, PBAT, a biodegradable polymer used in bioplastics, demonstrated increased sorption and desorption capacities. Therefore these bioplastic fragments transfer harmful contaminants such as phenanthrene across the ecosystem (Sajjad et al., 2022). As a result, further investigations and modifications are needed to address this challenge.

15.5 Future directions for mitigation strategies

7First, to improve consistency and relevance in MP sampling and analyses, developing standardized methodologies that focus not only on wastewater and marine environments but also on terrestrial environments has become integral (Lackner and Branka, 2024). The MP treatment technologies discussed above have several limitations, including affordability. Hence, in-depth analysis of these methods can promote the economic alleviation of MP pollution. A bioremediation tool that leverages oxic and anoxic microbes fosters a resource-efficient economy. These approaches curtail MP entry into both terrestrial and aquatic systems, reducing physiological impacts and safeguarding biodiversity (Ortiz et al., 2022). To determine the persistent consequences of infiltrated MPs across ecosystems, including the soil environment, longitudinal studies are paramount (Lackner and Branka, 2024). To date, most studies have been conducted on a limited scale for a short time, whereas the majority have focused on aquatic matrices (Sajjad et al., 2022). These findings emphasize the need for extensive studies involving all possible MP waste repositories. In addition to other interventions, such as combining soil amendment measures with nutrient-rich organic matter, compost, or biochar, bioremediation can subsequently be used to attenuate MPs in soil (Ortiz et al., 2022).

Furthermore, by prioritizing grant aid and legislation, attention should be given to facilitating the complete decomposition of biodegradable materials and to the optimal management of MP waste (Lackner and Branka, 2024). Another marked intervention to mitigate MP is the rational implementation and ensuring the recycling of products (Sajjad et al., 2022). Likewise, as upcycling represents a promising approach to waste management, greater recognition and strategic implementation are needed to fully harness its potential. Another crucial insight is to emphasize the socioeconomic effects of MP pollutants. Research in such fields will enable the augmented implementation of relevant policies and civic information initiatives, thereby minimizing MP waste and deposition (Lackner and Branka, 2024).

15.6 Conclusion

Given the escalation and dissemination of MPs across the environment and their ecological consequences for animals, MP pollution has become a priority requiring urgent attention. This chapter provides an overview of the major types and sources responsible for the extensive contamination of MPs, specifically highlighting the range of adverse outcomes they cause in animals. To minimize such adverse consequences, a set of mitigation strategies involving biological and technological advancements to ensure MP removal has been presented. Conventional and innovative treatment techniques have been summarized under the heading of technological advancements, with a focus mainly on MP removal from water matrices. MP pollution poses a significant threat to both terrestrial and aquatic ecosystems, with land as an essential source, accounting for 80% of the total MP waste in the ocean. This highlights the lack of treatment techniques or mitigation strategies that specifically target the accumulation of MP waste on land. Hence, effectively managing plastic debris, combined with mitigation strategies, will facilitate efficient MP removal and support the adoption of a circular economy. Implementing the 4-R principle, adopting waste-to-energy techniques, imposing bans, implementing ecolabeling, establishing treaties, and conducting global awareness campaigns would play a pivotal role in alleviating overall MP pollution. In addition, research on cost optimization of existing treatment processes or the introduction of more cost-effective alternatives would make these strategies accessible for implementation across all regions, regardless of financial barriers.

References

Agarwal, P., Prakash, S., Saini, G., 2024. Natural coagulants (Moringa oleifera and Benincasa hispida) based removal of microplastics. Clean. Water. 29502632. 1, 100010. doi:10.1016/j.clwat.2024.100010.

Arat, S.A., 2024. An overview of microplastic in marine waters: sources, abundance, characteristics and negative effects on various marine organisms. Desalination Water Treat. 19443986. 317, 100138. doi:10.1016/j.dwt.2024.100138.

Badawi, A.K., Hasan, R., Ismail, B., 2025. Sustainable coagulative removal of microplastic from aquatic systems: recent progress and outlook. RSC Adv. 20462069. 15 (31), 25256–25273. Available from http://pubs.rsc.org/en/journals/journal/ra. doi:10.1039/d5ra04074d.

Bhattacharjee, U., Baruah, K.N., Shah, M.P., 2025. Exploring sustainable strategies for mitigating microplastic contamination in soil, water, and the food chain: a comprehensive analysis. Environ. Chem. Ecotoxicol. 25901826. 7, 453–461. Available from https://www.sciencedirect.com/science/journal/25901826. doi:10.1016/j.enceco.2025.02.010.

Cao, B., Wan, S., Wang, Y., Guo, H., Ou, M., Zhong, Q., 2022. Highly-efficient visible-light-driven photocatalytic H2 evolution integrated with microplastic degradation over MXene/ZnxCd1-xS photocatalyst. J. Colloid Interface Sci. 10957103. 605, 311–319. Available from http://www.elsevier.com/inca/publications/store/6/2/2/8/6/1/index.htt. doi:10.1016/j.jcis.2021.07.113.

Chabi, K., Li, J., Ye, C., Kiki, C., Xiao, X., Li, X., Guo, L., Gad, M., Feng, M., Yu, X., 2024. Rapid sand filtration for <10 μm-sized microplastic removal in tap water treatment: efficiency and adsorption mechanisms. Sci. Total Environ. 912, 169074. doi:10.1016/j.scitotenv.2023.169074.

Ebrahimbabaie, P., Yousefi, K., Pichtel, J., 2022. Photocatalytic and biological technologies for elimination of microplastics in water: current status. Sci. Total Environ. 00489697. 806 (Pt 2), 150603. doi:10.1016/j.scitotenv.2021.150603.

FakhrHoseini, S.M., Dastanian, M., Sampedro, D., 2013. Predicting pyrolysis products of PE, PP, and PET using NRTL activity coefficient model. J. Chem. 2090-9063. 2013 (1). doi:10.1155/2013/487676.

Falco, G., Fenti, A., Galoppo, S., Chianese, S., Musmarra, D., Cocca, M., Mallardo, S., Iovino, P., 2024. Promoting removal of polystyrene microplastics from wastewater by electrochemical treatment. J. Water Process Eng. 22147144. 68, 106418. doi:10.1016/j.jwpe.2024.106418.

First Sentier MUFG. Sustainable Investment Institute, n.d. Sources of microplastics and their distribution in the environment. Available from: https://www.firstsentier-mufg-sustainability.com/insight/sources-of-microplastics-and-their-distribution-in-the-environment.html. (Accessed 19 June 2026).

Garcia, J.M., Robertson, M.L., 2017. The future of plastics recycling. Science. 10959203. 358 (6365), 870–872. Available from http://science.sciencemag.org/content/sci/358/6365/870.full.pdf. doi:10.1126/science.aaq0324.
Garfansa, M.P., Zalizar, L., Husen, S., Triwanto, J., Iswahyudi, I., Bakhtiar, A., Lasaksi, P., Ekalaturrahmah, Y.A.C., 2025. Addition of biochar based-activated carbon into sand filtration columns to improved removal efficiency microplastic from water. Bioresour. Technol. Rep. 2589014X. 30. Available from https://www.sciencedirect.com/science/journal/2589014X. doi:10.1016/j.biteb.2025.102099.
Grbic, J., Nguyen, B., Guo, E., You, J.B., Sinton, D., Rochman, C.M., 2019. Magnetic extraction of microplastics from environmental samples. Environ. Sci. Technol. Lett. 23288930. 6 (2), 68–72. Available from http://pubs.acs.org/page/estlcu/about.html. doi:10.1021/acs.estlett.8b00671.
Hechmi, S., Bhat, M.A., Kallel, A., Khiari, O., Louati, Z., Khelil, M.N., Zoghlami, R.I., Cherni, Y., Melki, S., Trabelsi, I., Jedidi, N., 2024. Soil contamination with microplastics (MPs) from treated wastewater and sewage sludge: risks and sustainable mitigation strategies. Discov. Environ.. 27319431. 2 (1). Available from https://link.springer.com/journal/44274. doi:10.1007/s44274-024-00135-0.
Hettiarachchi, H., Meegoda, J.N., 2023. Microplastic pollution prevention: the need for robust policy interventions to close the loopholes in current waste management practices. Int. J. Environ. Res. Public Health. 1660-4601. 20 (14), 6434. doi:10.3390/ijerph20146434.
Kumar, A., Indhur, R., Bux, F., Kumari, S., 2024. Recent advances in mechanistic insights into microplastics mitigation strategies via emerging advanced oxidation processes: legislation, challenges, and future direction. Sci. Total Environ. 00489697. 957, 177150. doi:10.1016/j.scitotenv.2024.177150.
Lackner, M., Branka, M., 2024. Microplastics in farmed animals—a review. Microplastics. 2673-8929. 3 (4), 559–588. doi:10.3390/microplastics3040035.
Liu, L., Mingjie, X., Yuheng, Y., Zhang, B., 2022. On the degradation of (micro)plastics: degradation methods, influencing factors, environmental impacts. Sci. Total Environ. 00489697. 806 (Pt 3), 151312. doi:10.1016/j.scitotenv.2021.151312.
Liu, Q., Khor, S.M., 2024. Emerging absorption-based techniques for removing microplastics and nanoplastics from actual water bodies. TrAC Trends Anal. Chem. 01659936. 170, 117465. doi:10.1016/j.trac.2023.117465.
Liu, Z., Xie, M., Zhou, T., Yang, J., Yang, Y., Liu, T., Dai, S., Huang, Q., Cen, Q., Xiao, P., Bin, L., Du, H., 2024. A review on liquid fuel produced from microwave-assisted pyrolysis of plastic waste. Process Saf. Environ. Prot. 09575820. 187, 833–844. doi:10.1016/j.psep.2024.05.035.
Mao, X., Yuchen, X., Cheng, Z., Yang, Y., Guan, Z., Jiang, L., Tang, K., 2022. The impact of microplastic pollution on ecological environment: a review. Front. Biosci.-Landmark. 2768-6701. 27 (2). doi:10.31083/j.fbl2702046.
Marthe Kiendrebeogo, M.R.K.E., Ouarda, Y., Patrick Drogui, R.D.T., 2022. Electrochemical degradation of nanoplastics in water: analysis of the role of reactive oxygen species. Sci. Total Environ. 00489697. 808, 151897. doi:10.1016/j.scitotenv.2021.151897.
Merrington, A., 2011. Recycling of Plastics. Elsevier BV, pp. 177–192. doi:10.1016/b978-1-4377-3514-7.10011-x.
Michielssen, M.R., Michielssen, E.R., Ni, J., Duhaime, M.B., 2016. Fate of microplastics and other small anthropogenic litter (SAL) in wastewater treatment plants depends on unit processes employed. Environ. Sci. Water Res. Technol. 20531419. 2 (6), 1064–1073. Available from http://pubs.rsc.org/en/journals/journalissues/ew#!recentarticles&adv. doi:10.1039/c6ew00207b.
Nafea, T.H., Chan, F.K.S., Xu, H., Wang, C., Xiao, H., He, J., 2024. Status of management and mitigation of microplastic pollution. Crit. Rev. Environ. Sci. Technol. 15476537. 54 (24), 1734–1756. Available from www.tandf.co.uk/journals/titles/10643389.asp. doi:10.1080/10643389.2024.2361502.
Nguyen, D., Nguyen, M.-K., Truong, Q.-M., Thai, V.-A., Minh-Thuan Pham, S.W.C., Nguyen, D.D., 2025. Microplastics and pharmaceuticals from water and wastewater: occurrence, impacts, and membrane bioreactor-based removal. Sep. Purif. Technol. 13835866. 361, 131489. doi:10.1016/j.seppur.2025.131489.
Ortiz, D., Munoz, M., Nieto-Sandoval, J., Romera-Castillo, C., de Pedro, Z.M., Casas, J.A., 2022. Insights into the degradation of microplastics by Fenton oxidation: from surface modification to mineralization. Chemosphere. 18791298. 309. Available from www.elsevier.com/locate/chemosphere. doi:10.1016/j.chemosphere.2022.136809.
Osman, A.I., Hosny, M., Eltaweil, A.S., Omar, S., Elgarahy, A.M., Farghali, M., Yap, P.S., Wu, Y.S., Nagandran, S., Batumalaie, K., Gopinath, S.C.B., John, O.D., Sekar, M., Saikia, T., Karunanithi, P., Hatta, M.H.M., Akinyede, K.A., 2023. Microplastic sources, formation, toxicity and remediation: a review. Environ. Chem. Lett. 16103661. 21 (4), 2129–2169. Available from https://www.springer.com/journal/10311. doi:10.1007/s10311-023-01593-3.

Prata, J.C., Patrício Silva, A.L., da Costa, J.P., Mouneyrac, C., Walker, T.R., Duarte, A.C., Rocha-Santos, T., 2019. Solutions and integrated strategies for the control and mitigation of plastic and microplastic pollution. Int. J. Environ. Res. Public Health. 16604601. 16 (13). Available from https://www.mdpi.com/1660-4601/16/13/2411/pdf. doi:10.3390/ijerph16132411.

Ramakrishnan, D., Sathiyamoorthy, M., 2025. Enhancing the remediation of polyamide microplastics: a comparative study of natural and synthetic coagulants. Front. Environ. Sci. 2296-665X. 13. doi:10.3389/fenvs.2025.1620074.

Sajjad, M., Huang, Q., Khan, S., Khan, M.A., Liu, Y., Wang, J., Lian, F., Wang, Q., Guo, G., 2022. Microplastics in the soil environment: a critical review. Environ. Technol. Innov. 23521864. 27, 102408. doi:10.1016/j.eti.2022.102408.

Senathirajah, K., Kandaiah, R., Logeshwaran Panneerselvan, C.I.S., Palanisami, T., 2023. Fate and transformation of microplastics due to electrocoagulation treatment: impacts of polymer type and shape. Environ. Pollut. 02697491. 334, 122159. doi:10.1016/j.envpol.2023.122159.

Sharma, S., Basu, S., Shetti, N.P., Nadagouda, M.N., Aminabhavi, T.M., 2021. Microplastics in the environment: occurrence, perils, and eradication. Chem. Eng. J. 13858947. 408. Available from www.elsevier.com/inca/publications/store/6/0/1/2/7/3/index.htt. doi:10.1016/j.cej.2020.127317.

Shuo, L., Yang, Y., Yang, S., Zheng, H., Zheng, Y., Jun, M., Nagarajan, D., Varjani, S., Chang, J.-S., 2023. Recent advances in biodegradation of emerging contaminants - microplastics (MPs): feasibility, mechanism, and future prospects. Chemosphere. 00456535. 331, 138776. doi:10.1016/j.chemosphere.2023.138776.

Talvitie, J., Mikola, A., Setälä, O., Heinonen, M., Koistinen, A., 2017a. How well is microlitter purified from wastewater? – a detailed study on the stepwise removal of microlitter in a tertiary level wastewater treatment plant. Water Res. 18792448. 109, 164–172. Available from www.elsevier.com/locate/watres. doi:10.1016/j.watres.2016.11.046.

Tiwari, E., Singh, N., Khandelwal, N., Monikh, F.A., Darbha, G.K., 2020. Application of Zn/Al layered double hydroxides for the removal of nano-scale plastic debris from aqueous systems. J. Hazard. Mater. 18733336. 397. Available from www.elsevier.com/locate/jhazmat. doi:10.1016/j.jhazmat.2020.122769.

Wulandari, M., Marpaung, K., Prasaningtyas, A., Yorika, R., Ma'arij Harfadli, M., Zulfikar, A., 2024. Performance of rapid sand filter dual media for microplastic removal in the water: the effect of microplastic size and effective size of filter media. J. Commun. Based Environ. Eng. Manage.. 2597-9736. 8 (8), 103–110. doi:10.23969/jcbeem.v8i1.12502.

Xing, X., Zhang, Y., Zhou, G., Zhang, Y., Yue, J., Wang, X., Yang, Z., Chen, J., Wang, Q., Zhang, J., 2023. Mechanisms of polystyrene nanoplastics adsorption onto activated carbon modified by ZnCl2. Sci. Total Environ. 00489697. 876, 162763. doi:10.1016/j.scitotenv.2023.162763.

Zhang, Y., Zhao, J., Liu, Z., Tian, S., Jingfang, L., Rong, M., Yuan, H., 2021. Coagulation removal of microplastics from wastewater by magnetic magnesium hydroxide and PAM. J. Water Process Eng. 22147144. 43, 102250. doi:10.1016/j.jwpe.2021.102250.

CHAPTER

16

Neurotoxicity and endocrine disruption induced by microplastic exposure

Younis Ahmad Hajam[a], *Neelam Bhatti*[a], *Ajmal Rashid Bhat*[b], and *Nisha Sharma*[c]

[a]Department of Life Sciences, Sant Baba Bhag Singh University, Jalandhar, India
[b]Department of Chemistry, Rashtrasant Tukadoji Maharaj Nagpur University, Nagpur, India
[c]Department of Physical Science, Sant Baba Bhag Singh University, Jalandhar, India

16.1 Introduction

Micro- and nanoplastics (MNPs) are generated from the breakdown of plastic waste, industrial activities, and the shedding of textile fibers. They persist in the environment and carry harmful substances, posing serious health risks worldwide (Ma et al., 2025). Plastics are formed through addition polymerization or polycondensation reactions using various monomers and additives. Their properties include insulation, stability, adaptability, and corrosion resistance. Increasing evidence suggests that exposure to MNPs leads to neurotoxicity, endocrine disorders, and other organ-level dysfunctions (Ma et al., 2025). MNPs are resistant to degradation, allowing them to persist in the environment. Their large surface area and small size increase their availability and interactions with organisms by disrupting various physiological processes. Studies have revealed that exposure to MNPs poses health and environmental risks, such as endocrine disorders (thyroid dysfunction, reproductive abnormalities, sex hormone imbalance, polycystic ovarian syndrome (PCOS)-like conditions, metabolic disorders, adrenal hormone dysregulation, neuroendocrine disturbances, growth and developmental abnormalities, pancreatic endocrine dysfunction, and autoimmune endocrine disorders), gastrointestinal problems, and reproductive toxicity affecting the gut microbiota, disruption of barrier integrity, a decrease in the quality of germ cells, and dysregulation of lipid metabolism.

Microplastics in Animal and Human Physiology
https://doi.org/10.1016/B978-0-443-45635-0.00017-3

Micronanoplastics (MNPs) have been detected in various human tissues, including cerebrospinal fluid and the frontal cortex. Moreover, MNPs have been found in aquatic, terrestrial, and atmospheric ecosystems. The understanding that microplastics (MPs) are not just inert litter but also active materials that interact with cellular systems has changed. In the current era, attention has shifted to their potential toxic effects on vital physiological systems, particularly the nervous and endocrine systems. Because MPs contain endocrine-disrupting substances such as bisphenols and phthalates, which increase their adverse biological impact and complicate toxic responses across various organ systems, the body can take in MNPs via ingestion, inhalation, and skin contact, leading to their distribution and accumulation throughout the whole body. Once inside, these particles might interact with biomolecules, cross physiological barriers, and accumulate in sensitive tissues. Moreover, particle size, surface chemistry, and biomolecular corona formation enable MPs to cross the blood–brain barrier or reach the brain via olfactory pathways. These findings indicate that direct exposure to MPs in the brain induces neurotoxicity. This finding is supported by both in vitro and in vivo studies, which indicate that MPs can penetrate through neural cells, causing cell death and compromising the integrity of the blood–brain barrier, potentially decreasing neuronal function and cognitive ability. MP-induced oxidative stress is another significant mechanism underlying neurotoxicity, causing neuronal injury, inflammation, and cell death by triggering excessive production of reactive oxygen species (ROS), disrupting antioxidant defenses, and harming mitochondrial function. These actions may affect synaptic transmission and brain balance, leading to behavioral changes and neurodegenerative effects. Therefore the neurotoxic effects of MPs arise not as isolated incidents but from a combination of redox imbalance, inflammatory responses, and chemical interactions within brain tissues.

In addition, MPs disrupt the neuroendocrine axis and compromise endocrine function. They can affect hormone signaling by binding to hormone receptors or altering hormone production and metabolism, as they can carry and release endocrine-disrupting chemicals (EDCs). This whole scenario occurs because MPs target the supercontroller of endocrine organs, such as the hypothalamus, pituitary gland, thyroid, gonads, and adrenal tissues. This can lead to imbalances in crucial hormonal systems, such as the hypothalamic–pituitary–gonadal (HPG), hypothalamic–pituitary–thyroid (HPT), and hypothalamic–pituitary–adrenal (HPA) axes. Such changes are associated with reproductive issues, metabolic disorders, immune problems, and developmental defects (Bossio et al., 2025).

Furthermore, oxidative stress plays a key role in linking endocrine disruption to MP exposure. This might be explained by the fact that increased ROS can impair the activity of steroidogenic enzymes, alter hormone receptor sensitivity, and disrupt hormone production in endocrine tissues. These effects may impair the endocrine system, leading to a self-perpetuating cycle of redox imbalance and hormonal disruption. This interrelated association between oxidative stress and endocrine disruption can help detect and trace the toxic effects of MPs in biological systems across a broader range (Bossio et al., 2025).

Because the neuroendocrine network regulates behavior, growth, reproduction, and homeostasis through tightly controlled feedback mechanisms, neuroinflammation and nerve damage can alter endocrine regulation. Moreover, disruptions in hormonal signaling can significantly affect brain growth and function (Manley et al., 2018). Experimental studies have reported that exposure to MPs during critical developmental stages can alter cytokine levels and thyroid hormone signaling in the fetal brain, suggesting long-term consequences for neurodevelopment and neuroendocrine function. Hence, scientific evidence suggests that MPs constitute a unique class of environmental toxins that can concurrently affect the brain and endocrine system via multiple molecular pathways. Owing to their widespread presence, potential bioaccumulation, and transport of harmful chemicals, this has raised concerns about long-term health effects on both humans and wildlife. Given the potentially toxic nature of MPs, the present chapter summarizes microinterpreted neurotoxicity and endocrine disruption.

MPs are conventionally defined as synthetic polymer particles <5 mm in at least one dimension, whereas NPs represent the nanosized fraction, typically <1 μm, although definitions vary across disciplines. MNPs are categorized as primary particles (manufactured at small sizes, such as microbeads and industrial pellets) or secondary particles (produced by the fragmentation and aging of larger plastics). Their polymer identity [e.g., polyethene (PE), polypropylene (PP), polystyrene (PS), polyethylene terephthalate (PET), and polyvinyl chloride (PVC)], morphology (fragment, fiber, film, and bead), size distribution, surface chemistry (oxidation and weathering), and associated chemicals (additives and sorbed environmental contaminants) strongly influence environmental fate, bioavailability, and toxicity. Harmonization of size bins, reporting units, and polymer identification is increasingly emphasized to facilitate cross-study comparability and robust risk assessment.

Since the 1950s, the production of plastic has increased significantly, reaching approximately 390.7 million metric tons in 2021, and is projected to exceed approximately 1100 million metric tons by 2050 worldwide (Dris et al., 2017). This surge has led to the broad accumulation of plastic waste in natural environments, where degradation processes generate MPs (particles <5 mm) and nanoplastics (NPs; <100 nm). In recent years, global plastic production has continued to increase, from approximately 250 million tons in 2009 (Alimba and Faggio, 2019) to 299 million tons in 2013 (Galgani et al., 2015) and reaching 335 million tons by 2016 (Alimba and Faggio, 2019). An estimated 10% of the plastic produced each year ultimately enters the marine environment as debris (Chen et al., 2025). Current assessments suggest that >5 trillion plastic particles spanning sizes from NPs to macroplastics, collectively weighing >250,000 tons, are present at the ocean surface alone (Eriksen et al., 2014). This figure does not account for the substantial quantities of plastic accumulated in freshwater systems, terrestrial environments, or on the seafloor, indicating that the true global burden of plastic pollution is considerably greater.

MPs and NPs are ubiquitously found in terrestrial and aquatic ecosystems. They are increasingly a concern because of their entry into and accumulation in the human body, which can cause various adverse health effects. Recent studies have reported that MPs and NPs are found in drinking water (both tap and bottled), seafood, salt, indoor and outdoor air, and even human tissues, including the lungs, colon, placenta, blood, and feces. A global survey revealed that bottled water can contain up to 10,000 MP particles per liter. In contrast, the concentrations of MP fibers in indoor air may reach hundreds to thousands of particles per cubic meter, particularly in urban environments. MNPs originate from multiple cross-linked sources, including microbeads from personal care products, synthetic textile fibers shed during laundering, tire and road-wear particles, industrial resin pellets, and paint fragments. Secondary particles are generated from larger plastic items such as packaging, fishing gear, and MPs through UV radiation, mechanical abrasion, and partial degradation. MPs are widely distributed in marine and freshwater systems, sediments, agricultural soils, indoor dust, and food webs. They are specifically present in wastewater treatment plant effluents and sewage sludge, which are the main vectors for contaminating terrestrial ecosystems. However, the spatial distribution varies from one component to another, with coastal sediments, urban soil, and indoor environments, particularly in homes and vehicles, exhibiting disproportionately higher concentrations because of airborne microfiber accumulation.

Humans are primarily exposed to MPs through ingestion and inhalation, with dermal contact contributing minimally, except in occupational settings or when skin barriers are compromised (Kannan and Vimalkumar, 2021). Moreover, current studies have reported that MPs are distributed across foodstuffs such as seafood, salt, sugar, honey, and infant formula, with dietary intake varying widely depending on particle size thresholds and analytical techniques (Pironti et al., 2021). Inhalation has been shown to increase through well-known pathways, as indoor aerosol studies indicate abundant airborne fibers and fragments, often at concentrations higher than those in outdoor air, suggesting significant daily respiratory intake, particularly in poorly ventilated or high-shedding indoor environments.

Advances in analytical techniques have enabled the detection of MNPs in multiple human tissues and fluids, including feces, blood, breast milk, placenta, semen, liver, kidneys, lungs, and even brain tissue. These studies indicate that MPs selectively accumulate in lipid-rich organs and suggest that body burdens may increase over time. However, quantitative interpretation remains constrained by small sample sizes, heterogeneous methodologies, variable particle size detection limits, and challenges in controlling contamination. Accurate exposure assessment, therefore, requires harmonized sampling and analytical protocols, validated methods for nanoparticle detection, and integrative models that combine ingestion and inhalation pathways. Coordinated biomonitoring across wastewater systems, indoor air, and food chains is now considered an urgent public health priority for quantifying cumulative human exposure and informing risk mitigation strategies.

16.2 Rationale for examining neurotoxic and endocrine-disrupting potential

Studies on experimental models and human tissues have shown that MNPs can be transported physically across important biological barriers, such as the intestinal epithelium and placental barrier, and, in animal studies, the blood–brain barrier (BBB), allowing them to enter systemic circulation and accumulate in sensitive organs such as the brain and endocrine tissues. Moreover, MNPs have been detected in the human prefrontal cortex and in mouse microvessels, indicating direct neurological effects and strengthening the mechanistic basis for the examination of neurotoxicity. Consequently, MPs act as carriers of chemical additives (e.g., phthalates, bisphenols, and flame retardants) and sorb environmental contaminants, many of which are established endocrine disruptors or neurotoxicants. When the matrix of polymers is itself inserted, these associated chemicals can leach from particles, causing localized or systemic toxicity. This pattern is well-documented in chemical–ecotoxicological surveys, which consistently show the co-occurrence of MPs with known hazardous compounds.

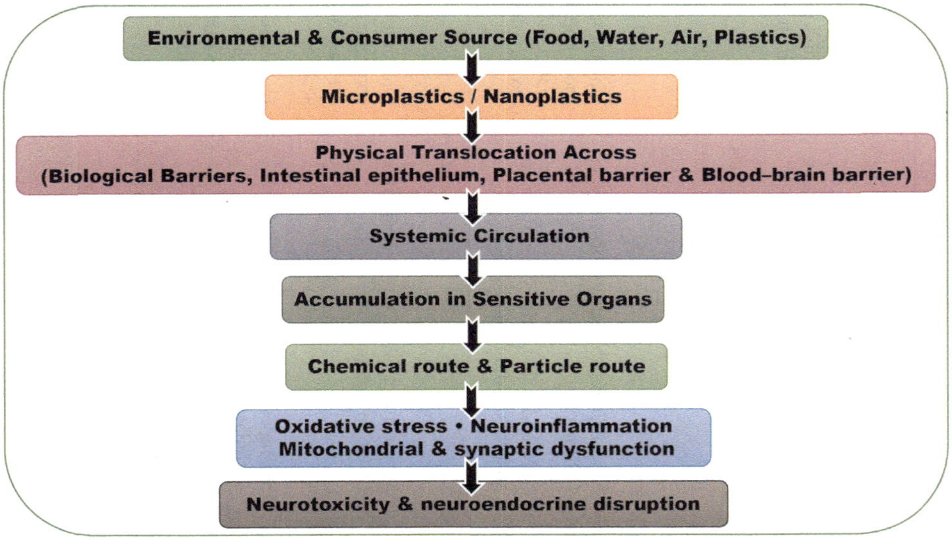

FIG. 16.1 Translocation and neurotoxic mechanisms of micro- and nanoplastics (MNPs): MNPs enter the body via ingestion, inhalation, or dermal exposure and are translocated across biological barriers, including the intestinal epithelium, pulmonary alveoli, and blood–brain barrier. Once in systemic circulation, MNPs accumulate in neural tissues, where they induce oxidative stress, neuroinflammation, and disruption of neuronal signaling, either directly or by acting as carriers for adsorbed toxic chemicals, ultimately leading to neurotoxicity.

In addition to chemical vectors, particle-driven biological mechanisms provide additional routes of harm. Various in vitro and in vivo studies have reported that MNP exposure induces oxidative stress, inflammation (including microglial activation in neural tissue), mitochondrial dysfunction, and disruptions to calcium signaling and synaptic protein pathways, which are fundamental to both neural health and endocrine regulation. High-sensitivity studies have also reported that NPs accumulate in the human brain and are spatially associated with neuropathological features, supporting the assumption that chronic low-level exposure could contribute to the development of neurodegenerative diseases. Complementary animal studies have revealed that MPs in circulation can be phagocytosed by immune cells and lodged in cerebral capillaries, leading to microvascular occlusions, decreased cerebral perfusion, and behavioral defects (Chowdhury et al., 2024). Interestingly, these studies reveal a mechanical route of neurotoxicity that functions independently of chemical leaching (Fig. 16.1).

16.3 Evidence of endocrine perturbation is equally compelling

In vivo studies in fish and rodents, along with receptor-based in vitro assays, have revealed that MP leachates and sorbed contaminants exhibit antiandrogenic and thyroid-disrupting effects (Jahedi et al., 2025). In addition, interactions between MPs and classical endocrine disruptors, including bisphenol A (BPA), can amplify biological responses by increasing internal doses or altering the bioavailability of contaminants (Tables 16.1 and 16.2).

TABLE 16.1 Classes of microplastics targeting the nervous and endocrine systems.

Class/polymer type	Key physicochemical characteristics	Nervous system toxicity (neurotoxicity)	Endocrine system toxicity (endocrine disruption)	Representative evidence/notes
Polystyrene (PS) micro- and nanoplastics	Spherical beads; available in charged forms (COOH, NH_2); high aromaticity; small NPs (<100 nm) cross membranes	Penetrate BBB via transcytosis Trigger microglial activation (NF-κB pathway) Induce oxidative stress, mitochondrial depolarization Reduce synaptic proteins (PSD-95, synapsin I) Impair learning and memory	Disrupt HPG and HPT axes Reduce estradiol/testosterone Inhibit steroidogenic enzymes (StAR, CYP19A1) Alter thyroid hormone (T3/T4) levels	Widely used model plastic; strong evidence in rodent and zebrafish models
Polyethylene (PE)	Hydrophobic; low-density; high sorption of organic pollutants	Indirect neurotoxicity via POP-loaded particles polychlorinated hydrocarbons (PAHs), polychlorinated biphenyls (PCBs) Increase neuroinflammation Alter microbiota–gut–brain axis	Modulate reproductive hormones via POP release Mild endocrine activity mainly from adsorbed contaminants	Ubiquitous in food, bottled water; indirect toxicity dominates
Polypropylene (PP)	Chemically inert; common in food packaging; fibrous fragments common	Induce systemic inflammation impacting CNS Behavioral abnormalities in fish/rodents	Minor direct ED; acts as a vector for EDCs and metals	One of the most common MPs in human stool studies
Polyvinyl chloride (PVC)	High chlorine content; contains phthalates and organotins; substantial additive leaching	Phthalate-mediated neurotoxicity Disruption of dopaminergic/GABAergic pathways Oxidative injury in brain tissues	Strong endocrine disruption due to phthalates Antiandrogenic, estrogenic activity Alter thyroid hormone synthesis and metabolic pathways	Highly hazardous MP class owing to endocrine-active additives
Polyethylene terephthalate (PET)	Polyester; releases antimony (Sb) and small oligomers; common in bottled water	Limited direct neurotoxicity; oxidative stress from Sb exposure	Altered thyroid axis (TSH/T3/T4 imbalance) Disrupted reproductive hormones Adrenal stress response activation	PET drinking water is the primary human exposure route

Polycarbonate (PC) (BPA-containing plastics)	Degrades to BPA and BPA derivatives; rigid polymer	BPA-related neurodevelopmental effects Disrupted neurotransmitter balance and synaptogenesis	Potent estrogen receptor agonism Thyroid suppression (↓TPO, ↓NIS) Metabolic disruption (glucose intolerance)	PC degradation products exert strong endocrine effects
Tire wear particles (TWP)/ synthetic rubber MPs	Mixture of styrene–butadiene rubber, carbon black, ZnO, benzothiazoles, and PAHs	Elevated neurotoxicity from Zn, PAHs• BBB permeability changes Behavioral deficits in aquatic organisms	Impair thyroid hormone synthesis (PAH-mediated) Alter cortisol/corticosterone via stress response Reduce androgen/estrogen levels in fish	Major contributor to urban MP exposure; complex chemical profile
Acrylics (PMMA), polyamide (Nylon)	Fibrous MPs; abrasive; may carry dyes/metals	Induce inflammation and oxidative stress affecting CNS indirectly	Leach dyes/metals with endocrine activity; reproductive disruption in aquatic species	Increasing prevalence in textiles and clothing
Biodegradable plastics (PLA, PHA)	Degraded into lactic acid and intermediates; may form nanofragments	Mild neurotoxicity from acidic degradation products; increased ROS	Limited direct ED; potential microbial dysbiosis influences estrogen metabolism	Often assumed safe but still produce harmful nanofragments
Weathered environmental MPs (mixed polymers)	UV-aged; oxidized surfaces; high pollutant sorption	Enhanced neural toxicity due to higher ROS generation Facilitated membrane penetration	Act as carriers for EDCs: BPA, phthalates, PCBs, dioxins Strong combined endocrine disruption through mixture effects	Environmental MPs are often more toxic than pristine ones

TABLE 16.2 Key cross-system mechanisms (nervous + endocrine overlap).

Shared mechanism	Relevance to neurotoxicity	Relevance to endocrine disruption
Oxidative stress and mitochondrial dysfunction	Neuronal apoptosis, impaired synaptic function	Hormone-producing cell damage (thyroid, gonads, pancreas)
Chronic inflammation	Microglial activation; neurodegeneration	Hypothalamic dysregulation; impaired hormone feedback
Leaching of additives (BPA, phthalates, organotins)	Neurodevelopmental effects, neurotransmitter imbalance	Estrogenic, anti-androgenic, thyroid-disrupting effects
Pollutant sorption (POPs, PAHs, PCBs)	Enhanced CNS toxicity	Synergistic endocrine disruption
Barrier penetration (gut, placenta, BBB)	Direct CNS accumulation	Transplacental endocrine effects; fetal programming

16.4 Physicochemical properties relevant to toxicity

The biological behavior and toxicity of MNPs are strongly influenced by their physicochemical characteristics, including particle size, shape, surface charge (zeta potential), polymer type, chemical additives, and adsorbed environmental pollutants. Among these factors, particle size is important: compared with larger particles, polystyrene nanoparticles (PS-NPs) smaller than 200 nm are internalized by cells more efficiently, penetrate tissues more readily, and localize within subcellular organelles to a greater extent (Luo et al., 2025). Moreover, particle shape further modulates biological interactions because most experimental studies use spherical particles, irregular fragments, and fibrous MNPs, which are increasingly detected in environmental samples. Owing to the elongated geometry of fibrous particles, they can deposit deeply in the alveolar regions of the lungs, resist clearance by macrophages, and promote persistent inflammation. In contrast, spherical and irregular fragments are more readily internalized via endocytosis (Gopallawa et al., 2023). Fibers may also induce frustrated phagocytosis, amplifying oxidative and inflammatory responses, emphasizing the importance of considering environmentally realistic morphologies in human health risk assessments.

In addition, surface chemistry and charge also play important roles by influencing the formation and composition of the protein corona, a dynamic layer of biomolecules that adsorbs onto particle surfaces when they come into contact with biological fluids. The corona affects cellular recognition, intracellular trafficking, and immune responses. This might be explained by the fact that, compared with neutral (PS) or carboxylated (PS-COOH) particles, amine-modified PS-NPs (PS-NH_2) become more cytotoxic and inflammatory. Moreover, positively charged PS-NPs destabilize cell membranes, increase permeability, induce mitochondrial damage, increase ROS generation, and trigger apoptotic signaling pathways. Furthermore, owing to their intrinsic properties, MNPs act as vectors for chemical additives such as phthalates, BPA, and brominated flame retardants. These compounds can leach upon particle internalization or upon interaction with physiological fluids, thereby increasing the toxicity of MNPs. Considering these characteristics, such as particle morphology, size, surface charge, and chemical payload, multiple interactions contribute to the diverse and complex toxicological outcomes observed with MNPs. Therefore there is a gap in the literature: a need exists for detailed physicochemical characterization and standardized particle profiling in experimental designs to accurately assess MNP safety and hazard potential.

Particle size is among the most important physicochemical factors that shape the biological activity and toxicity of MNPs. Owing to their high diffusion coefficients, high surface-to -volume ratios, and increased surface reactivity, NPs readily cross biological barriers via transcytosis, endocytosis, paracellular transport, or direct membrane insertion (Paul et al., 2020). NPs are translocated across the intestinal epithelium, placental barrier, and BBB, with preferential accumulation in lipid-rich organs such as the brain, liver, and adrenal glands (Pan and Umapathy, 2024). This expanded reactive surface area facilitates extensive protein corona formation, which, in turn, amplifies oxidative stress and inflammatory responses.

MPs (1–1000 μm) exhibit size-dependent interactions; however, the 20 μm particles are smaller and may be less likely to undergo cellular uptake, potentially becoming trapped within the microvasculature. In contrast, larger MPs (>100 μm) primarily cause local physical irritation in the gastrointestinal lumen and serve as reservoirs for leachable additives and sorbed pollutants. Among these plastic species, smaller particles consistently exhibit more severe toxicity, including increased neurotoxicity, endocrine disruption, mitochondrial dysfunction, and significantly higher cytokine production, indicating a clear inverse relationship between particle size and biological impact. In addition to size, particle shape has a stronger effect on toxicity. Fibers commonly produced by textile shedding, possess higher aspect ratios, which promote greater retention in the respiratory tract; their aerodynamic properties permit them to penetrate deep into pulmonary tissues, where they evade clearance, often eliciting asbestos-like frustrated phagocytosis, chronic inflammation, epithelial remodeling, and macrophage overload. Irregular fragments with sharp surfaces can mechanically disrupt epithelial barriers, trigger proinflammatory signaling, and exhibit increased sorption capacity, especially when weathered and roughened by environmental aging. In contrast, particles with spherical shapes, such as laboratory PS beads, are more predictable in behavior but fail to capture the complexity and increased toxicity of environmentally aged and irregular particles. Recent studies have reported that shape also governs the cellular uptake mechanism: elongated fibers are resistant to internalization, whereas smaller fragments with irregular geometries are more prone to macropinocytosis, receptor-mediated endocytosis, and phagocytosis.

Polymer chemistry further modulates MNP bioactivity. For example, PS, the most commonly used model polymer, readily forms NPs and is known to disrupt immune pathways, mitochondrial function, and calcium homeostasis. PE and PP, which dominate packaging waste, are highly hydrophobic and thus exhibit high sorption affinity for persistent organic pollutants (POPs); weathered PE fragments often acquire oxidized functional groups that enhance protein corona formation and cellular interactions (Peña et al., 2023). PVC mostly contains chlorine-based plasticizers and stabilizers and has a significantly greater endocrine-disrupting potential. Moreover, polyesters such as PET contribute significantly to airborne fiber exposure, have been linked to ROS production, and can cross the epithelial barrier via inhalation (Kudzin et al., 2024). These polymer-dependent differences in hydrophobicity, degradation patterns, and surface chemistry collectively shape bioavailability, pollutant loading, and toxicity profiles. Ultimately, surface charge (zeta potential) plays an important role in determining how MNPs interact with biomembranes and biological fluids. This might be explained by the fact that positively charged particles interact strongly with negatively charged phospholipid membranes, resulting in increased cellular uptake, membrane destabilization, and cytotoxicity and, additionally, generating increased oxidative and inflammatory responses. Moreover, negatively charged or neutral particles tend to be more colloidally stable. However, they eventually become less acutely toxic and form dynamic protein and lipid coronas that significantly alter biodistribution and cellular signaling.

Environmental weathering through UV exposure and photooxidation alters surface charge by introducing carbonyl and hydroxyl groups, which typically shift plastics toward more negative and hydrophilic states, increasing their mobility and biological accessibility (Khan and Jia, 2023). Finally, surface charge affects mucus penetration, immune recognition, endocytic pathways, and systemic transport. It interacts with size, shape, and polymer chemistry to determine the toxicological landscape of exposure to MNPs (Fig. 16.2).

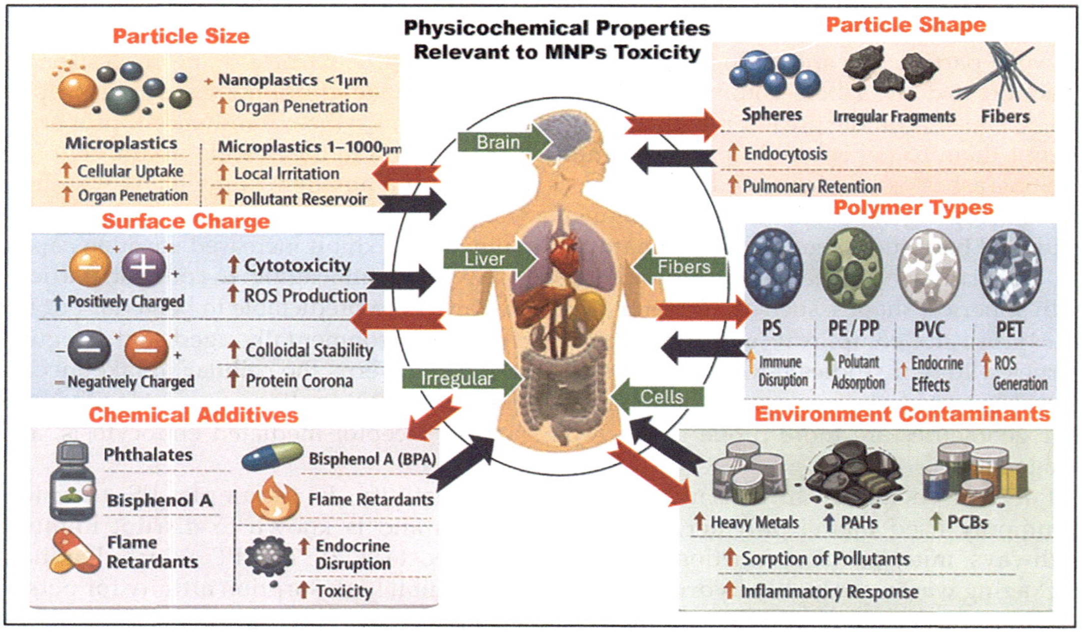

FIG. 16.2 Physicochemical determinants of micro- and nanoplastics (MNPs) toxicity and biological interactions. The key properties of MNPs, such as particle size, shape, surface charge, polymer composition, crystallinity, and surface functionalization, govern their environmental behavior, cellular uptake, biodistribution, and interactions with biomolecules. These physicochemical characteristics influence protein corona formation, membrane penetration, inflammatory responses, oxidative stress, and overall toxicological outcomes in biological systems.

16.5 Cellular uptake and translocation across biological barriers

MNPs can enter cells via various pathways, including clathrin- and caveolin-mediated endocytosis, macropinocytosis, and passive diffusion across lipid membranes (Khan and Jia, 2023). Their extremely small size, particularly at the nanoscale, enables them to penetrate cellular membranes and interfere with the fundamental processes of cells. This might be explained by the fact that PS-NPs smaller than 100 nm are more efficiently internalized by HeLa cells via active endocytic mechanisms, in contrast to these particles, which are larger (>250 nm), which mainly remain in the extracellular matrix and show lower cytotoxicity (Jiang et al., 2024; Ernhofer et al., 2025). Once MNPs are internalized, they can translocate across different epithelial and endothelial barriers, such as the intestinal mucosa, pulmonary alveoli, placental interface, and BBB, and may even penetrate dermal layers when the stratum corneum is absent. Experimental studies have reported that PS-NPs ranging in size from 50 to 100 nm can traverse gut and lung epithelial models, cross the placental barrier, and penetrate the BBB under both in vitro and in vivo conditions (Hwang et al., 2022). In human cerebral microvascular endothelial cells (HcCMECs/D3), exposure to 50-nm PS-NPs induces the generation of ROS, activates NF-κB signaling, disrupts tight junction proteins, and promotes necroptosis (Daniels et al., 2013). In addition, surface functionalization of PS-NPs, such as amine (PS-NH_2) or carboxyl (PS-COOH) modifications, intensifies these cytotoxic effects and indicates that particle chemistry mediates cellular toxicity.

16.6 Additives and sorbed pollutants (metals, POPs)

Plastics inherently comprise a complex mixture of chemical additives, such as plasticizers, stabilizers, flame retardants, pigments, and UV absorbers, which are important for polymer performance but pose significant toxicological risks when released from MNPs (Marturano et al., 2017). Phthalates, including Di (2-ethyhexyl) phthalate (DEHP) and Dibutyl phthalate (DBP), are the most concerning additives because they exhibit strong antiandrogenic and reproductive toxicity, and bisphenols [BPA and bisphenol S (BPS)] have well-known estrogenic and neuroactive properties. Polybrominated diphenyl ethers (PBDEs) contribute to thyroid hormone disruption and neurodevelopmental abnormalities, whereas organotins are associated with metabolic dysregulation, obesity, and immune dysfunction (Marques et al., 2018). Moreover, plastics frequently contain or sorb heavy metals such as cadmium, lead, and chromium, which exhibit potent neurotoxic and nephrotoxic effects (Mitra et al., 2022). The release potential of chemicals from MPs increases when particle size decreases, following UV-induced weathering or oxidative aging, or when higher temperatures and acidic environments are encountered, as is likewise found in the gastrointestinal tract (Dhaundiyal and Mittal, 2024). Following these processes, MPs act as slow-release chemical reservoirs, develop localized exposure at the tissue level that can occur independently of systemic chemical concentrations, and are not detected through conventional toxicant monitoring (Fig. 16.3).

In addition to being intrinsic additives, MPs work as efficient environmental sponges, concentrating POPs and metals from surrounding ecosystems by various orders of

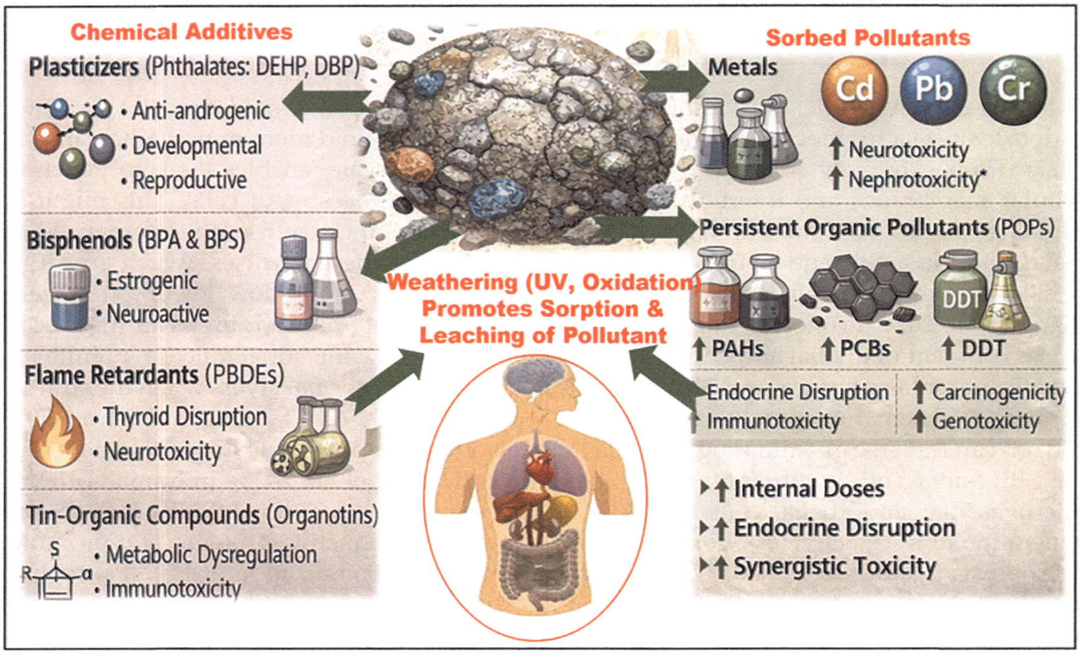

FIG. 16.3 Additives and sorbed pollutants associated with micro- and nanoplastics: metals, persistent organic pollutants, and synergistic toxicity. Additives and sorbed pollutants associated with micro- and nanoplastics. The figure illustrates the association of MNPs with plastic additives, heavy metals, and persistent organic pollutants (POPs), highlighting their role as vectors for cotransport into biological systems and the resulting synergistic toxicity through enhanced bioavailability, cellular uptake, and combined toxic effects.

magnitude. PCBs, dioxins, PAHs, and agricultural pesticides such as dichlorodiphenyltrichloroethane (DDT) metabolites, pyrethroids, and heavy metals (mercury, lead, cadmium, and nickel) are commonly sorbed contaminants (Giroux et al., 2024). This sorption is mediated by multiple chemical interactions, including hydrophobic interactions mediate this sorption, van der Waals forces, π–π interactions (particularly in aromatic polymers such as polystyrene), and coordination with oxygen-containing functional groups formed on the surfaces of weathered polyethylene (Agboola and Benson, 2021). Environmental weathering, which produces microcracks, surface roughness, and oxygen-containing groups, significantly enhances the sorption capacity of MPs, making these aged particles more hazardous than pristine particles. When ingested or inhaled, these pollutant-loaded particles can increase internal doses of EDCs, modify toxicokinetics, and lead to coexposure synergies in which MPs modify the bioavailability and metabolic fate of their associated contaminants. These interactions can intensify the disruption of the endocrine system, indicating synergistic toxicity between BPA and MPs, and alter the composition of the gut microbiota, which, in turn, increases the absorption of pollutants. Therefore the cumulative role of MPs as both sources and vectors of hazardous chemicals emphasizes their complex and versatile contribution to toxicological threats in exposed organisms, including humans.

16.7 Bioavailability and biopersistence

The bioavailability of MNPs is the fraction of particles that successfully interact with biological systems or enter systemic circulation and is governed by a combination of physicochemical and biological factors (Dube and Okuthe, 2023). Particle size is the primary determinant of the ability of NPs to readily traverse epithelial barriers, whereas only microparticles smaller than ~20 μm demonstrate significant potential for translocation beyond the gut lumen. Surface charge and hydrophobicity affect particle adhesion to mucus layers and cellular membranes, thus shaping uptake pathways. Moreover, the formation of a protein corona composed of albumin, apolipoproteins, immunoglobulins, and other biomolecules alters cellular recognition and internalization (Bilardo et al., 2022). Environmental weathering further increases bioavailability by inducing oxidation of the particle surface, increasing roughness, and enhancing the affinity for biological interfaces (Finlay et al., 2020). In the gastrointestinal tract, MNPs interact with mucus, bile acids, digestive enzymes, and the commensal microbiota; smaller or weathered particles can penetrate Peyer's patches and exploit M-cell-mediated transcytosis, thereby contributing to chronic, low-level systemic uptake. Similarly, during inhalation exposure, NPs can penetrate deep into the alveolar regions of the lungs, where impaired mucociliary clearance may prolong retention and increase the likelihood of systemic translocation.

Following the internalization of MPs, biopersistence becomes an important factor in shaping long-term toxicity, as most plastic polymers are chemically stable and degrade at very slow rates within biological environments (Khaki and Kumar, 2025). While intracellular NPs can undergo restricted oxidative modification or further fragmentation, they are generally not efficiently metabolized or cleared. Consequently, MNPs are deposited in multiple organs, including the liver, where they are sequestered by Kupffer cells; the spleen, through monocyte–macrophage uptake; and the brain, via BBB crossing or olfactory neuronal routes. Accumulation has been predominantly observed in placental and fetal tissues, raising concerns about developmental exposure and associated abnormalities. The persistent retention of MNPs drives chronic inflammation, oxidative stress, and cumulative endocrine dysregulation and may contribute to the development of neurodegenerative diseases through sustained microglial activation (Chrishtop et al., 2021). Recent studies in mice have demonstrated that persistent MPs can lodge in the cerebral microvasculature, causing focal blockage and impaired perfusion. Long-term sequestration within endocrine organs such as the thyroid, adrenal glands, and gonads may amplify hormonal disruption by prolonging local exposure to particles and their associated chemical additives. High bioavailability and biopersistence establish a foundation for chronic, cumulative, and multiorgan toxicity associated with MNP exposure.

16.8 Exposure routes and systemic translocation

MPs (<5 mm to ~1 μm) and NPs (<1 μm, often <100–1000 nm) exhibit size-dependent biological behavior, with particle dimensions serving as the primary determinant of whether they remain confined to the gastrointestinal lumen or traverse epithelial and vascular barriers. In addition to size, physicochemical characteristics such as particle shape (spherical, fragmented, fibrous), surface charge, hydrophobicity, and the composition of the biomolecular corona formed by adsorbed proteins, lipids, and cooccurring environmental contaminants [per- and polyfluoroalkyl substances (PFASs) and phthalates] collectively modulate cellular uptake pathways, immune detection, and systemic biodistribution (Dzierżyński et al., 2024). During ingestion, these features affect the efficiency of gastrointestinal absorption; smaller or nanoscale particles, particularly those with hydrophobic or functionally modified surfaces and a biologically conditioned corona, are more likely to undergo transcytosis or paracellular passage across the intestinal epithelium, thereby gaining access to circulation and contributing to systemic exposure (Fig. 16.4).

Nanosized MNPs (particularly <200 nm, especially <100 nm) can be transported across barriers via multiple size-dependent mechanisms, making ingestion a dominant exposure route. The American Chemical Society reported that transcellular endocytosis by enterocytes and M-cells is a primary pathway through which clathrin-/caveolin-mediated uptake and macropinocytosis involve nanoparticles and shuttle them through endosomal–lysosomal trafficking networks. When the gut barrier is compromised by inflammation, chemical coexposures, oxidative stress, or disease, tight junction dysfunction increases paracellular leakage, leading to decreased epithelial integrity and promoting particle passage between cells. Complementing these epithelial routes, Peyer's patches and GALT-associated M-cells provide a highly efficient pathway for nanoparticle translocation into the lymphatic circulation; these professional antigen-sampling cells readily transport nanoscale particles to immune compartments. In addition, MNPs often act as "Trojan horses," carrying adsorbed

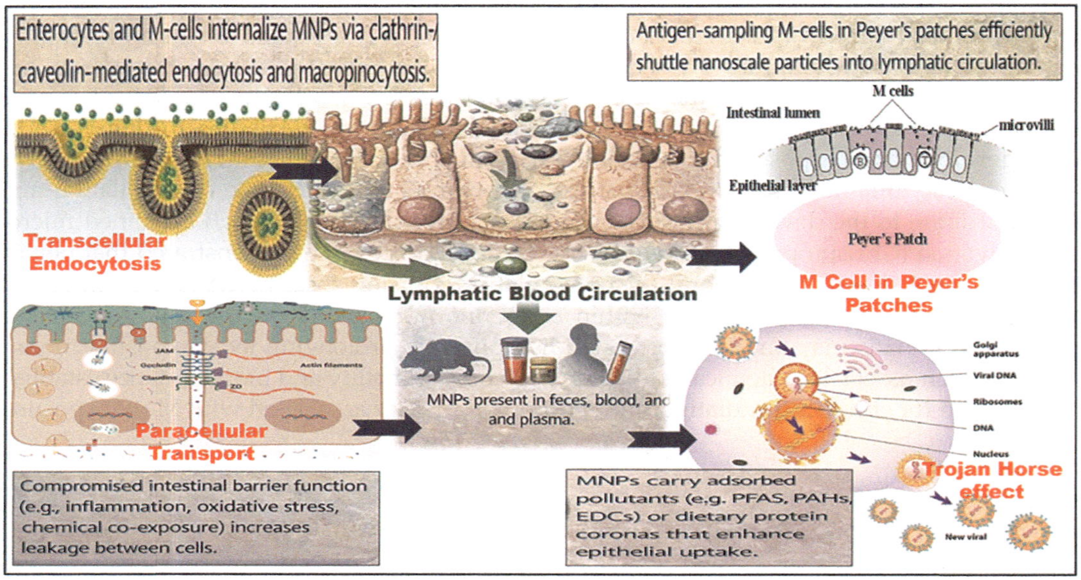

FIG. 16.4 Mechanisms of intestinal uptake of micro- and nanoplastics (MNPs). The figure illustrates the primary pathways through which MNPs cross the intestinal barrier, including transcellular endocytosis by enterocytes, uptake via M cells in Peyer's patches, paracellular transport through disrupted tight junctions, and subsequent translocation into systemic circulation.

lipophilic contaminants (e.g., PFAS, PAHs, and EDCs) or biomolecular coronas composed of dietary proteins, which can modulate cellular interactions and enhance epithelial uptake (Mahendran and Ramaswamy, 2024). Experimental animal and in vitro studies have consistently corroborated these mechanisms, reporting that the size-dependent intestinal translocation of PS and other NPs is amplified by barrier disruption (Kustra et al., 2025). Importantly, clinical studies have shown that these particles (MNPs) are present not only in feces, indicating widespread dietary exposure, but also in human blood and plasma. Collectively, these converging mechanistic and empirical findings indicate that ingestion is among the major pathways for internal exposure, with NPs (<200 nm) most capable of accessing lymphatic and blood compartments through endocytosis, M-cell transport, or paracellular leakage, especially under conditions of intestinal inflammation or compromised barrier function (Escher et al., 2017).

16.8.1 Deposition and interface with lung tissue

Inhalation is a quantitatively important route of exposure to MNPs, primarily because the aerodynamic diameter dictates how deeply particles penetrate the respiratory tract. Particles >10 μm tend to deposit in the upper airways, those 2.5–10 μm settle in the bronchi, and particles <2.5 μm, especially nanoscale particles <1 μm, reach the alveoli, where the thin gas-exchange epithelium and dense capillary network create an efficient interface for systemic uptake (Casella et al., 2025). Following deposition in the alveolar region, NP particles can cross into circulation via various pathways, including transcytosis by alveolar epithelial cells and uptake by pulmonary macrophages, followed by transport across the epithelial barrier into the interstitium and capillaries. When the tight junctions of the lung are weakened either by particle-induced oxidative stress and inflammation or by preexisting respiratory disease, paracellular leakage further increases particle passage. Moreover, cell-mediated mechanisms, such as macrophages engulfing particles, can result in migration, apoptosis, or the transfer of ingested material to neighboring immune cells. In contrast, dendritic cells may transport particles toward lymphatic vessels (Leikauf et al., 2020). Empirical evidence supports these mechanisms, as both ambient and indoor air monitoring consistently identify substantial loads of MP fibers and fragments, with sources ranging from urban dust and synthetic textiles to disposable masks, and experimental inhalation models demonstrate pulmonary inflammation, oxidative damage, and occasional detection of translocated particles in systemic tissues. Taken together, these findings show that inhaled MNPs can reach the deepest regions of the lung and that nanoscale particles have the greatest potential for translocation into blood, an effect markedly amplified under inflammatory or compromised pulmonary conditions.

16.8.2 Dermal exposure: Barriers, evidence, and limits

Although the intact stratum corneum is a highly effective physical and biochemical barrier, dermal exposure to MNPs remains plausible, particularly for nanoscale particles (<100 nm) and for particles incorporated into cosmetic or sunscreen formulations that enhance skin–particle interactions (Menichetti et al., 2025). The likelihood of penetration is strongly influenced by particle size, shape, surface chemistry, and the physiological state of the skin; compromised skin, including wounds, eczema, aging-related barrier thinning, or chemically irritated skin, exhibits significantly increased permeability, enabling deeper movement of nanosized particles (Menichetti et al., 2025). Mechanistically, hair follicles and sebaceous units provide a key follicular route that acts both as a reservoir and as an access point for nanoparticles. Moreover, transcellular endocytosis in keratinocytes has been demonstrated for various engineered nanomaterials in in vitro skin models, suggesting that similar processes may apply to NPs (Fang et al., 2014). When the epidermal barrier is physically breached, as in wounds, particles can bypass the stratum corneum completely and gain direct paracellular or open access to the dermis, increasing the potential for local immune interactions and possibly deeper tissue movement. Although clinical studies have reported that systemic dermal uptake of environmental MPs remains restricted, in vitro, ex vivo human skin, and cell culture studies have shown that nanoscale particles can penetrate or at minimum reach deeper epidermal layers, where they elicit oxidative stress, inflammatory signaling, and cytotoxic responses). Previous studies have reported that cosmetics and personal-care products are realistic sources of dermal MNP exposure; these findings underscore that while systemic translocation via skin is less documented than ingestion or inhalation, dermal penetration, especially under conditions of barrier compromise or with formulation-assisted delivery, is biologically plausible and represents an emerging research area (Bhardwaj et al., 2025; Pecorelli and Valacchi, 2023).

16.8.3 Crossing critical biological barriers

16.8.3.1 Intestinal epithelium: Uptake mechanisms, evidence, and consequences

The intestinal epithelium is the first major biological barrier that MPs/NPs encounter following oral exposure; particle size, shape, surface chemistry (charge, hydrophobicity), and the biomolecular/ecocorona acquired in the gut lumen determine the predominant uptake route. Large (>~1 μm) MPs are likely to accumulate in the lumen or become trapped in mucus, whereas NPs (<1000 nm and especially <100 nm) can interact directly with epithelial cells and use multiple entry paths, such as transcellular endocytosis/transcytosis (clathrin- or caveolin-mediated), paracellular passage when tight junctions are disrupted, M-cell transcytosis in Peyer's patches, and "persorption" through transient gaps created during epithelial turnover (Luo et al., 2025) (Fig. 16.5).

The corona (adsorbed proteins, lipids, and cocontaminants such as PFAS or phthalates) alters particle recognition and uptake; for example, certain corona compositions increase opsonization and receptor-mediated uptake, whereas others sterically hinder interactions. Evidence from human-relevant models includes human intestinal organoids and Caco-2 monolayers showing size- and charge-dependent uptake and limited but measurable translocation of nanosized PS particles; in vivo rodent gavage studies likewise report that nanosized particles can appear in mesenteric lymph, liver, and blood within hours. In experimental models, intestinal exposure to MPs/NPs is associated with epithelial oxidative stress, altered tight junction protein expression, low-grade inflammation, and perturbations of the gut microbiome, effects that may increase paracellular leakage and, consequently, systemic translocation (Fig. 16.6).

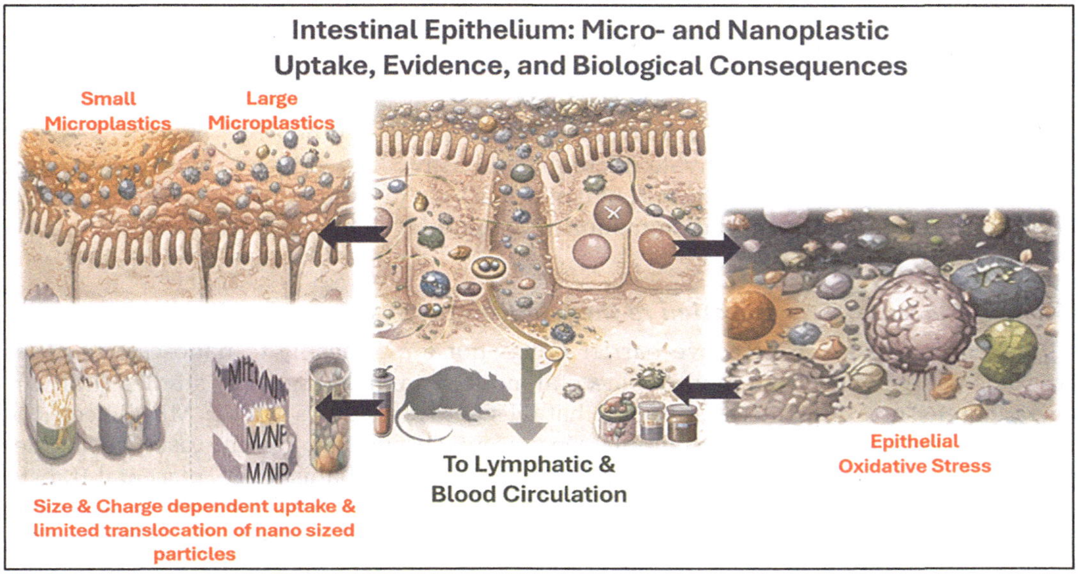

FIG. 16.5 Intestinal epithelium: micro- and nanoplastic uptake, evidence, and biological consequences. The figure summarizes experimental and observational evidence for MNP uptake across the intestinal epithelium and illustrates associated biological effects, including epithelial barrier disruption, inflammation, oxidative stress, and altered cellular and immune responses.

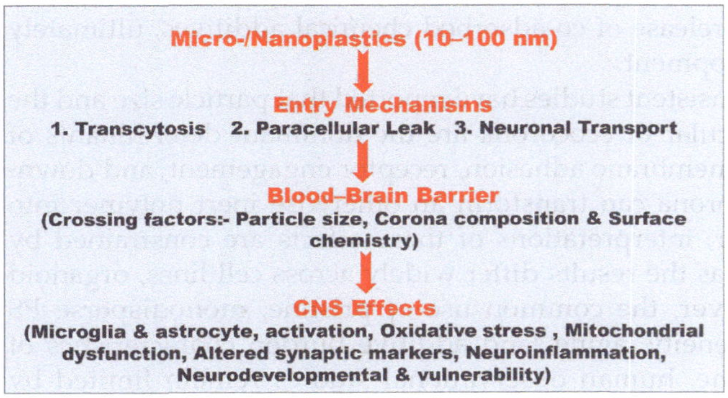

FIG. 16.6 Flowchart for blood–brain barrier (BBB): effects of micro- and nanoplastic entry and the CNS. The flowchart illustrating micro- and nanoplastic (MNP) entry across the blood–brain barrier (BBB) and associated central nervous system (CNS) effects. The figure outlines potential transport pathways across the BBB, the subsequent accumulation in brain tissues, and the resulting neurotoxic outcomes, including oxidative stress, neuroinflammation, and disruption of neuronal signaling.

16.8.3.2 BBB: How MPs/NPs reach the brain and what they do there

The BBB is highly selective, but NPs (typically tens of 10–100 nm) have been continually used in experimental studies to examine their effects on brain tissue under certain conditions (Ma et al., 2025). Three mechanisms have been proposed for this intervention: First, transcytosis across brain microvascular endothelial cells (receptor- or adsorptive-mediated), especially when a biomolecular corona presents lipoproteins or other ligands recognized by endothelial receptors; second, paracellular leakage when tight junctions are disrupted by inflammation or direct nanoparticle effects; and third, indirect routes such as neuronal transport via the olfactory nerve following nasal exposure. Various studies on rodents have reported that, when exposed to NPs via oral or systemic routes, their corona composition (cholesterol- or protein-enriched coronae) strongly alters membrane interactions and the ability of particles to cross lipid bilayers and BBB models. When NPs reach the central nervous system (CNS), they activate microglia and astrocytes, induce oxidative stress, and disturb mitochondrial functions and synaptic markers in animal models. These changes increase concerns about neuroinflammation and neurodevelopmental vulnerability.

16.8.3.3 *Placenta: Translocation to the fetal compartment and implications*

The placenta is a semiselective barrier but is not impermeable to nanoscale particles. Experimental evidence from in vitro placental models, animal pregnancy studies, and various human tissue examinations supports that NPs (and sometimes smaller MPs detected by sensitive analytic methods) can accumulate in placental tissue and, in some cases, be detected on both the maternal and fetal sides (Dusza et al., 2023). Mechanistically, placental transfer likely occurs through trophoblast endocytosis/transcytosis, paracellular passage when barrier integrity is compromised (e.g., inflammation and oxidative stress), and possibly trafficking within immune cells. Experimental studies in animals have shown that fetal deposition following maternal exposure is associated with altered fetal growth indices and tissue inflammation. Human observational work has detected diverse plastic polymers in term placentas and amniotic fluid; however, methods and contamination control are active areas. The potential consequences for fetal development include localized placental inflammation, impaired placental nutrient and oxygen transport, and endocrine disruption resulting from particle-induced inflammation or the release of co-adsorbed chemical additives, ultimately compromising fetal growth and development.

In terms of all biological barriers, consistent studies have reported that particle size and the composition of the acquired biomolecular or ecocorona are the dominant determinants of MNP interactions with cells, shaping membrane adhesion, receptor engagement, and downstream biodistribution. Indeed, the corona can transform an otherwise inert polymer into a biologically active ligand. However, interpretations of these effects are constrained by substantial limitations in the models, as the results differ widely across cell lines, organoid systems, and animal models. Moreover, the common use of pristine, monodisperse PS spheres fails to replicate the heterogeneity, aging, and additive burden characteristics of real-world particles. At the same time, human observational studies remain limited by contamination control and detection limits. Therefore the mere presence of MPs or NPs in human blood, placenta, or tissues, although evidence of exposure and translocation, does not establish causality for disease; instead, mechanistic animal studies demonstrating inflammation, oxidative stress, and endocrine disruption highlight plausible pathways requiring targeted human epidemiological investigation, especially for neurodevelopmental and perinatal outcomes. Finally, focusing on mechanistic endpoints such as tight junction integrity, transcytosis pathways, transport kinetics, and interactions with coexposures, such as PFAS and phthalates, will provide the most biologically informative insights into how MPs traverse critical barriers and exert potential toxicity.

16.9 Endocrine disruption

MNPs exert endocrine-disrupting effects through two primary mechanisms, viz., direct interference with hormone receptors and functioning as vectors for exogenous EDCs, such as BPA, BPS, and phthalates. The hydrophobic and porous surfaces of MNPs facilitate the accumulation and bioavailability of these persistent pollutants, which can disrupt estrogenic, androgenic, and thyroid hormone signaling. Experimental studies in rodents and zebrafish have consistently revealed that MNP exposure alters the HPG axis, suppresses sex hormone levels, and disturbs the expression of steroidogenic genes. Thyroid hormone synthesis and signaling are also affected by changes in circulating T3/T4 levels and altered expression of deiodinases and thyroid receptors. Oxidative stress further contributes to endocrine dysfunction by damaging endocrine tissues, impairing mitochondrial function, promoting apoptosis, and ultimately disrupting hormone biosynthesis.

Recent studies have reported that MNP exposure induces epigenetic modifications, such as DNA methylation and histone changes, in gonadal tissues, suggesting that MNPs are associated with transgenerational effects. MNPs have the potential to cross the BBB and influence neuroendocrine signaling, including the regulation of gonadotropin-releasing hormone (GnRH). The detection of MNPs in human placenta, breast milk, and fetal tissues raises concerns about perinatal hormonal disruption and long-term developmental impacts. Therefore these findings indicate that MNPs pose significant risks to reproductive health by interfering with gametogenesis, hormone production, placental function, embryogenesis, and fetal development. The upstream mechanisms of oxidative stress and endocrine disruption initiate a cascade of cellular dysfunctions in reproductive tissues across both sexes.

In females, MNP exposure disrupts folliculogenesis and oocyte maturation through oxidative stress, mitochondrial dysfunction, and ferroptosis. Placental tissues are also affected, with MNPs inducing autophagy inhibition, inflammasome activation, and necroptosis, thereby impairing hormone synthesis, disrupting cytokine signaling, reducing nutrient transport, and restricting fetal growth. In males, MNPs compromise the blood–testis barrier and damage Sertoli and Leydig cells, resulting in decreased testosterone synthesis and impaired sperm quality. These effects are mediated by mitochondrial and lysosomal damage, ROS overproduction, MAPK–Nrf2 pathway activation, apoptotic signaling, and lipid peroxidation, all of which exacerbate cellular injury and hormonal imbalance.

Beyond direct organ toxicity, MNPs can be translocated from parents to offspring, affecting the neurological, immune, metabolic, and reproductive systems, which raises serious concerns about transgenerational toxicity. While in vitro and organoid studies have shown that MNPs can harm human-derived germ cells and placental tissues, long-term and heritable effects remain insufficiently characterized (Miloradovic et al., 2021). Hence, the evidence emphasizes that MNP-induced reproductive toxicity results from ROS-driven cell death, endocrine disruption, and epigenetic modifications, compromising fertility and posing potential transgenerational risks in both males and females (Hernández et al., 2025).

Recurring findings across organ systems include barrier disruption (intestinal, vascular, and placental), oxidative and inflammatory injury in hepatic and cardiac tissues, reproductive dysfunction, and neurobehavioral alterations. Although limited, human biomonitoring studies have confirmed the presence of MNPs in multiple matrices, including the lungs, blood, placenta, feces, liver, kidney, and brain. Common polymers detected in the lungs, blood, placenta, feces, liver, kidney, and brain, such as PE, PP, and PET, indicate widespread exposure through inhalation and ingestion (Lechuga et al., 2023).

16.10 Nervous system

The neurotoxic potential of MNPs is being increasingly recognized because of their ability to cross biological barriers, accumulate in key brain regions such as the hippocampus and cortex, and impair CNS function (Prüst et al., 2020). Neurotoxicity arises through multiple interconnected mechanisms, including oxidative stress, mitochondrial dysfunction, neuroinflammation, and disrupted gene expression. MNPs generate excessive amounts of ROS, which lead to lipid peroxidation, mitochondrial injury, DNA damage, and subsequent neuronal apoptosis. In murine models, PS-NPs have been shown to impair spatial memory and synaptic integrity by inducing mitochondrial dysfunction via ROS. Moreover, zebrafish exposed to MNPs exhibit reduced locomotor activity, apoptosis, and altered neurotrophic signaling, including decreased expression of brain-derived neurotrophic factor (Chen et al., 2025).

In vitro studies using human SH-SY5Y neurons further demonstrated that MNPs activate AMPK/ULK1-mediated mitophagy, resulting in mitochondrial fragmentation and neuronal cell death (Kong et al., 2024). Disruption of autophagic balance is therefore implicated in neurodegeneration-like phenotypes. Neuroinflammatory pathways also contribute significantly to MNP-induced neurotoxicity, with microglial activation and NF-κB signaling leading to elevated proinflammatory cytokine production and depressive-like behaviors in murine models. Developmental exposure exacerbates these effects by interfering with neurotransmitter systems, including GABAergic and cholinergic signaling, resulting in behavioral abnormalities such as anxiety, depressive-like behavior, and impaired social interactions in offspring (Hernández et al., 2025).

Exposure to polypropylene nanoplastics (PP-NPs) in both mouse models and human brain organoids further impaired neurodevelopment by inhibiting neuronal differentiation, reducing neural stem cell proliferation, and compromising cognitive and motor functions. Mechanistic studies link these outcomes to the dysregulation of neurodevelopmental signaling pathways and the downregulation of key neuronal markers, including TUJ1, MAP2, and PAX6. Moreover, molecular assessment revealed CYSLTR1 and PTH1R as potential receptor-mediated targets of PP-NPs, suggesting that receptor-specific interactions may underlie some aspects of MNP neurotoxicity (Jha et al., 2022). Therefore these findings reveal the heightened vulnerability of the developing brain to NP exposure and emphasize the need for preventive strategies to safeguard fetal neurodevelopment during pregnancy, as well as targeted research to clarify molecular mechanisms, critical windows of susceptibility, and long-term cognitive and behavioral consequences (Preda et al., 2024).

16.10.1 Mechanistic pathways

16.10.1.1 *Oxidative stress and ROS generation*

Exposure to MNP particles consistently aggravates rapid bursts of ROS in exposed cells and tissues. ROS are produced from particle surface reactivity and from chemical cocontaminants (plastic monomers, plasticizers, adsorbed metals, and POPs) that catalyze redox cycling at the cellular level. MNPs stimulate membrane NADPH oxidase activity and disturb redox-sensitive pathways, resulting in the production of superoxide and hydrogen peroxide, which overwhelm antioxidant defenses reduced glutathione (GSH) and superoxide dismutase (SOD) and catalase (Dennis et al., 2025). Compensatory activation of the Nrf2 antioxidant response is often observed. Nevertheless, it is frequently insufficient for chronic exposure; thus oxidative damage accumulates in lipids (lipid peroxidation), proteins (carbonylation), and DNA (8-oxoG), which in turn sensitizes neurons and glia to downstream inflammatory and apoptotic signaling, revealing that oxidative stress is a primary and early event that links particle exposure to downstream neuronal dysfunction.

16.10.1.2 *Neuroinflammation and microglial activation*

ROS and particle recognition jointly trigger innate immune sensing in the brain. MNPs can cross or disrupt the BBB (or are transported via peripheral immune trafficking), are internalized by microglia and astrocytes, and engage pattern-recognition receptors (for example, Toll-like receptors (TLRs)) and endosomal danger sensors (Battaglini et al., 2024). This leads to microglial activation, morphological transformation, NF-κB signaling, and the production of proinflammatory cytokines (IL-1β, IL-6, TNF-α) and chemokines; various studies have also reported NLRP3 inflammasome assembly, caspase-1 activation, and IL-1β maturation (Biswas, 2023). Interestingly, microglial activation is not merely a transient defensive response; chronic or repeated exposures induce a sustained, primed microglial state that amplifies ROS production, impairs phagocytic clearance, and drives maladaptive synaptic pruning and oligodendrocyte stress, processes that are mechanistically implicated in neurodegeneration (Qin et al., 2023). The results of animal experiments and targeted cell studies replicate this pattern (particle uptake→cytokine release→chronic neuroinflammatory milieu) and link it to behavioral outcomes (Fig. 16.7).

16.10.1.3 *Mitochondrial dysfunction and impaired energy metabolism*

MNPs, especially NPs, readily interact with and localize to mitochondria or compromise mitochondrial membranes indirectly through lipid peroxidation and calcium dysregulation. Experimental studies have reported mitochondrial membrane depolarization, reduced activity of electron–transport chain complexes (notably complexes I and III), decreased ATP production, elevated mitochondrial ROS, and activation of mitophagy or apoptotic cascades (cytochrome *c* release, caspase-9). These defects both amplify cellular oxidative stress (a feed-forward loop) and deprive neurons of the ATP required for ion pump function and synaptic vesicle cycling. Transcriptomic and imaging studies in neuronal cultures and in vivo models corroborate mitochondrial pathways as key mechanistic nodes, explaining many electrophysiological abnormalities and the vulnerability of high-energy neurons (e.g., dopaminergic and hippocampal pyramidal cells) (Miao et al., 2025).

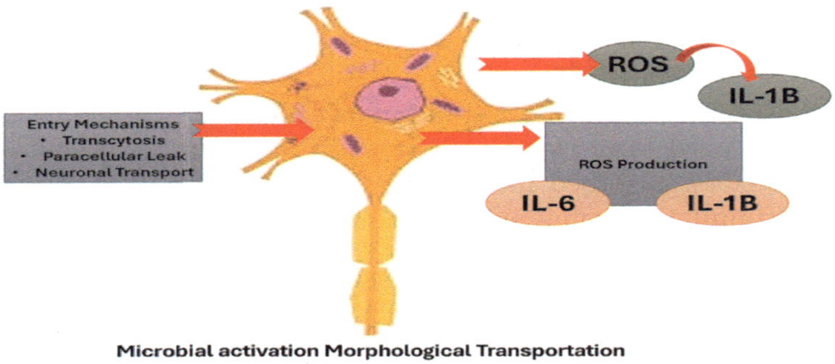

FIG. 16.7 Neuroinflammatory pathways and microglial activation triggered by micro- and nanoplastics. The figure depicts how MNP exposure activates microglia, stimulates inflammatory signaling cascades, increases proinflammatory cytokine release, and promotes oxidative stress, collectively contributing to neuronal dysfunction and neurotoxicity.

16.10.1.4 Disruption of synaptic plasticity and neurotransmission

The combined effects of ROS, chronic neuroinflammation, and mitochondrial energy failure converge on synapses. Oxidative modification of synaptic proteins and lipid membranes, inflammatory cytokine-mediated modulation of receptor trafficking, and ATP shortage for vesicle cycling reduce the probability of neurotransmitter release and alter the excitatory/inhibitory balance (often via glutamate–GABA dysregulation and cholinergic perturbation) (Pan and Yin, 2025). Molecular studies revealed that exposure to MNPs decreased the levels of synaptic scaffolding proteins (PSD-95 and synaptophysin), impaired N-Methyl-D-Aspartate (NMDA)/α-Amino-3-hydroxy-5-methyl-4-isoxazolepropionic acid (AMPA) receptor function, and suppressed the expression of molecular markers of long-term potentiation (LTP) (Pan and Yin, 2025). Behaviorally, these molecular and electrophysiological discrepancies map to learning/memory impairments, anxiety-like phenotypes, and motor anomalies observed in rodents after chronic MNP exposure. Notably, several recent in vivo studies directly linked microglial activation and particle-induced cerebrovascular obstruction to reduced cerebral perfusion and consequent synaptic dysfunction, underscoring that vascular and immune mechanisms can act in parallel. Hence, this evidence supports a multihit model in which oxidative stress primes inflammation, inflammation and ROS damage mitochondria, and mitochondrial/synaptic failure produces the cognitive and motor phenotypes observed experimentally. MNP exposure→early ROS surge and antioxidant insufficiency→microglial activation/inflammasome signaling→mitochondrial injury and ATP depletion→synaptic protein loss and neurotransmission failure→behavioral/cognitive deficits. This sequential but overlapping cascade highlights multiple measurable biomarkers (oxidative damage markers, cytokine panels, mitochondrial function assays, synaptic protein levels, electrophysiology, and behavior) that experimentalists can track to establish causality in future mechanistic work and suggests intervention points (antioxidants, inflammasome inhibitors, mitochondrial stabilizers, and BBB protection) to test mitigation strategies.

In cell-based, molecular, and whole-animal studies, a coherent mechanistic cascade in which exposure to MNPs initiates an early and disproportionate surge in ROS, overwhelming endogenous antioxidant defenses and establishing a state of persistent oxidative stress, has emerged. This redox imbalance acts as a primary trigger for microglial activation and inflammasome assembly, thereby promoting a sustained neuroinflammatory milieu that further amplifies ROS production and tissue damage (Birben et al., 2012). The combined oxidative–inflammatory burden subsequently impairs mitochondrial integrity, leading to membrane depolarization, disrupted electron–transport chain activity, and ATP depletion, changes that critically compromise the high energy demands of neuronal circuits (Jurcău et al., 2022). As mitochondrial dysfunction progresses, downstream synaptic consequences become evident, including the loss of key synaptic proteins, impaired receptor signaling, and disruptions in neurotransmitter dynamics, ultimately manifesting as deficits in cognition, memory, and behavior. This sequential but overlapping cascade provides a framework for identifying mechanistically relevant biomarkers, ranging from oxidative damage indices and cytokine signatures to mitochondrial bioenergetics profiles, synaptic protein quantification, electrophysiological parameters, and behavioral readouts, which can be leveraged to strengthen causal inference in experimental models. Moreover, delineation of this pathway highlights several potential intervention nodes, including antioxidant supplementation, inflammasome inhibition, mitochondrial stabilization, and strategies to preserve BBB integrity, all of which warrant systematic evaluation as candidate mitigation approaches against MNP-induced neurotoxicity (Ferrara, 2023).

16.11 Evidence from experimental studies

16.11.1 In vitro neuronal models

Studies at the cellular level using primary neurons, neuronal cell lines (e.g., SH-SY5Y, PC12), mixed neuron–glial cocultures, and brain-endothelial monolayers have demonstrated that MNP particles produce reproducible, mechanistically informative effects on neural cells. Most studies have explored how MNPs induce the rapid generation of ROS, lipid peroxidation, depletion of glutathione, and activation of redox-sensitive transcriptional programs (for example, partial induction of Nrf2 targets but inadequate compensatory antioxidant responses), together with mitochondrial membrane depolarization and reduced ATP production. In cultured microglia and mixed cultures, MNPs trigger TLR/NF-κB signaling and NLRP3 inflammasome activation with the release of IL-1β, IL-6, and TNFα (Lopez-Suarez et al., 2022). However, exposure to endothelial monolayers and in vitro BBB models revealed increased paracellular permeability and particle translocation mechanisms that plausibly link peripheral exposure to central effects. At the synaptic level, acute and subchronic exposures reduce the expression of synaptic protein markers (PSD-95 and synaptophysin) and impair the expression of chemically induced LTP surrogates in slice/cell systems, providing cellular correlates of later behavioral deficits (Saili et al., 2017). These in vitro observations provide both mechanistic anchors (oxidative stress→inflammation→mito-bioenergetic failure→ synaptic competition) and quantifiable endpoints for hypothesis testing.

16.11.2 In vivo rodent, fish, and invertebrate models

In rodents exposed to PS and other polymer particles via the oral route or inhalation, MNPs accumulate in peripheral organs and, in several studies, in brain tissue, and are associated with microglial activation, cytokine upregulation, and impaired performance on memory and motor tests. Advanced intravital imaging in mice has even demonstrated particle transit into the cerebral microvasculature during transient occlusions that reduce local perfusion and produce measurable behavioral deficits, supporting a vascular–immune mechanism in addition to direct neuronal effects.

Studies conducted on aquatic vertebrate models (zebrafish and medaka) have repeatedly reported that MP contamination disrupts neurodevelopment (altered neurogenesis markers and increased apoptosis in the developing brain), alters locomotor and predator avoidance behaviors, and disrupts cholinergic and monoaminergic signaling after embryonic or larval exposure. Invertebrate taxa (e.g., *Daphnia*, bivalves, and nematodes) provide complementary evidence that altered acetylcholinesterase (AChE) activity impairs sensorimotor behaviors and that population-level fitness decreases at environmentally relevant concentrations, indicating the neurofunctional sensitivity of phylogenetically broad nervous systems to particle exposure (Rojoni et al., 2024). Therefore the in vivo results confirm that the molecular and cellular endpoints observed in vitro translate into organism effects across taxa and identify conserved readouts (oxidative markers, cytokines, AChE changes, and behavioral assays) for environmental and toxicological assessment.

16.11.3 Dose–response patterns and nanoparticle-specific effects

Dose–response assessments in both cell and animal studies reveal nonlinearity and strong size and surface property dependence, particularly for smaller particles (NPs, <100–200 nm). Compared with larger MPs, larger microplastics typically elicit stronger ROS generation, greater cellular internalization, and more pronounced mitochondrial and synaptic perturbations at lower mass concentrations. Particle shape (spheres vs. fibers), surface charge, and the presence of sorbed contaminants or plastic additives modulate bioactivity. Several metaanalytical and systematic review studies reported that some "environmentally realistic" exposures in laboratory work still exceed typical ambient concentrations, complicating extrapolation (Balog et al., 2024). However, other studies now demonstrate neurobehavioral and biochemical changes at environmentally plausible burdens, particularly for chronic, low-dose exposures that allow bioaccumulation (Constantinescu et al., 2025). Importantly, cocontaminant effects (adsorbed metals, POPs, and plasticizers) frequently produce synergistic or additive toxicity, and protein or lipid corona formation in biological fluids can alter particle fate, BBB translocation, and immune recognition, thereby changing the effective dose at the target tissue (Rafa et al., 2024). These patterns indicate that hazard characterization must integrate particle size, surface chemistry, coexposures, and realistic exposure kinetics (chronic low dose, life-stage vulnerability) rather than rely solely on mass-based metrics.

16.11.4 Human health implications

16.11.4.1 *Cognitive deficits and behavioral alterations*

Experimental studies across cellular systems and multiple animal taxa consistently report that MNP exposure produces functional shortfalls that map onto impaired cognition and altered behavior (Lamoree et al., 2025). Experimental studies on rodent and fish models revealed that learning and memory impairments reduced exploratory or social behaviors and increased anxiety-like phenotypes and motor abnormalities after chronic or developmental exposure. These results cooccur with hallmark mechanistic changes (oxidative damage, microglial activation, synaptic protein loss, and impaired LTP-like measures), providing a plausible mechanistic bridge from particle exposure to altered neural circuit function. Translationally, early human biomonitoring studies reported that MNPs in blood, urine, and other tissues, together with initial epidemiological signals linking higher urinary MNP burden to poorer cognitive scores in school-aged children, raise concerns about their experimental relevance to humans.

16.11.4.2 *MNPs induce neurodevelopment*

MNPs induce unique patterns of brain development by increasing neurogenesis, synaptogenesis, and cellular proliferation, as well as the formation of immature barrier systems (the placenta and the neonatal BBB). The exposure of zebrafish and rodents to MNPs during the embryonic and perinatal stages results in altered neurodevelopmental trajectories, such as decreased production of neurogenesis markers, increased apoptotic signaling in developing brain regions, disrupted maturation of neurotransmitter systems (cholinergic and monoaminergic pathways), and persistent behavioral deficits that emerge later in life, supporting the biological plausibility that perinatal MNP exposure could perturb human neurodevelopment. In addition, particle translocation across the placenta and detection of MNPs in breast milk and neonatal tissues have provided mechanistic routes for early-life exposure; together, these data justify targeted epidemiological cohorts that measure prenatal/early-life exposure, neurodevelopmental biomarkers, and standardized cognitive/behavioral endpoints.

16.11.4.3 *Links to neurodegenerative risk*

Previous studies hypothesized that chronic, lifelong MNP accumulation could accelerate neurodegenerative processes by sustaining neuroinflammation, repeated oxidative insults, mitochondrial compromise, impaired proteostasis, and cerebrovascular dysfunction. Surprisingly, recent postmortem analyses reported measurable MNP burdens in human brain tissue and suggested higher loads in some individuals with dementia, but these observations are correlative and confounded by methodological and sampling limitations; however, they do not yet demonstrate that MNPs initiate or accelerate neurodegeneration in humans. Mechanistic animal and cellular data revealed that particles potentiate pathways implicated in the pathology of Alzheimer's disease and Parkinson's disease (inflammation, synaptic loss, and dopaminergic neuron vulnerability), providing a biologically plausible link; however, this requires longitudinal human studies and experimental validation of dose, kinetics, and thresholds for harm. Until stronger population-level evidence exists, the prudent interpretation is that MNPs are a plausible environmental modifier of neurodegenerative risk deserving prioritized, well-designed translational research rather than a proven causal factor.

16.12 Endocrine disruption induced by MPs

MPs (<5 mm) and NPs (<1 μm) are recognized as pervasive contaminants across different ecosystems, such as aquatic, terrestrial, and human environments, indicating that they function not only as inert particles but also as chemically active agents capable of disrupting endocrine systems. Their endocrine-disrupting potential might be due to two fundamental chemical roles: The first role is that MPs contain internal additives such as phthalates, bisphenols, brominated flame retardants, organotins, and various stabilizers integrated during polymer manufacturing, which gradually leach out of particles into surrounding environments; the second role is that MPs serve as vectors that adsorb and transport exogenous EDCs from water, sediments, and food into organisms. Recent reviews and experimental studies underscore the significance of these dual roles in shaping the toxicological profile of MPs. The release of internal additives is governed by physicochemical factors, including the additive–polymer affinity, polymer crystallinity, glass transition properties, particle size, and surface-area-to-volume ratio. Environmental drivers such as UV irradiation, temperature fluctuations, salinity, and dissolved organic matter accelerate surface oxidation and cracking, facilitating the diffusion of additives. Experimental leaching studies have consistently reported that the linear and long-term release of compounds, most notably phthalates from PVC and other polymers, is greater for UV-weathered and photooxidized MPs. Henkel and colleagues reported the multimonth, continuous release of DEHP and related PAEs, and corroborating studies revealed that leachates from aged agricultural films induce endocrine and developmental abnormalities in fish models, emphasizing that leachates, independent of particles, frequently account for observed endocrine toxicity (Bhaumik and Chakraborty, 2024).

In parallel, MPs act as effective sorbents for hydrophobic organic contaminants (HOCs), including PBDEs, PCBs, organochlorine pesticides, and other endocrine-active pollutants. Sorption occurs through hydrophobic partitioning, π–π interactions, and van der Waals forces. Moreover, aging introduces oxygenated functional groups that modulate the sorption capacity, often enhancing the uptake of semipolar compounds. The type of polymer strongly influences the sorption affinity. PE, PP, and PS have high sorption potential, while particle size, biofilm formation, and environmental chemistry also play decisive roles. These sorbed contaminants can desorb in the gastrointestinal environment, creating local hotspots of chemical exposure. Consequently, compared with other exposure pathways, MPs function as chemical concentrators and delivery vehicles, although the relative importance of MP-mediated transport varies contextually.

Recent findings on specific additive classes highlight clear endocrine implications. Phthalates, especially DEHP, constantly leach from MPs and are well-established antiandrogenic agents that disrupt testosterone synthesis and impair reproductive development. Similarly, bisphenols (BPA and its analogs) are frequently detected in plastic leachates and have been shown to elicit estrogenic responses, thyroid disruption, and adipogenic alterations, as observed in zebrafish and mammalian cell models. Brominated flame retardants and PBDEs, which readily sorb to MPs, pose risks for thyroid hormone dysregulation and neurodevelopmental toxicity (Mohajer, 2025). These chemical interactions are reflected in biological systems because in vitro studies have demonstrated that, compared with chemical-only exposure, coexposure to MPs and EDCs amplifies receptor activation, disrupts steroidogenic pathways, and increases cytotoxicity, likely because of localized desorption at cell surfaces. In vivo aquatic studies corroborate these mechanisms, reporting vitellogenin induction, reproductive impairment, thyroid hormone disruption, developmental delays, and behavioral changes in fish and bivalves exposed to MPs, MP leachates, or EDC-loaded particles. Notably, experimental designs that compare particle-only, leachate-only, and combined exposures repeatedly reveal synergistic, not merely additive, effects, with a 2025 zebrafish and MLTC-1 cell study demonstrating pronounced endocrine disruption when PE-MPs and BPA are coadministered.

Emerging mammalian studies similarly indicate that MPs and their leachates impair steroidogenesis, induce oxidative stress in endocrine tissues, and alter hormonal profiles. However, translating these doses to human relevance remains an ongoing challenge. In humans, the detection of MPs in stool, placenta, blood, and, most recently, in ovarian follicular fluid underscores the biological plausibility of direct exposure of endocrine organs to MP-associated chemicals. However, establishing causal epidemiological associations remains difficult because of confounding sources of EDCs (food packaging, indoor dust, and consumer products), analytical challenges in controlling contamination, and the lack of methods to differentiate MP-derived chemicals from other exposure pathways.)

16.12.1 MPs as endocrine-disrupting particles

MPs (<5 mm) and NPs (<1 μm) act as endocrine-disrupting particles through two intimately connected chemical pathways, such as the leaching of internally incorporated additives (phthalates, bisphenols, brominated flame retardants, organotins, and stabilizers) and the adsorption/transport (vectoring) of external EDCs from the environment. Additive release is controlled by intrinsic polymer–additive chemistry (partitioning, crystallinity, and glass transition temperature) and extrinsic environmental forcing (UV/photooxidation, temperature, salinity, and dissolved organic matter). Thus softer, low-crystallinity polymers (e.g., PVC) and smaller particles with large surface-to-volume ratios release plasticizers far more readily than highly crystalline polymers do (e.g., PET). Furthermore, photoaging dramatically increases surface oxidation, cracking, and diffusion, producing steady, long-term leaching profiles rather than single pulses and generating transformation products (e.g., MEHP from DEHP) that can themselves be biologically active.

Empirical work over the past few years has confirmed that these leachates are associated with toxicological impairments, such as controlled leaching experiments showing the continuous release of DEHP and related phthalates from PVC MPs over months, and that UV-weathered agricultural films produce leachates that induce developmental and endocrine perturbations in fish embryos, indicating that chemical fractions often explain much of the endocrine toxicity attributed to MPs in aquatic models. Moreover, MPs in the environment sorb HOCs such as PBDEs, PCBs, and many pesticides via hydrophobic partitioning, π–π interactions, and van der Waals forces, and aging increases their affinity or alters their selectivity by introducing oxygenated functional groups and surface roughness; polymer type (PE, PP, PS), biofilm formation, particle size, and ambient chemistry (salinity and dissolved organic carbon) further modulate their sorption and desorption behavior, indicating that MPs can concentrate low-level EDCs from water or sediments and deliver them into digestive tracts or tissues where desorption creates localized hotspots of exposure. Biologically, these two pathways combine and interact to produce complex endocrine outcomes. In vitro and in vivo studies have demonstrated that exposure to MP-bound chemicals or MP leachate alters steroidogenic enzyme expression, activates nuclear hormone receptors (e.g., estrogen receptors), and disrupts thyroid signaling (Prajapati et al., 2022).

Moreover, MNP exposure also disrupts the reproductive system, such as vitellogenin induction, gametogenesis, and offspring development; importantly, coexposure experiments increasingly report nonadditive (synergistic or potentiated) effects when MPs are present with bisphenols or phthalates, implying that MPs can both add their own chemical load and magnify the action of copresent pollutants by concentrating them at tissue interfaces. MPs and NPs have been detected in human tissues and reproductive fluids (placenta, blood, stool, and, recently, ovarian follicular fluid), suggesting plausible exposure routes to endocrine organs. However, the understanding of the underlying mechanism remains limited because EDCs are ubiquitous and originate from many sources, and methods for distinguishing MP-derived chemicals from other pathways are not yet mature.

16.12.2 Disruption of endocrine axes

MPs and NPs are now ubiquitous in aquatic, terrestrial, and human environments. They should be conceptualized not only as physical particles but also as combined particle–chemical stressors that disrupt endocrine homeostasis. Two chemical pathways principally drive endocrine outcomes: (1) the leaching of internally added additives (phthalates, bisphenols, brominated flame retardants, organotins, and stabilizers) that are not fully polymer-bound and thus migrate into surrounding media, especially from soft, low-crystallinity polymers such as PVC; (2) environmental endocrine-active pollutants (HOCs, PFASs, pesticides, and PBDEs) can be adsorbed/vectored onto weathered plastic surfaces, which can concentrate low-level contaminants and transport them into organisms. Photooxidative aging, small particle size (high surface-to-volume ratio), temperature, salinity, and dissolved organic matter all accelerate the release of additives and alter sorption selectivity, producing leachates and sorbed mixtures whose bioavailability differs markedly from that of bulk environmental reservoirs. Experimental leaching and modeling studies have revealed that the long-term release of phthalates (e.g., DEHP) from PVC MPs, with UV photoaging, substantially increases leaching rates and the generation of transformation products that may themselves be biologically active.

In organisms, dual chemicals produced through multiple routes can lead to endocrine dysfunction. In vitro and in vivo studies across taxa revealed that MP exposure, particularly to weathered particles, leachate, or MPs preloaded with EDCs, altered the expression of steroidogenic enzymes (aromatase, 17β-Hydroxysteroid dehydrogenase (17β-HSD), and cytochrome P450 enzymes (CYPs), activated or antagonized nuclear hormone receptors (ER, AR, and TR), and disturbed hormone synthesis/metabolism.

Studies in aquatic vertebrate and invertebrate models have consistently reported reproductive functions (vitellogenin induction, disrupted gametogenesis, and reduced fecundity), developmental deficits (growth retardation and neurodevelopmental changes), and thyroid dysfunction (altered T4/T3 ratios and dysregulated TRH/Thyroid-stimulating hormone beta subunit (TSHβ) expression). Many studies have shown that the chemical fraction (leachate or sorbed EDCs) explains a large share of observed endocrine toxicity, while simultaneous particle effects (inflammation, translocation) can potentiate those chemical effects (Ghosh et al., 2022). Recent mechanistic reviews and primary studies further report that mixed interactions between MPs and copollutants (PFASs, algal toxins, and BPA) often result in nonadditive and synergistic adverse outcomes.

16.12.3 HPA axis: Cortisol dysregulation and stress signaling

Exposure to MNPs disturbs the HPA axis by aggravating systemic stress responses, altering inflammatory signaling, and modifying adrenal output, with reported changes in circulating corticosterone/cortisol and upstream adrenocorticotropic hormone (ACTH) levels.

Moreover, particles and their chemical cargo (phthalates, bisphenols, and flame retardants) trigger gut and systemic inflammation, oxidative stress, and dysbiosis that communicate with the brain via immune and vagal pathways, activating hypothalamic corticotropin-releasing hormone (CRH) release and downstream HPA signaling. In parallel, some EDCs, which directly interfere with steroidogenic enzymes in the adrenal cortex, are associated with plastics and change the biosynthesis of cortisol. Moreover, it has been reported that both hyper and hypocortisolemic profiles depend on dose, particle type, aging status, and copollutant coexposures. Several rodent and fish studies revealed reduced serum cortisol levels, accompanied by compensatory ACTH elevation or altered stress-behavior phenotypes, after chronic MNP exposure, reflecting complex feedback disruption rather than simple up- or downregulation. These findings indicate that MNPs can reprogram stress axis homeostasis via combined immune, microbial, and direct endocrine actions, with likely downstream consequences for metabolism, immune competence, and behavior.

16.12.4 HPT axis: Thyroid hormone imbalance

MPs disrupt the HPT axis through multiple converging mechanisms, including direct interference with thyroid hormone synthesis and metabolism (inhibition/induction of thyroid peroxidase, deiodinases, and transport proteins). Endocrine disruption induced by sorbed additives, including brominated flame retardants and bisphenols, may alter thyroid hormone signaling through receptor agonism or antagonism. Furthermore, oxidative stress and gut microbiota dysbiosis can impair deiodinase-mediated peripheral conversion of thyroxine (T4) to the biologically active triiodothyronine (T3), thereby contributing to thyroid hormone imbalance. Recent experimental studies conducted in zebrafish, medaka, and rodent models have shown altered circulating T4/T3 ratios, dysregulated expression of hypothalamic and pituitary genes (TRH, TSHβ), and morphological thyroid changes after exposure to pristine and weathered MPs, MP leachates, or MPs preloaded with POPs. Several studies have also demonstrated that coexposure to algal toxins or other pollutants with MPs exacerbates thyroid hormone dysregulation. Interestingly, exposure during the early life and developmental phases disrupts embryonic/larval fish exposed to aged or biodegradable plastic leachates, which exhibit impaired neurodevelopment and thyroid-dependent endpoints (growth, metamorphosis, behavior), supporting a mode of action that links MP-associated chemicals to impaired thyroid homeostasis during sensitive windows.

16.12.5 HPG axis: Estrogenic/antiandrogenic effects and reproductive toxicity

MPs and their chemical mixtures disturb the HPG axis via direct receptor interactions (ER/AR agonism or antagonism), disruption of the expression of steroidogenic enzymes (aromatase, 17β-HSD, and CYPs), oxidative stress in gonadal tissues, and disruption of the blood–testis and ovarian follicular microenvironments. Laboratory evidence across taxa (fish, invertebrates, and rodents) revealed reduced sperm quality and counts. Moreover, they also alter sex steroid concentrations, disrupt gametogenesis, impair follicular development, and cause transgenerational reproductive deficits following exposure to MPs and MP leachates (phthalates, bisphenols). For MPs carrying BPA/other EDCs, coexposure studies demonstrate synergistic or additive effects (e.g., stronger suppression of steroidogenesis when MPs are combined with BPA) rather than simple additivity. Recent studies conducted on mammals and humans revealed ovarian dysfunction and reduced fertility parameters. MPs were also detected in reproductive fluids (placenta and follicular fluid), suggesting the biological plausibility of direct gonadal exposure. However, the causal epidemiological link to clinical infertility remains unproven (Haque and Tischkau, 2022).

16.13 Developmental and transgenerational effects

MNP exposure during early life stages results in a consistent pattern of developmental toxicity across taxa, mediated by overlapping physical, chemical, and indirect biological mechanisms. Small particles (especially NPs) translocate across the chorion and epithelial barriers and accumulate in embryonic tissues, where they cause mechanical interference and local oxidative stress. The chemical fraction (leached additives such as phthalates and bisphenols, and sorbed POPs/PFAS) disrupts endocrine signaling, which governs growth and differentiation. In teleost models (zebrafish and medaka), multiple independent studies reported dose-dependent morphogenesis defects (pericardial edema, delayed hatching, and reduced body length). Neurodevelopmental disturbances (impaired axon outgrowth, altered dopaminergic markers, and locomotor and anxiety-like behaviors), altered thyroid and steroid hormone profiles, and disrupted organogenesis were observed after exposure to pristine and weathered MPs/NPs or to their leachates. A robust experimental series demonstrated NP accumulation in embryos, with a consequent developmental delay and altered gene expression linked to apoptosis and oxidative pathways.

Developmental studies conducted in rodents revealed that maternal ingestion or inhalation of PS-NPs or realistic leachates during gestation/lactation alters offspring metabolic markers, immune status, and hepatic gene expression, and can impair neurobehavioral endpoints. Particle size and surface aging strongly increase developmental hazard, whereas chemical mixtures in leachates often account for much of the endocrine and growth-related toxicity observed in aquatic embryos. Coexposure to other environmental stressors (PFASs, algal toxins, and pathogens) frequently exacerbates developmental outcomes.

In addition to developmental dysfunctions, transgenerational and inherited effects occur following parental or ancestral exposure to MNPs, with DNA damage, epigenetic reprogramming, and germline abnormalities as key mechanisms. Several recent experimental studies have reported that parental exposure to PS-NPs induces reproducible reductions in fecundity, sperm quality, and embryonic viability that persist into F1 and, in some cases, F2 generations in invertebrate and fish models. Mechanistic studies have linked these phenotypes to increased germline DNA damage, activation of DNA damage checkpoint pathways (e.g., ATL-1 and CEP-1/p53 orthologues), impaired homologous recombination repair (reduced RAD-51), and increased germline apoptosis pathways, which are known to transmit compromised gamete integrity to progeny. Complementary epigenetic studies reported altered DNA methylation landscapes, dysregulated microRNA expression, and changes in chromatin-associated marks in the offspring of exposed parents. In addition to these transcriptional shifts in genes related to development, stress response, and metabolism, these findings are consistent with those of nongenetic inheritance.

Mammalian studies (mouse models) have reported that maternal gestational/lactational exposure to NPs and realistic plastic leachates produces hepatic, reproductive, and behavioral perturbations in F1 offspring, and some studies report persistent immune-epigenetic signatures that may be heritable. However, heterogeneity in exposure regimens, particle types/sizes, and endpoints means that the strength and persistence of transgenerational effects vary across studies. While various studies have provided mechanistic proof-of-principle for heritable effects via DNA damage and epigenetic reprogramming, other parental exposure experiments have revealed minimal or reversible effects in descendants, indicating context dependence (dose, timing, and repair capacity). Considering these findings, which provide evidence for a plausible and emerging paradigm, MNPs can induce a "toxic legacy" in offspring by compromising germline integrity and reprogramming regulatory epigenetic states. With important implications for population fitness and human risk assessment, definitive quantification of heritable risk for natural populations and humans requires standardized multigenerational designs, improved dosimetry for environmentally realistic exposures, and integrative molecular endpoints (DNA damage, methylome, small RNAs, and germline histone marks) to establish causality and persistence.

16.13.1 Maternal–fetal transfer

The maternal–fetal transfer of MPs and NPs has transitioned from a theoretical concern to a demonstrated phenomenon in both human samples and experimental animals. Multiple independent human studies have reported that MNPs in placental tissues and fetal compartments indicate that particles (and/or particle-derived chemical components) can reach the fetoplacental unit (Balali et al., 2024). Raman-guided analyses and complementary analytic approaches detected diverse polymer types within human placentas and reported associations with neonatal metrics in preliminary work. A pioneering study using multimodal detection (Raman spectroscopy, pyrolysis–gas chromatography–mass spectrometry (Py–GC–MS), and scanning electron microscopy (SEM)) reported the presence of MPs in umbilical cord blood, showing that particles were present in the fetal circulation and directly interacted with blood proteins, highlighting a plausible route to fetal tissues. Smaller cohorts and pilot mother–infant pair studies have confirmed the presence of MPs in the placenta and cord blood. Exploratory analyses of samples from mothers and early infants have begun to link higher placental burdens with altered cord hormone profiles and adverse birth outcomes. Mechanistically, maternal–fetal transfer depends strongly on particle size, surface chemistry, agglomeration state, and biological milieu (protein corona or biofilm).

NPs (<1 μm) have the greatest potential for translocation. Experimental studies conducted on animal models and in vitro placental perfusion have reported that NP particles can cross trophoblast layers via endocytic pathways, transcytosis, and paracellular leakage when barrier integrity is impaired or by hitchhiking within extracellular vesicles. Once these NPs cross the syncytiotrophoblast, the particles may access the fetal capillaries and thereby the circulation of the fetus. Maternal inflammation, oxidative stress, and placental pathology (e.g., altered trophoblast function and increased permeability) are commonly induced by MNPs or their leachates in animal models, further facilitating transfer and amplifying fetal exposure. Robust animal literature corroborates and extends human detection by demonstrating the functional consequences of maternal MNP exposure for the fetus. Rodent and fish experiments have revealed the maternal-to-offspring translocation of PS and other polymer particles, fetal accumulation, and downstream effects, including fetal growth restriction, altered organogenesis, disrupted endocrine signaling (thyroid and steroid hormones), neurodevelopmental deficits, and immune disturbance. Several studies have reported that compared with pristine particles, weathered particles or plastic leachates induce stronger placental inflammation and fetal effects and that maternal exposure during critical windows (gestation and lactation) yields the greatest developmental effects. Placental oxidative stress, upregulation of inflammatory cytokines, trophoblast apoptosis, and compromised nutrient transport pathways plausibly mediate the observed fetal growth and developmental endpoints. Environmental MNPs are ingested or inhaled by pregnant women and can reach and accumulate in placental tissues. A fraction of nanoparticles, especially NPs, can traverse the placental barrier into the fetal circulation, where they may interact with fetal proteins and tissues. These direct exposures, together with placental dysfunction induced by particles or their chemical cargo (phthalates, bisphenols, and PFAS cocontaminants), can disturb fetal endocrine milieus and developmental programs.

16.14 Epigenetic alterations (DNA methylation, miRNAs)

16.14.1 DNA methylation

MNP exposure alters DNA methylation patterns in exposed organisms, resulting in both global and locus-specific changes that affect development, stress response, and reproduction. Studies in fish and invertebrate models have reported that exposure to PE and PS particles or plastic leachates results in dose-dependent shifts in the methylation of genes involved in development, detoxification, and DNA repair pathways. This might be explained by laboratory work in fathead minnows and *Daphnia* showing altered methylomes associated with disrupted growth and reproductive endpoints. At the cellular level, NPs can induce oxidative stress and inflammatory signaling that activates DNA damage responses and the recruitment of DNA methyltransferases, leading to persistent promoter hyper or hypomethylation of regulatory genes (for instance, genes controlling apoptosis, antioxidant defenses, and endocrine signaling) (Aloisi and Poma, 2025). However, these effects are heterogeneous across studies, as some short-term human cell experiments have reported limited changes in global methylation after acute exposure, indicating that methylation responses are highly dependent on particle size (NPs>MPs), surface chemistry/aging, exposure duration, tissue/cell type, and the presence of copollutants in sorbed mixtures. Therefore MNPs can reprogram DNA methylation landscapes in an exposure- and context-dependent manner, providing a plausible mechanistic link to the persistent phenotypic changes observed during development and reproduction.

NMP-induced toxicity is very evident and concrete because noncoding RNA networks, especially those involving miRNAs, which act as rapid posttranscriptional regulators of stress, immune, and developmental pathways, are disturbed. Multiple aquatic and mammalian models exposed to PS or PE particles have shown differential expression of conserved miRNAs that regulate oxidative stress responses (members of the miR-30 and miR-21 families), apoptosis, and neurodevelopmental signaling. These miRNA shifts often precede or accompany transcriptomic and phenotypic changes such as altered locomotion, neurochemical imbalance, and impaired gametogenesis (Rim, 2025). Moreover, particle-induced inflammation, mitochondrial dysfunction, and altered extracellular vesicle signaling appear to drive miRNA reprogramming. MNPs induce the release and alter the cargo of extracellular vesicles in exposed tissues, alter intercellular miRNA messaging, and produce systemic regulatory effects. Importantly, miRNA responses are sensitive indicators of low-dose and short-term exposures and can integrate mixed chemical and particle signals, making them promising biomarkers; however, cross-study comparisons are currently limited by inconsistent profiling platforms, a lack of standardized exposure metrics, and sparse human data.

16.14.2 Integrated explanation and implications

Epigenetic effects from MNPs arise from two interacting pathways: direct particle–cell interactions (mechanical stress, ROS generation, membrane perturbation, and uptake) that trigger DNA damage and inflammatory signaling cascades. In addition, the chemical burden that particles carry (leached plasticizers, bisphenols, PBDEs, and sorbed POPs) acts as an endocrine and epigenetic modulator. Oxidative stress and DNA damage recruit repair and epigenetic machinery DNA Methyltransferases (DNMTs), Ten-Eleven Translocation enzymes (TETs), and histone modifiers) and reshape methylation and miRNA landscapes. Concurrently, sorbed EDCs can bind nuclear receptors and chromatin-associated regulators, thereby modifying transcription and epigenetic states. These epigenetic alterations plausibly explain observed developmental delays, reproductive deficits, and the emergence of transgenerational phenotypes because methylation and small-RNA changes can be maintained across cell divisions and, in some taxa, transmitted through the germline (Hoseini et al., 2025).

16.15 Integrated neuroendocrine toxicity

MNPs can cause integrated/synergistic neuroendocrine toxicity through the combination of physical particle effects (barrier crossing, local mechanical/oxidative stress, and inflammation) with chemical effects (leached plasticizers, bisphenols, flame retardants, and sorbed environmental EDCs) that target both neural tissue and endocrine control centers. This duality permits crosstalk between neurotoxic and endocrine-disrupting pathways and creates mechanistic routes for synergistic harm. NPs (<1 μm) and, in some cases, small MPs can be translocated from mucosal surfaces (the gut, respiratory epithelium, and olfactory mucosa) into systemic circulation. Following their penetration, they target the CNS via hematogenous routes, olfactory nerve transport, BBB penetration, and olfactory bulb deposition as realistic translocation routes. Once they accumulate in the systemic circulation or the brain parenchyma, particles provoke microglial activation, local cytokine release (IL-1β, TNFα), oxidative stress, and mitochondrial dysfunction, processes that have been repeatedly observed in in vivo fish and rodent models. Recent literature on neurotoxicity revealed that NPs/MPs impair neuronal function (synaptic deficits, neurotransmitter imbalance) and feed back to endocrine centers (hypothalamus and pituitary) that regulate the cortisol, thyroid, and gonadal axes.

In parallel, aquatic models (zebrafish, medaka) and terrestrial invertebrates provide high-resolution evidence that MNPs and their leachates alter behavior (locomotion, anxiety-like behavior, learning), neurotransmitter systems (dopaminergic, serotonergic signaling), and endocrine markers (cortisol/corticosterone changes, altered thyroid hormones, vitellogenin induction). These findings indicate that, compared with pristine particles alone, weathered particles or MP-bound EDCs produce stronger neurobehavioral and hormonal disturbances. Some complementary studies conducted in rodents revealed that exposure to PS-NPs during maternal life or at chronic doses modulates neurodevelopment and alters hypothalamic gene expression associated with stress and metabolic circuits. Moreover, they modify circulating cortisol and thyroid hormones associated with maternal particle exposure, thereby influencing both neural and endocrine developmental effects.

The gut–brain axis emerges as a key integrator of neuroendocrine toxicity from MNPs. Multiple recent studies have reported that ingested particles disturb gut barrier integrity, reshape the microbiome, and release various proinflammatory mediators that signal centrally through vagal and systemic immune routes. This microbial–immune signaling modulates HPA activity, alters serotonin and tryptophan metabolism, and can exacerbate neuroinflammation, thereby associating gut exposure with systemic stress-axis dysregulation and neurobehavioral outcomes. Experimental NPs exposures alter gut microbial composition in ways that correlate with anxiety-like behavior and HPA markers in fish and rodents, supporting microbiome-mediated signaling as a central conduit between peripheral particle exposure and central neuroendocrine disturbance (Bertollo et al., 2025).

Mixture effects and cocontaminant interactions critically amplify integrated neuroendocrine risk. Recent studies have reported that, compared with single agents, MPs can sorb PFAS, pesticides, and legacy POPs and that, compared with single agents, PFAS + MPs and photoaged MPs + pesticides synergistically increase oxidative stress, neurobehavioural impairment, and hormonal disruption in ecotoxicological models. This synergy arises because particles concentrate EDCs at tissue interfaces and because chemical additives (phthalates and bisphenols) leached from plastics directly disrupt steroidogenesis and thyroid signaling while simultaneously sensitizing neural tissue to inflammatory injury. Thus chemical and particle-mediated pathways converge to produce stronger, often nonadditive, neuroendocrine dysfunction. Taken together, these findings suggest that the integrated neuroendocrine toxicity of MNPs can be summarized as a cascade in which environmental particles (especially aged/weathered and nanoscale fractions) are translocated to or signal to the brain, where they can induce local and systemic inflammation and oxidative stress, delay endocrine-active chemicals to central and peripheral endocrine tissues, and disrupt neurotransmitter systems and hormone axes

(HPA, HPT, and HPG). These intersecting processes manifest as behavioral changes, impaired stress responses, thyroid and reproductive hormone imbalances, and developmental neuroendocrine deficits, often with heightened sensitivity during developmental windows or under conditions of coexposure. Therefore these findings support the treatment of MNPs as complex particle–chemical stressors whose neurotoxic and endocrine effects are mechanistically intertwined (Bertollo et al., 2025).

16.15.1 Crosstalk between endocrine signaling and neural function

MNPs disturb neural and endocrine systems in intimately connected ways, as the same particle- and chemical-mediated processes that damage neural tissue also dysregulate hormone signaling. This establishes bidirectional crosstalk in which neural injury alters endocrine set-points, and endocrine disruption alters neural development and function. Mechanistically, nanoscale particles and small MPs can be translocated from mucosal surfaces (gut, lung, and olfactory epithelium) into the circulation and reach central sites via hematogenous routes or olfactory nerve transport. Once there, they provoke microglial activation, local cytokine release (IL-1β, TNFα), mitochondrial dysfunction, and oxidative stress responses that both impair neurotransmission and feed back to hypothalamic controllers of endocrine axes (HPA, HPT, HPG). These processes have been summarized in several recent comprehensive reviews and mechanistic studies documenting BBB interactions, olfactory transport, and central inflammation after MNP exposure (Jahedi and Fard, 2025).

A central conduit for neuroendocrine crosstalk is the gut–brain–endocrine axis. Ingested MNPs disturb gut barrier integrity and the microbiome, promote local inflammation, and alter microbial metabolites (short-chain fatty acids and tryptophan metabolites), which, in turn, modulate vagal signaling, systemic cytokine levels, and hypothalamic circuits that set the tone of the HPA and HPT. Experimental studies conducted on rodents and fish have reported that MNP induces dysbiosis, which is correlated with anxiety-like behavior, altered cortisol/corticosterone responses, and changes in central serotonin and dopaminergic markers; critically, fecal microbiota-transfer experiments can reverse some behavioral deficits, supporting a causal microbiome→brain→endocrine pathway. This gut-mediated signaling hence associates peripheral particle exposure with central endocrine regulation.

Chemical payloads associated with MNPs (leached plasticizers such as phthalates and bisphenols, and sorbed POPs/PBDEs/PFAS) further compound crosstalk by directly engaging nuclear and membrane hormone receptors within brain tissue and at hypothalamic/pituitary targets. Bisphenols, PBDEs, and some phthalate metabolites act as estrogenic, thyroid-modulating, or antiandrogenic agents. They can alter the expression of neuropeptides (CRH, TRH, and GnRH) and steroidogenic enzymes, producing endocrine shifts that alter neuronal gene expression, synaptic plasticity, and neurodevelopmental trajectories.

Inflammation and oxidative stress provide a mechanistic bridge that commonly links neural dysfunction and endocrine dysregulation following MNP exposure. Central or systemic cytokine elevations (IL-6, IL-1β, and TNFα) induced by particle or leachate exposure alter hypothalamic sensitivity to feedback signals, disrupt glucocorticoid receptor function, and modify deiodinase activity and thyroid hormone availability. This explains the observed co-occurrence of altered stress hormones (cortisol/corticosterone), thyroid hormone imbalances (T4/T3 shifts), and neurobehavioral deficits in multiple models. Several recent animal experiments have revealed that adrenal and thyroid histopathology, along with microglial activation and impaired cognition after chronic MNP exposure, support a shared inflammation-mediated mechanism.

Photooxidative weathering of plastics increases the surface area and the abundance of oxygenated functional groups, enhancing the sorption of semipolar pollutants and accelerating the leaching of additives. Moreover, coexposure (PFAS, pesticides, and algal toxins) has been shown to yield synergistic neuroendocrine disruption in ecotoxicological models. In practice, this means that organisms (including humans) are exposed to complex particle–chemical combinations that concentrate at barrier interfaces (the gut, placenta, and BBB) and produce nonadditive effects on neural circuits and hormone axes. MNP detection in human nasal/olfactory tissues, placenta, blood, and even brain samples has established biological plausibility in which particles and their chemical cargos reach neural and endocrine organs. Some exploratory studies in humans have shown associations between increased placental or cord burdens and altered neonatal hormone profiles. Moreover, MNPs act as complex particle–chemical stressors that breach peripheral barriers and, sometimes, the BBB, provoke local and systemic inflammation and oxidative stress, alter hypothalamic set-points, deliver endocrine-active chemicals directly to neural and endocrine tissues, and perturb microbiome-mediated signaling pathways, further modulating neuroendocrine circuits. The result is bidirectional crosstalk, in which neural dysfunction alters endocrine regulation (stress, thyroid, and reproductive axes), and endocrine disturbance affects neural development and behavior, effects that are most pronounced during developmental windows and when exposures involve aged particles or copollutant mixtures.

16.15.2 Combined effects on neurodevelopment, metabolism, and behavior

MNPs impair neurodevelopment through a combination of direct particle translocation, inflammatory/oxidative mechanisms, and the delivery of endocrine-active chemical cargos. In particular, nanoscale particles cross epithelial barriers (placenta, gut, and olfactory mucosa) and access fetal and neonatal neural tissues, where they induce microglial activation, mitochondrial dysfunction, and altered neurogenesis. Recent spatial-transcriptomic and histological work in maternal-exposure models has shown that prenatal NP exposure remodels hippocampal gene expression programs (affecting synaptogenesis, neuronal survival, and ferroptosis pathways). In addition, NPs cause persistent learning and memory deficits in offspring. Moreover, zebrafish and rodent developmental studies have repeatedly documented dose-dependent morphological changes, impaired axon outgrowth, and neurochemical alterations after exposure to weathered MP or NP fractions. These neuronal and synaptic disruptions are mechanistically linked to both particle-mediated oxidative stress and plastic-associated chemicals (phthalates, bisphenols, and PBDEs) that disrupt steroid/thyroid signaling, which is critical for neurodevelopment, indicating that neural outcomes reflect a mix of physical injury, chemical endocrine interference, and altered developmental programming.

Metabolic disruption is an increasingly consistent outcome in MNP toxicology, documenting the effects on energy homeostasis, lipid and glucose metabolism, and adipose tissue function. At the mechanistic level, MNPs alter hepatic lipid metabolism and adipogenic gene networks (PPAR pathways and lipogenic enzymes), induce low-grade systemic inflammation, and perturb gut microbial communities that regulate nutrient absorption and host energy balance (Lolescu et al., 2025). Integrated multiomics and metabolomics studies in fish and mammalian models have revealed shifts in glycolipid pathways, altered bile acid metabolism, and changes in short-chain fatty acid profiles that plausibly drive obesogenic phenotypes. Several PS-MP studies now report increased adiposity, altered glucose tolerance, or transcriptomic signatures consistent with metabolic dysregulation after chronic ingestion, especially when particles are weathered or carry sorbed obesogenic chemicals. Moreover, MNP-induced dysbiosis alters the signaling of microbial metabolites (e.g., SCFAs and tryptophan metabolites) that control host appetite, insulin sensitivity, and hepatic lipid handling, linking gut perturbation to systemic metabolic phenotypes.

Behavioral consequences (locomotion, anxiety-like responses, social interaction, and cognitive performance) frequently cooccur with neurodevelopmental and metabolic effects. They are also associated with inflammation, neurotransmitter disruption, and endocrine disturbance. Experimental studies in zebrafish, rodents, and invertebrates have reported reductions in exploratory behavior, altered circadian activity, anxiety-like phenotypes, and learning deficits following exposure to Ps/NPs or MP leachate. Moreover, some studies have reported depression-like problems and impaired maze performance associated with hippocampal pathology or reduced cerebral blood flow in mammalian models. These behavioral changes can be driven by direct neural injury (synaptic dysfunction, neurotransmitter depletion), by endocrine axis dysregulation (altered cortisol/corticosterone and thyroid hormones that modify arousal and cognition), and by metabolic shifts (insulin resistance, altered energy supply to the brain) that together degrade neural processing and plasticity. Importantly, various studies have reported that behavioral impairments are more severe when the exposure material is aged/weathered particles or MPs preloaded with EDCs, indicating that chemical payload and particle aging critically amplify behavioral risk.

Neurodevelopmental effects alter hypothalamic circuits that regulate appetite, energy expenditure, and HPA responses; metabolic inflammation feeds back to the brain via cytokines and altered gut-derived metabolites (modifying mood, motivation, and cognition); and behavioral changes (reduced activity, altered feeding) further modify metabolic trajectories, creating self-reinforcing maladaptive loops. Manipulations of the microbiome in MNP-exposed animals can ameliorate both metabolic and anxiety-like phenotypes. Prenatal NP exposure alters offspring hippocampal transcriptomes and shifts glucose metabolism and stress-axis function, and coexposure to PFASs or pesticides results in synergistic impairments across neurodevelopmental, metabolic, and behavioral endpoints. Therefore MNPs function as complex particle–chemical stressors whose combined influence on brain development, systemic metabolism, and behavior emerges from overlapping pathways of inflammation, endocrine disruption (thyroid, steroid, and metabolic hormones), neurotransmitter imbalance, and microbiome-mediated signaling.

The implications of MPs and NPs for human health are significant but require cautious interpretation. Therefore the detection of MPs in the human placenta, blood, and brain and increasing experimental evidence of maternal-to-offspring transfer and persistent offspring deficits are needed. These adverse effects raise concerns that environmental MNP exposure could contribute to developmental neurobehavioral disorders, metabolic dysregulation, and later-life cognitive decline, particularly when exposures occur during sensitive windows or in combination with other chemical stressors. Nevertheless, translating laboratory doses to human exposures and disentangling MP-specific effects from the many cooccurring plastic-associated chemicals remain major challenges.

In the field, standardized dosimetry using environmentally relevant, weathered particle mixtures, integrative multiomics that link the particle burden to molecular signatures in brain and metabolic tissues, and longitudinal human cohorts with paired exposure and microbiome data are needed. Endocrine and neurobehavioral phenotyping is necessary to move from mechanistic plausibility to causal inference. In the meantime, the convergent experimental literature (2022–2025) supports treating MNPs as multifaceted stressors that can simultaneously impair neurodevelopment, rewire metabolic control, and degrade behavior through shared inflammatory, endocrine, and microbiome-mediated mechanisms (Ruprecht et al., 2024).

16.15.3 Potential synergism between MPs and cocontaminants

MPs act as chemical platforms that alter the fate, bioavailability, and biological activity of cocontaminants through sorption, desorption, and weathering-driven changes in surface chemistry. In addition, photooxidative aging, in particular, increases surface roughness and the formation of oxygenated functional groups, accelerating both additive leaching and the sorption of semipolar contaminants and often increasing particle toxicity when copollutants are present. Experimental studies have reported that UV-photoaged micro(nano)plastics increase the toxicity of cocontaminants (for example, photoaged PS increased the toxic effects of Tetrabromobisphenol A (TBBP A) on growth, behavior, and reproduction in aquatic models), demonstrating that aging converts MPs into more potent vectors and chemical sources.

Several classes of cocontaminants consistently show synergistic or additive effects when paired with MPs. PFASs, which frequently cooccur with MPs in aquatic systems and sediments, interact with plastics in ways that increase combined toxicity. Recent ecotoxicological studies (*Daphnia* and other sentinel species) and targeted mixture experiments reveal greater reductions in reproduction, growth, and survival from PFAS + MP mixtures than from either stressor alone, with charge and sorption interactions being implicated in the synergy. These findings have been independently summarized in reviews highlighting a "MP–PFAS nexus" that amplifies ecological risk.

Endocrine-active organics such as bisphenols (BPA and its analogs), phthalates, flame retardants (PBDEs/TBBPA), and pesticides also showed potentiated endocrine and developmental toxicity when presented with MPs. Controlled laboratory studies have reported that, compared with BPA alone, MPs preloaded with BPA or coadministered with BPA produce stronger steroidogenic disruption, apoptosis, and developmental defects in zebrafish and mammalian cell models, and photoaged MPs often magnify these effects by increasing leachate release and altering sorption/desorption dynamics. Similarly, compared with single agents, MPs have been shown to increase the developmental toxicity of common antibiotics and pesticides (e.g., tetracycline) in fish embryos, with combined exposures producing greater morphogenetic and behavioral deficits.

Interactions with metals and legacy hydrophobic organic pollutants create additional synergistic pathways. MPs sorb heavy metals and persistent hydrophobic organics (PCBs,

PBDEs) and can facilitate their trophic transfer. Mechanistic syntheses revealed that, compared with single-contaminant exposures, MP–metal and MP–POP mixtures can increase bioaccumulation and toxicity endpoints (oxidative stress, endocrine disruption, and immune impairment) in aquatic organisms. Photoaging and biofilm formation modulate these interactions and can convert MPs from sinks into secondary sources that pulse contaminants into biota under gut-like or digestive-fluid conditions.

MPs function not only as inert particles but also as dynamic carriers that modify the exposure, toxicokinetics, and biological effects of cooccurring environmental contaminants. Low-level pollutants such as plastic additives, POPs, metals, and emerging chemicals are concentrated on their surfaces. MPs can deliver disproportionately high local doses to epithelial interfaces, particularly in the gastrointestinal tract, where physicochemical conditions favor contaminant desorption. This vector effect is further amplified by environmental aging processes, including UV radiation, mechanical abrasion, and oxidative weathering, which increase surface roughness, introduce oxygen-containing functional groups, and enhance both the sorption capacity and additive leaching, thereby altering the chemical mixtures released upon ingestion or inhalation. Particle-mediated transport also alters toxicokinetics by shifting uptake routes from dissolved-phase exposure to particulate ingestion, increasing chemical persistence in the gut, facilitating translocation across biological barriers, and promoting trophic transfer within food webs. MPs simultaneously exert intrinsic biological stress by inducing inflammation, oxidative stress, epithelial damage, and barrier dysfunction, which can biologically prime tissues and sensitize them to coexposed chemicals, resulting in nonadditive or amplified toxic responses. At the molecular level, coadsorbed chemicals may interact through shared pharmacodynamic pathways, such as those involving endocrine receptors, detoxification enzymes, and redox systems, leading to synergistic receptor activation or inhibition, impaired biotransformation, and overwhelmed antioxidant defenses.

Because these physicochemical and biological mechanisms operate concurrently, MP–chemical coexposures frequently produce toxic effects that exceed the sum of those of individual agents, a prediction increasingly supported by empirical studies. Recent high-quality mixture experiments involving MPs combined with PFAS, BPA, pesticides, and other contaminants have demonstrated robust synergistic outcomes, including reduced fecundity, disrupted development, endocrine dysfunction, and elevated oxidative damage. However, the magnitude and direction of these interactions remain strongly context-dependent. Synergism varies with polymer composition, particle size (NPs often exhibit greater potency), degree of weathering, cocontaminant physicochemical properties, ambient environmental chemistry, organism feeding physiology, and exposure duration or life stage. Consequently, translating laboratory findings to field-relevant risk assessment remains challenging, as environmental concentrations, mixture compositions, and exposure routes are highly heterogeneous. In summary, converging mechanistic, experimental, and emerging human evidence indicates that MNPs act as vectors and sources of endocrine-active chemistry capable of integrated neuroendocrine toxicity, particularly under conditions of environmental aging and chemical coexposure. Standardized measurement, realistic mixture toxicology, and well-designed human studies are now essential to translate mechanistic plausibility into quantitative risk estimates and targeted regulatory action.

16.16 Conclusion and future perspectives

MNPs are being increasingly recognized as biologically active particle–chemical complexes with credible mechanistic links to neurotoxicity and endocrine disruption. Experimental evidence has demonstrated that small particles can translocate across epithelial, placental, and BBBs; induce oxidative stress and inflammation; disrupt microbiome composition; and deliver leached plastic additives (e.g., phthalates, bisphenols, and flame retardants), as well as sorbed copollutants to systemic and central compartments. These processes converge on endocrine and neural signaling pathways, including steroid, thyroid, and stress-axis regulation, yielding consistent organism-level outcomes across taxa, such as neuroinflammation, synaptic and neurotransmitter dysregulation, HPA/HPT/HPG axis disruption, metabolic alterations, and developmental deficits. Collectively, recent experimental syntheses place MNPs within the category of complex environmental stressors with plausible relevance to human neuroendocrine health.

Despite strong mechanistic coherence, quantitative risk translation remains constrained by key empirical gaps. These include the lack of standardized, contamination-controlled methods for quantifying MNP burdens in human tissues and fluids; limited ability to apportion EDC body burdens attributable specifically to MNP-mediated delivery; insufficient chronic low-dose studies using environmentally aged and chemically realistic particle mixtures; and a near absence of longitudinal human cohorts linking measured MNP biomarkers to endocrine and neurodevelopmental end points. Methodological heterogeneity in particle type, aging state, exposure route, and dose metrics further limits cross-study synthesis.

An important emerging insight is the central role of mixed biology. MNPs frequently cooccur with PFAS, POPs, and pesticides, and mixture experiments consistently show synergistic toxicity that exceeds the effects of the individual agents. These interactions arise from particle-mediated concentrations at tissue interfaces, aging-enhanced sorption and additive release, and biological priming through inflammation and barrier dysfunction. As a result, conventional single-chemical risk assessment frameworks are likely to underestimate real-world hazards associated with MNP exposure.

Advancing the field toward actionable inference requires a focused research agenda: (1) standardized reference materials and interlaboratory validation for MNP detection in biological matrices; (2) routine use of aged, biofilm-coated, and environmentally realistic particle mixtures in toxicology; (3) integration of toxicokinetics (particle translocation, protein corona formation, placental transfer) with multiomics endpoints to anchor mechanisms to endocrine-axis dysfunction; (4) longitudinal birth and life-course cohorts incorporating paired MNP/EDC measurements and neuroendocrine phenotyping; and (5) quantitative mixture experiments across life stages to derive interaction-aware dose–response models.

Acknowledgment

Younis Ahmad Hajam acknowledges the financial grant received from the Seed Money for Research through Research Project ref. no. SBBSU/R&D/SMP/011/2024-25 from Sant Baba Bhag Singh University, Jalandhar, Punjab.

AI disclosure

During the preparation of this work, the author(s) used ChatGPT as an AI tool to assist in the compilation of rough content and the designing of diagrams. However, the content was cross-checked, verified, rephrased, and aligned with scholarly literature. Only ideas were taken from AI sources for the creation of diagrams. After using this tool/service, the author(s) reviewed and edited the content as needed and take(s) full responsibility for the content of the publication.

References

Agboola, O.D., Benson, N.U., 2021. Physisorption and chemisorption mechanisms influencing micro (Nano) plastics-organic chemical contaminants interactions: a review. Front. Environ. Sci.. 2296665X. 9. Available from: journal.frontiersin.org/journal/environmental-science. doi:10.3389/fenvs.2021.678574.

Alimba, C.G., Faggio, C., 2019. Microplastics in the marine environment: current trends in environmental pollution and mechanisms of toxicological profile. Environ. Toxicol. Pharmacol. 13826689. 68, 61–74. doi:10.1016/j.etap.2019.03.001.

Aloisi, M., Poma, A.M.G., 2025. Nanoplastics as gene and epigenetic modulators of endocrine functions: a perspective. Int. J. Mol. Sci. 14220067. 26 (5). Available from: http://www.mdpi.com/journal/ijms. doi:10.3390/ijms26052071.

Balali, H., Morabbi, A., Karimian, M., 2024. Concerning influences of micro/nano plastics on female reproductive health: focusing on cellular and molecular pathways from animal models to human studies. Reproductive Biology and Endocrinology 22 (1), 141.

Balog, S., de Almeida, M.S., Taladriz-Blanco, P., Rothen-Rutishauser, B., Petri-Fink, A., 2024. Does the surface charge of the nanoparticles drive nanoparticle–cell membrane interactions? Curr. Opin. Biotechnol. 09581669. 87, 103128. doi:10.1016/j.copbio.2024.103128.

Battaglini, M., Marino, A., Montorsi, M., Carmignani, A., Ceccarelli, M.C., Ciofani, G., 2024. Nanomaterials as microglia modulators in the treatment of central nervous system disorders. Adv. Healthc. Mater. 21922659. 13 (12). Available from: http://onlinelibrary.wiley.com/journal/10.1002/(ISSN)2192-2659. doi:10.1002/adhm.202304180.

Bertollo, A.G., Santos, C.F., Bagatini, M.D., Ignácio, Z.M., 2025. Hypothalamus-pituitary-adrenal and gut-brain axes in biological interaction pathway of the depression. Front. Neurosci. 1662453X. 19. Available from: https://www.frontiersin.org/journals/neuroscience#. doi:10.3389/fnins.2025.1541075.

Bhardwaj, G., Abdulkadhim, M., Joshi, K., Wankhede, L., Das, R.K., Brar, S.K., 2025. Exposure pathways, systemic distribution, and health implications of micro- and nanoplastics in humans. Appl. Sci. 2076-3417. 15 (16), 8813. doi:10.3390/app15168813.

Bhaumik, S., Chakraborty, P., 2024. Interactions between microplastics (MPs) and trace/toxic metals in marine environments: implications and insights—a comprehensive review. Environ. Sci. Pollut. Res. 1614-7499. 31 (50), 59681–59699. doi:10.1007/s11356-024-34960-w.

Bilardo, R., Traldi, F., Vdovchenko, A., Resmini, M., 2022. Influence of surface chemistry and morphology of nanoparticles on protein Corona formation. WIREs Nanomed. Nanobiotechnol. 19390041. 14 (4). Available from: http://onlinelibrary.wiley.com/journal/10.1002/(ISSN)1939-0041. doi:10.1002/wnan.1788.

Birben, E., Sahiner, U.M., Sackesen, C., Erzurum, S., Kalayci, O., 2012. Oxidative stress and antioxidant defense. World Allergy Organ. J. 19394551. 5 (1), 9–19. Available from: https://www.journals.elsevier.com/world-allergy-organization-journal/. doi:10.1097/WOX.0b013e3182439613.

Biswas, K., 2023. Microglia mediated neuroinflammation in neurodegenerative diseases: a review on the cell signaling pathways involved in microglial activation. J. Neuroimmunol. 01655728. 383, 578180. doi:10.1016/j.jneuroim.2023.578180.

Bossio, S., Ruffolo, S.A., Lofaro, D., Perri, A., La Russa, M.F., 2025. Endocrine toxicity of micro-and nanoplastics, and advances in detection techniques for human tissues: a comprehensive review. Endocrines 6 (2), 23.

Casella, C., Cornelli, U., Ballaz, S., Zanoni, G., Merlo, G., Ramos-Guerrero, L., 2025. Plastic smell: a review of the hidden threat of airborne micro and nanoplastics to human health and the environment. Toxics. 2305-6304. 13 (5), 387. doi:10.3390/toxics13050387.

Chen, A., Zhong, H., Peng, J., Luo, T., Jun, S., 2025. Machine learning-based modeling for predicting hypopituitarism after cranial trauma. World Neurosurg. 18788750. 199, 124018. doi:10.1016/j.wneu.2025.124018.

Chowdhury, S.R., Dey, A., Mondal, S., Gautam, M.K., 2024. Environmental microplastics and nanoplastics: effects on cardiovascular system. Toxicologie Analytique Et Clinique. 23520078. 36 (2), 145–157. Available from: https://www.sciencedirect.com/science/journal/23520078. doi:10.1016/j.toxac.2023.11.006.

Chrishtop, V.V., Mironov, V.A., Prilepskii, A.Y., Nikonorova, V.G., Vinogradov, V.V., 2021. Organ-specific toxicity of magnetic iron oxide-based nanoparticles. Federation Nanotoxicol. 17435404. 15 (2), 167–204. Available from: http://www.tandfonline.com/loi/inan20. doi:10.1080/17435390.2020.1842934.

Constantinescu, A.M., Karzi, V.E., Docea, A.O., Tsitsimpikou, C., Nosyrev, A.E., Tsatsakis, A., Hernández, A.F., Bogdan, C., 2025. Neurobehavioral effects of low dose exposure to chemical mixtures: a review. Arch. Toxicol. 14320738. 99 (4), 1315–1331. Available from: https://www.springer.com/journal/204. doi:10.1007/s00204-025-04009-z.

Daniels, B.P., Cruz-Orengo, L., Pasieka, T.J., Couraud, P.O., Romero, I.A., Weksler, B., Cooper, J.A., Doering, T.L., Klein, R.S., 2013. Immortalized human cerebral microvascular endothelial cells maintain the properties of primary cells in an in vitro model of immune migration across the blood brain barrier. J. Neurosci. Methods. 1872678X. 212 (1), 173–179. doi:10.1016/j.jneumeth.2012.10.001.

Dennis, J., Arulraj, D., Mistri, T.K., 2025. Unseen toxins: exploring the human health consequences of micro and nanoplastics. Toxicol. Rep. 22147500. 14, 101955. doi:10.1016/j.toxrep.2025.101955.

Dhaundiyal, A., Mittal, A., 2024. Unveiling the microplastics: sources, distribution, toxicological impacts, extraction methods, degradational strategies, paving the path to a sustainable future. Water, Air, & Soil Pollution. 0049-6979. 235 (11). doi:10.1007/s11270-024-07506-6.

Dris, R., Gasperi, J., Mirande, C., Mandin, C., Guerrouache, M., Langlois, V., Tassin, B., 2017. A first overview of textile fibers, including microplastics, in indoor and outdoor environments. Environ. Pollut. 02697491. 221, 453–458. doi:10.1016/j.envpol.2016.12.013.

Dube, E., Okuthe, G.E., 2023. Plastics and micro/nano-plastics (MNPs) in the environment: occurrence, impact, and toxicity. Int. J. Environ. Res. Public Health. 1660-4601. 20 (17), 6667. doi:10.3390/ijerph20176667.

Dusza, H.M., van Boxel, J., van Duursen, M.B.M., Forsberg, M.M., Legler, J., Vähäkangas, K.H., 2023. Experimental human placental models for studying uptake, transport and toxicity of micro- and nanoplastics. Sci. Total Environ. 18791026. 860. Available from: www.elsevier.com/locate/scitotenv. doi:10.1016/j.scitotenv.2022.160403.

Dzierżyński, E., Gawlik, P.J., Puźniak, D., Flieger, W., Jóźwik, K., Teresiński, G., Forma, A., Wdowiak, P., Baj, J., Flieger, J., 2024. Microplastics in the human body: exposure, detection, and risk of carcinogenesis: a state-of-the-art review. Cancers. 2072-6694. 16 (21), 3703. doi:10.3390/cancers16213703.

Eriksen, M., Lebreton, L.C.M., Carson, H.S., Thiel, M., Moore, C.J., Borerro, J.C., Galgani, F., Ryan, P.G., Reisser, J., 2014. Plastic pollution in the world's oceans: more than 5 trillion plastic pieces weighing over 250,000 Tons afloat at sea. PLoS ONE. 19326203. 9 (12). Available from: http://www.plosone.org/article/fetchObject.action?uri=info%3Adoi%2F10.1371%2Fjournal.pone.0111913&representation=PDF. doi:10.1371/journal.pone.0111913.

Ernhofer, B., Spittler, A., Ferk, F., Mišík, M., Zylka, M.M., Glatt, L., Boettiger, K., Solta, A., Kirchhofer, D., Timelthaler, G., Megyesfalvi, Z., Kopatz, V., Kovar, H., Knasmueller, S., Aigner, C., Kenner, L., Döme, B., Schelch, K., 2025. Small particles, big problems: polystyrene nanoparticles induce dna damage, oxidative stress, migration, and mitogenic pathways predominantly in non-malignant lung cells. J. Hazard. Mater. 03043894. 495, 139129. doi:10.1016/j.jhazmat.2025.139129.

Escher, B.I., Hackermüller, J., Polte, T., Scholz, S., Aigner, A., Altenburger, R., Böhme, A., Bopp, S.K., Brack, W., Busch, W., Chadeau-Hyam, M., Covaci, A., Eisenträger, A., Galligan, J.J., Garcia-Reyero, N., Hartung, T., Hein, M., Herberth, G., Jahnke, A., Kleinjans, J., Klüver, N., Krauss, M., Lamoree, M., Lehmann, I., Luckenbach, T., Miller, G.W., Müller, A., Phillips, D.H., Reemtsma, T., Rolle-Kampczyk, U., Schüürmann, G., Schwikowski, B., Tan, Y.M., Trump, S., Walter-Rohde, S., Wambaugh, J.F., 2017. From the exposome to mechanistic understanding of chemical-induced adverse effects. Environ. Int. 18736750. 99, 97–106. Available from: www.elsevier.com/locate/envint. doi:10.1016/j.envint.2016.11.029.

Fang, C.L., Aljuffali, I.A., Li, Y.C., Fang, J.Y., 2014. Delivery and targeting of nanoparticles into hair follicles. Ther. Deliv. 20416008. 5 (9), 991–1006. Available from: http://www.future-science.com/loi/tde. doi:10.4155/TDE.14.61.

Ferrara, G., 2023. Abstracts of the fifth brainstorming research assembly for young neuroscientists (BraYn), Italy, 28–30 September 2022. Neurol. Int. 2035-8377. 15 (1), 415–496. doi:10.3390/neurolint15010028.

Finlay, R.D., Mahmood, S., Rosenstock, N., Bolou-Bi, E.B., Köhler, S.J., Fahad, Z., Rosling, A., Wallander, H., Belyazid, S., Bishop, K., Lian, B., 2020. Reviews and syntheses: biological weathering and its consequences at different spatial levels - From nanoscale to global scale. Biogeosciences. 17264189. 17 (6), 1507–1533. Available from: http://www.biogeosciences.net/volumes_and_issues.html. doi:10.5194/bg-17-1507-2020.

Galgani, F., Hanke, G., Maes, T., 2015. Global Distribution, Composition and Abundance of Marine Litter Marine Anthropogenic Litter. Springer International Publishing, France, pp. 29–56. doi:10.1007/978-3-319-16510-3_2.

Ghosh, A., Tripathy, A., Ghosh, D., 2022. Impact of endocrine disrupting chemicals (EDCs) on reproductive health of human. Proc. Zool. Soc. 09746919. 75 (1), 16–30. Available from: springer.com/journal/12595. doi:10.1007/s12595-021-00412-3.

Giroux, M.S., Siddiqui, S., Brander, S.M., 2024. Classic Contaminants in Aquatic Ecosystems: POPs, PFAS, Heavy Metals, and Microplastics Aquatic Ecotoxicology: Understanding Pollutants, Aquatic Organisms, and Their Environments. Springer International Publishing, United States, pp. 43–85. Available from: https://link.springer.com/book/9783031531309. doi:10.1007/978-3-031-53130-9_4.

Gopallawa, I., Dehinwal, R., Bhatia, V., Gujar, V., 2023. Narendra chirmule, a four-part guide to lung immunology: invasion, inflammation, immunity, and intervention. Front. Immunol. 1664-3224. 14. doi:10.3389/fimmu.2023.1119564.

Haque, N., Tischkau, S.A., 2022. Sexual dimorphism in adipose-hypothalamic crosstalk and the contribution of aryl hydrocarbon receptor to regulate energy homeostasis. Int. J. Mol. Sci. 1422-0067. 23 (14), 7679. doi:10.3390/ijms23147679.

Hernández, A.F., Lacasaña, M., Tsatsakis, A.M., Docea, A.O., 2025. Cellular and molecular mechanisms of micro- and nanoplastics driving adverse human health effects. Toxics. 23056304. 13 (11). Available from: http://www.mdpi.com/journal/toxics. doi:10.3390/toxics13110921.

Hoseini, Z.S., Rezaee, Z., Derakhshani, A., Zhang, S.Y., Tehrani, S.S.H., Taleb, M., Ghanbari, H., 2025. Nanoparticles and their impact on epigenetic mechanisms: insights and implications. Adv. Ther. 23663987. 8 (6). Available from: http://onlinelibrary.wiley.com/journal/10.1002/(ISSN)2366-3987. doi:10.1002/adtp.202500006.

Hwang, K.S., Son, Y., Kim, S.S., Shin, D.S., Lim, S.H., Yang, J.Y., Jeong, H.N., Lee, B.H., Bae, M.A., 2022. Size-dependent effects of polystyrene nanoparticles (PS-NPs) on behaviors and endogenous neurochemicals in zebrafish larvae. Int. J. Mol. Sci. 14220067. 23 (18). Available from: http://www.mdpi.com/journal/ijms. doi:10.3390/ijms231810682.

Jahedi, F., Fard, N.J.H., 2025. Micro-and nanoplastic toxicity in humans: exposure pathways, cellular effects, and mitigation strategies. Toxicol. Rep., 102043.

Jahedi, F., Frad, N.J.H., Khaksar, M.A., Rashidi, P., Safdari, F., Mansouri, Z., 2025. Nano and microplastics: unveiling their profound impact on endocrine health. Toxicol. Mech. Methods. 15376524. 35 (8), 865–893. Available from: http://www.tandfonline.com/loi/itxm20. doi:10.1080/15376516.2025.2509745.

Jha, N.K., Chen, W.C., Kumar, S., Dubey, R., Tsai, L.W., Kar, R., Jha, S.K., Gupta, P.K., Sharma, A., Gundamaraju, R., Pant, K., Mani, S., Singh, S.K., Maccioni, R.B., Datta, T., Singh, S.K., Gupta, G., Prasher, P., Dua, K., Dey, A., Sharma, C., Mughal, Y.H., Ruokolainen, J., Kesari, K.K., Ojha, S., 2022. Molecular mechanisms of developmental pathways in neurological disorders: a pharmacological and therapeutic review. Open Biol. 20462441. 12 (3). Available from: http://rsob.royalsocietypublishing.org/. doi:10.1098/rsob.210289.

Jiang, W., Liu, Y., Yuqi, W., Zhang, L., Zhang, B., Zhou, S., Zhang, P., Ting, X., Min, W., Lv, S., 2024. Polystyrene nanoplastics of different particle sizes regulate the polarization of pro-inflammatory macrophages. Sci. Rep. 2045-2322. 14 (1). doi:10.1038/s41598-024-67289-y.

Jurcău, M.C., Andronie-Cioara, F.L., Jurcău, A., Marcu, F., Ţiţ, D.M., Paşcalău, N., Nistor-Cseppentö, D.C., 2022. The link between oxidative stress, mitochondrial dysfunction and neuroinflammation in the pathophysiology of Alzheimer's disease: therapeutic implications and future perspectives. Antioxidants. 20763921. 11 (11). Available from: www.mdpi.com/journal/antioxidants. doi:10.3390/antiox11112167.

Kannan, K., Vimalkumar, K., 2021. A review of human exposure to microplastics and insights into microplastics as obesogens. Front. Endocrinol. 1664-2392. 12. doi:10.3389/fendo.2021.724989.

Khaki, Q.Z., Kumar, P., 2025. Ecological Impacts of Microplastics and Their Additives: Exposure Risk/Toxicity Assessment and Fate/Transport of Persistent, Bio-Accumulative and Toxic Substances Microplastics in the

Environment: Fate, Impacts, Removal, and Management. Wiley, India, pp. 259–282. Available from: https://onlinelibrary.wiley.com/doi/book/10.1002/9781394251100. doi:10.1002/9781394251100.ch10.

Khan, A., Jia, Z., 2023. Recent insights into uptake, toxicity, and molecular targets of microplastics and nanoplastics relevant to human health impacts. iScience. 25890042. 26 (2), 106061. doi:10.1016/j.isci.2023.106061.

Kong, M., Chen, Z., Lin, Z., Yin, P., Zhao, Q., 2024. SIGMAR1 targets AMPK/ULK1 pathway to inhibit SH-SY5Y cell apoptosis by regulating endoplasmic reticulum stress and autophagy. Funct. Integr. Genomics. 1438-793X. 24 (4). doi:10.1007/s10142-024-01414-y.

Kudzin, M.H., Piwowarska, D., Festinger, N., Chruściel, J.J., 2024. Risks associated with the presence of polyvinyl chloride in the environment and methods for its disposal and utilization. Materials. 19961944. 17 (1). Available from: http://www.mdpi.com/journal/materials. doi:10.3390/ma17010173.

Kustra, A., Maliszewska-Olejniczak, K., Sekrecka-Belniak, A., Kulawiak, B., Bednarczyk, P., 2025. Polystyrene nanoplastics in human gastrointestinal models—cellular and molecular mechanisms of toxicity. Int. J. Mol. Sci. 1422-0067. 26 (23), 11738. doi:10.3390/ijms262311738.

Lamoree, M.H., van Boxel, J., Nardella, F., Houthuijs, K.J., Brandsma, S.H., Béen, F., van Duursen, M.B.M., 2025. Health impacts of microplastic and nanoplastic exposure. Nat. Med. 1546170X. 31 (9), 2873–2887. Available from: https://www.nature.com/nm/. doi:10.1038/s41591-025-03902-5.

Lechuga, S., Braga-Neto, M.B., Naydenov, N.G., Rieder, F., Ivanov, A.I., 2023. Understanding disruption of the gut barrier during inflammation: should we abandon traditional epithelial cell lines and switch to intestinal organoids? Front. Immunol. 16643224. 14. Available from: https://www.frontiersin.org/journals/immunology#. doi:10.3389/fimmu.2023.1108289.

Leikauf, G.D., Kim, S.H., Jang, A.S., 2020. Mechanisms of ultrafine particle-induced respiratory health effects. Exp. Mol. Med. 20926413. 52 (3), 329–337. Available from: https://www.nature.com/emm/. doi:10.1038/s12276-020-0394-0.

Lolescu, B.M., Furdui-Linţa, A.V., Ilie, C.A., Sturza, A., Zară, F., Muntean, D.M., Blidişel, A., Creţu, O.M., 2025. Adipose tissue as target of environmental toxicants: focus on mitochondrial dysfunction and oxidative inflammation in metabolic dysfunction-associated steatotic liver disease. Mol. Cell. Biochem. 15734919. 480 (5), 2863–2879. Available from: https://www.springer.com/journal/11010. doi:10.1007/s11010-024-05165-z.

Lopez-Suarez, L., Awabdh, S.A., Coumoul, X., Chauvet, C., 2022. The SH-SY5Y human neuroblastoma cell line, a relevant in vitro cell model for investigating neurotoxicology in human: focus on organic pollutants. NeuroToxicology. 18729711. 92, 131–155. Available from: www.elsevier.com/inca/publications/store/6/2/1/3/5/5. doi:10.1016/j.neuro.2022.07.008.

Luo, W.Q., Cao, M.T., Sun, C.X., Wang, J.J., Gao, M.X., He, X.R., Dang, L.N., Geng, Y.Y., Li, B.Y., Li, J., Shi, Z.C., Yan, X.R., 2025. Size-dependent internalization of polystyrene microplastics as a key factor in macrophages and systemic toxicity. J. Hazard. Mater. 18733336. 490. Available from: https://www.sciencedirect.com/science/journal/03043894. doi:10.1016/j.jhazmat.2025.137701.

Ma, Y., Yang, H., Niu, S., Guo, M., Xue, Y., 2025. Mechanisms of micro- and nanoplastics on blood-brain barrier crossing and neurotoxicity: current evidence and future perspectives. NeuroToxicology. 18729711. 109, 92–107. Available from: https://www.sciencedirect.com/science/journal/0161813X. doi:10.1016/j.neuro.2025.06.003.

Mahendran, R., Ramaswamy, S.N., 2024. Nanoplastics as trojan horses: deciphering complex connections and environmental ramifications: a review. Chem. Afr. 2522-5758. 7 (5), 2265–2282. doi:10.1007/s42250-023-00865-3.

Manley, K., Han, W., Zelin, G., Lawrence, D.A., 2018. Crosstalk between the immune, endocrine, and nervous systems in immunotoxicology. Curr. Opin. Toxicol. 24682020. 10, 37–45. Available from: https://www.journals.elsevier.com/current-opinion-in-toxicology/. doi:10.1016/j.cotox.2017.12.003.

Marques, V.B., Faria, R.A., Santos, L.D., 2018. Overview of the pathophysiological implications of organotins on the endocrine system. Front. Endocrinol. 16642392. 9. Available from: https://www.frontiersin.org/articles/10.3389/fendo.2018.00101/full. doi:10.3389/fendo.2018.00101.

Marturano, V., Cerruti, P., Ambrogi, V., 2017. Polymer additives. Phys. Sci. Rev. 2365-659X. 2 (6). doi:10.1515/psr-2016-0130.

Menichetti, A., Mordini, D., Montalti, M., 2025. Penetration of microplastics and nanoparticles through skin: effects of size, shape, and surface chemistry. J. Xenobiotics. 20394713. 15 (1). Available from: https://www.mdpi.com/2039-4713/11. doi:10.3390/jox15010006.

Miao, X., Jiang, P., Wang, Z., Kong, W., Feng, L., 2025. Mitochondrial transplantation: a novel therapeutic approach for treating diseases. MedComm. 2688-2663. 6 (6). doi:10.1002/mco2.70253.

Miloradovic, D., Pavlovic, D., Jankovic, M.G., Nikolic, S., Papic, M., Milivojevic, N., Stojkovic, M., Ljujic, B., Embryos, H., 2021. Induced pluripotent stem cells, and organoids: models to assess the effects of environmental plastic pollution. Front. Cell Dev. Biol. 2296634X. 9. Available from: https://www.frontiersin.org/journals/cell-and-developmental-biology#. doi:10.3389/fcell.2021.709183.

Mitra, S., Chakraborty, A.J., Tareq, A.M., Emran, T.B., Nainu, F., Khusro, A., Idris, A.M., Khandaker, M.U., Osman, H., Alhumaydhi, F.A., Simal-Gandara, J., 2022. Impact of heavy metals on the environment and human health: novel therapeutic insights to counter the toxicity. J. King Saud Univ. Sci. 10183647. 34 (3). Available from: http://www.sciencedirect.com/science/journal/10183647. doi:10.1016/j.jksus.2022.101865.

Mohajer, C., 2025. Impact of real-life environmental exposures on reproduction: impact of human-relevant doses of endocrine-disrupting chemical and drug mixtures on testis development and function. Reproduction (Cambridge, England) 169(1).

Pan, H., Yin, F., 2025. Mitochondrial ROS sources steer neuroinflammation. Nat. Metab. 2522-5812. 7 (11), 2192–2194. doi:10.1038/s42255-025-01391-x.

Pan, I., Umapathy, S., 2024. Probiotics an emerging therapeutic approach towards gut-brain-axis oriented chronic health issues induced by microplastics: a comprehensive review. Heliyon. 24058440. 10 (11), e32004. doi:10.1016/j.heliyon.2024.e32004.

Paul, M.B., Stock, V., Cara-Carmona, J., Lisicki, E., Shopova, S., Fessard, V., Braeuning, A., Sieg, H., Böhmert, L., 2020. Micro- And nanoplastics-current state of knowledge with the focus on oral uptake and toxicity. Nanoscale Adv. 25160230. 2 (10), 4350–4367. Available from: pubs.rsc.org/en/journals/journalissues/na?_ga2.190536939.1555337663.1552312502-1364180372.1550481316#!issueidna001002&typecurrent&issnonline2516-0230. doi:10.1039/d0na00539h.

Pecorelli, A., Valacchi, G., 2023. Microplastics and Nanoplastics Contamination: An Emerging Environmental Issue for Skin Health Environmental Stressors and OxInflammatory Tissue Responses. CRC Press, Italy, pp. 136–152. Available from: http://www.tandfebooks.com/doi/book/9781000970098. doi:10.1201/9781003328100-15.

Peña, A., Rodríguez-Liébana, J.A., Delgado-Moreno, L., 2023. Interactions of microplastics with pesticides in soils and their ecotoxicological implications. Agronomy. 20734395. 13 (3). Available from: www.mdpi.com/journal/agronomy/. doi:10.3390/agronomy13030701.

Pironti, C., Ricciardi, M., Motta, O., Miele, Y., Proto, A., Montano, L., 2021. Microplastics in the environment: intake through the food web, human exposure and toxicological effects. Toxics. 2305-6304. 9 (9), 224. doi:10.3390/toxics9090224.

Prajapati, A., Vaidya, A.N., Kumar, A.R., 2022. Microplastic properties and their interaction with hydrophobic organic contaminants: a review. Environ. Sci. Pollut. Res. 16147499. 29 (33), 49490–49512. Available from: https://link.springer.com/journal/11356. doi:10.1007/s11356-022-20723-y.

Preda, O.T., Vlasceanu, A.M., Andreescu, C.V., Tsatsakis, A., Mezhuev, Y., Negrei, C., Baconi, D.L., 2024. Health implications of widespread micro- and nanoplastic exposure: environmental prevalence, mechanisms, and biological impact on humans. Toxics. 23056304. 12 (10). Available from: http://www.mdpi.com/journal/toxics. doi:10.3390/toxics12100730.

Prüst, M., Meijer, J., Westerink, R.H.S., 2020. The plastic brain: neurotoxicity of micro- and nanoplastics. Part. Fibre Toxicol. 1743-8977. 17 (1). doi:10.1186/s12989-020-00358-y.

Qin, J., Zhihui, M., Chen, X., Shu, S., 2023. Microglia activation in central nervous system disorders: a review of recent mechanistic investigations and development efforts. Front. Neurol. 1664-2295. 14. doi:10.3389/fneur.2023.1103416.

Rafa, N., Ahmed, B., Zohora, F., Bakya, J., Ahmed, S., Ahmed, S.F., Mofijur, M., Chowdhury, A.A., Almomani, F., 2024. Microplastics as carriers of toxic pollutants: source, transport, and toxicological effects. Environ. Pollut. 18736424. 343. Available from: https://www.sciencedirect.com/science/journal/02697491. doi:10.1016/j.envpol.2023.123190.

Rim, K.T., 2025. Literature review of the microRNA regulation of chemical toxicity. Toxicol. Environ. Health Sci. 22337784. 17 (2), 191–201. Available from: https://www.springer.com/journal/13530. doi:10.1007/s13530-025-00258-8.

Rojoni, S.A., Ahmed, M.T., Rahman, M., Hossain, M.M.M., Ali, M.S., Haq, M., 2024. Advances of microplastics ingestion on the morphological and behavioral conditions of model zebrafish: a review. Aquat. Toxicol. 18791514. 272. Available from: https://www.sciencedirect.com/science/journal/0166445X. doi:10.1016/j.aquatox.2024.106977.

Ruprecht, N.A., Singhal, S., Schaefer, K., Panda, O., Sens, D., Singhal, S.K., 2024. A review: multi-omics approach to studying the association between ionizing radiation effects on biological aging. Biology. 20797737. 13 (2). Available from: www.mdpi.com/journal/biology. doi:10.3390/biology13020098.

Saili, K.S., Zurlinden, T.J., Schwab, A.J., Silvin, A., Baker, N.C., Hunter, E.S., Ginhoux, F., Knudsen, T.B., 2017. Blood-brain barrier development: systems modeling and predictive toxicology. Birth Defects Res. 24721727. 109 (20), 1680–1710. Available from: http://onlinelibrary.wiley.com/journal/10.1002/%28ISSN%292472-1727. doi:10.1002/bdr2.1180.

CHAPTER

17

Microplastics and cardiac health

Divya Jyoti[a], *Ranbir Chander Sobti*[b], *and Arup Giri*[c]

[a]Lovely Professional University, School of Bioengineering and Biosciences, Phagwara, India
[b]Department of Biotechnology, Punjab University, Chandigarh, India
[c]Department of Zoology, Baba Mastnath University, Asthal Bohar, Rohtak, Haryana, India

17.1 Introduction

The term "plastic" is derived from the Greek word "plastikos," which means moldable. It is composed of synthetic organic polymers with high molecular masses, usually formed by the polymerization of gas-, coal-, or oil-derived monomers. Mass production began in the 1950s, leading to a plethora of versatile, durable, lightweight, corrosion-resistant, and flame-resistant polymers (An et al., 2021). Plastic production has increased over the past few decades, reaching approximately 445 million metric tons (mmt) globally in 2025, and is expected to rise to approximately 590 mmt by the year 2050. The ubiquity of plastic can be highlighted by the existence of approximately 45 different types of polymers, including polyethylene (PE), polypropylene (PP), polystyrene (PS), polyvinyl chloride, polyurethane, and polyethylene terephthalate (PET). The alarming situation of plastic pollution is underscored by the presence of approximately 12.7 mmt of plastics in the ocean. These large plastics can degrade into smaller particles due to UV radiation, mechanical abrasion, biological degradation, and wave action in the environment and are termed "microplastics" (MPs) (Sun et al., 2022). MPs ranging from 5 mm to 1 μm are chemically inert and persist indefinitely in both terrestrial and aquatic environments, leading to ecotoxicity (Zhou et al., 2021). These MPs are described as hazardous to health by many studies, as they can act as vectors for the distribution of contaminants, such as pathogens and heavy metals, leading to their bioaccumulation (Kumar et al., 2022; Menéndez-Pedriza and Jaumot, 2020). MPs can enter the food chain directly, as demonstrated by their ingestion and accumulation at relatively high trophic levels, either by ingesting contaminated plankton or algae or by ingesting contaminated prey (Diepens and Koelmans, 2018). Ingestion is among the major routes through which the human body is exposed to plastic. It is estimated that an average person consumes approximately 39,000–52,000 MPs every year via food and water, with sex and age affecting the total amount (Yee et al., 2021). Bottled water contains approximately 240,000 plastic particles, illustrating the distribution of these particles in everyday products

Microplastics in Animal and Human Physiology
https://doi.org/10.1016/B978-0-443-45635-0.00011-2

(Qian et al., 2024). A person who consumes bottled water is estimated to ingest approximately 90,000 additional plastic particles compared with someone who drinks tap water (Cox et al., 2019).

MPs have been found in human biological samples such as sputum, saliva, breast milk, liver, lungs, and feces (Enyoh et al., 2023; Jahedi et al., 2025; Kutralam-Muniasamy et al., 2020; Liu et al., 2023; Momeni et al., 2025). The finding of MPs in systemic tissues suggests that they can also be found in the blood, where they circulate throughout the body. Among the array of possible effects, MPs are considered potential risk factors for the progression of cardiovascular disease (CVD) (Zhu et al., 2023). The results from in vitro studies support the hypothesis that MPs activate pathophysiological pathways in cells, leading to the development of CVD. These include pathways through immune and endothelial cells and involve alterations such as immune activation and endothelial dysfunction. These findings are supported by various animal models and preliminary human studies (Marfella et al., 2024; Rotchell et al., 2023; Yang et al., 2023; Zhu et al., 2023).

The current review summarizes the various impacts on cardiac health resulting from the accumulation of MPs, including cellular damage, oxidative stress, disruption of lipid metabolism, induction of inflammation, cardiac mitochondrial dysfunction, and hemodynamic disruption.

17.2 Microplastic exposure routes

MPs can enter the food chain through many routes, raising concerns about their impacts on ecosystems and human health (Pironti et al., 2021). These particles are even detected in the foundation of the trophic level, i.e., plankton, which play a key role in aquatic life (Gouin, 2020; Pan et al., 2022; Siddiqui et al., 2024; Wang et al., 2019). Humans are exposed to MPs primarily through inhalation, dermal contact, ingestion, and iatrogenic sources (e.g., masks, PPE kits), each contributing to MP accumulation in the body; ingestion is the major route, followed by inhalation (Fig. 17.1).

17.2.1 Inhalation

Inhalation is a significant exposure route for humans, leading to potential health risks. Airborne MPs are found both indoors and outdoors. They can originate from industrial emissions, synthetic textiles, urban dust, construction materials, waste incineration, plastic abrasions, and indoor sources (furniture, textiles, etc.) (Acharya et al., 2021; Dris et al., 2017; Peng et al., 2023). Humans are estimated to inhale approximately 74,000–121,000 particles (Cox et al., 2019). It was also estimated that an individual could inhale approximately 130 airborne MPs per day, which could directly affect the lungs (Prata, 2018). Humans primarily breathe polyester (81%) indoors, where these particles contribute an average of approximately 4% of inhaled organic particles (Vianello et al., 2019). The aerodynamic size and shape of plastic particles affect the depth of penetration into fibers, and particles smaller than 5 μm are likely to reach the alveolar sacs upon inhalation. Plastic fibers (19.6 fibers/m^3) of PP were observed in indoor air by Dris and coworkers (Dris et al., 2017). These particles can be inhaled

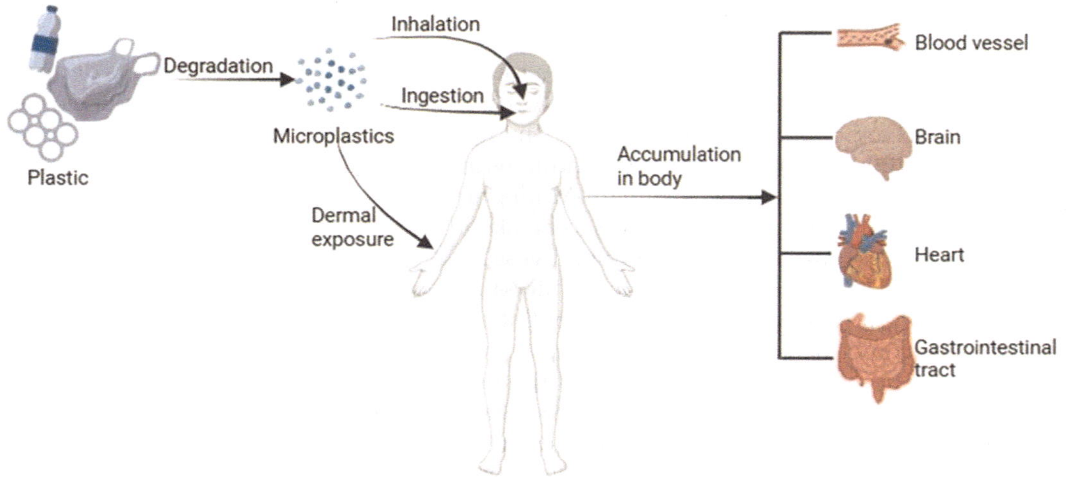

FIG. 17.1 Exposure routes of microplastics. MP accumulation in the body; ingestion is the major route, followed by inhalation.

and deposited in the respiratory tract. In addition to these studies, many studies have indicated that these inhaled plastic particles may cause inflammation, oxidative stress, and cell damage in pulmonary tissues (Dong et al., 2020; Jin et al., 2024; Kang et al., 2024; Lu et al., 2021). The biocompatibility and persistence of MPs raise concerns about their systemic and respiratory health effects, highlighting the need for regulatory measures and risk assessments to address airborne MP issues.

17.2.2 Ingestion

Ingestion is also among the primary routes of entry for MPs into the human body (Alqahtani et al., 2023). The source of MPs in humans via ingestion is commonly fish, mussels, commercial salt, tea bags, sugar, honey, and drinking water, leading to accumulation in various tissues (Jin et al., 2021; Kadac-Czapska et al., 2023; Lakshmayya et al., 2024; Pironti et al., 2021; Zhang et al., 2020). Once ingested, these MPs pass through the gastrointestinal tract, where they can interact with epithelial cells of the gut, disrupt the gut microbiota, and induce oxidative stress and inflammation (Hirt and Body-Malapel, 2020; Jia et al., 2023; Qiao et al., 2019; Shi et al., 2023; Zeng et al., 2024). Although it is very unlikely for MPs to penetrate the gastrointestinal tract at the paracellular level because of the fine size of the relevant pore, nanoplastics can be absorbed by the lymphatic tissue through phagocytosis or endocytosis by the cells of Peyer's patches (Ahmadi et al., 2025). In an in vitro study in mice, the intestinal absorption of MPs was low (0.04–0.3%), whereas that of nanoparticles <50 nm was greater (Walczak et al., 2015). Within the tract, plastic particles can interact with molecules such as lipids and proteins, affecting their absorption and ultimately posing serious health risks to both humans and animals through bioaccumulation and food chain transfer. Continuous exposure to dietary MPs raises concerns about their potential cumulative effects on human health.

17.3 Accumulation in the cardiovascular system

Many studies have demonstrated the ubiquitous presence of MPs across various domains, including wastewater, surface water, sediment, ice, air, food, and bottled and tap water (Feng et al., 2023; Horton et al., 2017; Prata, 2018). The presence of MPs in honey, milk, table salt, seafood, and drinking water suggests that humans are unavoidably exposed to MPs (Basaran et al., 2023; Katsara et al., 2024; Kutralam-Muniasamy et al., 2020; Zhang et al., 2023, 2020). Considering that MPs are small enough to be absorbed and found in blood, it can also be assumed that these particles can enter blood vessels. A study has shown that the detectable concentration of plastic particles in human blood is approximately 1.6 μg/mL (Leslie et al., 2022). However, it was anticipated that as pollution rose, the concentration of plastic in the bloodstream would increase over time.

The cardiovascular system, particularly the endothelium, is susceptible to substances present in the bloodstream. Rotchell and coworkers reported detectable levels of MPs in a preliminary study of human saphenous vein tissue. An average of 29.28 ± 34.88 MP/g of tissue was observed in the vein, with polytetrafluoroethane being the most abundant, followed by PET (Rotchell et al., 2023). In another study, the presence of MPs in patients undergoing cardiac surgery was explored via laser direct infrared spectroscopy and scanning electron microscopy. Pericardial, myocardial, pre- and postoperative venous blood, and pericardial and epicardial adipose tissues were analyzed. A range of nine types of MPs with diameters up to 469 μm were found in both the blood samples and cardiac tissues. Researchers confirmed the presence of polymethylmethacrylate in epicardial and pericardial adipose tissue and the left atrial appendage, suggesting the possible accumulation of MPs in the heart (Yang et al., 2023). Studies evaluating the detection of MPs in thrombi have also been conducted. In a recent study, evidence of MPs was found in thrombi from 30 patients who underwent thrombectomy due to myocardial infarction, deep vein thrombosis, and ischemic stroke (Wang et al., 2024). The thrombi were analyzed via pyrolysis–gas chromatography–mass spectrometry. Among the 30 assessed thrombi, 24 had detectable amounts of MPs, with PE being the most abundant among the 15 identified MP polymers. The concentration of MPs was greater in arteries (coronary and carotid) with atherosclerotic plaques than in those without atherosclerotic plaques, suggesting that MPs may preferentially accumulate at sites of atherosclerosis. Similarly, Marfella and coworkers reported that patients with carotid artery plaques in which MPs were found had a greater risk of stroke, myocardial infarction, or death than those in which no MPs were detected (Marfella et al., 2024). These findings collectively provide insights into the associations between MPs and cardiovascular health. The presence of MPs in thrombi, blood, cardiac tissues, and arterial plaques suggests their systemic distribution and potential role in cardiac health.

17.4 Impact on cardiac health

MPs have toxicological impacts on cardiac health, and these impacts are determined by various physicochemical factors, such as structure, type, surface features, exposure dose, duration, and size of MPs, as well as the species, sex, and life stage of the exposed organism

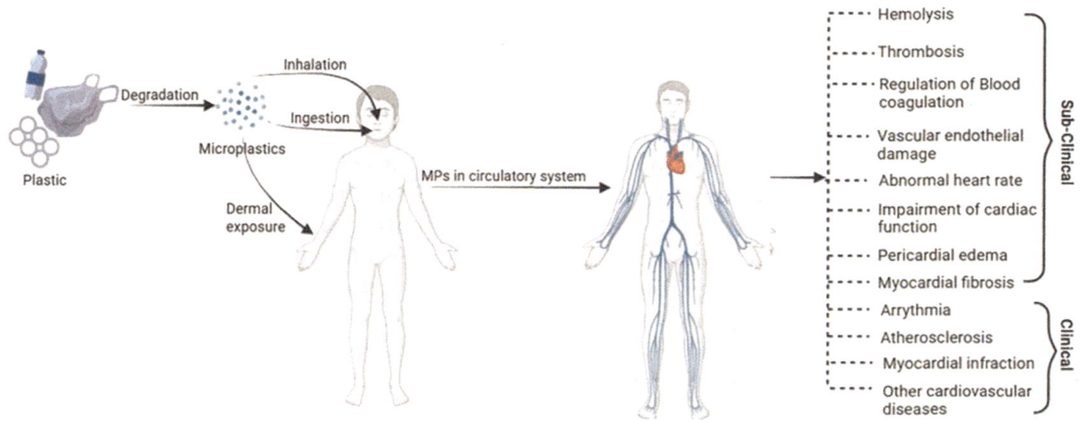

FIG. 17.2 Various impacts of MPs on the cardiovascular system. MPs can cause oxidative stress, inflammation, arrhythmia, atherosclerosis, myocardial fibrosis, myocardial infarction, cardiac fibrosis, cardiac mitochondrial dysfunction, and other effects.

(Zhu et al., 2023). Low doses of MPs can increase the heart beating rate, while higher doses in the blood can cause direct myocardial injury and brachycardia (Agathokleous, 2022). In addition, MPs can cause oxidative stress, inflammation, arrhythmia, atherosclerosis, myocardial fibrosis, myocardial infarction, cardiac fibrosis, cardiac mitochondrial dysfunction, and other effects (Fig. 17.2), as described in the following sections.

17.4.1 Cardiac mitochondrial dysfunction

Dysfunction of mitochondria is intricately associated with several cardiovascular disorders, such as ischemic heart disease, cardiomyopathy, and heart failure (Marín-García and Akhmedov, 2016). Mitochondrial metabolism is a fundamental process required for cellular homeostasis, signaling in eukaryotes, and energy transformation. Mitochondria play a crucial role in producing adenosine triphosphate (ATP) via oxidative phosphorylation within the electron transport chain (Liu et al., 2023). In addition, mitochondria are involved in several metabolic pathways, such as amino acid metabolism and pyruvate decarboxylation, which provide acetyl-CoA that enters the tricarboxylic acid cycle. The mitochondrial metabolic pathways also provide mediators for the biosynthesis of fatty acids, amino acids, nucleotides, cholesterol, heme, iron-sulfur clusters, and glucose and serve as signaling centers for regulating cellular processes such as ROS generation, calcium signaling, and apoptosis (Li et al., 2024). Owing to their central roles in metabolism, intracellular signaling, and bioenergetics, mitochondrial dysfunction can have negative consequences for cardiac physiology.

17.4.1.1 Cellular damage and oxidative stress

Oxidative stress is among the most frequently observed negative impacts of MPs. MPs evoke oxidative stress in two ways. First, MPs increase the ROS levels in cells and tissues.

In addition, they alter the actions of antioxidants [superoxide dismutase (SOD), catalase (CAT), and glutathione (GSH)], thereby preventing the efficient removal of ROS and leading to oxidative stress (An et al., 2021; Xie et al., 2020). The changes in the levels of antioxidant enzymes depend on the concentration, size, exposure time, and type of MPs, as well as the trophic level of the tissue and organism (Capó et al., 2022; Ding et al., 2018; Subaramaniyam et al., 2023). The antioxidant status of a cell is dictated by the balance between the production of reactive oxygen species (ROS) and the generation of the antioxidant system (e.g., SOD and GSH) within the cell (Bhattacharya, 2015). Under normal conditions, low levels of ROS are either removed by the antioxidant system, leading to redox balance, or act as signaling molecules that support cardiovascular function. However, toxins or pollutants such as heavy metals and plastics can disrupt the antioxidant-ROS balance, leading to increased ROS levels that can damage DNA, proteins, and lipids, causing inflammation, atherosclerosis, and endothelial dysfunction. Lipid peroxidation (LPO) by-products, such as malondialdehyde (MDA) and 4-hydroxynonenal, disrupt the cell membrane, promoting cardiomyocyte death and compensatory fibrosis (Yixiang et al., 2024). Environmental exposure can amplify these effects by directly increasing ROS production and impairing antioxidant capacity. Although mitochondria are a major source of ROS from complexes I and III, ROS also damage mitochondrial proteins, leading to mitochondrial dysfunction. The overproduction of ROS from dysfunctional mitochondria induces protein–protein and DNA damage, thereby impairing ATP synthesis (Werbner et al., 2023).

MPs can generate ROS through intra- and extracellular processes. The degradation of plastics is associated with extracellular ROS. Variables, including mechanical forces, light, chemical factors, temperature, and biological factors, contribute to the formation of MPs. These attributes can act independently or in combination to produce free radicals on the plastic surface via different mechanisms (Kadac-Czapska et al., 2023). In addition, MPs in intracellular processes can generate ROS (Qiao et al., 2019). MPs can be taken into cells by endocytosis, diffusion assisted by transport proteins, and passive transport across cell membranes. MPs are first transferred to lysosomes and then to mitochondria via the biofilm membrane, leading to alterations in the mitochondrial membrane potential (Lulu et al., 2021). These MPs are recognized as foreign substances by the body, triggering the innate immune system (Qiao et al., 2019).

ROS play crucial roles in regulating physiological functions and are major contributors to many cardiac-associated issues, such as cardiomyocyte cell cycle arrest (Pei et al., 2020). The increased production of ROS has been shown to upregulate the expression of nuclear factor-kappa B (NF-κB), which, in turn, facilitates the activation of the inflammasome (NLRP3, caspase-1) and its inflammatory factors [interleukin (IL)-1β and IL-18 precursors] (Liu et al., 2022; Jing et al., 2022). On the one hand, activated caspase-1 cleaves gasdermin-D (GSDMD) to produce a peptide containing the GSDMD-N active region, inducing cell membrane rupture and perforation and triggering pyroptosis (Sun et al., 2019). On the other hand, activated caspase-1 cleaves IL-1β and the precursors of IL-18, releasing them into the extracellular space, recruiting inflammatory cells, and enhancing the inflammatory response (Dai et al., 2021). Wei and colleagues reported that administering PS-MPs could trigger oxidative stress, as indicated by elevated MDA levels and reduced activities of SOD, CAT, and glutathione peroxidase (GSH-Px) in the hearts of Wistar rats (Wei et al., 2021). In another

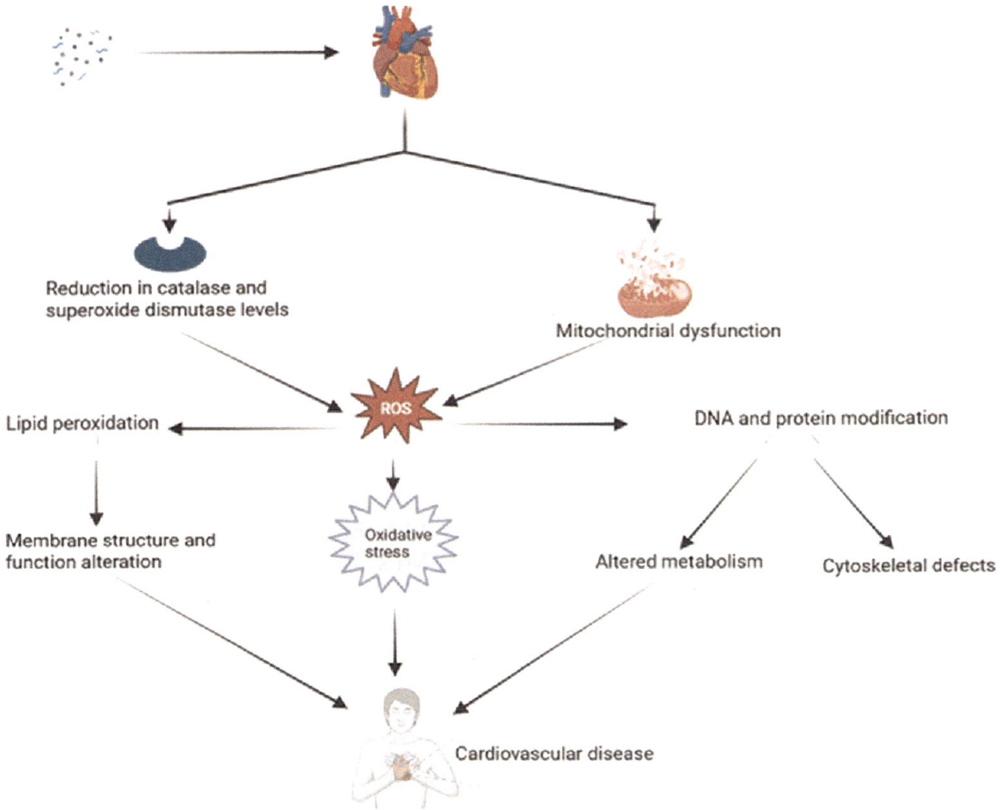

FIG. 17.3 Schematic representation of the process by which reactive oxygen species affect cardiovascular health. The accumulation of ROS in H9C2 cells was observed, along with elevated levels of MDA in myocardial tissues and reduced activities of SOD, CAT, and GSH-Px. This oxidative stress, along with the activation of a mitochondrial-mediated apoptotic pathway, ultimately leads to cardiomyocyte death and causes cardiac fibrosis.

study, mice were exposed to PET-MPs at concentrations of 0, 25, 50, and 75 μg/ml via oral ingestion. The accumulation of ROS in H9C2 cells was observed, along with elevated levels of MDA in myocardial tissues and reduced activities of SOD, CAT, and GSH-Px. This oxidative stress, along with the activation of a mitochondrial-mediated apoptotic pathway, ultimately leads to cardiomyocyte death and causes cardiac fibrosis (Tao et al., 2024) (Fig. 17.3).

17.4.1.2 *Lipid metabolism disruption and inflammation induction*

MPs can act as causative agents in many cardiovascular and metabolic diseases, such as obesity, diabetes, and atherosclerosis (Oraee et al., 2025; Zheng et al., 2024). These particles play crucial roles in the pathogenesis of atherosclerosis, affecting macrophages, lipid homeostasis, and gene expression related to the inflammatory response. Atherosclerosis is a vascular inflammatory disorder characterized by the contracting lumen of blood vessels due to lipid accumulation, vascular smooth muscle cells (VSMCs), inflammatory cells, and platelets activated by LPO, VSMC migration, inflammatory mediator release by macrophage activation, and intimal layer proliferation (Gusev and Sarapultsev, 2023; Rudijanto, 2007). Atherosclerotic lesions may progress with increased accumulation of macrophages and the release of inflammatory mediators, including tumor necrosis factor, inducible nitric oxide synthase, IL, and cyclooxygenase-2, followed by the migration of VSMCs and their proliferation from the tunica media to the tunica intima due to the activation of mitogen-activated protein kinase pathways, lesion calcification, and collagen production (Kim and Jeon, 2022). In addition, MPs can induce histological changes. These changes include aberrant cytoskeletal organization, increased spacing of microfilament proteins, and increased expression of fibronectin and type I and III collagens (Roshanzadeh et al., 2021; Zhou et al., 2023).

In a previous study, human vascular endothelial (HVEC) EA.hy926 cells were utilized to assess the toxicity of PS-MPs at realistic blood concentrations (4×10^{-6}–40 μg/mL), with a mean concentration of 1.6 μg/mL reflecting realistic levels of human exposure (Leslie et al., 2022). At the end of the study, no acute cytotoxicity to vascular cells was observed. A similar study also evaluated the potential toxicity of PS-MPs on HVECs, in which the cells were exposed to PS-MPs at concentrations ranging from 4×10^{-6} μg/mL to 40 μg/mL for 24 h. A significant reduction in cell viability was observed at concentrations of 0.04, 4, and 40 μg/mL. PS-MPs induced apoptosis, as evidenced by the loss of the mitochondrial membrane potential in cells exposed to 0.04 μg/mL PS-MPs. Further analysis revealed that PS-MPs induced oxidative stress, as evidenced by the accumulation of ROS in HVECs. Compared with MPs in a previous study, MPs compromised the integrity of the vascular barrier by reducing ZO-1 expression, potentially promoting atherosclerosis (Chen et al., 2023). At the molecular level, human vascular cell inflammation is associated with the inhibition of critical inflammatory response mediators, such as NF-κB p65 proteins and Rho-linked Rho-kinase (Gao et al., 2023; Xia et al., 2025). This inhibition also led to apoptosis and the formation of atherosclerotic plaques. The discrepancy in the results may be due to the use of different MP sizes in the experiments. The laboratory-prepared MPs are mostly uniform in size and are likely to penetrate cells.

In contrast, in real scenarios, MPs degrade unevenly, producing a wide array of particle sizes, i.e., from nanometers to micrometers. These varying sizes might coexist, accumulate differently in the body, and contribute to differences between real-world and laboratory exposure scenarios. In one study, a human pluripotent stem cell-derived three-dimensional cardiac organoid was used to assess the effects of PS-MPs (0.025, 0.25, and 2.5 μg/mL) on the human heart. The organoids exhibited an inflammatory response, oxidative stress, collagen accumulation, and apoptosis, which were consistent with observations in vivo in terms of increased septal thickness (intraventricular). PS-MPs can also induce cardiac hypertrophy both in vitro and in vivo, as evidenced by alterations in the expression of genes related to hypertrophy (atrial natriuretic peptide, *MYH7B, COL1A1* and B-type natriuretic peptide) and increased cardiac-specific markers (CX43, MYL2, and MYL4), indicating that MPs are potential risk factors for cardiac health (Zhou et al., 2023).

Exposure to MPs has been found to disrupt lipid metabolism. Zhao and coworkers examined lipid homeostasis disruption and the proinflammatory effects of PS-MP (bead) exposure. The mice were exposed to PS beads of two sizes, i.e., 0.5 and 5 μm, for 12 weeks at concentrations of 0.1 and 1 μg/ml through drinking water. By week three, the weight and body fat percentage of the mice exposed to 0.5 μm at 1 μg/ml were significantly greater than those of the control mice, suggesting that the ingestion of PS beads promoted adiposity. In addition, alterations in the gene expression of perivascular adipose tissue, which is crucial for cardiovascular health, were observed. Genes related to inflammation and adipogenesis, such as MCP-1 and IL-6, were also significantly upregulated, leading to adipose tissue dysfunction, a prominent risk factor for CVD (Zhao et al., 2022). In another study by Song and colleagues, Wistar rats were exposed to PS-MPs (0, 0.5, 5, and 50 mg/kg/d). At the end of the experiments, the rats exposed to higher MPs presented a significant increase in myocardial enzyme levels. In addition, MPs induced lipid metabolic disorders in rats and increased inflammation and oxidative stress in aortic and myocardial tissues. Untargeted metabolomics revealed the upregulation of metabolites with antiinflammatory and antioxidant effects, such as 4-hydroxybenzoic acid and equol (Song et al., 2024).

17.4.2 Hemodynamic disruption

Red blood cells are among the longest-lived and most abundant cells in the blood, and the internalization of plastic particles has been consistently observed within these cells (Rajendran and Chandrasekaran, 2023; Rubin et al., 2012). This might activate the adhesion of RBCs to the vascular endothelium, a crucial feature of thrombus pathogenesis. Thrombosis has been identified as a factor in thrombotic and cardiovascular complications (Falk and Fernández-Ortiz, 1995; Page and Ariëns, 2021). MPs disrupt homeostasis, leading to increased thrombin and thrombus formation, and they increase the risk of thrombotic diseases related to their exposure (Oraee et al., 2025). Globally, thrombosis is recognized as a pathological factor in thrombotic complications and CVDs.

Chen and coworkers investigated the adverse effects of MPs on thrombosis and blood clotting both in vitro (a human thrombosis model simulating blood vessel conditions for 30 min) and in vivo (a mouse model for 16 days). In both models, 1-μm fluorescent PS-MPs were administered via two methods: intraperitoneal injection targeting the abdominal mucosa and continuous exposure via drinking water, which targets the intestine. The findings in mice revealed that MPs remained in the bloodstream, suggesting their potential for long-term accumulation. In an in vitro model, MPs invade thrombi, disrupt platelets, and disrupt fibrinogen colocalization, leading to a greater risk of thrombus shedding. The structural integrity of MPs invading thrombi is compromised, promoting the aggregation of blood cells and the additional formation of thrombi, which worsens the shear rate induced by MPs. These impacts are hazardous, especially in the case of endothelial damage, increasing the risk of thrombus shedding under real blood flow conditions (Chen et al., 2022). However, no definitive evidence indicates whether high flow rates cause MPs to disrupt thrombi or whether the MPs themselves are responsible for clot disruption. Another study investigated the effects of carboxylated, aminated, and nonfunctionalized PS particles (50, 100, and 500 nm) at different concentrations (25, 100, and 250 μg/mL) in an in vitro model (human whole blood). The results suggested that the R-time (clotting cascade activation time) was related to the particle size; the smaller the particle size, the more accelerated the R-time. Similarly, the aminated PS particles significantly enhanced clotting. The carboxylated PS particles produced more pronounced effects, including increased clot strength and an increased fibrin polymerization rate. The molecular mechanisms for these impacts may vary. The nonfunctionalized PS may activate the intrinsic pathway, whereas the aminated

PS may increase clotting via the extrinsic pathway by activating factor VII. The carboxylated PS serves as a scaffold for kallikrein, factor XII, and kininogen (high molecular weight) in the intrinsic coagulation pathway.

In a cross-sectional study by Lee and coworkers, human blood was quantitatively analyzed for MPs via μ-FTIR. They reported MPs in 88.9% of the participants, with an average concentration of 4.2 MPs/mL. MPs were significantly more common in adults with greater use of plastic food containers, and the most prevalent polymers were PS and PP. Blood samples were also analyzed for inflammatory markers, including prothrombin time, activated partial thromboplastin time (aPTT), antithrombin III, C-reactive protein, platelet count, fibrinogen, and erythrocyte sedimentation rate. A higher load of MPs (≥3 MPs/mL) in the blood was significantly correlated with increased C-reactive protein, fibrinogen, and aPTT. This underscores the association of specific lifestyles with alterations in coagulation markers (Lee et al., 2024)

17.4.3 Histopathological alterations in the heart

Exposure to MPs induces histopathological changes in myocardial tissue, including damage to fibers, swollen cells, disrupted tissue architecture, and an increased number of apoptotic cells, ultimately leading to cardiac fibrosis (Zhang et al., 2023). The damage is further accelerated by a significant reduction in cardiac Na^+-K^+-ATPase activity, a condition linked to ischemic heart disease (Zhang et al., 2023).

Zekang et al. (2020) reported that PS-MPs might activate the mechanism of cardiotoxicity in the hearts of rats. They observed that exposure to a high concentration (50 mg/L) of PS-MPs increased the levels of creatine kinase-MB and troponin I (two crucial biomarkers of myocardial damage), leading to structural damage and myocardial apoptosis, ultimately resulting in collagen proliferation in the heart. Transmission electron microscopy was used to visualize the PS-MPs internalized by cardiomyocytes, confirming their translocation to the heart via the circulatory system. They also reported that PS-MPs could induce oxidative stress (reduced levels of antioxidant molecules), which activated the Wnt/β-catenin signaling pathway associated with cardiac fibrosis (Zekang et al., 2020). These findings suggest that PS-MPs lead to cardiotoxicity by inducing apoptosis, which is driven by Wnt/β-catenin-mediated signaling, and causing Wnt/β-catenin-mediated fibrosis. In addition to PS-MPs, PET-MPs exacerbate cardiac fibrosis and postmyocardial infarction inflammation by activating the inflammasome (NLRP3) pathway (Xin et al., 2025). A study by Wei et al. revealed that exposure to MPs causes notable alterations in morphology, such as fractures in myocardial fibers and capillary congestion. In addition to the internalization of MPs in cardiomyocytes, the mitochondrial cristae disappeared (Wei et al., 2021). Li and colleagues examined the effects of PS-MPs on rat hearts and observed structural changes, including vascular congestion, interstitial hyperplasia, and increased cardiac fibrosis (Zekang et al., 2020). Similarly, in another study, cardiac fibrosis was observed in the hearts of mice exposed to PET-MPs (Tao et al., 2024). In another study, the effects of airborne PS particles were studied in human cardiac tissues, and histological assessments suggested toxic effects on the heart's structure, including myocardial necrosis, fibrosis, and ventricular hypertrophy.

The negative effects of MPs on the heart provide a foundation for a better understanding of their impact on human cardiac health, emphasizing the need to increase vigilance in the use of plastics and to implement mitigation strategies.

17.5 Conclusion and future perspectives

MPs represent a novel class of environmental pollutants that are distributed ubiquitously. Exposure to MPs via different routes could affect the cells of the blood immune system, leading to genotoxicity and altering heart physiology. This review presents growing evidence from animal experiments, human research, and laboratory studies showing that MPs can enter the body through ingestion, inhalation, and dermal contact and can be transported through the blood to the heart, triggering inflammation, damaging cells, increasing the clotting rate, and disrupting blood vessel function and the normal heart rate. Many studies have identified MPs in plaques of human arteries, associated them with strokes, heart attacks, and increased death rates. MPs can affect factors that play crucial roles in heart disease and blood pressure, such as gut health and hormonal balance, suggesting that MPs may contribute to CVD.

Future studies of the cardiovascular toxicity of MPs should consider factors such as particle size, exposure duration and dose, particle structure, surface modification, and the life stage of the organism to obtain a unified standard of procedure. In addition, most of the related research has focused on PS. It is important to study other plastics, such as PET, PE, and PP, given their widespread environmental distribution, thereby bringing the results closer to real-world applications. Furthermore, research is needed on the mechanisms of vascular absorption and cardiac bioaccumulation, as well as on the identification of early indicators of MP-induced cardiovascular toxicity. An appropriate deliberation is also needed to determine the relationship between inflammation and the oxidative stress response triggered by MPs and coronary atherosclerosis, arrhythmia, CVDs, and myocardial infarction.

References

Acharya, S., Rumi, S.S., Hu, Y., Abidi, N., 2021. Microfibers from synthetic textiles as a major source of microplastics in the environment: a review. Text. Res. J. 17467748. 91 (17-18), 2136–2156. Available from: https://journals.sagepub.com/home/TRJ. doi:10.1177/0040517521991244.

Agathokleous, E., 2022. The hormetic response of heart rate of fish embryos to contaminants – implications for research and policy. Sci. Total Environ. 00489697. 815, 152911. doi:10.1016/j.scitotenv.2021.152911.

Ahmadi, P., Doyle, D., Mojarad, N., Taherkhani, S., Janzadeh, A., Honardoost, M., Gholami, M., 2025. Effects of micro- and nanoplastic exposure on macrophages: a review of molecular and cellular mechanisms. Toxicol. Mech. Methods. 15376524. 35 (8), 823–846. Available from: http://www.tandfonline.com/loi/itxm20. doi:10.1080/15376516.2025.2500546.

Alqahtani, S., Alqahtani, S., Saquib, Q., Mohiddin, F., 2023. Toxicological impact of microplastics and nanoplastics on humans: understanding the mechanistic aspect of the interaction. Front. Toxicol. 2673-3080. 5. doi:10.3389/ftox.2023.1193386.

An, R., Wang, X., Yang, L., Zhang, J., Wang, N., Feibo, X., Hou, Y., Zhang, H., Zhang, L., 2021. Polystyrene microplastics cause granulosa cells apoptosis and fibrosis in ovary through oxidative stress in rats. Toxicology. 0300483X. 449, 152665. doi:10.1016/j.tox.2020.152665.

Basaran, B., Özçifçi, Z., Türker Akcay, H., Aytan, Ü., 2023. Microplastics in branded milk: dietary exposure and risk assessment. J. Food Compos. Anal. 08891575. 123, 105611. doi:10.1016/j.jfca.2023.105611.

Bhattacharya, S., 2015. Reactive Oxygen Species and Cellular Defense System Free Radicals in Human Health and Disease. Springer, India, pp. 17–29. doi:10.1007/978-81-322-2035-0_2.

Capó, X., Alomar, C., Compa, M., Sole, M., Sanahuja, I., Rojas, D.L.S., González, G.P., Garcinuño Martínez, R.M., Deudero, S., 2022. Quantification of differential tissue biomarker responses to microplastic ingestion and plasticizer bioaccumulation in aquaculture reared sea bream Sparus aurata. Environ. Res. 10960953. 211. Available from: http://www.elsevier.com/inca/publications/store/6/2/2/8/2/1/index.htt. doi:10.1016/j.envres.2022.113063.

Chen, L., Zheng, Y., Liu, Y., Tian, P., Yu, L., Bai, L., Zhou, F., Yang, Y., Cheng, Y., Wang, F., Zheng, L., Jiang, F., Zhu, Y., 2022. Microfluidic-based in vitro thrombosis model for studying microplastics toxicity. Lab Chip. 1473-0197. 22 (7), 1344–1353. doi:10.1039/d1lc00989c.

Chen, Y.C., Chen, K.F., Lin, K.Y.A., Su, H.P., Wu, D.N., Lin, C.H., 2023. Evaluation of toxicity of polystyrene microplastics under realistic exposure levels in human vascular endothelial EA.hy926 cells. Chemosphere. 18791298. 313. Available from: www.elsevier.com/locate/chemosphere. doi:10.1016/j.chemosphere.2022.137582.

Cox, K.D., Covernton, G.A., Davies, H.L., Dower, J.F., Juanes, F., Dudas, S.E., 2019. Human consumption of microplastics. Environ. Sci. Technol. 15205851. 53 (12), 7068–7074. Available from: http://pubs.acs.org/journal/esthag. doi:10.1021/acs.est.9b01517.

Dai, F., Xuan, L., Xia, L., Ding, Z., Ran, X., Yin, P., Wang, S., Junbo, G., Jian, W., Zou, Y., 2021. Caspase-1 abrogates the salutary effects of hypertrophic preconditioning in pressure overload hearts via IL-1β and IL-18. Front. Mol. Biosci. 2296-889X. 8. doi:10.3389/fmolb.2021.641585.

Diepens, N.J., Koelmans, A.A., 2018. Accumulation of plastic debris and associated contaminants in aquatic food webs. Environ. Sci. Technol. 15205851. 52 (15), 8510–8520. Available from: http://pubs.acs.org/journal/esthag. doi:10.1021/acs.est.8b02515.

Ding, J., Zhang, S., Razanajatovo, R.M., Zou, H., Zhu, W., 2018. Accumulation, tissue distribution, and biochemical effects of polystyrene microplastics in the freshwater fish red tilapia (Oreochromis niloticus). Environ. Pollut. 18736424. 238, 1–9. Available from: www.elsevier.com/inca/publications/store/4/0/5/8/5/6. doi:10.1016/j.envpol.2018.03.001.

Dong, C.D., Chen, C.W., Chen, Y.C., Chen, H.H., Lee, J.S., Lin, C.H., 2020. Polystyrene microplastic particles: in vitro pulmonary toxicity assessment. J. Hazard. Mater. 18733336. 385. Available from: www.elsevier.com/locate/jhazmat. doi:10.1016/j.jhazmat.2019.121575.

Dris, R., Gasperi, J., Mirande, C., Mandin, C., Guerrouache, M., Langlois, V., Tassin, B., 2017. A first overview of textile fibers, including microplastics, in indoor and outdoor environments. Environ. Pollut. 02697491. 221, 453–458. doi:10.1016/j.envpol.2016.12.013.

Enyoh, C.E., Devi, A., Kadono, H., Wang, Q., Rabin, M.H., 2023. The plastic within: microplastics invading human organs and bodily fluids systems. Environments. 2076-3298. 10 (11), 194. doi:10.3390/environments10110194.

Falk, E., Fernández-Ortiz, A., 1995. Role of thrombosis in atherosclerosis and its complications. Am. J. Cardiol. 00029149. 75 (6), 5B. doi:10.1016/0002-9149(95)80003-b.

Feng, Y., Chen, T., Ruijie, L., Wu, D., Yang, J., Xia, Y., Peijnenburg, W.J.G.M., Luo, Y., 2023. A systematic review of the impacts of exposure to micro- and nano-plastics on human tissue accumulation and health. Eco-Environ. Health. 27729850. 2 (4), 195–207. doi:10.1016/j.eehl.2023.08.002.

Gao, J., Zhang, Z., Yan, J.Y., Ge, Y.X., Gao, Y., 2023. Inflammation and coagulation abnormalities via the activation of the HMGB1-RAGE/NF-κB and F2/Rho pathways in lung injury induced by acute hypoxia. Int. J. Mol. Med. 1791244X. 52 (2). Available from: https://www.spandidos-publications.com/10.3892/ijmm.2023.5270. doi:10.3892/ijmm.2023.5270.

Gouin, T., 2020. Toward an improved understanding of the ingestion and trophic transfer of microplastic particles: critical review and implications for future research. Environ. Toxicol. Chem. 0730-7268. 39 (6), 1119–1137. doi:10.1002/etc.4718.

Gusev, E., Sarapultsev, A., 2023. Atherosclerosis and inflammation: insights from the theory of general pathological processes. Int. J. Mol. Sci. 1422-0067. 24 (9), 7910. doi:10.3390/ijms24097910.

Hirt, N., Body-Malapel, M., 2020. Immunotoxicity and intestinal effects of nano- and microplastics: a review of the literature. Part. Fibre Toxicol. 1743-8977. 17 (1). doi:10.1186/s12989-020-00387-7.

Horton, A.A., Walton, A., Spurgeon, D.J., Lahive, E., Svendsen, C., 2017. Microplastics in freshwater and terrestrial environments: evaluating the current understanding to identify the knowledge gaps and future research priorities. Sci. Total Environ. 18791026. 586, 127–141. Available from: www.elsevier.com/locate/scitotenv. doi:10.1016/j.scitotenv.2017.01.190.

Jahedi, F., Fard, N.J.H., Ahmadi, M., Takdastan, A., Shoushtari, M.H., Dehbandi, R., Turner, A., 2025. Microplastics in urine, sputum and lung lavage fluid from patients with respiratory illnesses. Environ. Res. 10960953. 274. Available from: https://www.sciencedirect.com/science/journal/00139351. doi:10.1016/j.envres.2025.121278.

Jia, R., Han, J., Liu, X., Kang, L., Lai, W., Bian, L., Yan, J., Zhuge, X., 2023. Exposure to polypropylene microplastics via oral ingestion induces colonic apoptosis and intestinal barrier damage through oxidative stress and inflammation in mice. Toxics. 2305-6304. 11 (2), 127. doi:10.3390/toxics11020127.

Jin, M., Wang, X., Ren, T., Wang, J., Shan, J., 2021. Microplastics contamination in food and beverages: direct exposure to humans. J. Food Sci. 0022-1147. 86 (7), 2816–2837. doi:10.1111/1750-3841.15802.

Jin, W., Zhang, W., Tang, H., Wang, P., Zhang, Y., Liu, S., Qiu, J., Chen, H., Wang, L., Wang, R., Sun, Y., Liu, P., Tang, H., Zhu, Y., 2024. Microplastics exposure causes the senescence of human lung epithelial cells and mouse lungs by inducing ROS signaling. Environ. Int. 01604120. 185, 108489. doi:10.1016/j.envint.2024.108489.

Jing, H., Wang, F., Gao, X., 2022. Lithium intoxication induced pyroptosis via ROS/NF-κB/NLRP3 inflammasome regulatory networks in kidney of mice. Environ. Toxicol. 37 (4), 825–835. doi:10.1002/tox.23446.

Kadac-Czapska, K., Knez, E., Gierszewska, M., Olewnik-Kruszkowska, E., Grembecka, M., 2023. Microplastics derived from food packaging waste—their origin and health risks. Materials. 1996-1944. 16 (2), 674. doi:10.3390/ma16020674.

Kang, H., Huang, D., Zhang, W., Wang, J., Liu, Z., Wang, Z., Jiang, G., Gao, A., 2024. Inhaled polystyrene microplastics impaired lung function through pulmonary flora/TLR4-mediated iron homeostasis imbalance. Sci. Total Environ. 00489697. 946, 174300. doi:10.1016/j.scitotenv.2024.174300.

Katsara, K., Viskadourakis, Z., Alissandrakis, E., Kountourakis, N., Kenanakis, G., Papadakis, V.M., 2024. Microplastics' detection in honey: development of protocols in a simulation. Appl. Sci. 2076-3417. 14 (11), 4720. doi:10.3390/app14114720.

Kim, M.J., Jeon, J.H., 2022. Recent advances in understanding nrf2 agonism and its potential clinical application to metabolic and inflammatory diseases. Int. J. Mol. Sci. 14220067. 23 (5). Available from: https://www.mdpi.com/1422-0067/23/5/2846/pdf. doi:10.3390/ijms23052846.

Kumar, R., Ivy, N., Bhattacharya, S., Dey, A., Sharma, P., 2022. Coupled effects of microplastics and heavy metals on plants: uptake, bioaccumulation, and environmental health perspectives. Sci. Total Environ. 00489697. 836, 155619. doi:10.1016/j.scitotenv.2022.155619.

Kutralam-Muniasamy, G., Pérez-Guevara, F., Elizalde-Martínez, I., Shruti, V.C., 2020. An overview of recent advances in micro/nano beads and microfibers research: critical assessment and promoting the less known. Sci. Total Environ. 00489697. 740, 139991. doi:10.1016/j.scitotenv.2020.139991.

Lakshmayya, N.S.V., Panday, A., Yadavalli, R., Reddy, C.N., Mandal, S.K., Agrawal, D.C., Mishra, B., 2024. Food contamination with micro-plastics: occurrences, bioavailability, human vulnerability, and prevention. Curr. Nutr. Food Sci. 22123881. 20 (7), 797–810. Available from: https://www.eurekaselect.com/article/134628. doi:10.2174/1573401319666230915164116.

Lee, D.W., Jung, J., Park, S.A., Lee, Y., Kim, J., Han, C., Kim, H.C., Lee, J.H., Hong, Y.C., 2024. Microplastic particles in human blood and their association with coagulation markers. Sci. Rep. 20452322. 14 (1). Available from: https://www.nature.com/srep/. doi:10.1038/s41598-024-81931-9.

Leslie, H.A., van Velzen, M.J.M., Brandsma, S.H., Vethaak, A.D., Garcia-Vallejo, J.J., Lamoree, M.H., 2022. Discovery and quantification of plastic particle pollution in human blood. Environ. Int. 18736750. 163. Available from: www.elsevier.com/locate/envint. doi:10.1016/j.envint.2022.107199.

Li, Y., Zhang, H., Chunjuan, Y., Dong, X., Yang, F., Wang, M., Wen, Z., Mohan, S., Bing, L., Yang, L., 2024. New insights into mitochondria in health and diseases. Int. J. Mol. Sci. 1422-0067. 25 (18), 9975. doi:10.3390/ijms25189975.

Liu, X.J., Wang, Y.Q., Shang, S.Q., Xu, S., Guo, M., 2022. TMT induces apoptosis and necroptosis in mouse kidneys through oxidative stress-induced activation of the NLRP3 inflammasome. Ecotoxicol. Environ. Saf. 10902414. 230. Available from: http://www.elsevier.com/inca/publications/store/6/2/2/8/1/9/index.htt. doi:10.1016/j.ecoenv.2022.113167.

Liu, Y., Sun, Y., Guo, Y., Shi, X., Chen, X., Feng, W., Wu, L.L., Zhang, J., Yu, S., Wang, Y., Shi, Y., 2023. An overview: the diversified role of mitochondria in cancer metabolism. Int. J. Biol. Sci. 14492288. 19 (3), 897–915. Available from: https://www.ijbs.com/v19p0897.pdf. doi:10.7150/ijbs.81609.

Lu, K., Lai, K.P., Stoeger, T., Ji, S., Lin, Z., Lin, X., Chan, T.F., Fang, J.K.H., Lo, M., Gao, L., Qiu, C., Chen, S., Chen, G., Li, L., Wang, L., 2021. Detrimental effects of microplastic exposure on normal and asthmatic pulmonary physiology. J. Hazard. Mater. 18733336. 416. Available from: www.elsevier.com/locate/jhazmat. doi:10.1016/j.jhazmat.2021.126069.

Lulu, L., Yan, X., Feng, L., Jiang, S., Zifan, L., Xie, H., Sun, S., Chen, J., Chengyong, L., 2021. Challenge for the detection of microplastics in the environment. Water Environ. Res. 1061-4303. 93 (1), 5–15. doi:10.1002/wer.1281.

Marfella, R., Prattichizzo, F., Sardu, C., Fulgenzi, G., Graciotti, L., Spadoni, T., D'Onofrio, N., Scisciola, L., Grotta, R. L., Frigé, C., Pellegrini, V., Municinò, M., Siniscalchi, M., Spinetti, F., Vigliotti, G., Vecchione, C., Carrizzo, A., Accarino, G., Squillante, A., Spaziano, G., Mirra, D., Esposito, R., Altieri, S., Falco, G., Fenti, A., Galoppo, S., Canzano, S., Sasso, F.C., Matacchione, G., Olivieri, F., Ferraraccio, F., Panarese, I., Paolisso, P., Barbato, E., Lubritto, C., Balestrieri, M.L., Mauro, C., Caballero, A.E., Rajagopalan, S., Ceriello, A., D'Agostino, B., Iovino, P., Paolisso, G., 2024. Microplastics and nanoplastics in atheromas and cardiovascular events. N. Engl. J. Med. 15334406. 390 (10), 900–910. Available from: http://www.nejm.org/medical-index. doi:10.1056/NEJMoa2309822.

Marín-García, J., Akhmedov, A.T., 2016. Mitochondrial dynamics and cell death in heart failure. Heart Fail. Rev. 15737322. 21 (2), 123–136. Available from: http://www.springer.com/medicine/cardiology/journal/10741. doi:10.1007/s10741-016-9530-2.

Menéndez-Pedriza, A., Jaumot, J., 2020. Interaction of environmental pollutants with microplastics: a critical review of sorption factors, bioaccumulation and ecotoxicological effects. Toxics. 2305-6304. 8 (2), 40. doi:10.3390/toxics8020040.

Momeni, M.K., Taghipour, H., Ghayebzadeh, M., Mohammadi, M., Keikhaee, R., 2025. Isolation and characterization of microplastics from the human respiratory system: sputum, broncho-alveolar lavage fluid, and pleural fluid simultaneously. Environ. Pollut. 02697491. 365, 125389. doi:10.1016/j.envpol.2024.125389.

Oraee, S., Alinejadfard, M., Mahdavi, S., Firouzabadi, S.R., Hatami, S., Mohammadi, I., Zarinfar, Y., Khaheshi, I., 2025. The association between micro- and nano-plastics and cardiovascular outcomes: a comprehensive review. Toxicol. Environ. Health Sci. 2005-9752. 17 (2), 203–218. doi:10.1007/s13530-025-00248-w.

Page, E.M., Ariëns, R.A.S., 2021. Mechanisms of thrombosis and cardiovascular complications in COVID-19. Thromb. Res. 18792472. 200, 1–8. Available from: www.elsevier.com/locate/thromres. doi:10.1016/j.thromres.2021.01.005.

Pan, Y., Long, Y., Hui, J., Xiao, W., Yin, J., Li, Y., Liu, D., Tian, Q., Chen, L., 2022. Microplastics can affect the trophic cascade strength and stability of plankton ecosystems via behavior-mediated indirect interactions. J. Hazard. Mater. 03043894. 430, 128415. doi:10.1016/j.jhazmat.2022.128415.

Pei, J., Wang, F., Pei, S., Bai, R., Cong, X., Nie, Y., Chen, X., 2020. Hydrogen sulfide promotes cardiomyocyte proliferation and heart regeneration via ROS scavenging. Oxid. Med. Cell. Longev. 1942-0900. 2020, 1–11. doi:10.1155/2020/1412696.

Peng, C., Zhang, X., Mengxi, L., Yuan, L., Liu, C., Wang, L., 2023. Source apportionment of microplastics in indoor dust: two strategies based on shape and composition. Environ. Pollut. 02697491. 334, 122178. doi:10.1016/j.envpol.2023.122178.

Pironti, C., Ricciardi, M., Motta, O., Miele, Y., Proto, A., Montano, L., 2021. Microplastics in the environment: intake through the food web, human exposure and toxicological effects. Toxics. 2305-6304. 9 (9), 224. doi:10.3390/toxics9090224.

Prata, J.C., 2018. Airborne microplastics: consequences to human health? Environ. Pollut. 18736424. 234, 115–126. Available from: www.elsevier.com/inca/publications/store/4/0/5/8/5/6. doi:10.1016/j.envpol.2017.11.043.

Qian, N., Gao, X., Lang, X., Deng, H., Bratu, T.M., Chen, Q., Stapleton, P., Yan, B., Min, W., 2024. Rapid single-particle chemical imaging of nanoplastics by SRS microscopy. Proc. Natl. Acad. Sci. U. S. A. 0027-8424. 121 (3). doi:10.1073/pnas.2300582121.

Qiao, R., Sheng, C., Yifeng, L., Zhang, Y., Ren, H., Lemos, B., 2019. Microplastics induce intestinal inflammation, oxidative stress, and disorders of metabolome and microbiome in zebrafish. Sci. Total Environ. 00489697. 662, 246–253. doi:10.1016/j.scitotenv.2019.01.245.

Rajendran, D., Chandrasekaran, N., 2023. Journey of micronanoplastics with blood components. RSC Adv. 2046-2069. 13 (45), 31435–31459. doi:10.1039/d3ra05620a.

Roshanzadeh, A., Oyunbaatar, N.E., Ganjbakhsh, S.E., Park, S., Kim, D.S., Kanade, P.P., Lee, S., Lee, D.W., Kim, E.S., 2021. Exposure to nanoplastics impairs collective contractility of neonatal cardiomyocytes under electrical synchronization. Biomaterials. 18785905. 278. Available from: http://www.journals.elsevier.com/biomaterials/. doi:10.1016/j.biomaterials.2021.121175.

Rotchell, J.M., Jenner, L.C., Chapman, E., Bennett, R.T., Bolanle, I.O., Loubani, M., Sadofsky, L., Palmer, T.M., 2023. Detection of microplastics in human saphenous vein tissue using μFTIR: a pilot study. PLoS ONE. 19326203. 18 (2). Available from: https://journals.plos.org/plosone/article/file?id=10.1371/journal.pone.0280594&type=printable. doi:10.1371/journal.pone.0280594.

Rubin, O., Canellini, G., Delobel, J., Lion, N., Tissot, J.D., 2012. Red blood cell microparticles: clinical relevance. Transfus. Med. Hemother. 16603818. 39 (5), 342–347. doi:10.1159/000342228.

Rudijanto, A., 2007. The role of vascular smooth muscle cells on the pathogenesis of atherosclerosis. Acta Medica Indonesiana. 01259326. 39 (2), 86–93.

Shi, Y., Chen, C., Han, Z., Chen, K., Xiangyang, W., Qiu, X., 2023. Combined exposure to microplastics and amitriptyline caused intestinal damage, oxidative stress and gut microbiota dysbiosis in zebrafish (Danio rerio). Aquat. Toxicol. 0166445X. 260, 106589. doi:10.1016/j.aquatox.2023.106589.

Siddiqui, S.A., Ucak, İ., Afreen, M., Sasidharan, A., Yunusa, B.M., Bhowmik, S., Pandiselvam, R., Ambartsumov, T. G., Shah, M.A., 2024. Microalgae as a potential raw material for plant-based seafood alternatives: a comprehensive review. Food Sci. Nutr. 20487177. 12 (11), 8559–8593. Available from: onlinelibrary.wiley.com/journal/10.1002/%28ISSN%292048-7177. doi:10.1002/fsn3.4313.

Song, Z., Haidi, W., Fang, X., Feng, X., Zhou, L., 2024. The cardiovascular toxicity of polystyrene microplastics in rats: based on untargeted metabolomics analysis. Front. Pharmacol. 1663-9812. 15. doi:10.3389/fphar.2024.1336369.

Subaramaniyam, U., Allimuthu, R.S., Vappu, S., Ramalingam, D., Balan, R., Paital, B., Panda, N., Rath, P.K., Ramalingam, N., Sahoo, D.K., 2023. Effects of microplastics, pesticides and nano-materials on fish health, oxidative stress and antioxidant defense mechanism. Front. Physiol. 1664042X. 14. Available from: http://www.frontiersin.org/Physiology/archive/. doi:10.3389/fphys.2023.1217666.

Sun, J., Zheng, H., Xiang, H., Fan, J., Jiang, H., 2022. The surface degradation and release of microplastics from plastic films studied by UV radiation and mechanical abrasion. Sci. Total Environ. 00489697. 838 (Pt 3), 156369. doi:10.1016/j.scitotenv.2022.156369.

Sun, L., Ma, W., Gao, W., Xing, Y., Chen, L., Xia, Z., Zhang, Z., Dai, Z., 2019. Propofol directly induces caspase-1-dependent macrophage pyroptosis through the NLRP3-ASC inflammasome. Cell Death Dis. 20414889. 10 (8). Available from: https://www.nature.com/cddis/. doi:10.1038/s41419-019-1761-4.

Tao, L., Desheng, L., Yuan, X., Wang, Z., Shao, Z., Feng, X., Yang, C., Liu, H., Zhang, G., Wang, Y., Liu, X., Zhou, L., Maolei, X., 2024. Potential effects of orally ingesting polyethylene terephthalate microplastics on the mouse heart. Cardiovasc. Toxicol. 1530-7905. 24 (3), 291–301. doi:10.1007/s12012-024-09837-6.

Vianello, A., Jensen, R.L., Liu, L., Vollertsen, J., 2019. Simulating human exposure to indoor airborne microplastics using a Breathing Thermal Manikin. Sci. Rep. 20452322. 9 (1). Available from: www.nature.com/srep/index.html. doi:10.1038/s41598-019-45054-w.

Walczak, A.P., Kramer, E., Hendriksen, P.J.M., Tromp, P., Helsper, J.P.F.G., Zande, M.V.D., Rietjens, I.M.C.M., Bouwmeester, H., 2015. Translocation of differently sized and charged polystyrene nanoparticles in in vitro intestinal cell models of increasing complexity. Nanotoxicology. 17435404. 9 (4), 453–461. doi:10.3109/17435390.2014.944599.

Wang, T., Zhiheng, Y., Liu, X., Cai, Y., Huang, X., Fang, J., Shen, R., Weikun, L., Xiao, Y., Zhuang, W., Guo, S., 2024. Multimodal detection and analysis of microplastics in human thrombi from multiple anatomically distinct sites. eBioMedicine. 23523964. 103, 105118. doi:10.1016/j.ebiom.2024.105118.

Wang, W., Gao, H., Jin, S., Ruijing, L., Guangshui, N., 2019. The ecotoxicological effects of microplastics on aquatic food web, from primary producer to human: a review. Ecotoxicol. Environ. Saf. 01476513. 173, 110–117. doi:10.1016/j.ecoenv.2019.01.113.

Wei, J., Wang, X., Liu, Q., Zhou, N., Zhu, S., Zekang, L., Xiaoli, L., Yao, J., Zhang, L., 2021. The impact of polystyrene microplastics on cardiomyocytes pyroptosis through NLRP3 /Caspase-1 signaling pathway and oxidative stress in Wistar rats. Environ. Toxicol. 1520-4081. 36 (5), 935–944. doi:10.1002/tox.23095.

Werbner, B., Tavakoli-Rouzbehani, O.M., Fatahian, A.N., Boudina, S., 2023. The dynamic interplay between cardiac mitochondrial health and myocardial structural remodeling in metabolic heart disease, aging, and heart failure. J. Cardiovasc. Aging. 27685993. 3 (1). Available from: https://oaepublishstorage.blob.core.windows.net/255e1f7c-1f2f-42d0-a855-6f3e3f566de8/5367.pdf. doi:10.20517/jca.2022.42.

Xia, T., Jiachi, Y., Meng, D., Chen, X., Wang, C., Ruibing, L., 2025. Vascular endothelial cell injury: causes, molecular mechanisms, and treatments. MedComm. 2688-2663. 6 (2). doi:10.1002/mco2.70057.

Xie, X., Deng, T., Duan, J., Xie, J., Yuan, J., Chen, M., 2020. Exposure to polystyrene microplastics causes reproductive toxicity through oxidative stress and activation of the p38 MAPK signaling pathway. Ecotoxicol. Environ. Saf. 01476513. 190, 110133. doi:10.1016/j.ecoenv.2019.110133.

Xin, G., Zhang, Z., Zhao, L., Lijie, L., Xin, L., Yafei, L., Tianya, G., Huang, X., Huang, G., Liang, Y., Meng, H., Baihong, L., Zhang, X., Zhang, J., Wang, X., Yingqiang, D., 2025. Exposure to polyethylene terephthalate micro-(nano)plastics exacerbates inflammation and fibrosis after myocardial infarction by reprogramming the gut and lung microbiota and metabolome. J. Hazard. Mater. 03043894. 488, 137410. doi:10.1016/j.jhazmat.2025.137410.

Yang, Y., Xie, E., Zhiyong, D., Peng, Z., Han, Z., Linyi, L., Zhao, R., Qin, Y., Xue, M., Fengwang, L., Hua, K., Yang, X., 2023. Detection of various microplastics in patients undergoing cardiac surgery. Environ. Sci. Technol. 0013-936X. 57 (30), 10911–10918. doi:10.1021/acs.est.2c07179.

Yee, M.S.L., Hii, L.W., Looi, C.K., Lim, W.M., Wong, S.F., Kok, Y.Y., Tan, B.K., Wong, C.Y., Leong, C.O., 2021. Impact of microplastics and nanoplastics on human health. Nanomaterials. 20794991. 11 (2), 1–23. Available from: https://www.mdpi.com/2079-4991/11/2/496/pdf. doi:10.3390/nano11020496.

Yixiang, H., Huang, Y., Zong, L., Lin, J., Liu, X., Ning, S., 2024. Emerging roles of ferroptosis in pulmonary fibrosis: current perspectives, opportunities and challenges. Cell Death Discov. 2058-7716. 10 (1). doi:10.1038/s41420-024-02078-0.

Zekang, L., Zhu, S., Liu, Q., Wei, J., Jin, Y., Wang, X., Zhang, L., 2020. Polystyrene microplastics cause cardiac fibrosis by activating Wnt/β-catenin signaling pathway and promoting cardiomyocyte apoptosis in rats. Environ. Pollut. 02697491. 265 (Pt A), 115025. doi:10.1016/j.envpol.2020.115025.

Zeng, G., Jingyi, L., Wang, Y., Jingran, S., Zhongbing, L., Zhang, F., Ding, W., 2024. Polystyrene microplastic-induced oxidative stress triggers intestinal barrier dysfunction via the NF-κB/NLRP3/IL-1β/MCLK pathway. Environ. Pollut. 02697491. 345, 123473. doi:10.1016/j.envpol.2024.123473.

Zhang, Q., Liu, L., Jiang, Y., Zhang, Y., Fan, Y., Rao, W., Qian, X., 2023. Microplastics in infant milk powder. Environ. Pollut. 18736424. 323. Available from: https://www.journals.elsevier.com/environmental-pollution. doi:10.1016/j.envpol.2023.121225.

Zhang, Q., Xu, E.G., Li, J., Chen, Q., Ma, L., Zeng, E.Y., Shi, H., 2020. A review of microplastics in table salt, drinking water, and air: direct human exposure. Environ. Sci. Technol. 15205851. 54 (7), 3740–3751. Available from: http://pubs.acs.org/journal/esthag. doi:10.1021/acs.est.9b04535.

Zhao, J., Gomes, D., Jin, L., Mathis, S.P., Li, X., Rouchka, E.C., Bodduluri, H., Conklin, D.J., O'Toole, T.E., 2022. Polystyrene bead ingestion promotes adiposity and cardiometabolic disease in mice. Ecotoxicol. Environ. Saf. 10902414. 232. Available from: http://www.elsevier.com/inca/publications/store/6/2/2/8/1/9/index.htt. doi:10.1016/j.ecoenv.2022.113239.

Zheng, P.C., Li, R., Lai, K.P., Zhang, X.X., 2024. Biological exposure to microplastics and nanoplastics and plastic additives: impairment of glycolipid metabolism and adverse effects on metabolic diseases. Environ. Sci. Pollut. Res. 16147499. 31 (51), 60778–60791. Available from: https://www.springer.com/journal/11356. doi:10.1007/s11356-024-35220-7.

Zhou, J., Gao, L., Lin, Y., Pan, B., Ming, L., 2021. Micrometer scale polystyrene plastics of varying concentrations and particle sizes inhibit growth and upregulate microcystin-related gene expression in Microcystis aeruginosa. J. Hazard. Mater. 03043894. 420, 126591. doi:10.1016/j.jhazmat.2021.126591.

Zhou, Y., Qian, W., Yan, L., Feng, Y., Wang, Y., Cheng, W., 2023. Low-dose of polystyrene microplastics induce cardiotoxicity in mice and human-originated cardiac organoids. Environ. Int. 01604120. 179, 108171. doi:10.1016/j.envint.2023.108171.

Zhu, X., Wang, C., Duan, X., Liang, B., Elvis Genbo, X., Huang, Z., 2023. Micro- and nanoplastics: a new cardiovascular risk factor? Environ. Int. 01604120. 171, 107662. doi:10.1016/j.envint.2022.107662.

CHAPTER

18

From plastic to microplastic in the food chain: Toxicity and mitigation

Nguyen Thi Nhan[a], *Nguyen Tuan Anh*[b], *Huynh Do Hieu Nhu*[a], *and Nguyen Minh Ky*[c]

[a]**Master Program in Water Technology, Reuse, and Management, Vietnamese German University, Ben Cat, Binh Duong Province, Vietnam**
[b]**Department of Environment and Natural Resources, Nong Lam University of Ho Chi Minh City – Gialai Campus, Pleiku City, Vietnam**
[c]**Faculty of Environment and Natural Resources, Nong Lam University, Ho Chi Minh City, Vietnam**

18.1 Introduction

Plastics have become indispensable materials in recent decades due to their convenience, low purchase cost, and long-term durability (Mohajan, 2025). Plastics are used not only for daily activities but also across various sectors, including packaging and materials for consumer goods and construction (Islam, 2025). According to statistics, global plastic production currently reaches approximately 430 MMT annually, with only approximately 9% being recycled (Barone et al., 2025). While plastics provide undeniable socioeconomic benefits, their persistence in the environment has created one of the most pressing challenges to date. Unlike biodegradable materials, which account for approximately 0.5% of the 400 million tons of plastic produced annually, plastics resist natural degradation and can persist in ecosystems for many years, leading to widespread environmental contamination.

Microplastics (MPs), generally defined as plastic items <5 mm, are now regarded as emerging environmental contaminants of global concern (Hossain and Engelhardt, 2025). MPs can be classified into two categories: primary MPs (manufactured at microscopic size, such as microbeads in cosmetics or industrial materials) and secondary MPs (resulting from the degradation of large plastic items under the effects of physical, chemical, or biological conditions in the environment). To date, MPs have been detected in diverse environments. MPs can be found in oceans, freshwater systems (both surface and groundwater), soil, air, and even remote regions, such as deep-sea sediments (Hossain and Engelhardt, 2025; Hoang et al., 2025). As a result, the impacts of MPs on ecosystems and human health need to be urgently addressed.

Microplastics in Animal and Human Physiology
https://doi.org/10.1016/B978-0-443-45635-0.00013-6

The significant impact of MPs is evident not only in their widespread distribution but also in their complex interactions with surrounding environmental systems. Owing to their small size, MPs are easily ingested or inhaled by organisms. This allows these particles to enter food webs and become harmful components within food chains (Nguyen et al., 2023). Moreover, MPs are also considered effective carriers of pollutants when they are distributed in environmental systems. Owing to their large specific surface area, MPs adsorb heavy metals and persistent organic pollutants (POPs). Therefore once they enter biological systems, MPs can cause harmful effects through several pathways: (1) physical impacts, such as tissue abrasion or gut blockage; (2) chemical impacts, such as the release of additives or the adsorption and transfer of environmental contaminants; and (3) biological impacts, such as oxidative stress, immune disruption, or reproductive toxicity (Rathour et al., 2025). In addition, the detection of MPs in human tissues and aquatic organisms further raises the potential risk to both ecosystems and human health.

Although numerous studies have explored the implications of MPs regarding their distribution, degradation, and environmental and human health effects, significant knowledge gaps remain. Problems related to their synergistic fragmentation interactions, specific characteristics in the food chain, cytotoxicity, and mitigation strategies remain unclear. This chapter addresses these gaps by addressing several issues related to MPs in animal and human physiology, including (1) degradation mechanisms to form MPs; (2) the existence of MPs in the food chain through sources, bioaccumulation, and biomagnification; (3) physical, chemical, biological, and health toxicity effects related to MPs; and (4) MP mitigation strategies from source to consumer. Understanding these aspects is crucial not only for scientists but also for proposing and implementing approaches to eliminate this emerging global pollutant to protect ecosystems and human health.

18.2 Plastic-to-MP mechanism

When entering ecosystems, large plastic items take several years to degrade due to their durability. The fragmentation of plastics in natural environments is driven by synergistic interactions among physical, chemical, and biological degradation processes, as described in Fig. 18.1.

18.2.1 Physical degradation

Physical degradation is one of the primary pathways by which plastics transform into MPs and is strongly influenced by mechanical and physical forces present in natural conditions. The fragmentation of large plastics can be caused by mechanical stress, including shear forces from shorelines, sediments, and the sea surface. Under various hydrodynamic conditions, such as wave action, wind, or tides, plastics can move back and forth repeatedly. These movements across hard sand or rock facilitate the fragmentation of material, leading to the further formation of different MPs of various shapes and sizes (Chubarenko et al., 2020). To confirm these mechanisms in the environment, Bullard et al. (2023) demonstrated that MPs can be degraded during eolian saltation, during which wind-driven abrasion significantly alters MP size, mass, and surface characteristics, potentially facilitating long-distance transport.

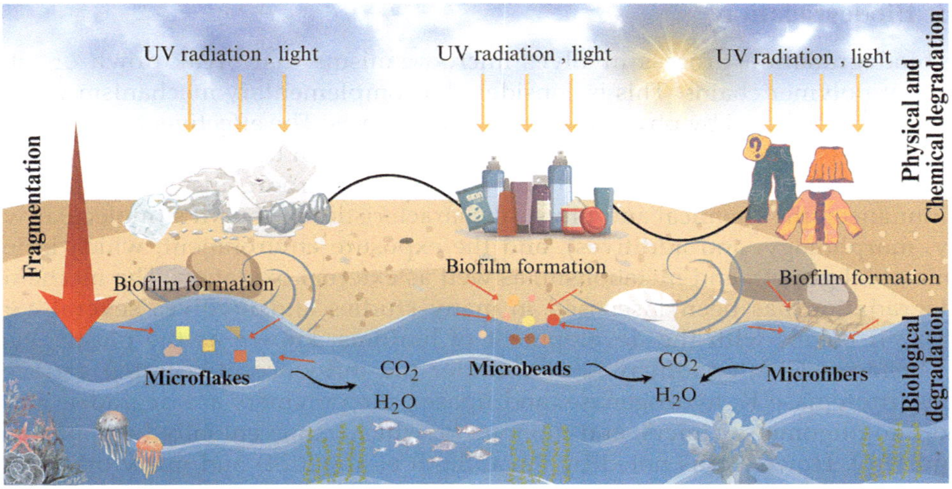

FIG. 18.1 Mechanisms of plastic degradation into microplastics. Under natural conditions, including sunlight or UV radiation, plastic waste undergoes physical, chemical, and biological degradation, forming different types of microplastics. Biofilm formation on particle surfaces and subsequent biodegradation release CO_2 and H_2O into the ecosystem.

In addition, synergistic effects among multiple factors have been identified. As reported by Song et al. (2017), the combined effects of mechanical abrasion and UV radiation (as mentioned in the following) not only weaken polymer structures but also produce unaccounted MP particles. Thus physical degradation provides the initial driving force, converting large plastics into weakened MPs that continue to undergo further transformations in the environment.

18.2.2 Chemical degradation

Under the facilitation of UV radiation and visible light, plastic fragmentation can occur through chemical and photochemical pathways (Manzoor et al., 2022). When exposed to light, photons can excite additives or chromophores within the polymer, generating reactive radicals, such as hydroxyl radicals. The generated radicals then attack the polymer backbones, leading to chain scission, crosslinking, and the incorporation of oxygen-containing functional groups. Such modifications reduce molecular weight, decrease extensibility, and increase the plastic's vulnerability to fragmentation. Factors such as polymer components, additives used, and depositional environments (water, soil, and sediment) are recognized as influencing the chemical degradation rates of plastics (Yousafzai et al., 2025). As proposed by Xia et al. (2025), plastics in surface water are exposed to more sunlight or UV radiation than those distributed in the water column or sediment, thereby fostering their degradation. Owing to their weakened structure (resulting from the hydrolysis of ester linkages) and altered surface properties (higher surface roughness and porosity), plastics can absorb and interact with other pollutants and microorganisms. In addition, aged plastics increase environmental and ecological risks by creating smaller fragments, carrying adsorbed contaminants, and releasing additives (Zeng et al., 2025).

18.2.3 Biodegradation

The biodegradation process involves microorganisms that break down or alter the structure of polymer chains. This is considered a complementary mechanism when plastics are partially affected by physical and chemical effects. The effectiveness of the biodegradation process depends on various conditions. These can be classified into three main groups: the conditions for the growth of the involved microorganisms (substrate, strain, and abundance), the chemical and physical characteristics of the MPs (molecular weight, density, morphology, and additives), and the exposure environment, which influences both microbial growth and metabolism, as well as external oxidation. The mechanism of MPs degradation has been discussed in many studies and involves several complex processes. It begins with the development of biofilms, known as the plastisphere, on plastic surfaces (Chattopadhyay, 2022). Under the influence of secreted extracellular enzymes (proteinase K, hydrogenase, and lipase), the polymers are degraded and fragmented into oligomers, dimers, and monomers, which can eventually be mineralized to CO_2 and H_2O. However, as noted by Payanthoth et al. (2024) and many other reviews, biodegradation often does not result in complete mineralization (i.e., incomplete degradation). As a result, further fragmentation is facilitated, releasing smaller polymer fragments, including nanoplastics (NPs).

18.3 MPs in the food chain

MPs in the food chain represent an emerging environmental and public health concern due to their widespread presence across trophic levels. These particles can be ingested by organisms ranging from plankton to higher consumers, leading to bioaccumulation and potential trophic transfer. In addition, MPs may act as vectors for POPs and additives, thereby increasing the risk of toxic effects on ecosystems and human health.

18.3.1 Sources

MPs are categorized into two groups based on their entry into the environment: primary and secondary MPs. Small primary MPs are made directly and are released into the air at the end of a product's use phase or during industrial production processes (Le et al., 2023). Microbeads have been used in cosmetics, such as facial cleansers and exfoliating creams, as skin abrasives instead of ground oats or almonds. They are then released into the environment through wastewater treatment systems. During the washing process, the textile industry also releases thousands of broken synthetic fibers into the water. Solar radiation, waves, wind, mechanical forces, or biological activity can break down larger plastic debris, such as bottles, bags, nets, and packaging, into smaller pieces called secondary MPs. When these items are not handled properly and disposed of incorrectly, they break down over time into particles known as MPs, which are believed to be the primary cause of MP pollution (Lebreton et al., 2017). In addition, paint fragments have become a significant source of

MP pollution in the environment (Thuan et al., 2025). More than one-third of all plastic produced globally is used for packaging, primarily in the food industry. MPs can easily migrate through air, soil, or water due to their small size and specific physical properties. Continuous human activities in the environment, including industrial and agricultural processes, wastewater discharge, emissions, and waste disposal, have contributed to the increasing detection of MPs in the food chain over time (Mamun et al., 2023).

18.3.2 Pathways of MPs entry into the food chain

The food chain can become contaminated with MPs in many ways, for example, during the production, use, or disposal of consumer, industrial, agricultural, and fishing products, as well as throughout food processing, handling, and distribution. All of these factors can increase the likelihood that MP particles will move from lower to higher trophic levels (Cverenkárová et al., 2021). Fig. 18.2 shows an example of how MPs can infiltrate nearly all parts of the environment and ecosystems, both in water and on land.

MPs and other types of plastic in the environment can be ingested directly through the digestive tract or mistaken for food by bacteria, fungi, plankton, larvae, and invertebrates, especially if the particles are colored. Larger organisms subsequently consume these contaminated prey, facilitating the trophic transfer of MPs (Saeedi, 2024). Studies of plastic debris in nature have revealed bite marks from larger animals and signs that smaller organisms also ingest it. Several investigations also demonstrated that MPs can be transferred through aquatic food chains. For example, zooplankton consumed algae with adhered MPs, which were subsequently ingested by fish, enabling transfer along the food chain (Ky et al., 2023).

MP particles have been detected in fish species commonly consumed by people, such as tuna, mackerel, sardines, cod, and herring (Liboiron et al., 2016). In addition, organisms at various trophic levels in aquatic ecosystems have been examined to assess their function as indicators of MP pollution. For example, benthic fish can indicate pollution in sediment. In contrast, sardines and anchovies can reveal pollution in the open sea and pose a risk of human exposure because they are caught in large numbers and consumed together (Mercogliano et al., 2020). MPs also tend to accumulate on the seabed, where benthic organisms consume them, leading to bioaccumulation in the food chain. Studies indicate that MPs remain in the organs of predatory animals for a duration exceeding the interval between ingestion and excretion, implying an increased risk of MP accumulation in predators within the food chain. This shows how MPs move up and down the food chain. As a result, concerns about the trophic transfer of MPs in aquatic, benthic, and marine food webs are increasing. This is because predators may indirectly accumulate them by eating contaminated prey (Lusher et al., 2017; Lusher et al., 2017).

Researchers have also investigated MP contamination in other important food sources, including milk, salt, honey, spices, and vegetables. Pathways encompass the utilization of plastic packaging materials, interactions with plastic equipment during production and processing, and exposure to contaminated water and animal feed. A total of 19 honey samples were analyzed, mainly from Germany, with additional samples from France,

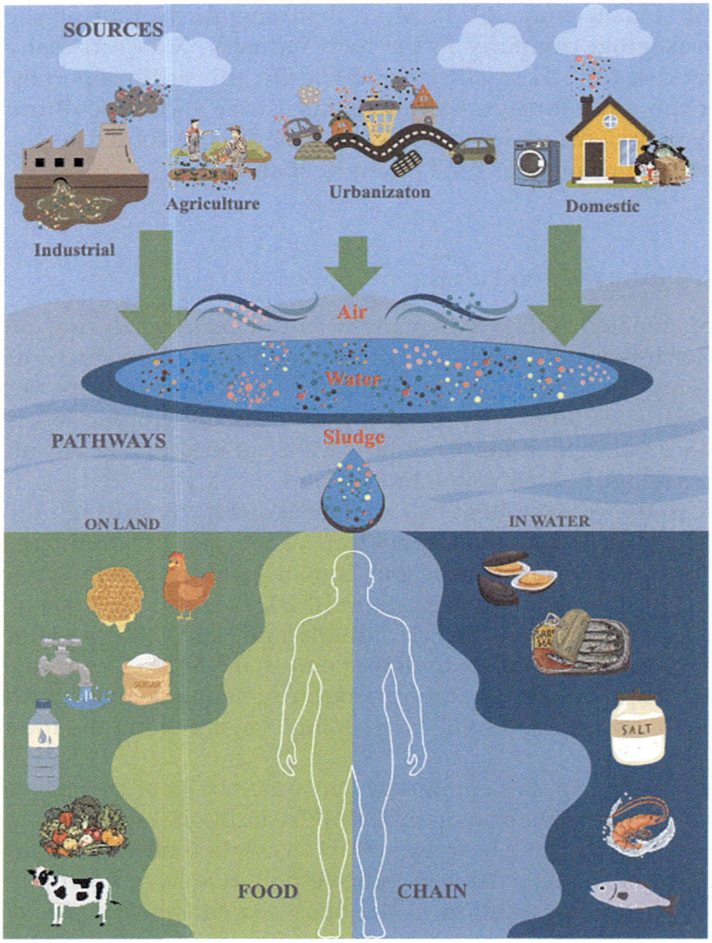

FIG. 18.2 Sources and pathways of microplastics into the food chain. Industrial, agricultural, urban, and domestic activities are among the main sources of microplastic release into the air, water, and soil. These particles accumulate through water, sludge, and land systems, contaminating food and water resources. Eventually, microplastics reach humans through the consumption of contaminated food sources.

Spain, Italy, and other countries. The analyses revealed that the average MP concentration in honey was 166 ± 147 fibers/kg and 9 ± 9 fragments/kg (Liebezeit and Liebezeit, 2013). MPs were also found in 23 milk samples from Mexico, with particle counts ranging from 1 to 14 particles/L (Kutralam-Muniasamy et al., 2020). MPs have also been found in several processed and dry foods. Some examples include traditional medicines, seaweed, tea bags, and dried fish. Experiments in the laboratory with lettuce and carrot plants showed that MPs can be absorbed through both roots and leaves. Smaller particles are more likely to enter plant tissues.

MPs can also absorb other pollutants because they are hydrophobic, which could make them more toxic to both living organisms and the environments where they reside. Many studies have shown that MPs contain harmful chemical additives and POPs. These pollutants can then enter marine food chains when aquatic animals consume contaminated particles (Bouwmeester et al., 2015). This illustrates the severity and prevalence of MP pollution in the food chain, highlighting its potential impact on human health.

18.3.3 Bioaccumulation and biomagnification

Research indicates that MPs infiltrate the food chain and are exchanged among trophic levels, suggesting bioaccumulation and biomagnification. Bioaccumulation is the process by which MPs or chemical additives are taken up from the environment through contact, ingestion, or respiration, resulting from sources such as water, sediments, and prey. Biomagnification refers to the increase in the concentration of MPs or additives as they move up the trophic levels, from lower to higher organisms (Miller et al., 2020).

One of the first studies on bioaccumulation in coastal ecosystems was conducted in Port Blair Bay, Andaman and Nicobar Islands. It examined samples of water, sediments, zooplankton, finfish, and shellfish. The findings indicated average concentrations of 0.93 ± 0.59 particles/m^3, 45.17 ± 25.23 particles/kg, 0.12 ± 0.07 fragments per zooplankton, and 10.65 ± 7.83 particles per specimen, with notably elevated retention in zooplankton populations (Goswami et al., 2020). Bioaccumulation has been substantiated in bivalves, crustaceans, fish, reptiles, seabirds, and marine mammals (Pennati et al., 2022). Baleen whales (*Balaenoptera physalus* L.) can filter approximately 70,000 liters of water at once. They probably do not consume MPs directly from seawater but acquire them indirectly through trophic transfer from prey (Lusher et al., 2017).

Extensive examinations of marine organisms have validated that bioaccumulation occurs at every trophic level within the marine food chain. Certain studies suggest that fish, especially those that feed on zooplankton, are at an increased risk of ingesting and accumulating MPs. In addition, because fish live longer than invertebrates do and are more prevalent in the food chain, they are thought to be more prone to plastic pollution. Numerous studies have identified MPs in the intestines, gills, and muscle tissues of both pelagic and demersal fish, indicating slow elimination rates that facilitate the accumulation and biomagnification of MPs within the food chain. These results suggest that organisms may experience various detrimental effects from the accumulation of substantial amounts of plastic in their bodies, such as modified behavior and impaired physiological regulation, decreased foraging efficiency, impaired predator evasion, diminished survival rates in the face of anthropogenic threats, slowed growth rates, and compromised reproductive success (Nelms et al., 2016).

18.4 Toxicity of MPs

The toxicity of MPs arises from their physical characteristics, chemical composition, and ability to interact with co-contaminants. They can induce oxidative stress, inflammation, and cellular damage after ingestion or exposure to organisms. Moreover, they may act as carriers for additives and environmental pollutants, enhancing their bioavailability and associated toxic effects.

18.4.1 Physical effects

The physical toxicity of MPs is associated mainly with their physical properties, such as shape, size, surface characteristics, and concentration. When MPs enter the body through ingestion or inhalation, they can cause mechanical blockage, tissue abrasion, or disruption of physiological processes (Witczak et al., 2024). These impacts have been observed not only in aquatic organisms but also at the cellular level. Numerous studies have evaluated various aspects of MPs' physical toxicity, providing important data for comprehensive assessments. These effects have been studied in both cellular systems and animal models, yielding valuable insights. In particular, Choi et al. (2021) conducted an in vitro study on high-density polyethylene (HDPE) and low-density polyethylene that considered two factors, namely, particle shape and concentration, to better understand the physical toxicity and demonstrated the hypothesis that the physical characteristics of MPs could impact immunity, lysis, and viability of cells. The results indicated that cytotoxicity, hemolysis, and inflammatory cytokine release were associated with higher concentrations and rougher morphologies of the MPs. While no severe cytotoxicity was observed at the tested doses, irregularly shaped particles (sharp edges) with high curvature had significant harmful effects. These findings suggest that such effects have potential for human toxicity under real-life conditions. Other physical effects related to different shapes were also observed in a recent study. The authors compared the toxicity profile and mechanistic pathways of irregularly shaped and fibrous MPs. The investigation combined in vivo animal exposures (mice) with in vitro cell culture models. The experimental results revealed that both types of testis MPs caused ventilatory dysfunction and airway remodeling in mice. Exposure also led to epithelial–mesenchymal transition, characterized by increased cell migration, reduced adhesion, and cytoskeletal changes. In addition, fibrous MPs caused more severe airway toxicity and uniquely increased cilia formation (reduction in Piezo1 inhibition). Thus research indicates that MPs' shape influences both their toxicity and their underlying biological mechanisms.

In addition to shape, the size of the MPs also plays an important role in causing physical damage to the experimental subjects. In vitro experiments were performed to compare the effects of polystyrene (PS, 1 μm) and nanoparticles (100 nm) at various concentrations (5, 25, and 75 μg/mL) on mouse HT22 cells (Liu et al., 2022). Both sizes of MPs were efficiently taken up by cells, but smaller particles at higher concentrations caused greater reactive oxygen species (ROS) production, S-phase arrest, and mild cytotoxicity. Moreover, a microfluidic device was used to simulate in vivo conditions, and such dynamic exposure further reduced cell viability compared with static culture. These findings highlight that particle size and exposure conditions strongly influence cytotoxicity, thereby informing risk assessments of PS particles. Ding et al. (2024) examined the size-dependent gastric epithelial toxicity of PS-MPs and NPs using in vivo and in vitro models. Experiments demonstrated that exposure to PS-MPs caused gastric barrier injury and mitochondrial damage in rats, whereas exposure to smaller PS-NPs induced greater oxidative stress by reducing antioxidant enzyme activity and increasing markers of lipid peroxidation, DNA damage, and oxidative stress (MDA, 8-OHdG, and γ-H2AX). In addition, PS-MPs activated the β-catenin/YAP pathway through redox regulation, revealing a size-dependent toxic mechanism with potential carcinogenic implications.

In addition to studies on cellular systems and animals, aquatic organisms, such as benthic crustaceans, are also considered. A recent representative study on *Artemia franciscana* was also investigated (Kim et al., 2025). In this study, different polyethylene terephthalate (PET) fragment sizes (<20 μm, 45–75 μm, and >125 μm) and fibers (200 μm and 3 mm) were used, and *A. franciscana* was exposed for 48 h. The effects on growth rate, apoptosis, survival, gut damage, and swimming ability were tested. These findings indicated that small PET fragments (<75 μm) caused gut damage and movement inhibition, whereas larger fragments and fibers led to apoptosis due to physical stress. These findings demonstrate that both size and shape significantly affect the ecotoxicity of PET MPs to marine benthic organisms.

In summary, MPs have been shown to impair essential functions in both humans and aquatic organisms. Different shapes and sizes determine varying levels of risk. They can cause not only mechanical blockages and tissue abrasions but also cellular penetration.

18.4.2 Chemical effects

The chemical toxicity of MPs results from both the leaching of additives incorporated during plastic manufacturing and the adsorption of environmental pollutants onto their surfaces, given their high surface area (Agboola and Benson, 2021). These additives include plasticizers, stabilizers, and flame retardants, whereas the adsorbed pollutants often include heavy metals, polycyclic aromatic hydrocarbons, pesticides, and POPs. When these chemicals enter biological systems, they can be gradually released, thereby causing oxidative stress, endocrine disruption, and organ toxicity (Ullah et al., 2023).

To illustrate these effects, many studies have been conducted. In particular, (Savva et al., 2023) studied additives in bioplastics, including polylactic acid (PLA) and polyhydroxybutyrate (PHB), as well as single-use materials commonly found in daily life. By using HPLC-HESI(+/−)-HRMS, 203 additives were tentatively identified, with PLA and PHB containing, on average, 123 and 121 additives, respectively. Plasticizers, particularly phthalates, were the most abundant, and 63 additives were confirmed and quantified. Moreover, cytotoxicity assays revealed that plastic particles caused only a slight reduction in cell viability but induced significant oxidative stress, with PLA producing the strongest effect. Extracts also cause oxidative stress, especially from PLA, despite the limited migration of toxicants. These findings highlight that bioplastics may contain numerous potentially toxic additives, underscoring the need for further investigations. An essential product used daily by humans is masks. Since the COVID-19 pandemic, mask use has increased significantly. Therefore many studies have been conducted to determine the presence of hazardous substances. In the study by Chang and Wang (2023), the authors assessed the leaching of phthalate acid esters (PAEs), metals, and MPs from six common mask types and their cytotoxic effects. MPs (<3 μm) were quantified via Py–GC–MS, with reusable masks releasing up to 1504 μg/g. PAEs were prevalent, reaching a peak at 42 μg/g. In addition, antimicrobial cloth masks released 173 μg of Cu and 4.5 μg of Ag. The experiments on cells revealed that mask leachates induced DNA damage, oxidative stress, mitochondrial fragmentation, and cell death, with PAEs identified as the primary contributors. Masks containing Cu or Ag also disrupted lysosomal and cellular pH.

Another in vitro study by Choi et al. (2020) investigated the chemical effects of MPs. These findings suggest that chemical toxicity may originate from the chemical reagents used in MPs. PS was used as the research material. This study revealed that, upon chemical agent release, MPs increased acute inflammation 20-fold compared with the control. In addition, ROS generation and increased death of both fibroblasts and cancer cells were observed.

MPs are known as strong adsorbents of pollutants in the environment due to their unique physicochemical properties. In the study by Yuan et al. (2020), the adsorption ability of various heavy metal ions (Pb^{2+}, Cu^{2+}, Cd^{2+}, and Ni^{2+}) on PS-MPs (10 μm and 50 μm) was investigated. The combined effects of these two factors on *Daphnia magna* (water fleas) were evaluated. The results indicate that, compared with larger MPs, smaller MPs had greater adsorption abilities for heavy metals (0.261–0.579 mg/g and 0.243–0.525 mg/g), followed by the order $Pb^{2+} > Cu^{2+} > Cd^{2+} > Ni^{2+}$. Furthermore, the authors confirmed that in *D. magna*, the combined effects of MPs and heavy metals shifted from antagonistic to additive with increasing MP concentration, with smaller MPs posing greater toxicological risks than larger MPs. In-depth case study evidence has demonstrated that MPs act as effective vectors for toxicants, not only increasing the bioavailability of hazardous compounds but also amplifying their toxicological effects in organisms and ecosystems.

18.4.3 Biological and health effects

The biological and health effects of MPs result from both their physical and chemical interactions (Osman et al., 2023). Once MPs enter the body, they can cause inflammation, increase oxidative stress, and weaken the immune system. They disrupt the gut microbiome by altering the balance of bacteria, which can reduce digestion efficiency and lead to chronic inflammation. Reproductive and developmental toxicity has been reported in multiple animal models. For example, it can reduce fertility, slow growth, or cause abnormal development in embryos. At the molecular level, MPs induce oxidative DNA damage and change the normal cell cycle, increasing the risk of long-term disease progression (Joseph et al., 2025). In rodents, MP exposure has been linked to liver inflammation and weakens the protective barrier of the gut. In humans, these particles have already been detected in the blood, lungs, liver, and placenta, indicating they can cross biological barriers and accumulate in sensitive organs. These findings raise serious concerns about the links between MP exposure and long-term diseases such as metabolic disorders, heart problems, and cancer (Bora et al., 2024). Table 18.1 summarizes recent studies on the biological and health effects of long-term exposure to MPs. The experiments have been carried out in animal models and cell cultures, with PS being the most commonly studied MP type. These efforts have contributed to a better understanding of the mechanisms of action and the harmful effects caused by MP exposure.

DNA—Deoxyribonucleic acid; ATP—Adenosine triphosphate; ROS—Reactive oxygen species; NO—Nitric oxide; EROD—7-ethoxyresorufin-O-deethylase; TCA—Tricarboxylic acid; 5mC—5-methylcytosine; EA.hy926—Human endothelial hybrid cell line; ESCs—Embryonic stem cells; THP-1—Human leukemic cell line; U937—Human monoblast leukemia cell line; Caco-2—Human colon cancer cell line; HEK001—Human keratinocyte cell line; MRC-5—Human lung fibroblast cell line; HMEC-1—Human microvascular endothelial cells; HMC-3—Human microglial cell line; T47D—Human breast cancer cell line; A549—Human lung cancer cell line; HEK293—Human embryonic kidney cell line.

TABLE 18.1 Recent studies related to the biological and health effects caused by MP exposure.

Subjects		Types of MPs	Toxic effects	References
Mouse model	Ovary	Polystyrene	Oxidative stress, DNA oxidative damage, ovarian damage	(Hao et al., 2023)
Human pluripotent stem cell	Liver organoids	Polystyrene	Hepatotoxicity, ATP production, hepatocytotoxicity, lipid metabolism, ROS generation, oxidative stress, inflammatory response, lipotoxicity	(Cheng et al., 2022)
Mouse model	Liver, kidney, spleen, and pancreas	Polystyrene	Hepatic function damage, lipid metabolism abnormality	(Fan et al., 2022)
Human cell lines	THP-1, Jurkat, hacat, U937	Polyethylene	Cell viability, intracellular reactive oxygen species (ROS), nitric oxide (NO), and cytokines	(Gautam et al., 2022)
Human cell lines	Caco-2, HEK001, MRC-5, HMEC-1, heparg, HMC-3, and T47D	Polystyrene	Cell viability, DNA fragmentation	(Janiga-MacNelly et al., 2025)
Human cell lines	A549, HEK293, HeLa	Polystyrene Polymethylmethacrylate	No major effect on metabolic activity, viability, and cell cycle Effect on intracellular thiol content and cytokine secretion	(da Silva Brito et al., 2023)
Human cell	Human dermal fibroblast-derived spheroids	Polystyrene	Cytotoxicity increased, weakening the cytoskeleton and cell-matrix adhesion	(Eom et al., 2024)
Human cell	Vascular endothelial EA.hy926 cells	Polystyrene	Oxidative stress, reducing the expression of antioxidants, apoptotic cytotoxicity in EA.hy926 cells, and vascular barrier dysfunction	(Chen et al., 2023)
Marine organism	Rotifer (*Brachionus koreanus)*	Polyethylene terephthalate	Increased intracellular ROS levels, antioxidant enzyme activity, immunotoxicity, and metabolic disorder	(Yoo et al., 2024)
Marine organism	Microcrustacean (*Artemia salina)*	Polypropylene	Oxidative stress, homeostasis changes, reduced enzyme activities, and epithelial cell damage	(Jeyavani et al., 2022)
Marine organism	Fish embryos	Polystyrene Polypropylene Polyethylene	Decreased head/body ratios, increased EROD activity, and DNA breaks	(Pannetier et al., 2020)
Marine organism	Rainbow trout (*Oncorhynchus mykiss*)	Polyethylene terephthalate Polyvinyl chloride	Oxidative stress, cell viability, ROS generation	(Boháčková et al., 2023)

Continued

TABLE 18.1 Recent studies related to the biological and health effects caused by MP exposure—cont'd

Subjects		Types of MPs	Toxic effects	References
Marine organism	Mussel (*Mytilus edulis)*	Polyethylene	Hemocyte mortality, DNA damage	(Roman et al., 2023)
Marine organism	Mussel (*Unio delicatus)*	Polystyrene	Decreases glutathione, advanced oxidation protein products, gills, digestive gland, and disrupts cellular structure	(Arslan Yüce et al., 2025)
Marine organism	Bivalve (*Ruditapes philippinarum)*	Polystyrene	Growth inhibition, oxidative stress, lipid peroxidation, DNA damage	(Zhou et al., 2024)
Mouse model	Mouse eye	Low-density polyethylene	Oxidative stress, cellular damage, diminished tear secretion, and increased intraocular pressure	(Yang et al., 2024)
Human cell	Endometrial stromal cells	Polystyrene	Cell death, reduced cell proliferation, and induced apoptosis	(Kim et al., 2025)
Human cell	Placental explants	Polystyrene	Cytotoxicity, oxidative stress, TCA cycle alterations, folate, amino acid, energy metabolism	(de Sousa et al., 2024)
Human stem cells	Human embryonic stem cells (ESCs)	Polystyrene	5-mC expression reduction, impairing cell proliferation and differentiation processes, and the disruption of DNA methylation	(Jeong et al., 2025)

18.5 Mitigation strategies to reduce MP pollution

MP pollution poses a significant threat to ecosystems and human health (Nguyen et al., 2023). A comprehensive approach, based on source reduction, effective waste management, and the application of advanced processing techniques to prevent MPs from entering various environments and accumulating in ecosystems, is needed to mitigate plastic pollution.

Fig. 18.3 presents the sustainable waste management hierarchy used to mitigate plastic and MP pollution. The pyramid emphasizes the transition to a circular economy, aiming to reduce waste generation and environmental impacts, and ranks strategies from the most environmentally beneficial ("Prevention") to the least preferred ("Disposal"). This approach aligns with international initiatives to reduce plastic pollution, which involve the generation of primary MPs from direct sources, such as cosmetics and textiles, or secondary MPs from the breakdown of larger plastics.

Prevention is the most effective strategy, focusing on reducing plastic waste at its source to reduce overall pollution. For plastics, this involves reducing the consumption of single-use items, redesigning products to eliminate unnecessary plastic, and implementing policies such as banning nonbiodegradable plastics. Examples include removing intentionally added microbeads from personal care and cosmetic products, such as facial scrubs, toothpaste, shower gels, and exfoliating creams (one of the major sources of MPs), and promoting

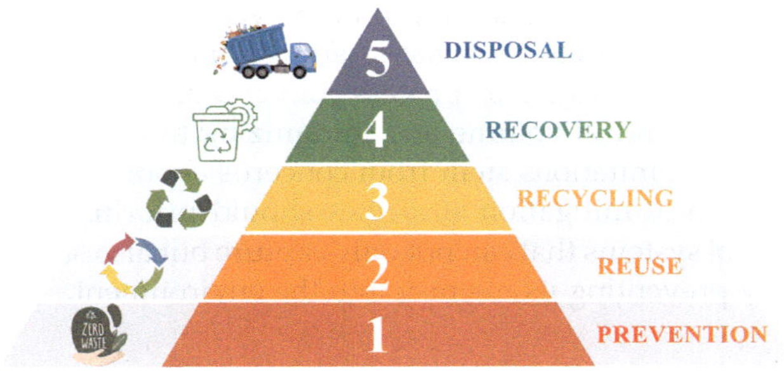

FIG. 18.3 Hierarchy of plastic waste management interventions. The plastic waste management interventions illustrate the waste hierarchy pyramid with five different colors. At the base, prevention represents avoiding waste generation. Above it are reuse, recycling, and recovery, emphasizing resource efficiency. The lowest top-level disposal indicates the least preferred waste management option.

alternatives, such as biodegradable materials or bulk purchases, to reduce packaging. For MPs, prevention targets sources such as synthetic textiles by encouraging the use of natural fibers or reducing washing frequency (Ghosh et al., 2023). Policies banning or reducing plastic waste hold promising potential in many regions worldwide. Regulatory actions, such as the EPA's Objective A (reducing pollution from plastic production) and Objective B (innovating material design), support this by advocating for lower virgin plastic output and eco-friendly substitutes. Reuse extends the lifespan of plastic products, reduces the demand for new products, and prevents their fragmentation into MPs. Specifically, this can be achieved through refill programs for bottles and containers and deposit–refund systems for bottles and jars. In addition, Zheng et al. (2022), in Cleaner Waste Systems, showed through a cost–benefit analysis of deposit–refund systems in Sweden that reusing PET trays for food-grade applications not only results in a higher benefit–cost ratio (1.67) than recycling (0.73) but also significantly reduces plastic waste leakage into ecosystems (Zheng et al., 2022).

Public awareness campaigns and consumer education are essential to encourage the reuse or return of reusable items. These strategies require the collaboration of multiple stakeholders, including governments, manufacturers, and consumers (Prata et al., 2019). When prevention and reuse are not feasible, recycling transforms waste into new materials, conserving resources and limiting landfill inputs that could leach MPs. Key strategies include improving collection systems, standardizing recyclable polymers (e.g., PET and HDPE), and advancing chemical recycling technologies. For MPs, enhanced recycling reduces environmental leakage from mismanaged waste. Challenges include low global recycling rates (approximately 9% for plastics), so solutions emphasize education, infrastructure expansion, and policies to increase demand for recycled content.

In addition, public education campaigns will inform communities about the impacts of plastic pollution and the importance of reducing plastic use. The next step involves enhancing waste management and increasing awareness through plastic recycling initiatives. In today's fast-paced society, people often discard items rather than reuse them. Therefore governments should promote participation in waste collection by improving recycling infrastructure, thereby reducing the amount of plastic waste entering water bodies. Encouraging involvement in recycling programs can be achieved through deposit-return schemes for waste. To ensure effective plastic waste management, governments worldwide should establish stricter regulations and stronger enforcement systems (Thacharodi et al., 2024).

Recently, recovery has extracted value from waste that cannot be recycled (i.e., plastic waste), often through energy recovery methods such as waste-to-energy incineration or advanced processes such as pyrolysis (Luu et al., 2025). This process converts nonrecyclable items into fuel or chemicals, thereby reducing landfill volume and minimizing the potential release of MPs from degradation. However, its limitations stem from concerns about secondary emissions. To address this challenge, critical mitigation strategies should prioritize the implementation of advanced pollution control systems that can not only capture but also safely manage or transform plastic waste, thereby preventing its re-entry into the environment.

18.6 Conclusion

Plastics entering the environment gradually fragment into smaller particles, eventually becoming MPs that can infiltrate food chains. These particles are ingested by aquatic organisms such as plankton, fish, and shellfish, and can accumulate at relatively high trophic levels, including humans. MPs pose multiple risks by causing physical damage, blocking nutrient uptake, and carrying toxic additives or adsorbed pollutants. Their persistence in biological systems raises concerns about long-term ecological and human health effects, including inflammation and oxidative stress.

Addressing this issue requires both preventive and remedial actions. Preventive strategies include reducing plastic production and banning intentionally added MPs, such as microbeads, in personal care and cosmetic products. The remaining measures focus on advanced wastewater treatment technologies, nature-based solutions, and innovations in plastic alternatives. Together, these approaches offer a pathway to mitigate MP pollution and protect food security and ecosystem health.

References

Agboola, O.D., Benson, N.U., 2021. Physisorption and chemisorption mechanisms influencing micro (Nano) plastics-organic chemical contaminants interactions: a review. Front. Environ. Sci. 2296665X. 9. Available from: journal.frontiersin.org/journal/environmental-science. doi:10.3389/fenvs.2021.678574.

Arslan Yüce, P., Günal, A.Ç., Erkmen, B., Yurdakok-Dikmen, B., Çağan, A.S., Çırak, T., Başaran Kankılıç, G., Seyfe, M., Filazi, A., Tavşanoğlu, Ü.N., 2025. In vitro and in vivo effects of commercial and environmental microplastics on Unio delicatus. Ecotoxicology. 15733017. 34 (1), 22–37. Available from: https://www.springer.com/journal/10646. doi:10.1007/s10646-024-02807-2.

Barone, A.S., Maragoni-Santos, C., Farias, P.M.D., Cortat, C.M.G., Maniglia, B.C., Ongaratto, R.S., Ferreira, S., Fai, A.E.C., 2025. Rethinking single-use plastics: innovations, polices, consumer awareness and market shaping biodegradable solutions in the packaging industry. Trends Food Sci. Technol. 09242244. 158. Available from: https://www.sciencedirect.com/science/journal/09242244. doi:10.1016/j.tifs.2025.104906.

Boháčková, J., Havlíčková, L., Semerád, J., Titov, I., Trhlíková, O., Beneš, H., Cajthaml, T., 2023. In vitro toxicity assessment of polyethylene terephthalate and polyvinyl chloride microplastics using three cell lines from rainbow trout (Oncorhynchus mykiss). Chemosphere. 00456535. 312 (Pt 1), 136996. doi:10.1016/j.chemosphere.2022.136996.

Bora, S.S., Gogoi, R., Sharma, M.R., Anshu, M.P.B., Deka, P., Bora, J., Naorem, R.S., Das, J., Teli, A.B., 2024. Microplastics and human health: unveiling the gut microbiome disruption and chronic disease risks. Front. Cell. Infect. Microbiol. 22352988. 14. Available from: http://www.frontiersin.org/Cellular_and_Infection_Microbiology/archive. doi:10.3389/fcimb.2024.1492759.

Bouwmeester, H., Hollman, P.C.H., Peters, R.J.B., 2015. Potential health impact of environmentally released micro- and nanoplastics in the human food production chain: experiences from nanotoxicology. Environ. Sci. Technol. 15205851. 49 (15), 8932–8947. Available from: http://pubs.acs.org/journal/esthag. doi:10.1021/acs.est.5b01090.

Bullard, J.E., Zhou, Z., Davis, S., Fowler, S., 2023. Breakdown and modification of microplastic beads by Aeolian Abrasion. Environ. Sci. Technol. 15205851. 57 (1), 76–84. Available from: http://pubs.acs.org/journal/esthag. doi:10.1021/acs.est.2c05396.

Chang, X., Wang, W.-X., 2023. Phthalate acid esters contribute to the cytotoxicity of mask leachate: cell-based assay for toxicity assessment. J. Hazard. Mater. 03043894. 459, 132093. doi:10.1016/j.jhazmat.2023.132093.

Chattopadhyay, I., 2022. Role of microbiome and biofilm in environmental plastic degradation. Biocatal. Agric. Biotechnol. 18788181. 39, 102263. doi:10.1016/j.bcab.2021.102263.

Chen, Y.C., Chen, K.F., Andrew Lin, K.Y., Su, H.P., Wu, D.N., Lin, C.H., 2023. Evaluation of toxicity of polystyrene microplastics under realistic exposure levels in human vascular endothelial EA.hy926 cells. Chemosphere. 18791298. 313. Available from: www.elsevier.com/locate/chemosphere. doi:10.1016/j.chemosphere.2022.137582.

Cheng, W., Xiaolan, L., Zhou, Y., Hengyi, Y., Xie, Y., Guo, H., Wang, H., Yan, L., Feng, Y., Wang, Y., 2022. Polystyrene microplastics induce hepatotoxicity and disrupt lipid metabolism in the liver organoids. Sci. Total Environ. 00489697. 806 (Pt 1), 150328. doi:10.1016/j.scitotenv.2021.150328.

Choi, D., Bang, J., Kim, T., Yoogyeong, O., Hwang, Y., Hong, J., 2020. In vitro chemical and physical toxicities of polystyrene microfragments in human-derived cells. J. Hazard. Mater. 03043894. 400, 123308. doi:10.1016/j.jhazmat.2020.123308.

Choi, D., Hwang, J., Bang, J., Han, S., Kim, T., Yoogyeong, O., Hwang, Y., Choi, J., Hong, J., 2021. In vitro toxicity from a physical perspective of polyethylene microplastics based on statistical curvature change analysis. Sci. Total Environ. 00489697. 752, 142242. doi:10.1016/j.scitotenv.2020.142242.

Chubarenko, I., Efimova, I., Bagaeva, M., Bagaev, A., Isachenko, I., 2020. On mechanical fragmentation of single-use plastics in the sea swash zone with different types of bottom sediments: insights from laboratory experiments. Mar. Pollut. Bull. 0025326X. 150, 110726. doi:10.1016/j.marpolbul.2019.110726.

Cverenkárová, K., Valachovičová, M., Mackuľak, T., Žemlička, L., Bírošová, L., 2021. Microplastics in the Food Chain. Life. 2075-1729. 11 (12), 1349. doi:10.3390/life11121349.

da Silva Brito, W.A., Singer, D., Miebach, L., Saadati, F., Wende, K., Schmidt, A., Bekeschus, S., 2023. Comprehensive in vitro polymer type, concentration, and size correlation analysis to microplastic toxicity and inflammation. Sci. Total Environ. 00489697. 854, 158731. doi:10.1016/j.scitotenv.2022.158731.

de Sousa, A.K.A., Pires, K.S.N., Cavalcante, I.H., Cavalcante, I.C.L., Santos, J.D., Queiroz, M.I.C., Leite, A.C.R., Crispim, A.C., da Rocha Junior, E.R., Aquino, T.M., Weingrill, R.B., Urschitz, J., Ospina-Prieto, S., Borbely, A. U., 2024. Polystyrene microplastics exposition on human placental explants induces time-dependent cytotoxicity, oxidative stress and metabolic alterations. Front. Endocrinol. 16642392. 15. Available from: https://www.frontiersin.org/journals/endocrinology#. doi:10.3389/fendo.2024.1481014.

Ding, R., Chen, Y., Shi, X., Yang, L., Yang, Y., Sun, Z., Duan, J., 2024. Size-dependent toxicity of polystyrene microplastics on the gastrointestinal tract: oxidative stress related-DNA damage and potential carcinogenicity. Sci. Total Environ. 00489697. 912, 169514. doi:10.1016/j.scitotenv.2023.169514.

Eom, S., Shim, W., Choi, I., 2024. Microplastic-induced inhibition of cell adhesion and toxicity evaluation using human dermal fibroblast-derived spheroids. J. Hazard. Mater. 03043894. 465, 133359. doi:10.1016/j.jhazmat.2023.133359.

Fan, X., Wei, X., Hailong, H., Zhang, B., Yang, D., Haining, D., Zhu, R., Sun, X., Yuri, O., Ning, G., 2022. Effects of oral administration of polystyrene nanoplastics on plasma glucose metabolism in mice. Chemosphere. 00456535. 288 (Pt 3), 132607. doi:10.1016/j.chemosphere.2021.132607.

Gautam, R., JiHun, J., Acharya, M., Maharjan, A., Lee, D., Pramod Bahadur, K.C., Kim, C., Kim, K., Kim, H., Heo, Y., 2022. Evaluation of potential toxicity of polyethylene microplastics on human derived cell lines. Sci. Total Environ. 00489697. 838 (Pt 2), 156089. doi:10.1016/j.scitotenv.2022.156089.

Ghosh, S., Sinha, J.K., Ghosh, S., Vashisth, K., Han, S., Bhaskar, R., 2023. Microplastics as an emerging threat to the global environment and human health. Sustainability. 2071-1050. 15 (14), 10821. doi:10.3390/su151410821.

Goswami, P., Vinithkumar, N.V., Dharani, G., 2020. First evidence of microplastics bioaccumulation by marine organisms in the Port Blair Bay, Andaman Islands. Mar. Pollut. Bull. 0025326X. 155, 111163. doi:10.1016/j.marpolbul.2020.111163.

Hao, W., Liu, Q., Yang, N., Shiwen, X., 2023. Polystyrene-microplastics and DEHP co-exposure induced DNA damage, cell cycle arrest and necroptosis of ovarian granulosa cells in mice by promoting ROS production. Sci. Total Environ. 00489697. 871, 161962. doi:10.1016/j.scitotenv.2023.161962.

Hoang, V.H., Nguyen, M.K., Hoang, T.D., Rangel-Buitrago, N., Lin, C., Pham, M.T., Ha, M.C., Nguyen, T.P., Shaaban, M., Chang, S.W., Nguyen, D.D., 2025. Microplastic characteristics, transport, risks, and remediation in groundwater: a review. Environ. Chem. Lett. 16103661. 23 (3), 817–837. Available from: https://www.springer.com/journal/10311. doi:10.1007/s10311-025-01825-8.

Hossain, M., Engelhardt, I., 2025. Global plastic footprint: unveiling property trends, environmental fate, and emerging threats of microplastic and nanoplastics pollution across ecosystems. Energy, Ecology and Environment. 2363-7692. 10 (6), 637–674. doi:10.1007/s40974-025-00383-7.

Islam, F.A.S., 2025. The impact of plastic waste on ecosystems and human health and strategies for managing it for a sustainable environment. Int. J. Latest Technol. Eng. Manag. Appl. Sci. 2278-2540. 14 (3), 706–723. doi:10.51583/ijltemas.2025.140300075.

Janiga-MacNelly, A., Hoang, T.C., Lavado, R., 2025. Comparative toxicity of microplastics obtained from human consumer products on human cell-based models. Food Chem. Toxicol. 02786915. 196, 115194. doi:10.1016/j.fct.2024.115194.

Jeong, J., Kang, K., Kim, H., Chong, C., Jeongeun, I., Park, J.-S., Cho, M., Choi, J., 2025. Developmental toxicity of microplastics in human stem cells using adverse outcome pathway based integrated approaches to testing and assessment approach. Toxicology. 0300483X. 518, 154268. doi:10.1016/j.tox.2025.154268.

Jeyavani, J., Sibiya, A., Bhavaniramya, S., Mahboob, S., Al-Ghanim, K.A., Nisa, Z.U., Riaz, M.N., Nicoletti, M., Govindarajan, M., Vaseeharan, B., 2022. Toxicity evaluation of polypropylene microplastic on marine microcrustacean Artemia salina: an analysis of implications and vulnerability. Chemosphere. 18791298. 296. Available from: www.elsevier.com/locate/chemosphere. doi:10.1016/j.chemosphere.2022.133990.

Joseph, M.M., Nair, J.B., Joseph, A.M., 2025. Microscopic menace: exploring the link between microplastics and cancer pathogenesis. Royal Society of Chemistry, India Environmental Science: Processes and Impacts. 20507895. 27 (7), 1768–1795. Available from: http://pubs.rsc.org/en/journals/journal/em. doi:10.1039/d5em00232j.

Kim, L., Kim, H., Kim, T.H., An, Y.J., 2025. Size- and shape-dependent effects of polyethylene terephthalate microplastics on the benthic crustacean Artemia franciscana. Mar. Pollut. Bull. 18793363. 211. Available from: https://www.sciencedirect.com/science/journal/0025326X. doi:10.1016/j.marpolbul.2024.117391.

Kutralam-Muniasamy, G., Pérez-Guevara, F., Elizalde-Martínez, I., Shruti, V.C., 2020. Branded milks – are they immune from microplastics contamination? Sci. Total Environ. 00489697. 714, 136823. doi:10.1016/j.scitotenv.2020.136823.

Ky, N.M., Lin, C., Nguyen, H.L., Le, V.G., Haddout, S., Um, M.J., Nguyen, D.D., 2023. Ecotoxicity of micro-and nanoplastics on aquatic algae: facts, challenges, and future opportunities. J. Environ. Manag. 346, 118982.

Le, V.G., Nguyen, M.K., Nguyen, H.L., Lin, C., Hadi, M., Hung, N.T.Q., Bolan, N.S., 2023. A comprehensive review of micro-and nano-plastics in the atmosphere: occurrence, fate, toxicity, and strategies for risk reduction. Sci. Total Environ. 904, 166649.

Lebreton, L.C.M., Zwet, J.V.D., Damsteeg, J.W., Slat, B., Andrady, A., Reisser, J., 2017. River plastic emissions to the world's oceans. Nat. Commun. 20411723. 8. Available from: http://www.nature.com/ncomms/index.html. doi:10.1038/ncomms15611.

Liboiron, M., Liboiron, F., Wells, E., Richárd, N., Zahara, A., Mather, C., Bradshaw, H., Murichi, J., 2016. Low plastic ingestion rate in Atlantic cod (Gadus morhua) from Newfoundland destined for human consumption collected through citizen science methods. Mar. Pollut. Bull. 0025326X. 113 (1-2), 428–437. doi:10.1016/j.marpolbul.2016.10.043.

Liebezeit, G., Liebezeit, E., 2013. Non-pollen particulates in honey and sugar. Food Addit. Contam. Part A. 1944-0049. 30 (12), 2136–2140. doi:10.1080/19440049.2013.843025.

Liu, S., Yuewu, L., Shang, L., Yin, J., Qian, Z., Chen, C., Yang, Y., 2022. Size-dependent neurotoxicity of micro- and nanoplastics in flowing condition based on an in vitro microfluidic study. Chemosphere. 00456535. 303 (Pt 3), 135280. doi:10.1016/j.chemosphere.2022.135280.

Lusher, A., Hollman, P., Mendoza-Hill, J., 2017. Microplastics in Fisheries and Aquaculture: Status of Knowledge on Their Occurrence and Implications for Aquatic Organisms and Food Safety. Vol. 1. Food and agriculture organization of the united nations, Rome. Available from: https://oceanrep.geomar.de/id/eprint/49179/1/Microplastics%20in%20fisheries%20and%20aquaculture.pdf.

Luu, T.A., Le, V.G., Thai, V.A., Nguyen, M.K., Lin, C., Pham, M.T., Chang, S.W., Nguyen, D.D., 2025. Plastic wastes for carbon-based materials: investigations on recent applications towards environmentally sustainable, carbon dioxide capture and green energy. Environ. Pollut. 18736424. 370. Available from: https://www.sciencedirect.com/science/journal/02697491. doi:10.1016/j.envpol.2025.125899.

Mamun, A.A., Prasetya, T.A.E., Dewi, I.R., Ahmad, M., 2023. Microplastics in human food chains: food becoming a threat to health safety. Sci. Total Environ. 18791026. 858. Available from: www.elsevier.com/locate/scitotenv. doi:10.1016/j.scitotenv.2022.159834.

Manzoor, S., Naqash, N., Rashid, G., Singh, R., 2022. Plastic material degradation and formation of microplastic in the environment: a review. Mater. Today Proc. 22147853. 56, 3254–3260. doi:10.1016/j.matpr.2021.09.379.

Mercogliano, R., Avio, C.G., Regoli, F., Anastasio, A., Colavita, G., Santonicola, S., 2020. Occurrence of microplastics in commercial seafood under the perspective of the human food chain. A review. J. Agric. Food Chem. 15205118. 68 (19), 5296–5301. Available from: http://pubs.acs.org/journal/jafcau. doi:10.1021/acs.jafc.0c01209.

Miller, M.E., Hamann, M., Kroon, F.J., 2020. Bioaccumulation and biomagnification of microplastics in marine organisms: a review and meta-analysis of current data. PLoS ONE. 19326203. 15 (10). Available from: https://journals.plos.org/plosone/article/file?id=10.1371/journal.pone.0240792&type=printable. doi:10.1371/journal.pone.0240792.

Mohajan, H.K., 2025. Plastic pollution: a potential threat on health and environment. Stud. Soc. Sci. Humanities. 27097862. 4 (2), 25–30. doi:10.56397/sssh.2025.03.04.

Nelms, S.E., Duncan, E.M., Broderick, A.C., Galloway, T.S., Godfrey, M.H., Hamann, M., Lindeque, P.K., Godley, B. J., 2016. Plastic and marine turtles: a review and call for research. ICES J. Mar. Sci. 10959289. 73 (2), 165–181. Available from: http://icesjms.oxfordjournals.org/. doi:10.1093/icesjms/fsv165.

Nguyen, M.K., Lin, C., Nguyen, H.L., Le, V.R., KL, P., Singh, J., Chang, S.W., Um, M.J., Nguyen, D.D., 2023. Emergence of microplastics in the aquatic ecosystem and their potential effects on health risks: the insights into Vietnam. J. Environ. Manag. 10958630. 344. Available from: https://www.sciencedirect.com/journal/journal-of-environmental-management. doi:10.1016/j.jenvman.2023.118499.

Osman, A.I., Hosny, M., Eltaweil, A.S., Omar, S., Elgarahy, A.M., Farghali, M., Yap, P.S., Wu, Y.S., Nagandran, S., Batumalaie, K., Gopinath, S.C.B., John, O.D., Sekar, M., Saikia, T., Karunanithi, P., Hatta, M.H.M., Akinyede, K. A., 2023. Microplastic sources, formation, toxicity and remediation: a review. Environ. Chem. Lett. 16103661. 21 (4), 2129–2169. Available from: https://www.springer.com/journal/10311. doi:10.1007/s10311-023-01593-3.

Pannetier, P., Morin, B., Bihanic, F.L., Dubreil, L., Clérandeau, C., Chouvellon, F., Van Arkel, K., Danion, M., Cachot, J., 2020. Environmental samples of microplastics induce significant toxic effects in fish larvae. Environ. Int. 01604120. 134, 105047. doi:10.1016/j.envint.2019.105047.

Payanthoth, N.S., Mut, N.N.N., Samanta, P., Li, G., Jung, J., 2024. A review of biodegradation and formation of biodegradable microplastics in soil and freshwater environments. Appl. Biol. Chem. 24680842. 67 (1). Available from: https://www.springer.com/journal/13765. doi:10.1186/s13765-024-00959-7.

Pennati, R., Castelletti, C., Parolini, M., Scarì, G., Mercurio, S., 2022. Mixotrophic flagellate ingestion boosts microplastic accumulation in ascidians. J. Exp. Zool. A Ecol. Integr. Physiol. 2471-5638. 337 (6), 639–644. doi:10.1002/jez.2596.

Prata, J.C., Patrício Silva, A.L., da Costa, J.P., Mouneyrac, C., Walker, T.R., Duarte, A.C., Rocha-Santos, T., 2019. Solutions and integrated strategies for the control and mitigation of plastic and microplastic pollution. Int. J. Environ. Res. Public Health. 16604601. 16 (13). Available from: https://www.mdpi.com/1660-4601/16/13/2411/pdf. doi:10.3390/ijerph16132411.

Rathour, A., Dhar, A., Pathania, S., 2025. Impact of microplastics pollution on human health and aquatic life: a review. Polym. Int. 0959–8103. doi:10.1002/pi.6782.

Roman, C., Mahé, P., Latchere, O., Catrouillet, C., Gigault, J., Métais, I., Châtel, A., 2023. Effect of size continuum from nanoplastics to microplastics on marine mussel Mytilus edulis: comparison in vitro/in vivo exposure scenarios. Comp. Biochem. Physiol. C Toxicol. Pharmacol. 15320456. 264, 109512. doi:10.1016/j.cbpc.2022.109512.

Saeedi, M., 2024. How microplastics interact with food chain: a short overview of fate and impacts. J. Food Sci. Technol. 0022-1155. 61 (3), 403–413. doi:10.1007/s13197-023-05720-4.

Savva, K., Borrell, X., Moreno, T., Pérez-Pomeda, I., Barata, C., Llorca, M., Farré, M., 2023. Cytotoxicity assessment and suspected screening of PLASTIC ADDITIVES in bioplastics of single-use household items. Chemosphere. 00456535. 313, 137494. doi:10.1016/j.chemosphere.2022.137494.

Song, Y.K., Hong, S.H., Jang, M., Han, G.M., Jung, S.W., Shim, W.J., 2017. Combined Effects of UV Exposure Duration and Mechanical Abrasion on Microplastic Fragmentation by Polymer Type. Environ. Sci. Technol. 15205851. 51 (8), 4368–4376. Available from: http://pubs.acs.org/journal/esthag. doi:10.1021/acs.est.6b06155.

Thacharodi, A., Hassan, S., Meenatchi, R., Bhat, M.A., Hussain, N., Arockiaraj, J., Ngo, H.H., Sharma, A., Nguyen, H. T., Pugazhendhi, A., 2024. Mitigating microplastic pollution: a critical review on the effects, remediation, and utilization strategies of microplastics. J. Environ. Manag. 10958630. 351. Available from: https://www.sciencedirect.com/science/journal/03014797. doi:10.1016/j.jenvman.2023.119988.

Thuan, P.M., Nguyen, M.K., Nguyen, D.D., 2025. The potential release of microplastics from paint fragments: characterizing sources, occurrence and ecological impacts. Environ. Geochem. Health. 15732983. 47 (6). Available from: https://www.springer.com/journal/10653. doi:10.1007/s10653-025-02525-6.

Ullah, S., Ahmad, S., Guo, X., Ullah, S., Ullah, S., Nabi, G., Wanghe, K., 2023. A review of the endocrine disrupting effects of micro and nano plastic and their associated chemicals in mammals. Front. Endocrinol. 1664-2392. 13. doi:10.3389/fendo.2022.1084236.

Witczak, A., Przedpełska, L., Pokorska-Niewiada, K., Cybulski, J., 2024. Microplastics as a threat to aquatic ecosystems and human health. Toxics. 2305-6304. 12 (8), 571. doi:10.3390/toxics12080571.

Xia, F., Yang, W., Zhao, H., Cai, Y., Tan, Q., 2025. Occurrence characteristics and transport processes of riverine microplastics in different connectivity contexts. Npj Clean Water. 2059-7037. 8 (1). doi:10.1038/s41545-024-00424-4.

Yang, Y., Wang, J., Shi, Y., Cao, H., Wei, L., Gao, L., Liu, M., 2024. Oxidation enhances the toxicity of polyethylene microplastics to mouse eye: perspective from in vitro and in vivo. Environ. Pollut. 02697491. 360, 124633. doi:10.1016/j.envpol.2024.124633.

Yoo, J.W., Park, J.S., Lee, Y.H., Choi, T.J., Kim, C.B., Jeong, T.Y., Kim, C.H., Kim, T.H., Lee, Y.M., 2024. Toxic effects of fragmented polyethylene terephthalate particles on the marine rotifer Brachionus koreanus: based on ingestion and egestion assay, in vivo toxicity test, and multi-omics analysis. J. Hazard. Mater. 18733336. 472. Available from: https://www.sciencedirect.com/science/journal/03043894. doi:10.1016/j.jhazmat.2024.134448.

Yousafzai, S., Farid, M., Zubair, M., Naeem, N., Zafar, W., Asam, Z.U.Z., Farid, S., Ali, S., 2025. Detection and degradation of microplastics in the environment: a review. Environ. Sci. Adv. 27547000. 4 (8), 1142–1165. Available from: www.rsc.org/journals-books-databases/about-journals/environmental-science-advances. doi:10.1039/d5va00064e.

Yuan, J., Jie, M., Sun, Y., Zhou, T., Zhao, Y., Fei, Y., 2020. Microbial degradation and other environmental aspects of microplastics/plastics. Sci. Total Environ. 00489697. 715, 136968. doi:10.1016/j.scitotenv.2020.136968.

Zeng, L., Yang, H., Tong, T., Chen, J., Huang, Y., Yang, J., Xiao, Y., Yang, F., Chen, Y., 2025. Study on the toxic effect of seawater-aged microplastics on Philippine curtain clams. Sci. Rep. 20452322. 15 (1). Available from: https://www.nature.com/srep/. doi:10.1038/s41598-025-02823-0.

Zheng, L., Hasselström, L., Finnveden, G., Johansson, N., 2022. Cost-benefit analysis of two possible deposit-refund systems for reuse and recycling of plastic packaging in Sweden. Cleaner Waste Syst. 27729125. 3, 100048. doi:10.1016/j.clwas.2022.100048.

Zhou, Y., Zhou, X.X., Jiang, H., Liu, W., Chen, F., Gardea-Torresdey, J.L., Yan, B., 2024. In vitro toxicity and modeling reveal nanoplastic effects on marine bivalves. ACS Nano. 1936086X. 18 (26), 17228–17239. Available from: http://pubs.acs.org/journal/ancac3. doi:10.1021/acsnano.4c04607.

Index

Note: Page numbers followed by "*f*" indicate figures, "*t*" indicate tables.

P

R

9780443456350